Applied Optimal Control

Applied Optimal Control

OPTIMIZATION, ESTIMATION, AND CONTROL

Arthur E. Bryson, Jr.
Stanford University

Yu-Chi Ho
Harvard University

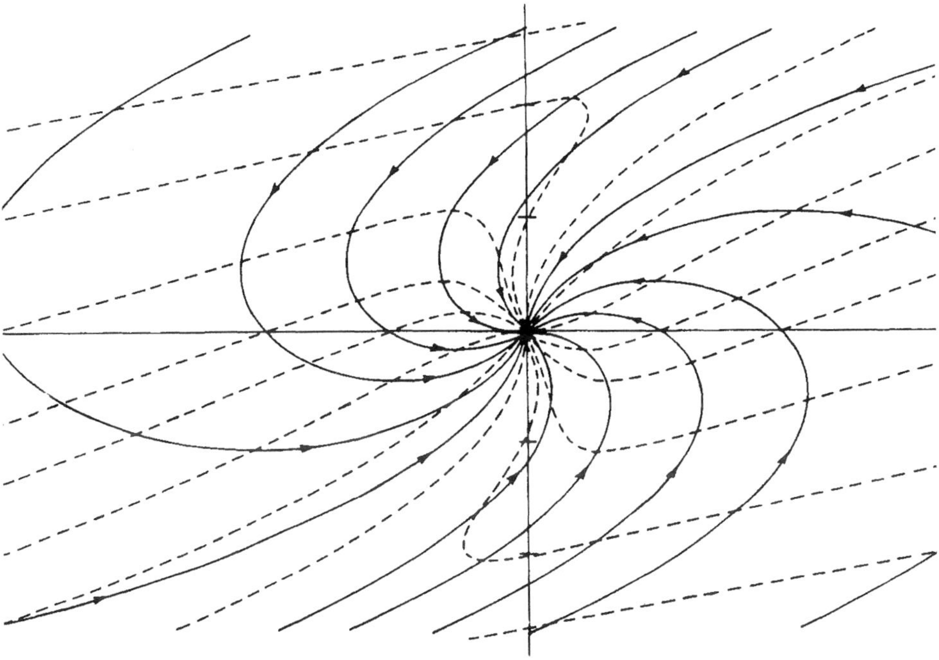

REVISED PRINTING

Taylor & Francis
Taylor & Francis Group
New York London

Taylor & Francis is an imprint of the
Taylor & Francis Group, an informa business

To Harvard University, its high standards, and its great tradition of seeking the truth, and to our teachers, colleagues, and students who have all taught and inspired us.

Published in 1975 by
Taylor & Francis Group
270 Madison Avenue
New York, NY 10016

Published in Great Britain by
Taylor & Francis Group
2 Park Square
Milton Park, Abingdon
Oxon OX14 4RN

20 19 18 17 16 15 14 13 12 11

International Standard Book Number-10: 0-89116-228-3 (Softcover)
International Standard Book Number-13: 978-0-89116-228-5 (Softcover)
Library of Congress catalog number: 75-16114

Library of Congress Cataloging-in-Publication Data

Catalog record is available from the Library of Congress

Visit the Taylor & Francis Web site at
http://www.taylorandfrancis.com

Preface to Revised Printing

After publication of the original Ginn and Company edition, we accumulated a large number of modifications, including corrections of typographical errors and misprints, some minor improvements in clarity and other changes in graphical illustrations as a result of errata accumulated from such sources as students, teachers and other readers.

We are happy to have Hemisphere Publishing Corporation continue publication of our book and are grateful for the support and fine reception our colleagues have given it.

A. E. B.
Y.-C. H.

Preface

This text is intended for use at the senior-graduate level for university courses on the analysis and design of dynamic systems and for independent study by engineers and applied mathematicians. An elementary knowledge of mechanics and ordinary differential equations is presumed. Some acquaintance with matrix algebra and the properties of linear systems is desirable, but the required familiarity can be obtained by studying the two appendices. The book grew out of a set of lecture notes for a Harvard University summer school program on optimization of dynamic systems given in 1963. These notes were rewritten and extended for graduate courses given at Harvard from 1963 to 1968 and at M.I.T. in 1966.

The book is concerned with the analysis and design of complicated dynamic systems. Particular emphasis is placed on determining the "best" way to guide and/or control such systems. Over the past twenty-five years, a great body of knowledge has been built up on the subject of feedback control systems for linear, time-invariant dynamic systems. This knowledge plays an important role in our technology, and almost every engineering school recognizes this fact by teaching courses in this area. However, many dynamic systems—such as aerospace systems—are nonlinear and/or time-varying, and the techniques for analysis and design of linear, time-invariant systems are, in general, not applicable to these more complicated systems.

The appearance of practical, high-speed digital computers in the 1950's provided an essential tool for dealing with nonlinear and time-varying systems. Engineers were quick to take advantage of these remarkable computers to do extensive cut-and-try design work on paper instead of in the development laboratory. In many instances, particularly when designing guidance and control systems, it became clear that a more systematic approach was desirable. This led to a renewed interest in an old subject, the *calculus of variations*, and to the discovery of an interesting extension of this subject, *dynamic programming*. The application of these techniques to deterministic, nonlinear and time-varying systems forms the basis of Chapters 1 through 9.

The first part of the book assumes precise knowledge of the structure and parameters of the dynamic systems investigated and precise measurements for feedback control. In practice, such precise knowledge is seldom available. Thus, it is important to be able to predict the sensitivity of controlled systems to *random fluctuations* in the environment and in the measurement system. Chapters 10 through 14 are concerned with this topic, starting with a review of the fundamentals of probability and random processes and proceeding to the design of the "best" control system, on the average, using noisy measurements and taking into account random perturbations of the system by the environment.

Our main objective has been to produce useful results which can be easily translated into digital computer programs. Several versions of the book in the form of lecture notes have been scrutinized by our competent and critical students and colleagues, and we therefore hope that most of the errors have been caught. However, we take responsibility for any that may still exist.

This book can be, and has been, used for both one-semester and two-semester courses in modern control theory. It can also be taught in either the deterministic–stochastic or the introductory–advanced sequence. The logical dependence of various chapters as well as the breakdown of chapters into semester–long parts are shown at the end of the table of contents.

The exercises form an integral part of the text. They either illustrate the subject matter covered or extend it and in some cases constitute a semi-research problem. Serious students should pursue them with diligence.

The authors are indebted to many persons for material. Special thanks go to John V. Breakwell, Henry J. Kelley, R. E. Kalman, George E. Leitmann, Placido Cicala, Colin C. Blaydon, Sheldon Baron, Rangasami L. Kashyap, Walter F. Denham, Stuart E. Dreyfus, Donald E. Johansen, Robert C. K. Lee, Stephen R. McReynolds, Jason L. Speyer, Kevin S. Tait, Laurie J. Henrikson, Raman Mehra, Mukund Desai, Robert Behn, and David Jacobson for papers and/or discussions.

We are also indebted to Marion Remillard, Sandra Nagy, and Skippi Torrance for typing the successive versions of the manuscript.

A. E. B.

Y.-C. H.

Contents

6. *Neighboring extremals and the second variation*

7. *Numerical solution of optimal programming and control problems*

8. *Singular solutions of optimization and control problems*

Appendix B—Properties of linear systems

LOGICAL DEPENDENCE OF CHAPTERS

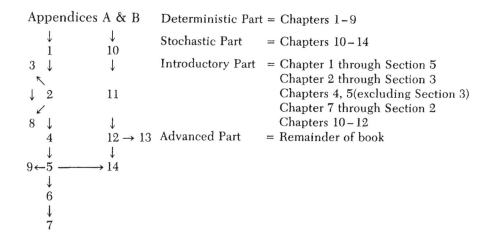

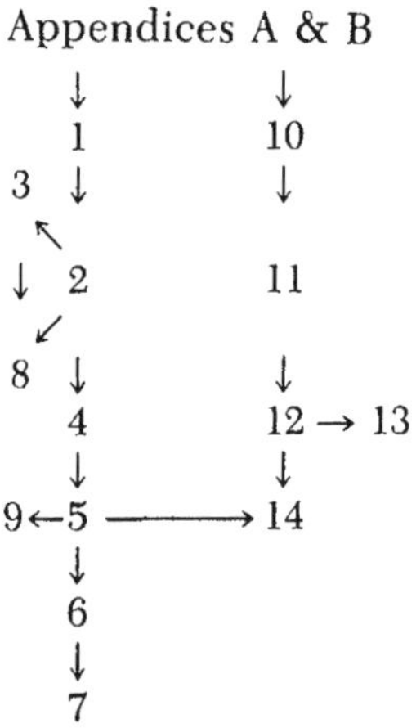

Deterministic Part = Chapters 1–9

Stochastic Part = Chapters 10–14

Introductory Part = Chapter 1 through Section 5
Chapter 2 through Section 3
Chapters 4, 5(excluding Section 3)
Chapter 7 through Section 2
Chapters 10–12

Advanced Part = Remainder of book

Parameter optimization problems

1

1.1 Problems without constraints

The simplest class of parameter optimization problems involves finding the values of m parameters $u_1, \ldots, u_m$ that minimize a performance index which is a function of these parameters,

$$L(u_1, \ldots, u_m) .$$

For convenience, we shall use a more compact nomenclature; let

$$u = \begin{bmatrix} u_1 \\ \cdot \\ \cdot \\ \cdot \\ u_m \end{bmatrix} = \text{decision vector} \tag{1.1.1}$$

and write the performance index then as

$$L(u) . \tag{1.1.2}$$

If there are no constraints on possible values of u and if the function $L(u)$ has first and second partial derivatives everywhere, *necessary conditions for a minimum* are

$$\frac{\partial L}{\partial u} = 0 , \tag{1.1.3}$$

by which we mean that $\partial L/\partial u_i = 0$, $i = 1, \ldots, m$, and

$$\frac{\partial^2 L}{\partial u^2} \geq 0 , \tag{1.1.4}$$

by which we mean that the $(m \times m)$-matrix whose components are $\partial^2 L/\partial u_i \, \partial u_j$ must be positive semidefinite, i.e., have eigenvalues that are zero or positive.

All points that satisfy (1.1.3) are called *stationary* points.

Sufficient conditions for a local minimum are (1.1.3) and

$$\frac{\partial^2 L}{\partial u^2} > 0\,; \tag{1.1.5}$$

that is, all the eigenvalues must be positive.

If (1.1.3) is satisfied but $\partial^2 L/\partial u^2 = 0$, that is, the determinant of the matrix is zero (meaning that one or more of its eigenvalues is zero), additional information is needed to establish whether or not the point is a minimum. Such a point is called a *singular* point. Note that, if L is a linear function of u, then $\partial^2 L/\partial u^2 = 0$ everywhere, and, in general, a minimum does not exist.

Examples *for* $L = L(u_1,u_2)$.

(a) *minimum*: both eigenvalues of $\partial^2 L/\partial u_i\, \partial u_j > 0$

$$L = [u_1 u_2]\begin{bmatrix} 1, -1 \\ -1, \ 4 \end{bmatrix}\begin{bmatrix} u_1 \\ u_2 \end{bmatrix}$$

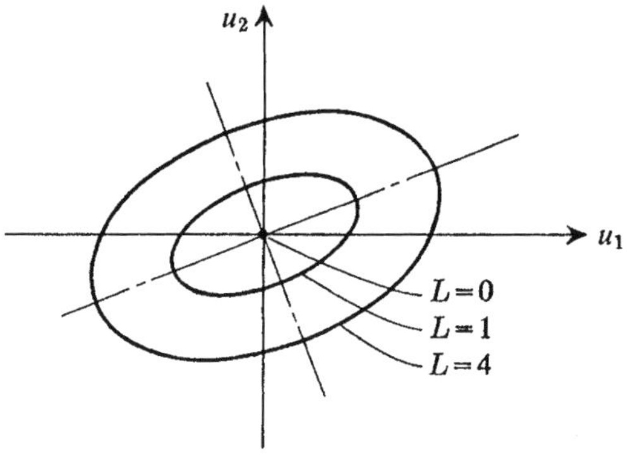

Figure 1.1.1. A minimum point.

(b) *saddlepoint*: one positive eigenvalue, one negative eigenvalue of $\partial^2 L/\partial u_i\, \partial u_j$

$$L = [u_1 u_2]\begin{bmatrix} -1,1 \\ 1,3 \end{bmatrix}\begin{bmatrix} u_1 \\ u_2 \end{bmatrix}$$

(c) *singular point*: one positive eigenvalue, one zero eigenvalue

$$L = (u_1 - u_2^2)(u_1 - 3u_2^2).$$

1.2 Problems with equality constraints; necessary conditions for a stationary point

A more general class of parameter optimization problems involves finding the values of m decision parameters $u_1, \ldots, u_m$ that minimize

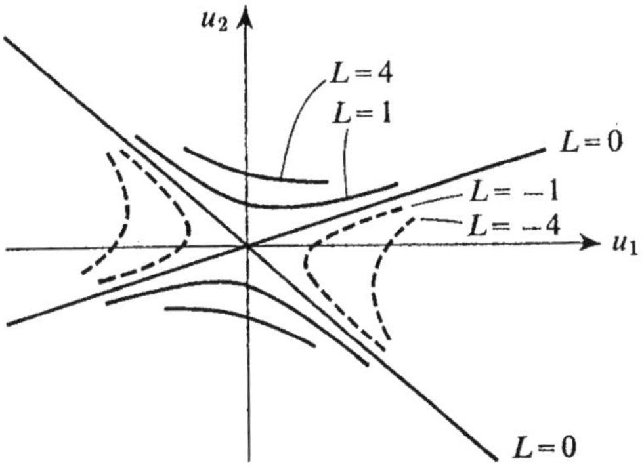

Figure 1.1.2. A saddle point.

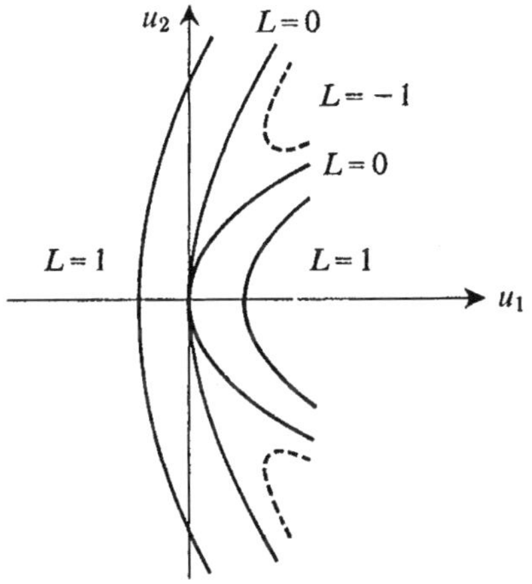

Figure 1.1.3. A singular point.

a performance index which is a scalar function of $n + m$ parameters,

$$L(x_1, \ldots x_n ; u_1, \ldots u_m) ,$$

where the n *state parameters* $x_1, \ldots, x_n$ are determined by the decision parameters through a set of n *constraint relations*,

$$f_1(x_1, \ldots, x_n ; u_1, \ldots, u_m) = 0 ,$$

$$\vdots$$

$$f_n(x_1, \ldots, x_n ; u_1, \ldots, u_m) = 0 .$$

For convenience, we shall again use a more compact nomenclature. Let

$$u = \begin{bmatrix} u_1 \\ \cdot \\ \cdot \\ \cdot \\ u_m \end{bmatrix} = \text{decision vector}, \qquad x = \begin{bmatrix} x_1 \\ \cdot \\ \cdot \\ \cdot \\ x_n \end{bmatrix} = \text{state vector},$$

$$f = \begin{bmatrix} f_1 \\ \cdot \\ \cdot \\ \cdot \\ f_n \end{bmatrix} = \text{constraint vector}.$$

In this nomenclature, the problem may be stated as follows.

Find the decision vector u that minimizes

$$L(x,u), \tag{1.2.1}$$

where the state vector x is determined by the decision vector through the constraint relations

$$f(x,u) = 0 \qquad (n \text{ equations}). \tag{1.2.2}$$

For a given parameter optimization problem, the choice of which parameters to designate as decision parameters is not unique; it is only a matter of convenience to make a distinction between decision and state parameters. However, the choice must be such that u determines x through the constraint relations (1.2.2).

If the relations (1.2.1) and (1.2.2) are *linear* in both x and u, then, in general, a minimum does not exist. Inequality constraints on the magnitudes of x and/or u are necessary to make the problem meaningful; such problems, treated in later sections of this chapter, are called *linear programming problems* if the inequality constraints are also linear in x and u.

In the first part of this chapter we shall discuss problems that have some nonlinearity in (1.2.1) and (1.2.2). Of course, the presence of nonlinearity does not, in itself, insure that a minimum exists.

A *stationary* point is one where $dL = 0$, for arbitrary du, while holding $df = 0$ (letting dx change as it will). Now we have

$$dL = L_x\, dx + L_u\, du, \tag{1.2.3}$$

$$df = f_x\, dx + f_u\, du. \tag{1.2.4}$$

When we require that $df = 0$, then, if f_x is nonsingular (and it should be if u determines x from (1.2.2)), (1.2.4) may be solved for dx

$$dx = -f_x^{-1} f_u\, du. \tag{1.2.5}$$

Substituting (1.2.5) into (1.2.3) yields

$$dL = (L_u - L_x f_x^{-1} f_u)\, du\,. \tag{1.2.6}$$

Hence, if dL is to be zero for arbitrary du, *it is necessary that*

$$\boxed{L_u - L_x f_x^{-1} f_u = 0} \qquad (m \text{ equations})\,. \tag{1.2.7}$$

These m equations together with the n equations (1.2.2), determine the m quantities u and the n quantities x at stationary points. Note that (1.2.7) represents the partial derivative of L with respect to u, *holding f constant*, whereas L_u represents the partial derivative of L with respect to u, *holding x constant.*

Another (equivalent) approach to Equation (1.2.7) is to notice that (1.2.3) and (1.2.4), with $dL = 0$, $df = 0$, must be *consistent* linear equations in dx and du at a stationary point. If they are consistent, we should be able to find a set of n constants $\lambda_1, \ldots, \lambda_n$ such that

$$L_y + \sum_{i=1}^{n} \lambda_i \frac{\partial f_i}{\partial_y} = 0, \tag{1.2.8}$$

where

$$y^T = (x_1, \ldots, x_n, u_1, \ldots u_m) \tag{1.2.9}$$

that is, a linear combination of rows of f_y must equal L_y.† For convenience, let

$$\lambda = \begin{bmatrix} \lambda_1 \\ \cdot \\ \cdot \\ \cdot \\ \lambda_n \end{bmatrix} \qquad \text{and} \qquad \lambda^T = [\lambda_1, \ldots, \lambda_n]\,.$$

Then we may write (1.2.8) and (1.2.9) as

$$L_x + \lambda^T f_x = 0 \qquad (n \text{ equations})\,, \tag{1.2.10}$$

$$L_u + \lambda^T f_u = 0 \qquad (m \text{ equations})\,. \tag{1.2.11}$$

Equation (1.2.10) may be solved for λ^T (since f_x must be nonsingular):

$$\lambda^T = -L_x (f_x)^{-1}\,, \tag{1.2.12}$$

which, in turn, may be substituted into (1.2.11) to yield (1.2.7).

---

†More generally, consistency requires that the rank of the $[(n + 1) \times (n + m)]$-matrix

$$\begin{bmatrix} f_x \,.\, f_u \\ L_x \,.\, L_u \end{bmatrix}$$

be less than $(n + 1)$.

The interpretation of λ may be inferred from (1.2.3) and (1.2.4) by placing $du = 0$ and eliminating dx

$$-\lambda^T = L_x(f_x)^{-1} = (\partial L/\partial f)_u ;$$

that is, the λ's are partial derivatives of L with respect to f, holding u constant, letting x change as required. This will have a special significance in optimization problems with *inequality* constraints (Section 1.7).

Still another (equivalent) approach, which we shall use many times throughout the remainder of the book, is to "adjoin" the constraints (1.2.2) to the performance index (1.2.1) by a set of n "undetermined multipliers," $\lambda_1, \ldots, \lambda_n$,* as follows:

$$H(x,u,\lambda) = L(x,u) + \sum_{i=1}^{n} \lambda_i f_i(x,u) = L(x,u) + \lambda^T f(x,u) . \quad (1.2.13)$$

Suppose we have chosen some nominal values of u and determined the corresponding values of x from (1.2.2) so that $L = H$. Differential changes in H due to differential changes in x and u are given by

$$dH = \frac{\partial H}{\partial x} dx + \frac{\partial H}{\partial u} du . \quad (1.2.14)$$

Since we are interested in how H (and, hence, L) changes as the control vector u changes, it is convenient to *choose* the λ vector so that

$$\frac{\partial H}{\partial x} = \frac{\partial L}{\partial x} + \lambda^T \frac{\partial f}{\partial x} = 0 \Rightarrow \lambda^T = -\frac{\partial L}{\partial x}\left(\frac{\partial f}{\partial x}\right)^{-1} , \quad (1.2.15)$$

which is, of course, the same as (1.2.12).

Since x was found from (1.2.2), it follows that

$$dL \equiv dH = \frac{\partial H}{\partial u} du . \quad (1.2.16)$$

Thus, $\partial H/\partial u$ is the gradient of L with respect to u *while holding* $f(x,u) = 0$.

At a stationary point in the u-space, dL vanishes for arbitrary du; this can happen only if

$$\frac{\partial H}{\partial u} \equiv \frac{\partial L}{\partial u} + \lambda^T \frac{\partial f}{\partial u} = 0 . \quad (1.2.17)$$

Thus, *necessary conditions for a stationary value of* $L(x,u)$ are

$$f(x,u) = 0 ; \quad (1.2.18)$$

*The constants $\lambda_1, \ldots, \lambda_n$ are often referred to as *Lagrange multipliers*.

$$\frac{\partial H}{\partial x} = 0\,, \qquad \text{where} \qquad H = L(x,u) + \lambda^T f(x,u)\,, \tag{1.2.19}$$

$$\frac{\partial H}{\partial u} = 0\,, \tag{1.2.20}$$

which are $2n + m$ equations for the $2n + m$ quantities x, λ, and u.

Example 1. Find the scalar parameter u that yields a stationary value of

$$L = \frac{1}{2}\left(\frac{x^2}{a^2} + \frac{u^2}{b^2}\right),$$

with the linear constraint

$$f(x,u) = x + mu - c = 0\,,$$

where x is a scalar parameter and a, b, m, and c are constants.

The curves of constant L are ellipses, with L increasing with the size of the ellipse, whereas $x + mu - c = 0$ is a fixed straight line.

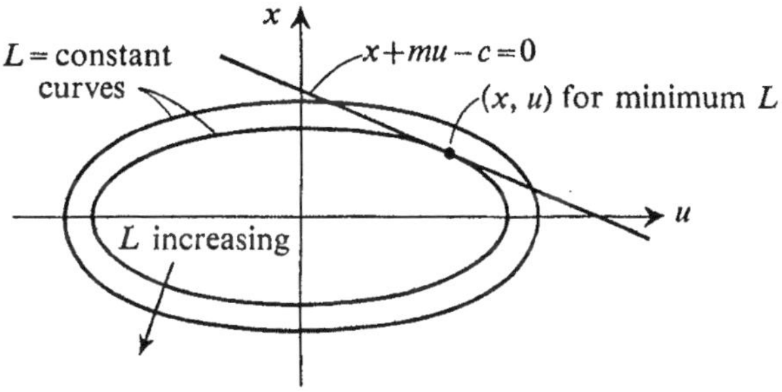

Figure 1.2.1. Example of minimization subject to a constraint.

Clearly, the minimum value of L satisfying the constraint is obtained when the ellipse is just tangent to the straight line. Analytically, the H function is

$$H = \frac{1}{2}\left(\frac{x^2}{a^2} + \frac{u^2}{b^2}\right) + \lambda(x + mu - c)\,,$$

and necessary conditions for a stationary value are

$$x + mu - c = 0\,, \qquad \frac{\partial H}{\partial x} = \frac{x}{a^2} + \lambda = 0\,, \qquad \frac{\partial H}{\partial u} = \frac{u}{b^2} + \lambda m = 0\,.$$

These three equations for the three unknowns, x, u, λ, have a simple unique solution:

$$x = \frac{a^2c}{a^2 + m^2b^2}, \qquad u = \frac{b^2mc}{a^2 + m^2b^2}, \qquad \lambda = -\frac{c}{a^2 + m^2b^2},$$

and the minimum value of L is thus given by

$$J = L_{\min} = \frac{c^2}{2(a^2 + m^2b^2)}.$$

Note that $-\lambda = \partial J/\partial c \equiv \partial J/\partial f$.

Example 2. *Maximum steady rate of climb for aircraft.* The net force on an aircraft maintaining a steady rate of climb must be zero. If we choose force components parallel and perpendicular to the flight path (see Figure 1.2.2), this requires that

$$f_1(V,\gamma,\alpha) = T\cos(\alpha + \epsilon) - D - mg\sin\gamma = 0,$$

$$f_2(V,\gamma,\alpha) = T\sin(\alpha + \epsilon) + L - mg\cos\gamma = 0,$$

where

V = velocity,
γ = flight path angle to horizontal,
α = angle-of-attack,
m = mass of aircraft,
g = gravitational force per unit mass,
ϵ = angle between thrust axis and zero-lift axis,

and, at a given altitude,

$L = L(V,\alpha)$ = lift force,
$D = D(V,\alpha)$ = drag force,
$T = T(V)$ = thrust of engine.

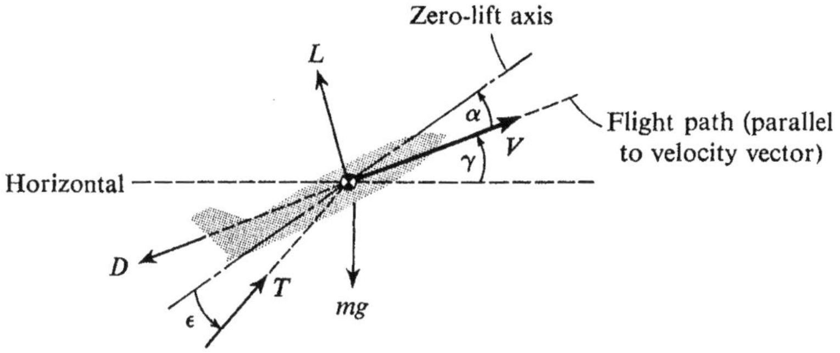

Figure 1.2.2. Force equilibrium of climbing aircraft.

The rate of climb is simply

$$V \sin \gamma .$$

We choose V and γ as state parameters and α as the control parameter since, at a given altitude, a choice of α determines V, γ from the two force equilibrium relations.

The H function is

$$H = V \sin \gamma + \lambda_1 (T \cos (\alpha + \epsilon) - D - mg \sin \gamma) + \lambda_2 (T \sin (\alpha + \epsilon) + L - mg \cos \gamma) .$$

Hence, the necessary conditions for a stationary value of rate of climb are:

$$f_1 = T(V) \cos (\alpha + \epsilon) - D(V,\alpha) - mg \sin \gamma = 0 ,$$

$$f_2 = T(V) \sin (\alpha + \epsilon) - L(V,\alpha) - mg \cos \gamma = 0 ,$$

$$\frac{\partial H}{\partial V} = \sin \gamma + \lambda_1 \left[\frac{\partial T}{\partial V} \cos (\alpha + \epsilon) - \frac{\partial D}{\partial V} \right] + \lambda_2 \left[\frac{\partial T}{\partial V} \sin (\alpha + \epsilon) + \frac{\partial L}{\partial V} \right] = 0 ,$$

$$\frac{\partial H}{\partial \gamma} = V \cos \gamma - \lambda_1 \, mg \cos \gamma + \lambda_2 \, mg \sin \gamma = 0 ,$$

$$\frac{\partial H}{\partial \alpha} = \lambda_1 \left[-T \sin (\alpha + \epsilon) - \frac{\partial D}{\partial \alpha} \right] + \lambda_2 \left[T \cos (\alpha + \epsilon) + \frac{\partial L}{\partial \alpha} \right] = 0 .$$

These five equations for the five unknowns, V, γ, α, λ_1, and λ_2, will, in general, have to be solved numerically for realistic lift, drag, and thrust functions.

1.3 Problems with equality constraints; sufficient conditions for a local minimum

To second order, the differential changes of $L(x,u)$ and $f(x,u)$ away from a nominal point (x,u) are

$$dL = (L_x, L_u) \begin{pmatrix} dx \\ du \end{pmatrix} + \frac{1}{2} (dx^T, du^T) \begin{pmatrix} L_{xx}, L_{xu} \\ L_{ux}, L_{uu} \end{pmatrix} \begin{pmatrix} dx \\ du \end{pmatrix} , \tag{1.3.1}$$

$$df = (f_x, f_u) \begin{pmatrix} dx \\ du \end{pmatrix} + \frac{1}{2} (dx^T, du^T) \begin{pmatrix} f_{xx}, f_{xu} \\ f_{ux}, f_{uu} \end{pmatrix} \begin{pmatrix} dx \\ du \end{pmatrix} , \qquad (1.3.2)\dagger$$

where

$$L_{xu} = \frac{\partial}{\partial u} \left(\frac{\partial L}{\partial x} \right)^T , \qquad L_{xx} = \frac{\partial}{\partial x} \left(\frac{\partial L}{\partial x} \right)^T , \qquad \text{etc.}$$

---

†This equation must be interpreted as applying to each component of f.

If we multiply (1.3.2) by λ^T, determined from Equation (1.2.19), where $(H_x = 0)$, and add the result to (1.3.1), we obtain

$$dL = (0, H_u)\begin{pmatrix} dx \\ du \end{pmatrix} + \frac{1}{2}(dx^T, du^T)\begin{pmatrix} H_{xx}, & H_{xu} \\ H_{ux}, & H_{uu} \end{pmatrix}\begin{pmatrix} dx \\ du \end{pmatrix} - \lambda^T df, \quad (1.3.3)$$

where

$$H = L(x,u) + \lambda^T f(x,u). \quad (1.3.4)$$

Let us assume that the nominal point (x,u) satisfies the constraints $f(x,u) = 0$. We wish to examine the behavior of $L(x,u)$ in an infinitesimal neighborhood of this nominal point, keeping $f(x,u) = 0$ to second order.

If we place $df = 0$ in (1.3.2), we can write for dx

$$dx = -f_x^{-1} f_u \, du + \text{second and higher-order terms in components of } du \text{ and } dx. \quad (1.3.5)$$

Now, if the nominal point is a *stationary point*, then $H_u = 0$, and (1.3.3) becomes, with $df = 0$,

$$dL = \frac{1}{2} du^T[-f_u^T(f_x^T)^{-1}, I]\begin{bmatrix} H_{xx}, & H_{xu} \\ H_{ux}, & H_{uu} \end{bmatrix}\begin{bmatrix} -f_x^{-1} f_u \\ I \end{bmatrix} du \quad (1.3.6)$$

$$+ \text{third and higher-order terms in components of } du \text{ and } dx.$$

It follows that

$$\left(\frac{\partial^2 L}{\partial u^2}\right)_{f=0} = H_{uu} - H_{ux} f_x^{-1} f_u - f_u^T (f_x^T)^{-1} H_{xu} + f_u^T (f_x^T)^{-1} H_{xx} f_x^{-1} f_u. \quad (1.3.7)$$

Thus, *sufficient conditions for a local minimum* are the stationarity conditions (1.2.18) through (1.2.20) and the *positive-definiteness* of the matrix in (1.3.7). Clearly, a *necessary condition* for a local minimum is that the matrix in (1.3.7) be *positive semidefinite.* Note that (1.3.6) could have been obtained directly by considering the augmented criterion H to *second* order while considering the constraint $f = 0$ to *first* order only.

Furthermore, it is *not* necessary that H_{uu} be positive semidefinite.

Example. Find the scalar quantity u that minimizes

$$L = \frac{1}{2}\left(\frac{x^2}{a^2} + \frac{u^2}{b^2}\right)$$

with the quadratic constraint

$$f(x,u) = c - xu = 0,$$

where x is a scalar parameter and a, b, and c are positive constants. The curves of constant L are ellipses, with L increasing with the size of the ellipse, whereas $c - xu = 0$ is a hyperbola with two branches.

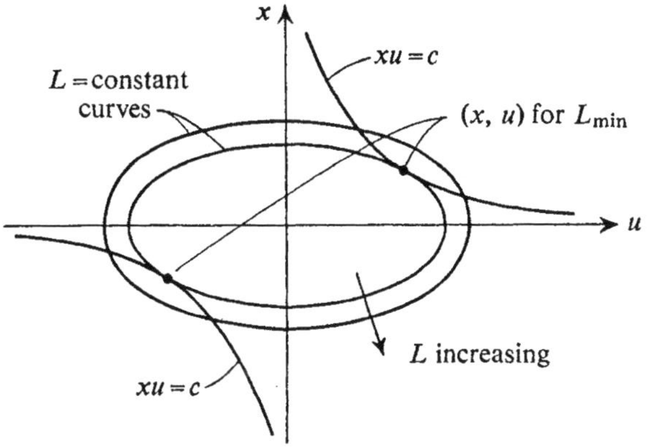

Figure 1.3.1. Example of minimization subject to a nonlinear constraint.

The minimum value of L satisfying the constraint is obtained when the ellipse is just tangent to the hyperbola. Analytically, the H function is

$$H = \frac{1}{2}\left(\frac{x^2}{a^2} + \frac{u^2}{b^2}\right) + \lambda(c - xu),$$

and necessary conditions for a stationary value are

$$c - xu = 0, \qquad \frac{\partial H}{\partial x} = \frac{x}{a^2} - \lambda u = 0, \qquad \frac{\partial H}{\partial u} = \frac{u}{b^2} - \lambda x = 0.$$

Using these relations, we find that

$$x = \pm\sqrt{\frac{ac}{b}}, \qquad u = \pm\sqrt{\frac{bc}{a}}, \qquad \lambda = \frac{1}{ab}, \qquad J = L_{\min} = \frac{c}{ab}.$$

The sufficient condition (1.3.7) for this problem is

$$\left(-\frac{a}{b}, 1\right)\begin{bmatrix} \frac{1}{a^2}, & -\frac{1}{ab} \\ -\frac{1}{ab}, & \frac{1}{b^2} \end{bmatrix}\begin{bmatrix} -\frac{a}{b} \\ 1 \end{bmatrix} = \frac{4}{b^2} > 0,$$

which is clearly satisfied. Note that there are two points at which the same minimum value of L occurs. Note, also, that

$$\lambda = \frac{\partial J}{\partial c}.$$

Problem 1. Find the point nearest the origin on the line

$$x + 2y + 3z = 10\,, \qquad x - y + 2z = 1\,,$$

where x, y, z are rectangular coordinates; i.e., minimize

$$L = x^2 + y^2 + z^2$$

subject to the two linear constraints.

Problem 2. Find the rectangle of maximum perimeter that can be inscribed in an ellipse; i.e., maximize

$$P = 4(x + y)$$

with the constraint

$$\frac{x^2}{a^2} + \frac{y^2}{b^2} = 1\,.$$

Problem 3. Find the rectangular parallelopiped of maximum volume that can be contained in a given ellipsoid; i.e., maximize

$$V = 8xyz$$

with the constraint

$$\frac{x^2}{a^2} + \frac{y^2}{b^2} + \frac{z^2}{c^2} = 1\,.$$

Problem 4. *Quadratic performance index with linear constraints.* Show that the control vector u that minimizes the nonnegative definite quadratic form

$$L = \frac{1}{2}x^T Q x + \frac{1}{2}u^T R u\,,$$

with the linear constraints

$$f(x,u) = x + Gu + c = 0\,,$$

is

$$u = -(R + G^T Q G)^{-1} G^T Q c\,.$$

Show, also, that the minimum value of L is

$$J = L_{\min} = \tfrac{1}{2} c^T \left(Q - QG(R + G^T Q G)^{-1} G^T Q\right) c$$

and that

$$\begin{aligned} \lambda &= \left(Q - QG(R + G^T Q G)^{-1} G^T Q\right) c \\ &\equiv (Q^{-1} + GR^{-1}G^T)^{-1} c \qquad \text{if} \quad Q^{-1} \text{ exists;}† \\ x &= -\left(I - G(R + G^T Q G)^{-1} G^T Q\right) c\,. \end{aligned}$$

Note, also, that

$$\lambda^T = \frac{\partial J}{\partial c}.$$

Problem 5. *Sail setting and heading for maximum upwind velocity.* A simplified model of a sailboat moving at constant velocity is shown in Figure 1.3.2.

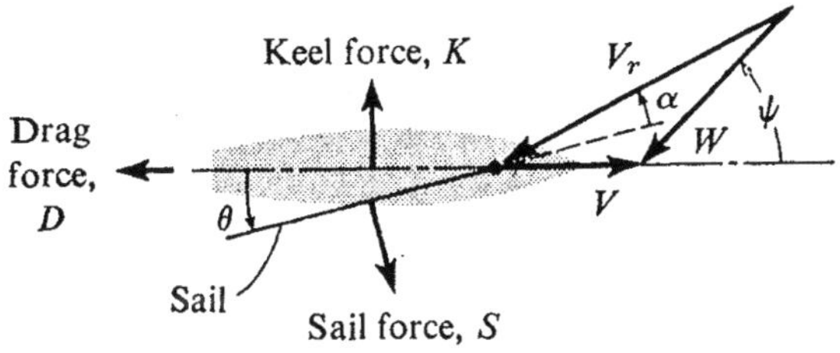

Figure 1.3.2. Force equilibrium of sailboat.

The sailboat's velocity relative to the water is V, at an angle ψ to the wind, which is blowing with velocity W relative to the water. The sail is set at an angle θ to the centerline of the boat, and the aerodynamic force S is assumed to act normal to the sail. The hydrodynamic force on the hull is resolved into components perpendicular to the centerline K and parallel to the centerline D. The magnitude of S is assumed to vary with the square of the relative wind, V_r, and the sine of the sail angle of attack, α:

$$S = C_1 V_r^2 \sin\alpha,$$

where C_1 is a constant and V_r and α are as defined in Figure 1.3.2. The drag is assumed to vary with the square of the boat velocity, V:

$$D = C_2 V^2,$$

where C_2 is a constant. For equilibrium of forces parallel to the centerline, we have

$$D = S\sin\theta.$$

Show that: (a) For given ψ, maximum V is obtained when $\alpha = \theta$. (b) The maximum velocity for $\psi = 180°$ (running before the wind) is $W\mu/(1+\mu)$ and is obtained when $\theta = 90°$, where $\mu^2 = C_1/C_2$. (c) The maximum upwind velocity, $V\cos\psi$, is equal to $W\mu/4$ and is obtained when the sail setting and the heading are chosen to be

---

†This is known as the "matrix inversion lemma." See Section 12.2 for discussion of its importance.

$$\theta \cong [(\mu+2)^2 + 4]^{-1/2}, \qquad \psi \cong 45^\circ .$$

Assume for this part of the problem that α and θ are small angles so that $\sin\alpha \cong \alpha$, $\sin\theta \cong \theta$, $\cos\alpha \cong 1$, $\cos\theta \cong 1$.

Problem 6. *Angle of attack and bank angle for maximum lateral range glide.* A quasisteady approximation for gliding turns of a low-speed (subsonic) glider, made with constant angle of attack and constant bank angle, gives lateral gliding range, y_f, as

$$y_f = r(1 - \cos\beta_f),$$

where

$$r = \frac{\ell}{\alpha}\frac{\cos^2\gamma}{\sin\sigma} = \text{radius of the helix},$$

$$\beta_f = \frac{z_0}{\ell}\frac{\alpha\sin\sigma}{\sin\gamma\cos\gamma} = \text{final heading angle},$$

$$\gamma = \tan^{-1}\left[\left(\alpha + \frac{\delta^2}{4\alpha}\right)\sec\sigma\right] = \text{gliding helix angle},$$

and

$\alpha = \eta\bar{\alpha}$; $\bar{\alpha}$ = angle of attack, } (decision parameters),
σ = bank angle }
z_0 = initial altitude,
$\ell = \dfrac{2m\eta}{\rho S C_{L_\alpha}}$ = characteristic length ($\cong$ 10 ft for typical sailplane),
$\delta = 2(\eta C_{D_0}/C_{L_\alpha})^{1/2}$ = minimum drag to lift ratio ($\cong \frac{1}{30}$ for typical sailplane),

η = efficiency factor ($0 < \eta < 1$).

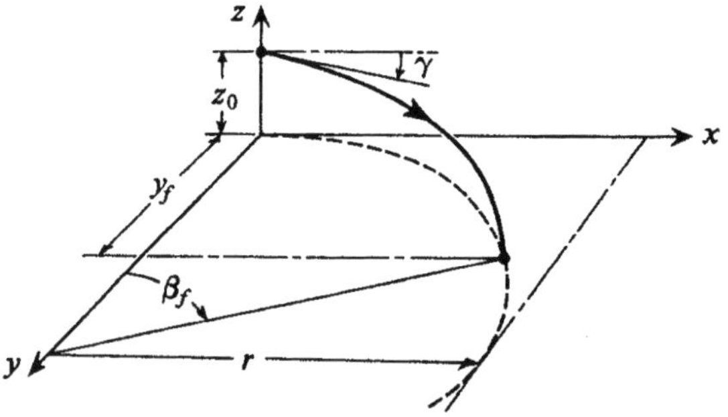

Figure 1.3.3. Geometry of flight path for lateral turn.

Show that the maximum value of y_f for a given z_o is obtained when we have

$$\tan\frac{\beta_f}{2} = \frac{\beta_f}{1 + (4\beta_f^2/\zeta^2)},$$

which may be regarded as a transcendental equation for β_f as a function of $\zeta = z_o/\ell$. The corresponding values of σ, α, and γ are obtained from

$$\tan\sigma = \frac{2\beta_f}{\zeta}, \qquad \alpha = \frac{\delta}{2\sqrt{\cos 2\sigma}}, \qquad \gamma = 2\,\alpha\cos\sigma\,.$$

Assume that α, γ, δ are $<< 1$.

[NOTE 1. Within this same approximation, the maximum value of x_f for given z_o is

$$x_f = z_o/\delta$$

and is obtained with

$$\alpha = (1/2)\delta\,, \qquad \sigma = 0 \Rightarrow \tan\gamma = \delta\,.]$$

[NOTE 2: Further definition of symbols:

m = mass of glider, V = velocity,
ρ = density of the atmosphere (approximated as constant in this problem),
C_{L_α} = lift coefficient slope,
C_{D_0} = zero-lift drag coefficient
S = reference area for coefficients,

$$\text{Lift} = C_{L_\alpha}\bar{\alpha}\frac{\rho V^2}{2}S\,, \qquad \text{Drag} = (C_{D_0} + \eta C_{L_\alpha}\bar{\alpha}^2)\frac{\rho V^2}{2}S\,.]$$

Problem 7. ***Maximum steady rate of climb for an aircraft.*** For the problem stated in Example 2 of Section 1.2, find the maximum steady rate of climb at sea level and at altitudes of 10,000 ft, 20,000 ft, 30,000 ft, and 40,000 ft for an airplane with weight mg = 34,000 lbs and wing area S = 530 ft². The lift, drag, and thrust characteristics are given below:

$$L = C_{L_\alpha}\alpha\frac{\rho V^2}{2}S\,, \qquad D = (C_{D_0} + \eta C_{L_\alpha}\alpha^2)\frac{\rho V^2}{2}S\,.$$

Here C_{L_α}, C_{D_0}, and η are functions of Mach number $M \equiv V/c$, as shown in Figures 1.3.4 and 1.3.5; c = speed of sound and ρ = density of the air, both of which are functions of altitude, that is, $c = c(h)$, $\rho = \rho(h)$. These functions are given in Table 1.3.1. The thrust, T, at full throttle, is a function of Mach number and altitude, as shown in Figure 1.3.5. Use $\epsilon = 3°$.

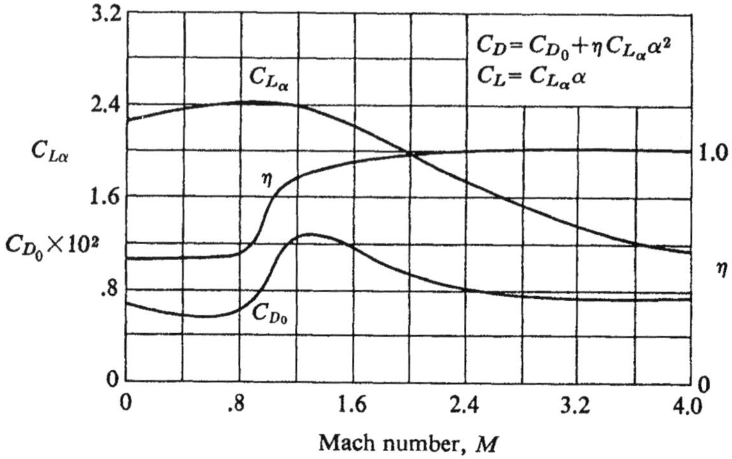

Figure 1.3.4. Drag and lift coefficients as function of Mach number.

Find, also, the altitude at which the maximum rate of climb is zero. This is called the "ceiling" of the airplane.

Table 1.3.1. Air density and speed of sound variation with altitude

Altitude $\sim h$, ft	Speed of sound $\sim C$, ft/sec	Air density $\sim \rho$, slugs/ft^3
0	1,116	2,377 × 10^{-6}
5,000	1,097	2,048
10,000	1,077	1,755
15,000	1,057	1,496
20,000	1,037	1,266
25,000	1,016	1,065
30,000	994.7	889.3
36,090	968.1	706.1
40,000	968.1	585.1
45,000	968.1	460.1
50,000	968.1	361.8
55,000	968.1	284.5
60,000	968.1	223.8
70,000	968.1	138.4
80,000	968.1	85.56
82,020	968.1	77.64
90,000	984.2	51.51
100,000	1,004	31.38

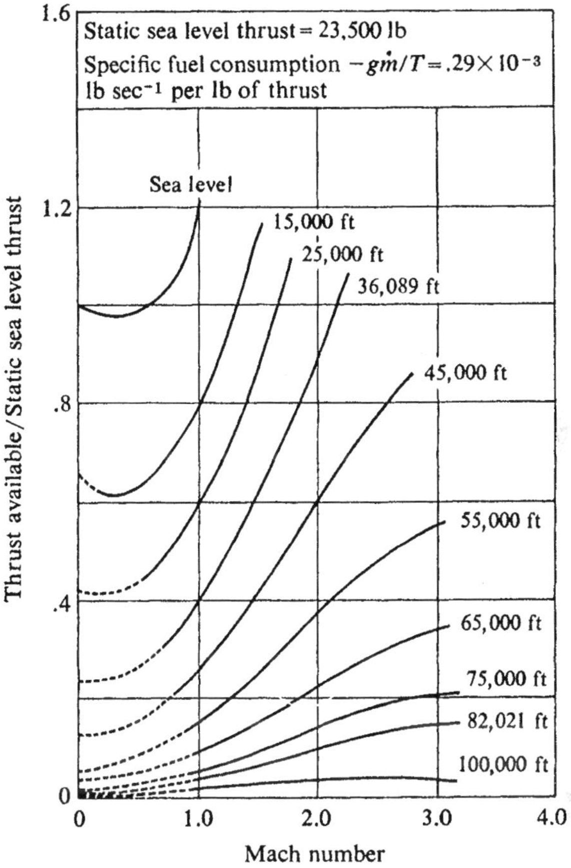

Figure 1.3.5. Thrust as function of Mach number and altitude at full throttle.

Problem 8. *Minimum fuel turn at constant altitude.* A steady turn ($\dot{V} = 0$, $\dot{r} = 0$) at constant altitude is described by

$$(C_{D_0} + \eta C_{L_\alpha} \bar{\alpha}^2)\frac{\rho V^2}{2} S = T \qquad \text{(drag = thrust)},$$

$$mg = C_{L_\alpha} \bar{\alpha} \frac{\rho V^2}{2} S (\cos \sigma) \qquad \text{(weight = vertical component of lift)},$$

$$mV\dot{\beta} = C_{L_\alpha} \bar{\alpha} \frac{\rho V^2}{2} S (\sin \sigma) \qquad \text{(turn rate} \approx \text{horizontal component of lift)},$$

where

$$\left.\begin{array}{l}\bar{\alpha} = \text{angle-of-attack,}\\ \sigma : \text{bank angle}\end{array}\right\} \quad \text{decision parameters}$$

and the rest of the symbols are as defined in Problem 6.

Find $\alpha \triangleq \eta\bar{\alpha}$ and σ to minimize the fuel in making a turn from $\beta = \beta_o$ to $\beta = \beta_f$ where fuel is proportional to

$$J \triangleq \int_0^{t_f} T\,dt = \int_{\beta_o}^{\beta_f} \frac{T\,d\beta}{\dot{\beta}} \equiv \frac{T}{\dot{\beta}}(\beta_f - \beta_o)\,;$$

that is, minimize

$$\frac{T}{\dot{\beta}} = \frac{(C_{D_0} + \eta C_{L_\alpha}\bar{\alpha}^2)\,mV}{C_{L_\alpha}\bar{\alpha}\sin\sigma}$$

subject to $mg = C_{L_\alpha}\bar{\alpha}\left(\dfrac{\rho V^2}{2}\right) S\cos\sigma$.

ANSWER. $\alpha = (\sqrt{3}/2)\delta\,,\ \sigma = \cos^{-1}(1/\sqrt{3}) = 54.7°\,,$
where $\delta = 2\sqrt{\eta C_{D_0}/C_{L_\alpha}}$.

Note that this implies $V = \sqrt{2gl/\delta}\,, \dfrac{L}{D} = (\sqrt{3}/2)\,(1/\delta)$ and $T = 2mg\,\delta$, where $\ell = \dfrac{2m\eta}{C_{L_\alpha}\rho S}$.

1.4. Neighboring optimum solutions and the interpretation of the Lagrange multipliers

Occasionally we wish to find how the optimum solution changes if some of the constants in the constraint equations are changed by small amounts.

Let us suppose that the constraints (1.2.2) are increased by infinitesimal amounts so that we have $f(x,u) = df$, where df is an infinitesimal constant vector. Then, assuming that the values of x and u for the minimal solution will be changed by infinitesimal amounts dx and du, we have, from (1.2.19), (1.2.20), (1.2.18),

$$dH_x^T = H_{xx}\,dx + H_{xu}\,du + f_x^T\,d\lambda = 0\,, \tag{1.4.1}$$

$$dH_u^T = H_{ux}\,dx + H_{uu}\,du + f_u^T\,d\lambda = 0\,, \tag{1.4.2}$$

$$df = f_x\,dx + f_u\,du\,, \tag{1.4.3}$$

where the partial derivatives are evaluated at the point corresponding to the original optimum solution.

The $2n + m$ relations (1.4.1), (1.4.2), and (1.4.3) determine the $2n + m$ parameters dx, du, and $d\lambda$. Since f_x must be nonsingular for du to determine dx, it follows from (1.4.3) and (1.4.1) that

$$dx = f_x^{-1}\, df - f_x^{-1} f_u\, du\,, \tag{1.4.4}$$

$$d\lambda = -(f_x^T)^{-1}(H_{xx}\, dx + H_{xu}\, du)\,. \tag{1.4.5}$$

Substituting these relations into (1.4.2) and solving for *du* yields

$$du = -\,C\, df\,, \tag{1.4.6}$$

where

$$C = \left(\frac{\partial^2 L}{\partial u^2}\right)_{f=0}^{-1} [H_{ux} - f_u^T (f_x^T)^{-1} H_{xx}] f_x^{-1}, \tag{1.4.7}$$

and $(\partial^2 L/\partial u^2)_{f=0}$ is as defined in Equation (1.3.7). Thus, the existence of neighboring optimum solutions is guaranteed if the original stationary point was a local minimum; that is, if $(\partial^2 L/\partial u^2)_{f=0} > 0$.

By substituting (1.4.4) into (1.3.3), with $H_u = 0$, we obtain dL correct to second order. If, in addition, we substitute the expression for *du* from (1.4.6) into (1.3.3), we have, after some manipulation,

$$dL = -\lambda^T df + (1/2)\, df^T [f_x^{-T} H_{xx} f_x^{-1} - C^T L_{uu} C]\, df + \cdots, \tag{1.4.8}$$

where

$$L_{uu} \equiv \left(\frac{\partial^2 L}{\partial u^2}\right)_{f=0},$$

which is given by (1.3.7).

Thus, we have

$$\frac{\partial L_{\min}}{\partial f} = -\lambda^T, \tag{1.4.9}$$

$$\frac{\partial^2 L_{\min}}{\partial f^2} = f_x^{-T} H_{xx} f_x^{-1} - C^T L_{uu} C\,. \tag{1.4.10}$$

1.5 Numerical solution by a first-order gradient method†

Unless the relations for $L(x,u)$ and $f(x,u)$ in Section 1.2 are quite simple, numerical methods must be used to determine the values of u that minimize H. A straightforward numerical method, in common use for many years, is that of ***steepest descent*** for finding minima (or steepest ascent for finding maxima).

Steepest descent or gradient methods are characterized by iterative algorithms for improving estimates of the control parameters, u, so as to come closer to satisfying the stationarity conditions $\partial H/\partial u = 0$.

The steps in using the gradient method are as follows:

†Grateful acknowledgment is made to Walter F. Denham for his assistance in writing this section.

(a) Guess a set of values for u.
(b) Determine the values of x from $f(x,u) = 0$.
(c) Determine the values of λ from $\lambda^T = -(\partial L/\partial x)(\partial f/\partial x)^{-1}$.
(d) Determine the values of $\partial H/\partial u = (\partial L/\partial u) + \lambda^T(\partial f/\partial u)$, which, in general, will *not* be zero.
(e) Interpreting $\partial H/\partial u$ as a gradient vector, change the estimates of u by amounts $\Delta u = -K(\partial H/\partial u)^T$, where K is a positive scalar constant. The predicted change in the criterion, ΔJ, is thus $-K(\partial H/\partial u)(\partial H/\partial u)^T$. (The "−" is replaced by a "+" if a maximum is being sought.)
(f) Repeat steps (a) through (f), using the revised estimates of u, until $(\partial H/\partial u)(\partial H/\partial u)^T$ is very small.

There are many variations of this approach, and we will consider one of them in the next section. Graphically, the gradient method (for finding a maximum) is a hill-climbing technique in the u-space; if u is a two-component vector, we can show contours of constant J in the u-plane (see Figure 1.5.1). Starting with an initial guess of u, a sequence of changes Δu is made. At each step, Δu is in the direction of the gradient $\partial H/\partial u$ whose magnitude gives the steepest slope at that point on the hill. The choice of K, which determines the magnitude of Δu, involves judging the extent of nonlinearity so that the linearized prediction will be reasonably accurate, while at the same time trying to keep the number of steps in the sequence from becoming excessive. K should almost always be varied in the sequence.

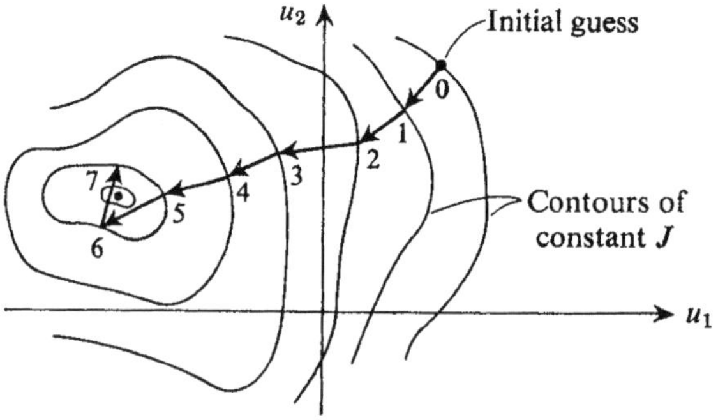

Figure 1.5.1. Typical numerical procedure for first-order gradient method.

Usually this will be done to decrease the magnitude of Δu when it is thought the minimum (or maximum) is near. As Figure 1.5.1 suggests, it is easy to overshoot the extremal point. In problems of higher dimension, the geometrical concept of hypersurfaces of constant J in the u-hyperspace provides valuable insight.

Gradient methods usually show substantial improvements in the first few iterations but have poor convergence characteristics as the optimal solution is approached. Second-order gradient methods that use the "curvature" as well as the "slope" at the nominal point are discussed in the next section; they have excellent convergence characteristics as the optimal solution is approached but may have starting difficulties associated with picking a "convex" nominal solution.†

1.6 Numerical solution by a second-order gradient method

Second-order gradient methods†† use information on the curvature as well as the slope at the nominal point in the u-space. If u is a scalar, we can sketch a simple description of the second-order gradient method as in Figure 1.6.1.

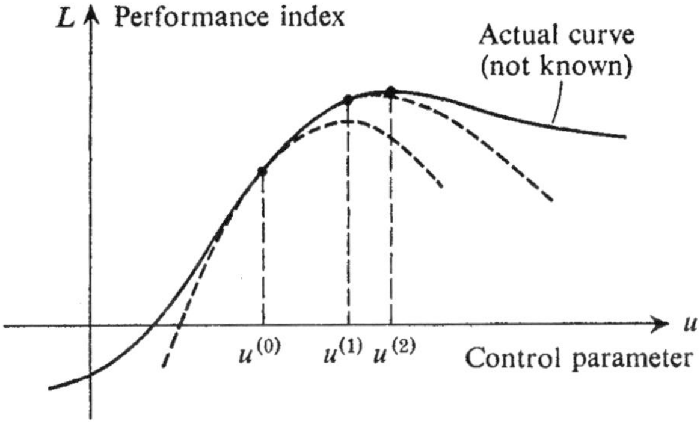

Figure 1.6.1. Typical numerical procedure for second-order ascent method.

The actual curve of performance index L vs. control parameter u could, in principle, be calculated and the maximum picked out. However, this may involve an enormous amount of computation. Instead, using the second-order gradient method, we *guess* a control parameter $u^{(o)}$ and determine $x^{(o)}$ from $f(x^{(o)},u^{(o)})=0$, and then $L(x^{(o)},u^{(o)})$. We then determine the first and second derivatives of L with respect to u, holding $f(x,u)=0$ using (1.2.6) and (1.3.7) and approximate the (L vs. u)-curve by a quadratic curve (a parabola):

$$L \cong L(x^{(o)}, u^{(o)}) + \left(\frac{\partial L}{\partial u}\right)^0_{f=0} (u-u^o) + \frac{1}{2}\left(\frac{\partial^2 L}{\partial u^2}\right)^0_{f=0} (u-u^o)^2 .$$

†By this we mean that the nominal approximating quadratic surface has a minimum.
††These are often called Newton-Raphson methods.

The value of u that yields the maximum of this approximate curve is then easily determined; call it $u^{(1)}$. This value is taken as an improved guess and the process is repeated. In Figure 1.6.1 it is apparent that two steps in the iterative procedure already yield a good approximation to the maximizing value of u. In more complicated problems, more steps may be required. Also, if the initial guess $u^{(o)}$ is too far away from the maximizing value, we may find that $(\partial^2 L/\partial u^2)^o_{f=o} > 0$; i.e., the curvature has the wrong sign. In this case, the method fails completely; note, however, that the first-order gradient method may still converge.

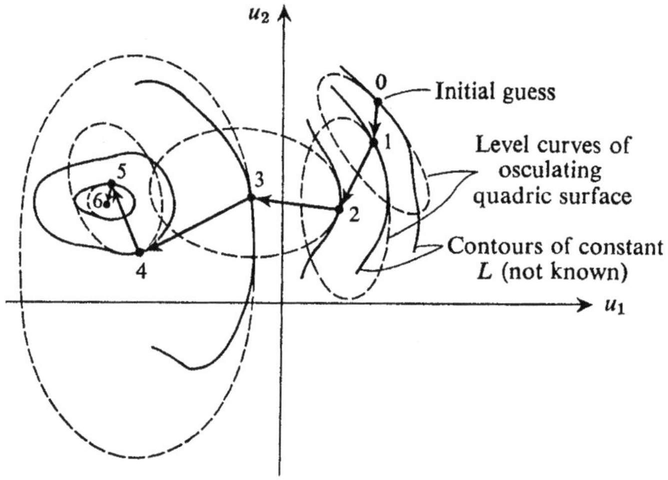

Figure 1.6.2. Two-dimensional illustration of second-order gradient method.

Figure 1.6.2 shows a two-parameter maximization problem with contours of constant performance index L, holding $f(x,u) = 0$ (unknown to the optimizer). An initial guess is made at point 0, and an *osculating quadric surface* is fitted locally to the region around 0 by determining the first and second derivatives of L, holding $f(x,u) = 0$, from (1.2.6) and (1.3.7). If this quadric surface turns out to be an elliptic paraboloid with a maximum (that is, the matrix of second derivatives is negative definite), the location of that maximum is taken as the next guess (point 1).† The procedure is repeated until we have $(\partial L/\partial u)_{f=o} = 0$, hopefully with $(\partial^2 L/\partial u^2)_{f=o} < 0$ all the way. Figure 1.6.2 shows the maximum being achieved in six steps.

†If the matrix of second derivatives is positive definite or indefinite, the procedure fails.

The constraint relations $f(x,u) = 0$ are often so complicated that numerical methods are needed just to determine x, given u. In this case a slightly more general version of the second-order gradient method may be used. Recall that necessary conditions for a stationary value of $L(x,u)$ are

$$H_x = 0\,, \tag{1.6.2}$$

$$H_u = 0\,, \tag{1.6.3}$$

$$f = 0\,, \quad \text{where} \quad H(x,u,\lambda) = L(x,u) + \lambda^T f(x,u). \tag{1.6.4}$$

The steps in the generalized second-order gradient method are:

(a) Guess a set of values for x, u, and λ; call them x^o, u^o, and λ^o.
(b) Determine the values of

$$H_x(x^o,u^o,\lambda^o) = H_x^o\,, \tag{1.6.5}$$

$$H_u(x^o,u^o,\lambda^o) = H_u^o\,, \tag{1.6.6}$$

$$f(x^o,u^o) = f^o\,. \tag{1.6.7}$$

(c) Linearize the relations (1.6.2), (1.6.3), and (1.6.4) about x^o, u^o, and λ^o:

$$H_x^o + H_{xx}^o\,dx + H_{xu}^o\,du + (f_x^T)^o\,d\lambda = 0\,, \tag{1.6.8}$$

$$H_u^o + H_{ux}^o\,dx + H_{uu}^o\,du + (f_u^T)^o\,d\lambda = 0\,, \tag{1.6.9}$$

$$f^o + f_x^o\,dx + f_u^o\,du = 0\,. \tag{1.6.10}$$

(d) Solve Equations (1.6.8), (1.6.9), and (1.6.10) for dx, du, and $d\lambda$ in terms of H_x^o, H_u^o, and f^o†,
(e) Repeat (a) through (d), using, as improved guesses,

$$x^1 = x^o + dx\,, \qquad u^1 = u^o + du\,, \qquad \lambda^1 = \lambda^o + d\lambda\,.$$

The process is repeated until the necessary conditions (1.6.2), (1.6.3), and (1.6.4) are satisfied to the desired degree of accuracy.

If the method converges at all, it may converge on a minimum, a maximum, or a saddle point. To determine which of these it is, we must examine the curvature matrix given in Equation (1.3.7). If the matrix is positive definite, the point is a minimum; if the matrix is negative definite, the point is a maximum; if the matrix is not singular but is neither positive definite nor negative definite, the point is a saddle point; if the matrix is singular, we do not know the nature of the point without going to higher derivatives.

---

†If values of H_x^o, H_u^o, f^o are such that dx, du, $d\lambda$ obtained from step (d) are very large then ϵH_x^o, ϵH_u^o, ϵf^o may be used instead, where $0 < \epsilon < 1$.

Problem. Another variant of the second-order gradient method would be to guess only x and u, determining λ from $H_x = 0$. Work out the procedure for this variant.

1.7 Problems with inequality constraints

Parameter optimization problems involving inequality constraints require extension of the methods treated in the previous sections. An important class of problems of this type involves minimizing

$$L(y) \tag{1.7.1}$$

subject to

$$f(y) \leq 0\,, \tag{1.7.2}$$

where, in general, f and y are vectors of different dimension.†

Consider first the case in which y and f are scalars. There are two possibilities for the optimal value of y, $y^o : f(y^o) < 0$ or $f(y^o) = 0$. In the former case, the constraint is not effective and can be ignored. The situation remains the same as in Section 1.1. In the latter case, consider small perturbations about y^o; if $L(y^o)$ is a minimum, then we have

$$dL = \left.\frac{\partial L}{\partial y}\right|_{y^o} dy \geq 0 \tag{1.7.3}$$

for all *admissible* values of dy, which must satisfy

$$df = \frac{\partial f}{\partial y}\, dy \leq 0\,. \tag{1.7.4}$$

In order that Equations (1.7.3) and (1.7.4) be consistent, it is clearly necessary that either

$$\operatorname{sgn}\frac{\partial L}{\partial y} = -\operatorname{sgn}\frac{\partial f}{\partial y} \qquad \text{or} \qquad \frac{\partial L}{\partial y} = 0\,.$$

These two possibilities may be expressed in one relation as

$$\frac{\partial L}{\partial y} + \lambda\frac{\partial f}{\partial y} = 0\,, \qquad \lambda \geq 0\,. \tag{1.7.5}$$

†Such problems, referred to as *nonlinear programming problems*, have been treated extensively in the technical literature. We give only brief treatment in this section. We will not distinguish between state and control variables as was done in Section 1.2. Indeed, in many nonlinear programming problems, the dimension of f is greater than the dimension of y, so that it is not possible to decompose y into state and decision variables.

The two situations are shown geometrically in Figures 1.7.1(a) and (b).

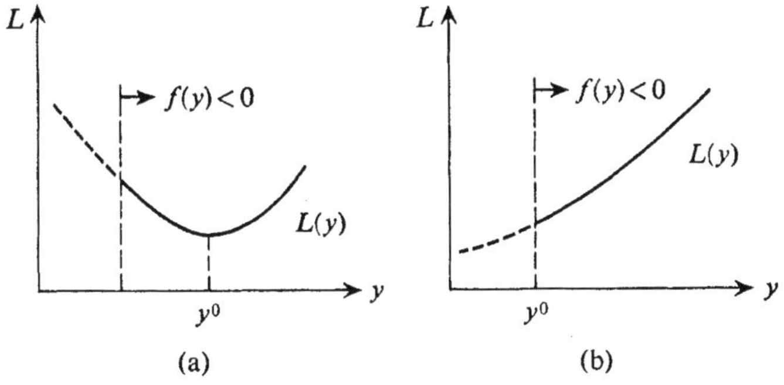

Figure 1.7.1. One-dimensional illustration of two possible types of minimum with inequality constraints.

The two cases may be treated analytically by adjoining (1.7.2) to (1.7.1):

$$H(y,\lambda) = L(y) + \lambda f(y)\,. \tag{1.7.6}$$

The necessary conditions become

$$\frac{\partial H}{\partial y} = 0 \tag{1.7.7}$$

and

$$f(y) \leq 0\,, \tag{1.7.8}$$

where

$$\lambda \begin{cases} \geq 0\,, & f(y) = 0\,, \\ = 0\,, & f(y) < 0\,. \end{cases} \tag{1.7.9}$$

When y is a vector but f is still a scalar, Equations (1.7.3), (1.7.4), and (1.7.5) remain applicable if we interpret the symbols in vector notation. We should interpret conclusion (1.7.5) to mean

$$\frac{\partial L}{\partial y} \text{ parallel to } \frac{\partial f}{\partial y} \text{ and pointing in opposite directions.} \tag{1.7.10}$$

The necessity of (1.7.10) is easily established by contradiction. Let us suppose that (1.7.10) is not true as illustrated in two dimensions in Figure 1.7.2. Then the cross-hatched region represents a region of admissible y which will yield smaller L.

This and the other situation (namely, when $f(y^o) < 0$) again can be

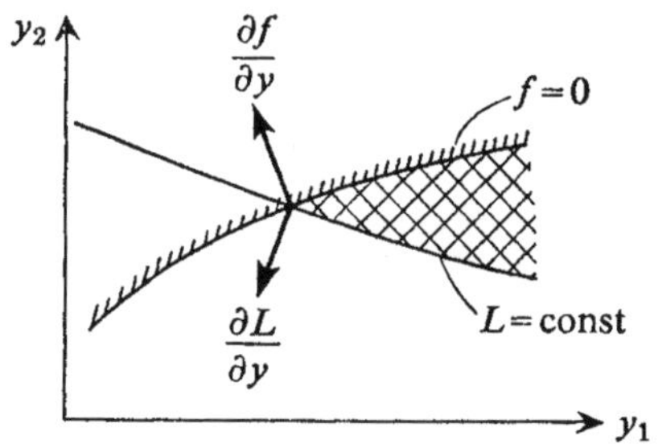

Figure 1.7.2. Two-dimensional illustration showing the necessity of Equation (1.7.10).

summarized by the necessary conditions stated by (1.7.7), (1.7.8), and (1.7.9).

In the more general case, when f itself is a vector, we can still employ (1.7.4) and (1.7.5), noting that $\partial f/\partial y$ is now a matrix. If only one component of f is effective, the problem is the same as that just treated. If two components of f are effective, the situation, in two dimensions, is as shown in Figure 1.7.3.

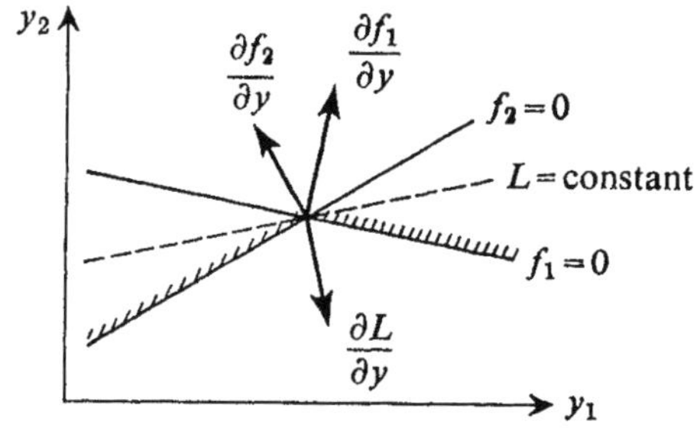

Figure 1.7.3. Two-dimensional illustration of minimization subject to two inequality constraints.

It is clear that, if y^o is to be an optimal point on $f_1 = f_2 = 0$, then $\partial L/\partial y$ must lie between the negative gradients of f_1 and f_2.† Analytically, this means that $\partial L/\partial y$ can be expressed as a *negative* linear combination of $\partial f_1/\partial y$ and $\partial f_2/\partial y$. In general, when q components are effective at a boundary optimal point, we must have

$$\frac{\partial L}{\partial y} + \lambda_1 \frac{\partial f_1}{\partial y} + \cdots + \lambda_q \frac{\partial f_q}{\partial y} = 0 \qquad (1.7.11)$$

†Recall the parallelogram construction of a resultant from two components.

or

$$\frac{\partial L}{\partial y} + \lambda^T \frac{\partial f}{\partial y} = 0\,, \tag{1.7.12}$$

where

$$\lambda \begin{cases} \geq 0\,, & f(y) = 0\,, \\ = 0\,, & f(y) < 0\,. \end{cases} \tag{1.7.13†}$$

Hence, as in Section 1.2, we may define a quantity $H \equiv L + \lambda^T f$ and write (1.7.12) as $\partial H/\partial y = 0$. Equations (1.7.12) and (1.7.13) are necessary conditions for minimality. For a maximum, the sign of λ must be changed in (1.7.13). In words, *the gradient of L with respect to y at a minimum must be pointed in such a way that decrease of L can only come by violating the constraints.*

Let us suppose that y has p components and that n components of the inequality constraint are "effective," that is,

$$f_i(y) = 0\,, \qquad i = 1, \ldots, n\,. \tag{1.7.14}$$

The "ineffective" constraints, $f_i(y) < 0$, $i = n+1, \ldots$, may be disregarded. It is clear that $p \geq n$. Next, take n of the components of y and call them x; let the remaining $p - n$ components be called u; that is,

$$y^T = (x_1, \ldots, x_n\,; u_1, \ldots, u_{p-n}) \triangleq (x^T, u^T)\,.$$

The choice must be such that

$$f_i(x,u) = 0\,, \qquad i = 1, \ldots, n \tag{1.7.15}$$

determines x when u is given. Then *sufficient conditions* for a local minimum of $L(y)$, with $f(y) \leq 0$, are given in Section 1.3, to which we must add the condition that $\lambda_1, \ldots, \lambda_n$ *all* be positive.‡ The latter condition is easily interpreted from Equation (1.4.8) since $-\lambda_i$ is equal to $(\partial L/\partial f_i)_u$, which must be negative (that is, $dL > 0$ for $df_i < 0$).

Example. Consider $L(y_1,y_2)$ with $f_i(y_1,y_2) \leq 0$, $i = 1,2$, and suppose the level curves are as shown in Figure 1.7.4.
It is clear that $f_2 < 0$ is "ineffective" and $f_1 = 0$. From Figure 1.7.4, we have

$$\frac{\partial L}{\partial y} + \lambda_1 \frac{\partial f_1}{\partial y} = 0\,, \qquad \lambda_1 > 0\,;$$

---

†Equation (1.7.13) is understood, of course, to be in the component-by-component sense.

‡For a precise statement, see G. McCormick, "Second Order Sufficient Conditions for Constrained Minimum," *SIAM Journal on Appl. Math.*, Vol. 15, No. 3, May 1967.

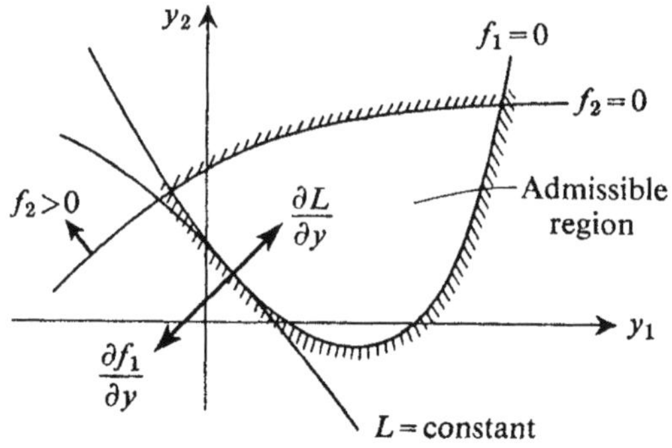

Figure 1.7.4. Example of minimization subject to inequality constraints.

that is, grad L is parallel and in the opposite direction to grad f_1. Also, the "curvature" of L along $f_1 = 0$ is such that L increases on $f_1 = 0$ away from the minimum; to show this analytically, we may let $y_1 = x$ and $y_2 = u$. Then we have

$$f_1(x,u) = 0 \Rightarrow x, \qquad \text{given } u,$$

and, from (1.3.7), we can compute $(\partial^2 L/\partial u^2)_{f_1=0}$ which, as we can see from Figure 1.7.4, is positive.

Equations (1.7.12) and (1.7.13) are the essence of the Kuhn-Tucker Theorem in nonlinear programming. The precise statement of the condition requires the assumption of the so-called "constraint qualification" on the set $f(y) \leq 0$ (Kuhn-Tucker, p. 483). This qualification is designed to rule out geometric situations as shown in Figure 1.7.5. At the minimum point we see that $(y_1,y_2) = (1,0)$ and $(\partial L/\partial y)$ is *not* equal to any finite linear combination $(\partial f_1/\partial y)$, $(\partial f_2/\partial y)$.

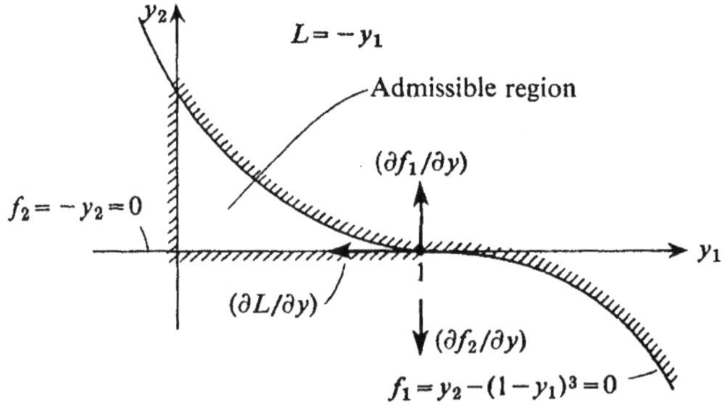

Figure 1.7.5. Example of Kuhn-Tucker constraints qualification.

Another approach to sufficiency is the *saddle-point theorem* of nonlinear programming. It is more elegant (but usually harder to apply) than the conditions given above since it does not require the arbitrary separation of y into x and u. Consider the function $H(y,\lambda) = L + \lambda^T f$. Suppose that it is possible to find y^o and λ^o such that they constitute a saddle point for H; that is,

$$H(y^o,\lambda) \leq H(y^o,\lambda^o) \leq H(y,\lambda^o) \qquad (1.7.16)$$

for all $\lambda \geq 0$ and $f(y) \leq 0$. Then we may conclude that y^o is a minimum point for $L(y)$ subject to $f(y) \leq 0$, regardless of the nature of L and f.

Problem 1. Prove the saddle-point theorem. [HINT: The left-hand inequality in (1.7.16) implies that $\lambda_i^o f_i(y^o) = 0$ for all i.]

Problem 2. *Aircraft cruise condition for minimum fuel consumption.* For the airplane described in Example 2, Section 1.2, and in Problem 7, Section 1.3, find the steady level-flight ($\gamma = 0$) condition for minimum fuel consumption per unit distance. Assume constant specific fuel consumption, $\sigma = .29 \times 10^{-3}$ lb sec^{-1} per lb of thrust, so that fuel consumption per unit distance is given by

$$J = \frac{\sigma T}{V},$$

where

$$T \leq T_{max}(V,h)$$

and $T_{max}(V,h)$ is as given graphically in Problem 7, Section 1.3.
The constraint equations are

$$L - mg + T \sin(\alpha + \epsilon) = 0, \qquad D - T\cos(\alpha + \epsilon) = 0,$$

where $L = L(V,h,\alpha)$, $D = D(V,h,\alpha)$ are as given in Problem 7, Section 1.3.

Problem 3. Write out a mathematical proof of the geometrical argument of Figure 1.7.2. In particular, show why $\lambda \geq 0$.

1.8 Linear programming problems

If both the performance index and the inequality constraints are linear functions of y, the problem is called a *linear programming problem*. Clearly, in this case, the minimum, if it exists, *must* occur on the boundary since the curvature of L is zero everywhere. Let the problem be to choose y to minimize

$$L = b^T y\,, \tag{1.8.1}$$

subject to

$$A^T y + c \leq 0\,, \tag{1.8.2}$$

where y is an n-vector and c is an m-vector, $m > n$. If A is of rank n and b^T is not collinear with any of the rows of A^T or any negative linear combinations of $n + 1$ rows of A^T, the minimum, if it exists, must occur at a *point* determined by the simultaneous satisfaction of n of the constraints $A^T y + c = 0$. This result is not surprising to anyone with geometric intuition; it is the *fundamental theorem of linear programming.*

Example 1. Minimize $L = -5y_1 - y_2$ subject to

$$f_1 = -y_1 \leq 0\,, \quad f_2 = -y_2 \leq 0\,, \quad f_3 = y_1 + y_2 - 6 \leq 0\,,$$
$$f_4 = 3y_1 + y_2 - 12 \leq 0\,, \quad f_5 = y_1 - 2y_2 - 2 \leq 0\,.$$

Figure 1.8.1 shows the admissible region, with contours of constant L.

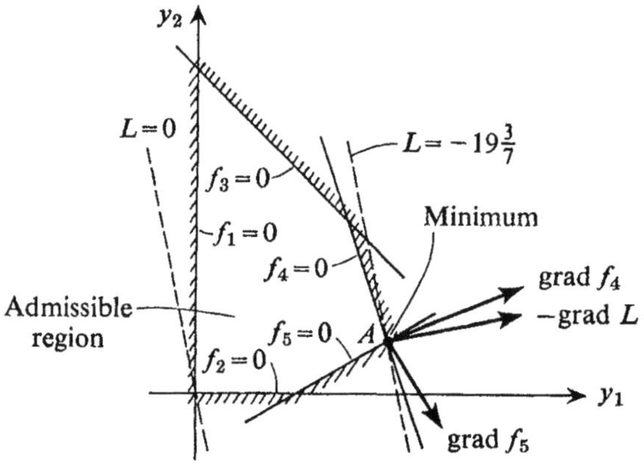

Figure 1.8.1. Solution of the linear programming problem of Example 1.

Obviously, the minimum occurs at point A, where we have $3y_1 + y_2 - 12 = 0$ and $y_1 - 2y_2 - 2 = 0 \Rightarrow y_1 = 3\frac{5}{7}\,, y_2 = \frac{6}{7} \Rightarrow L_{\text{min}} = -19\frac{3}{7}$, and grad L can be expressed as a negative linear combination of n (but not $n - 1$) rows of A^T (namely, grad f_4 and grad f_5), as is obvious from Figure 1.8.1.

The implication of the fundamental theorem of linear programming for the *numerical solution of linear programming problems* is clear. Take n constraints at a time and treat them as equalities. Solving the equalities yields one solution (assuming admissibility), which is either optimal or nonoptimal. In the latter case, we can discard one of the constraints and, substituting another, repeat the process with the requirement that the new solution be admissible and better. Since there are only a finite number of such possibilities, this process must eventually arrive at the optimal combination (if it exists). The method that accomplishes this is known as the *simplex* method. We shall say more about this in the next section.

Problem. Show that the necessary conditions for a maximum of

$$L = c^T\lambda ,$$

subject to

$$A\lambda + b = 0 , \qquad \lambda \geq 0 ,$$

are simply (1.8.2) of the minimization problem discussed in this section. These two problems are called *duals* of each other.

Example 2. *A blending problem.* There are many *blending problems* that involve finding the cheapest mixture of several materials that contains at least a certain fraction of specified ingredients. A typical problem is to find the cheapest mixture of several feeds that contains at least certain specified amounts of nutrients (proteins, fats, vitamins, etc.).† Suppose that we are considering a mixture of three feeds and we have three inequality specifications on nutrients. Table 1.8.1, below, shows the fraction of each of the three nutrients contained in each of the three feeds and the cost per unit amount of each of the three feeds.

Table 1.8.1

Feed	Fraction of nutrient in each feed 1	2	3	Cost
1	.06	.02	.09	15
2	.03	.04	.05	12
3	.04	.01	.03	8

Our problem is to find the cheapest mixture of the three feeds such that the fraction of nutrients one, two, three in the mixture is greater than or equal to .04, .02, and .07, respectively.

Let F_j = the fraction of feed j in the mixture, where j = 1,2, or 3; these are the quantities we are trying to find (our design parameters).

†Other blending problems occur in mixing fuel oils and fertilizers.

Let N_i = fraction of nutrient i in the mixture, where i = 1,2, or 3; then we have

$$N_i = n_{i1}F_1 + n_{i2}F_2 + n_{i3}F_3 ,$$

where n_{ij} = fraction of nutrient i in feed j (the data given in Table 1.8.1). For this problem we have $N_1 \geq .04$, $N_2 \geq .02$, $N_3 \geq .07$.

Let C = cost per unit amount of the mixture and c_j = cost per unit amount of feed j (also given in Table 1.8.1). Then we have

$$C = c_1F_1 + c_2F_2 + c_3F_3 .$$

Of course, the fractions of the three feeds in the mixture must add up to one:

$$F_1 + F_2 + F_3 = 1 .$$

The problem, then, is to find the *two* quantities F_1 and F_2 (using $F_3 = 1 - F_1 - F_2$) to minimize C and satisfy the inequalities.†

$$N_i \geq \overline{N}_i , \qquad i = 1,2,3, \qquad 0 \leq F_j \leq 1 , \qquad j = 1,2$$

where $\overline{N}_i$ is the minimum allowable fraction of nutrient i in the mixture.

We draw a graph, using F_1 and F_2 as coordinates (see Figure 1.8.2). The inequalities are shown as lines with arrows perpendicular to them pointing in the "allowable" directions. In this problem, the inequalities are

(a) $N_1 = .06F_1 + .03F_2 + .04(1-F_1-F_2) \geq .04$ or $2F_1 - F_2 \geq 0$;
(b) $N_2 = .02F_1 + .04F_2 + .01(1-F_1-F_2) \geq .02$ or $F_1 + 3F_2 \geq 1$;
(c) $N_3 = .09F_1 + .05F_2 + .03(1-F_1-F_2) \geq .07$ or $3F_1 + F_2 \geq 2$;
(d) $F_3 = 1 - F_1 - F_2 \geq 0$ or $F_1 + F_2 \leq 1$;
(e) $0 \leq F_1 \leq 1$;
(f) $0 \leq F_2 \leq 1$.

Note that the inequality $F_3 = 1 - F_1 - F_2 \leq 1$ or $F_1 + F_2 \geq 0$ is redundant since we have $F_1 \geq 0$ and $F_2 \geq 0$. Which of the other inequalities are redundant? (See the graph, Figure 1.8.2.) The *feasible region* is the region in the graph where all of the inequalities are satisfied; it is clearly marked in Figure 1.8.2 and is surrounded by extra-heavy lines.

The lines of constant cost are given by setting C = different constants in

$$C = 15F_1 + 12F_2 + 8(1-F_1-F_2) \qquad \text{or} \qquad C = 8 + 7F_1 + 4F_2 .$$

†Several of these turn out to be redundant; i.e., other inequalities automatically cause them to be satisfied.

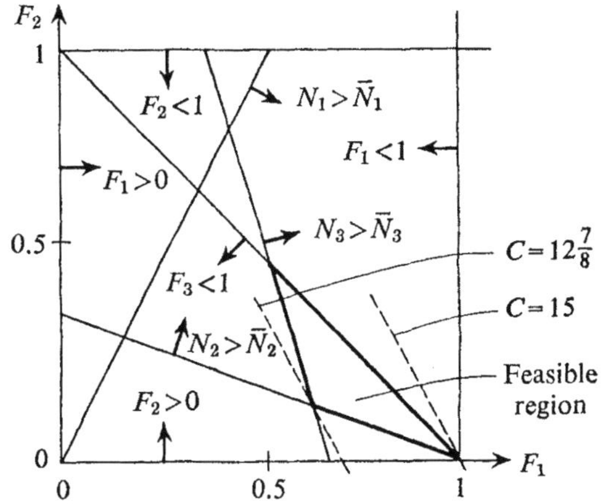

Figure 1.8.2. Solution of the linear programming problem of Example 2.

From Figure 1.8.2 it is clear that the cheapest feasible solution is one that occurs at the corner where $N_2 = \bar{N}_2$ and $N_3 = \bar{N}_3$. From above, this requires that

$$F_1 + 3F_2 = 1, \qquad 3F_1 + F_2 = 2.$$

These are two linear equations in two unknowns (F_1 and F_2); they are easily solved to yield

$$F_1 = \tfrac{5}{8} \qquad \text{and} \qquad F_2 = \tfrac{1}{8}.$$

The amount of feed number three is obtained by substituting the results above into $F_3 = 1 - F_1 - F_2$. This yields

$$F_3 = \tfrac{1}{4}.$$

The minimum cost is given by

$$C = 15(\tfrac{5}{8}) + 12(\tfrac{1}{8}) + 8(\tfrac{1}{4}) = 12\tfrac{7}{8} \quad \text{per unit amount of mixture.}$$

The amount of nutrient one is *above* the minimum required fraction:

$$N_1 = .06(\tfrac{5}{8}) + .03(\tfrac{1}{8}) + .04(\tfrac{1}{4}) = .05125 > .04.$$

Notice, also, that the *most expensive feasible solution* is to use feed number one all by itself.

Example 3. *A transportation planning problem.* A grain dealer owns 50,000 bushels of wheat in Grand Forks, North Dakota, and 40,000 bushels in Chicago. He has sold 20,000 bushels to a customer in Denver, 36,000 bushels to a customer in Miami, and the remaining

34,000 bushels to a customer in New York. He wishes to determine the minimum-cost shipping schedule, given the following freight rates in cents per bushel:

Table 1.8.2

	Denver	Miami	New York
Grand Forks	42	55	60
Chicago	36	47	51

Different modes of shipment cause the rates *not* to be proportional to the distance between the cities. For convenience, we can combine our data into a table, leaving space for the answer, as follows:

Table 1.8.3

Destination / Origin	Denver	Miami	New York	
Grand Forks	42	55	60	50,000
Chicago	36	47	51	40,000
	20,000	36,000	34,000	

The figure in the upper right-hand corner of each square is the freight rate between the two cities.

Our problem is to find a *nonnegative amount* in each of the six squares so that (a) the amounts in the first row add up to 50,000 and the amounts in the second row add up to 40,000; (b) the amounts in the first, second, and third columns add up to 20,000, 36,000, and 34,000, respectively; (c) the total freight cost is a minimum; this cost is obtained by multiplying the amount in each square by the rate in the upper right-hand corner and adding these numbers together.

This problem is a little bit like a crossword puzzle, only harder, since it is not sufficient just to get the rows and columns to add up properly (a feasible solution); we must, in addition, minimize the total cost. By "cut and try" we might be able to find the solution. However, a systematic approach is apt to take less time, and, for problems with more shipping points and more destinations, a systematic approach (an algorithm) and a computer are essential.

Suppose we designate the amount from Grand Forks to Denver as "x" in thousands of bushels. Then, clearly, the amount from Chicago to Denver must be $20 - x$ (see Table 1.8.4). Similarly, let us designate the amount shipped from Grand Forks to Miami as "y"; then the amount from Chicago to Miami must be $36 - y$. Now the amount from Grand Forks to New York must be $50 - x - y$, and the amount from Chicago to New York must be $40 - (20-x) - (36-y) = x + y - 16$. (We automatically satisfy the requirement that the total shipped to New York be 34,000 since the total amount sold equals the amount owned.)

Table 1.8.4

42 x	55 y	60 $50 - x$ $- y$	50
36 $20 - x$	47 $36 - y$	51 $x + y$ $- 16$	40
20	36	34	

We have reduced the number of unknowns to two, x and y, which must satisfy six inequalities:

$$x \geq 0, \quad y \geq 0, \quad 50 - x - y \geq 0, \quad 20 - x \geq 0,$$
$$36 - y \geq 0, \quad x + y - 16 \geq 0.$$

We can conveniently plot all these inequalities on an (x versus y)-graph as in Figure 1.8.3. Again, as in the previous section, there is a *feasible region* where all the inequalities are satisfied.

Next, we calculate the cost in terms of x and y:

$$C = 1000/100 \times [42x + 55y + 60(50-x-y) + 36(20-x) + 47(36-y) + 51(x+y-16)],$$

or

$$C = 45{,}960 - 30x - 10y \quad \text{(in dollars)}.$$

Lines of constant cost are shown in Figure 1.8.3 as dashed lines; clearly, the feasible solution with minimum cost is at $x = 20$, $y = 30$ in Figure 1.8.3. This minimum cost solution is shown in Table 1.8.5.

Note that *no* wheat should be shipped from Chicago to Denver, even though such shipment would involve the lowest rate per bushel.

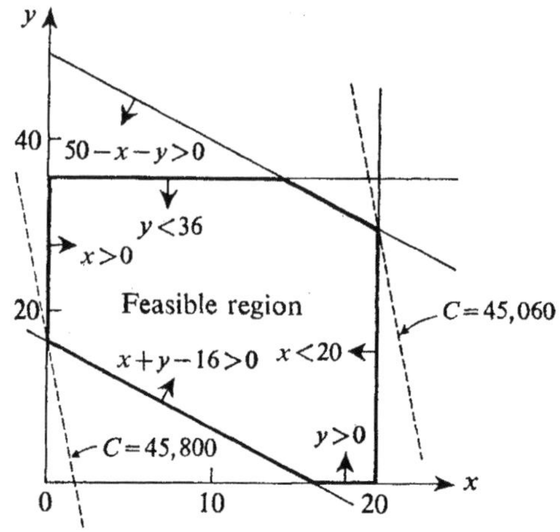

Figure 1.8.3. Solution to minimum cost shipping problem.

Table 1.8.5

42 20	55 30	60 0	50
36 0	47 6	51 34	40
20	36	34	

The difference between the best and the worst feasible solutions is only \$740 out of about \$45,000. However, this 1.6% difference could be a substantial per cent of the profit involved in the sales.

1.9 Numerical solution of problems with inequality constraints

Numerical solution of optimization problems with inequality constraints is one of the major concerns of the field of "mathematical programming." Numerous texts exist on the subject (e.g., Zoutendijk), and we shall describe only the main features of the "method of feasible directions" or "gradient projection method." This method is divided into two separate but related steps:

STEP 1. *Finding a feasible solution.* With reference to Section 1.7, locating a value of y such that $f(y) \leq 0$ is often *not* a trivial task. In problems with equality constraints, such as those treated in Sections 1.5 and 1.6, finding a feasible solution is usually straightforward since there are more variables (x and u) than there are constraint equations

($f(x,u) = 0$). In problems with inequality constraints, there are often more constraint equations (components of f) than variables (components of y). Finding a feasible solution may be approached by guessing a value for y, then considering a small perturbation, dy, which will change f according to

$$df = \frac{\partial f}{\partial y} dy . \tag{1.9.1}$$

If certain components of $f(y)$ are greater than zero, i.e., not feasible, we require a dy such that the corresponding components of df are less than zero. In other words, $f(y + dy)$ should be an improvement toward a feasible solution, that is,

$$F \, dy \leq 0 , \tag{1.9.2}$$

where F contains only rows of $\partial f/\partial y$ corresponding to infeasible values of f. The problem is thus reduced to finding feasible solutions to successive linear inequalities (instead of nonlinear inequalities).

STEP 2. *Finding a feasible improvement.* If a feasible y can be found, the next step is to find a dy which is not only feasible but which also improves the performance index; that is, we must have $f(y + dy) \leq 0$ and $L(y + dy) < L(y)$. This gives rise to another set of linear inequalities like (1.9.2) above:

$$\begin{bmatrix} \dfrac{\partial L}{\partial y} \\ \hline \dfrac{\partial f}{\partial y} \end{bmatrix} dy \equiv H \, dy \leq 0 . \tag{1.9.3}$$

Example. *Quadratic performance index with linear inequality constraints.* Minimize

$$L = \frac{(y_1 - 2)^2}{4} + (y_2 - 1)^2$$

subject to

$$3y_1 + 2y_2 - 6 \leq 0 , \qquad y_1 > 0 , \qquad y_2 > 0 .$$

A sketch of the admissible region with contours of constant L is shown in Figure 1.9.1. If we guess $y_1 = y_2 = \frac{1}{2}$ to start with, we find that

$$\frac{\partial L}{\partial y_1} = -\frac{3}{4} , \frac{\partial L}{\partial y_2} = -1 .$$

Since we are minimizing, the greatest improvement will be in the direction of the *negative gradient* that is shown at $A(y_1 = y_2 = \frac{1}{2})$. We

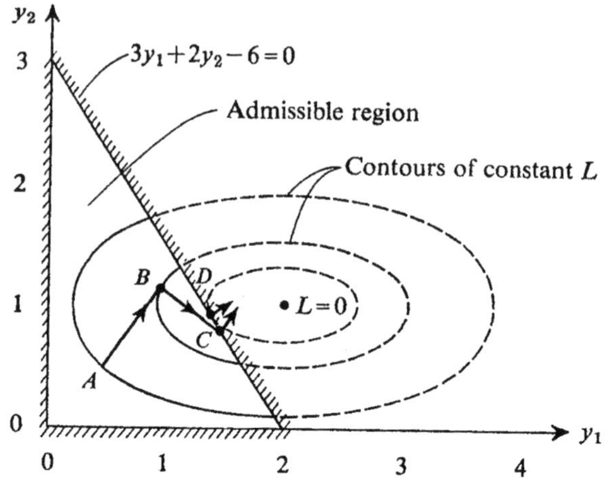

Figure 1.9.1. Solution of the quadratic programming problem of preceding example.

proceed in that direction until we reach a minimum (which may be on a boundary). In this case we reach a minimum at point B in the *interior* of the admissible region. Following the direction of the negative gradient at B, we reach point C on the constraint $3y_1 + 2y_2 - 6 = 0$. Here the negative gradient points *out of* the admissible region, so we take the component of the negative gradient *along the constraint boundary*, which is *up* in this case. Moving along the constraint, we finally arrive at the minimum at D, where the negative gradient points out of the admissible region and is perpendicular to the boundary.

As discussed in Section 1.8, *linear programming problems* have gained a great importance in recent years. Thus, it seems worthwhile to discuss briefly the special procedures applicable to them. Consider again the problem of minimizing

$$L = b^T y \tag{1.9.4}$$

subject to

$$A^T y + c \leq 0\,, \tag{1.9.5}$$

where y is an n-vector, A^T is an $(m \times n)$-matrix, $m > n$. We know that the minimum must occur at a point intersection of n hyperplanes† whose normals are in the direction of rows of A^T. We start the solution, then, by selecting n equations out of (1.9.5) and solving them

†There are abnormal situations in which the minimum lies on an "edge" rather than at a "point"; see conditions specified in Section 1.8 for a "point" solution.

(set *equal* to zero). If this point is feasible, we then examine the n "edges" leading away from this point (formed by the intersection of the n sets of $n-1$ hyperplanes we have chosen); these "edges" will have directions, away from the point, of $e^1, e^2, \ldots, e^n$, where e^i is a *unit* n-vector along the ith edge. The gradient of L is simply b^T, so we consider the projection of the edge directions on b^T; that is, we consider the scalar products b^Te^i, $i = 1, \ldots, n$. If all the scalar products are positive, it is not possible to move along any edge to obtain an improvement (i.e., a smaller value of L); we have the optimal solution. On the other hand, if some of the scalar products are negative, let us choose the one with the largest magnitude and move along the corresponding edge until we encounter another constraint. This new constraint and the $n-1$ old constraints that form the "edge" determine a new point at which the value of L is necessarily smaller since we moved along a *projected gradient* $b^Te^i < 0$. The process is then repeated over and over until a point is found at which we have all $b^Te^i > 0$; that is, no improvement is possible. This is the basis of the *simplex algorithm* proposed by Dantzig (1963), which uses, essentially, a method of feasible directions.

Problem. Perform one step of the above-described process for Example 2 of Section 1.8. Why are we allowed to move as far as the next constraint boundary during each step of the simplex method?

1.10 The penalty function method

Another method for the handling of equality constraints as well as inequality constraints is the so-called penalty function method. The idea is quite simple. Suppose we wish to minimize $L(y)$ subject to

$$f(y) = 0 . \tag{1.10.1}$$

Instead of solving the desired problem directly, we consider the minimization of

$$\overline{L} = L(y) + K\|f(y)\|^2 \tag{1.10.2}$$

subject to *no* constraints, where K is very large.† If $\overline{L}$ attains a minimum at y^*, it is reasonable to expect that

$$f(y^*) \approx 0 \tag{1.10.3}$$

and

$$L(y^*) \approx L(y^o) , \tag{1.10.4}$$

†Other functions of $f(y)$ that are zero when $f(y) = 0$ and positive when $f(y) \neq 0$ are, of course, possible.

where y^o is the value of y that minimizes L subject to $f(y) = 0$. In fact, it is possible to show that, in some cases,

$$\lim_{K\to\infty} y^* \to y^o\,, \lim_{K\to\infty} L(y^*) \to L(y^o)\,. \tag{1.10.5}$$

Computationally, the penalty function method is appealing and has been used both in parameter optimization problems and in function optimization problems (see Chapters 2, 3, and 4).

Nevertheless, it is important to note that, *in practice*, the penalty function method occasionally does *not* come very close to the proper limit indicated in (1.10.5). One reason for this is as follows: The augmented performance index (1.10.2), with large K, has a long narrow "valley" containing the point y^* at the "bottom" (see Figure 1.10.1). Gradient procedures for finding this point tend to go back and forth, from one side of the narrow valley to the other side, instead of down the "long" direction of the valley. Even worse, if K is very large, the "width" of the valley becomes comparable to the numerical accuracy of the computation, and the gradient procedure breaks down completely.

Another potential source of difficulty is the creation of artificial minima that are not present in the original problem.

Example. Find y_1 and y_2 to minimize

$$L = (y_1 - 2)^2 + y_2^2$$

subject to $y_1 = 0$. Now, this is a trivial problem with the obvious answer $y_1 = y_2 = 0$. However, if we use the penalty function approach, we minimize

$$\overline{L} = (y_1 - 2)^2 + y_2^2 + K\,y_1^2 = y_2^2 + \left[\left(y_1 - \frac{2}{1+K}\right)^2 \Big/ \frac{1}{1+K}\right] + \frac{4K}{1+K}\,.$$

Contours of constant $\overline{L}$ are ellipses with centers at $y_1 = (2/1 + K)$, $y_2 = 0$ and semi-axes in a ratio of $\sqrt{1+K}$. Figure 1.10.1 shows contours of constant $\overline{L}$ for $K = 35$. Note the long "valley" created by the penalty function.

Inequality constraints can also be handled by the penalty function method. Suppose that, instead of (1.10.1), we have

$$f(y) \le 0\,. \tag{1.10.6}$$

Then we may consider minimizing

$$\overline{L} = L(y) + K[f(y)]^2\, \mathbf{1}[f(y)]\,, \tag{1.10.7}$$

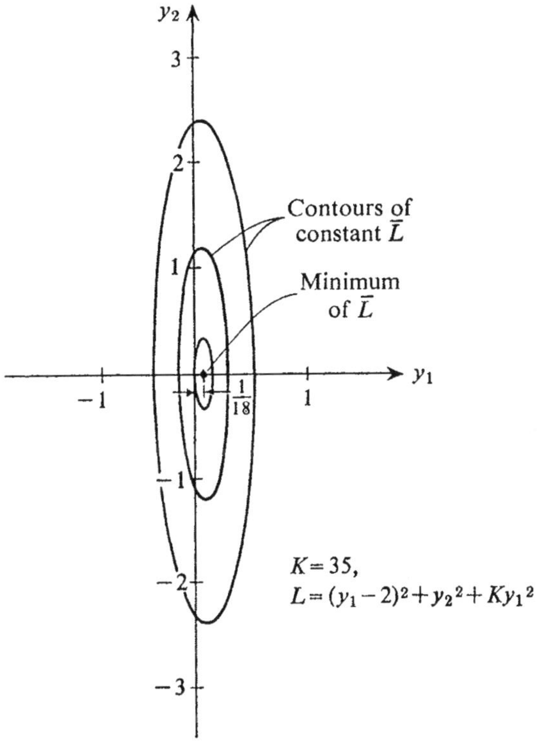

Figure 1.10.1 Cost contours created by penalty functions in the preceding example.

where $1(f)$ is the unit step function defined as

$$1(f) = \begin{cases} 1, & f > 0, \\ 0, & f < 0. \end{cases} \tag{1.10.8}$$

The use of penalty functions is often quite helpful during the initial stages of numerical computation on problems with complex constraints.

Optimization problems for dynamic systems

2

2.1 Single-stage systems

As an introduction to multistage systems, let us consider the simplest nontrivial multistage system, namely, the single-stage system.

A system is initially in a known state described by $x(0)$, an n-dimensional state vector. Choice of an m-dimensional control vector $u(0)$ determines a transition to a state described by $x(1)$ through the relation

$$x(1) = f^o[x(0), u(0)] , \tag{2.1.1}$$

which is shown schematically in Figure 2.1.1.

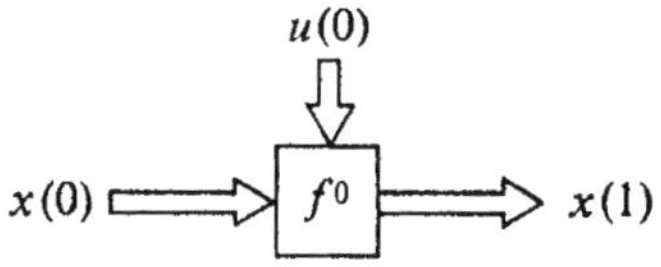

Figure 2.1.1. Flow chart for a single-stage system.

We wish to choose $u(0)$ to minimize a performance index of the form

$$J = \phi[x(1)] + L^o[x(0), u(0)] . \tag{2.1.2}$$

This is a parameter optimization problem with equality constraints, exactly like the ones considered in Section 1.2. We shall treat it in the same way, differing only slightly in our choice of an H-function; we adjoin the constraints (2.1.1) to (2.1.2) with undetermined multipliers $\lambda(1)$:

$$\bar{J} = \phi[x(1)] + L^o[x(0), u(0)] + \lambda^T(1)\{f^o[x(0), u(0)] - x(1)\} . \tag{2.1.3}$$

Now, let

$$H^o = L^o[x(0), u(0)] + \lambda^T(1) f^o[x(0), u(0)], \tag{2.1.4}$$

so that (2.1.3) may be written as

$$\bar{J} = \phi[x(1)] + H^o[x(0), u(0), \lambda(1)] - \lambda^T(1)x(1). \tag{2.1.5}$$

Next, consider infinitesimal changes in $\bar{J}$ due to infinitesimal changes in $u(0)$, $x(1)$, and $x(0)$:

$$d\bar{J} = \left[\frac{\partial \phi}{\partial x(1)} - \lambda^T(1)\right] dx(1) + \frac{\partial H^o}{\partial u(0)} du(0) + \frac{\partial H^o}{\partial x(0)} dx(0). \tag{2.1.6}$$

An expedient choice of $\lambda(1)$ is apparent from (2.1.6); to avoid determining $dx(1)$ in terms of $du(0)$ by differentiating (2.1.1), we *choose*

$$\lambda^T(1) = \frac{\partial \phi}{\partial x(1)}. \tag{2.1.7}$$

As a result of this choice, (2.1.6) becomes

$$d\bar{J} = \frac{\partial H^o}{\partial u(0)} du(0) + \frac{\partial H^o}{\partial x(0)} dx(0). \tag{2.1.8}$$

Thus $\partial H^o/\partial u(0)$ is the gradient of J with respect to $u(0)$, holding $x(0)$ constant and satisfying (2.1.1), and $\partial H^o/\partial x(0)$ is the gradient of J with respect to $x(0)$, holding $u(0)$ constant and satisfying (2.1.1). If $x(0)$ is given, we have $dx(0) = 0$.

Clearly, a stationary value of $\bar{J}$ and, hence, J, for given values of $x(0)$, will be obtained if

$$\frac{\partial H^o}{\partial u(0)} = 0. \tag{2.1.9}$$

Note that (2.1.1), (2.1.7), and (2.1.9) constitute $n + n + m$ equations for determining the $n + n + m$ quantities $x(1)$, $\lambda(1)$, and $u(0)$.

2.2 Multistage systems; no terminal constraints, fixed number of stages

Optimal programming problems for multistage systems are also parameter optimization problems. Consider the multistage system described by the nonlinear difference equations:

$$x(i+1) = f^i[x(i), u(i)]; \quad x(0) \text{ given}, \quad i = 0, \ldots, N-1, \tag{2.2.1}$$

which is nothing but a sequential set of equality constraints, where $x(i)$, a sequence of n-vectors, is determined by $u(i)$, a sequence of m-vectors. This is shown schematically in Figure 2.2.1.

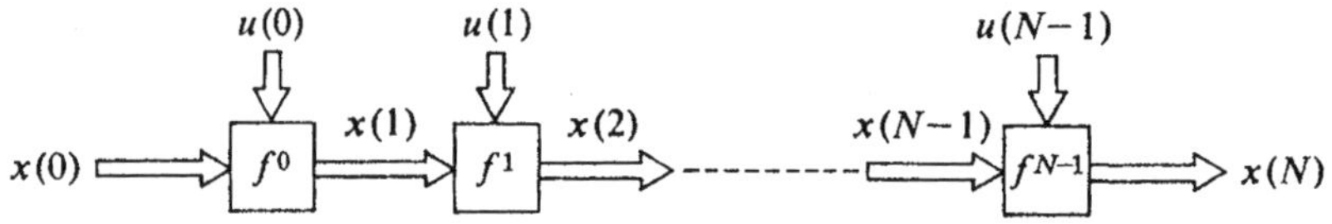

Figure 2.2.1. Flow chart for a multistage system.

Consider a performance index of the form

$$J = \phi[x(N)] + \sum_{i=0}^{N-1} L^i[x(i), u(i)]. \tag{2.2.2}$$

The problem is to find the sequence $u(i)$ that minimizes (or maximizes) J. Adjoin the system equations (2.2.1) to J with a multiplier sequence $\lambda(i)$:

$$\bar{J} = \phi[x(N)] + \sum_{i=0}^{N-1} [L^i[x(i), u(i)] + \lambda^T(i+1)\{f^i[x(i), u(i)] - x(i+1)\}]. \tag{2.2.3}$$

For convenience, define a scalar sequence H^i:

$$H^i = L^i[x(i), u(i)] + \lambda^T(i+1) f^i[x(i), u(i)], \qquad i=0, \ldots, N-1. \tag{2.2.4}$$

Also, change indices of summation on the last term in (2.2.3), obtaining

$$\bar{J} = \phi[x(N)] - \lambda^T(N)x(N) + \sum_{i=1}^{N-1} [H^i - \lambda^T(i)x(i)] + H^o. \tag{2.2.5}$$

Now consider differential changes in $\bar{J}$ due to differential changes in $u(i)$:

$$d\bar{J} = \left[\frac{\partial \phi}{\partial x(N)} - \lambda^T(N)\right] dx(N) + \sum_{i=1}^{N-1} \left\{\left[\frac{\partial H^i}{\partial x(i)} - \lambda^T(i)\right] dx(i) + \frac{\partial H^i}{\partial u(i)} du(i)\right\} + \frac{\partial H^o}{\partial x(0)} dx(0) + \frac{\partial H^o}{\partial u(0)} du(0). \tag{2.2.6}$$

It would be tedious to determine the differential changes $dx(i)$ produced by a given $du(i)$ sequence, so we *choose* the multiplier sequence $\lambda(i)$ so that we have

$$\lambda^T(i) - \frac{\partial H^i}{\partial x(i)} = 0 \Rightarrow \lambda^T(i) = \frac{\partial L^i}{\partial x(i)} + \lambda^T(i+1)\frac{\partial f^i}{\partial x(i)}; \qquad i = 0, \ldots, N-1, \tag{2.2.7}$$

with boundary conditions

$$\lambda^T(N) = \frac{\partial \phi}{\partial x(N)} . \tag{2.2.8}$$

Equation (2.2.6) then becomes

$$d\bar{J} = \sum_{i=0}^{N-1} \frac{\partial H^i}{\partial u(i)} du(i) + \lambda^T(0)\, dx(0) . \tag{2.2.9}$$

Thus, $\partial H^i/\partial u(i)$ is the *gradient* of J with respect to $u(i)$ while holding $x(0)$ constant and satisfying (2.2.1), and $\lambda^T(0) = \partial H^0/\partial x(0)$ is the gradient of J with respect to $x(0)$ while holding $u(i)$ constant and satisfying (2.2.1). If $x(0)$ is given, we have $dx(0) = 0$.

For an extremum, $d\bar{J}$ must be zero for arbitrary $du(i)$; this can only happen if we have

$$\frac{\partial H^i}{\partial u(i)} = 0 , \qquad i = 0, \ldots, N-1 . \tag{2.2.10}$$

In summary, to find a control vector sequence $u(i)$ that produces a stationary value of the performance index J, we must solve the following difference equations:

$$x(i+1) = f^i[x(i), u(i)] , \tag{2.2.11}$$

$$\lambda(i) = \left[\frac{\partial f^i}{\partial x(i)}\right]^T \lambda(i+1) + \left[\frac{\partial L^i}{\partial x(i)}\right]^T , \tag{2.2.12}$$

where $u(i)$ is determined by seeking a stationary point of H^i, which requires that we have

$$\frac{\partial H^i}{\partial u(i)} = \frac{\partial L^i}{\partial u(i)} + \lambda^T(i+1) \frac{\partial f^i}{\partial u(i)} = 0 , \qquad i = 0, \ldots, N-1 . \tag{2.2.13}$$

The boundary conditions for (2.2.11) and (2.2.12) are *split*; i.e., some are given for $i = 0$, and some are given for $i = N$

$$x(0) \quad \text{given} \tag{2.2.14}$$

$$\lambda(N) = \left[\frac{\partial \varphi}{\partial x(N)}\right]^T . \tag{2.2.15}$$

Such problems are called *two-point boundary-value problems*, and they are sometimes rather difficult to solve, even with a high-speed computer. Notice that the difference equations (2.2.11) and (2.2.12) are *coupled* since $u(i)$ depends on $\lambda(i)$ through (2.2.13), and the coefficients of (2.2.12) depend, in general, on $x(i)$ and $u(i)$.

In order that J be a local minimum, not only must we have $\partial H^i/\partial u(i) = 0$, but, in addition, the second-order expression for dJ

with the constraint (2.2.1) must be nonnegative for all (infinitesimal) values of $du(i)$; that is, we must have $dJ \geq 0$, where (from (2.2.3))

$$dJ = \frac{1}{2}\, dx^T(N) \frac{\partial^2 \phi}{\partial x(N)\, \partial x(N)}\, dx(N)$$

$$+ \frac{1}{2} \sum_{i=0}^{N-1} [dx^T(i)\, , du^T(i)] \begin{bmatrix} \dfrac{\partial^2 H^i}{\partial x(i)\, \partial x(i)}, & \dfrac{\partial^2 H^i}{\partial x(i)\, \partial u(i)} \\ \dfrac{\partial^2 H^i}{\partial u(i)\, \partial x(i)}, & \dfrac{\partial^2 H^i}{\partial u(i)\, \partial u(i)} \end{bmatrix} \begin{bmatrix} dx(i) \\ du(i) \end{bmatrix}. \quad (2.2.16)$$

The values of $dx(i)$ are determined by the sequence $du(i)$ from the differential of (2.2.1).

$$dx(i+1) = \frac{\partial f}{\partial x(i)}\, dx(i) + \frac{\partial f}{\partial u(i)}\, du(i)\, , \qquad dx(0) = 0\, . \quad (2.2.17)$$

Methods for checking this criterion are given in Chapter 6.

Example. *Quadratic performance index with linear system equations. Find* the control vector sequence $u(i)$, $i = 0, \ldots, N-1$ that minimizes the quadratic form

$$J = \tfrac{1}{2} x^T(N)\, A(N)\, x(N) + \sum_{i=0}^{N-1} [\tfrac{1}{2} x^T(i)\, A(i)\, x(i) + \tfrac{1}{2} u^T(i)\, B(i)\, u(i)]\, , \quad (2.2.18)$$

where $A(i)$ and $B(i)$ are given positive definite matrices, with the linear system equations

$$x(i+1) = \Phi(i) x(i) + \Gamma(i) u(i)\, , \qquad x(0) \text{ given}\, . \quad (2.2.19)$$

SOLUTION. The H^i sequence for the problem is

$$H^i = \tfrac{1}{2} x^T(i)\, A(i)\, x(i) + \tfrac{1}{2} u^T(i)\, B(i)\, u(i) + \lambda^T(i+1)\, [\Phi(i)\, x(i) + \Gamma(i)\, u(i)]\, , \quad (2.2.20)$$

where

$$\lambda^T(i) = \lambda^T(i+1)\, \Phi(i) + x^T(i)\, A(i) \quad \text{and} \quad \lambda^T(N) = x^T(N)\, A(N)\, . \quad (2.2.21)$$

A stationary value of H^i with respect to $u(i)$ will occur where we have

$$\frac{\partial H^i}{\partial u(i)} = u^T(i)\, B(i) + \lambda^T(i+1)\, \Gamma(i) = 0\, , \quad (2.2.22)$$

$$\Rightarrow u(i) = -[B(i)]^{-1}\, \Gamma^T(i)\, \lambda(i+1)\, . \quad (2.2.23)$$

Hence, we obtain

$$x(i+1) = \Phi(i)\, x(i) - \Gamma(i)\, [B(i)]^{-1}\, \Gamma^T(i)\, \lambda(i+1)\, ; \quad (2.2.24)$$

$$\lambda(i) = \Phi^T(i)\, \lambda(i+1) + A(i)\, x(i)\, ; \qquad i = 0, \ldots, N-1\, , \quad (2.2.25)$$

with boundary conditions $\lambda(N) = A(N)x(N)$ and $x(0)$ given. These are coupled sets of *linear* difference equations with two-point boundary values; the solution to this boundary-value problem yields the minimizing sequence $u(i)$ from (2.2.23).

Problem 1. Show that the two-point boundary-value problem of the Example can be solved by placing

$$\lambda(i) = S(i)x(i)$$

and determining $S(i)$ from the backward recursive relations:

$$\left.\begin{aligned} S(i) &= \Phi^T(i)\, M(i+1)\Phi(i) + A(i)\,, \\ M(i+1) &= [S^{-1}(i+1) + \Gamma(i)B^{-1}(i)\,\Gamma^T(i)]^{-1}\,; \end{aligned}\right. \qquad i = N-1,\ldots,0\,;$$

or

$$\begin{aligned} M(i+1) \equiv S(i+1) - S(i+1)\,\Gamma(i)\,[B(i) \\ + \Gamma^T(i)\,S(i+1)\,\Gamma(i)]^{-1}\,\Gamma^T(i)S(i+1)\,, \end{aligned}$$

where

$$S(N) = A(N)\,.$$

Having determined $S(i)$, $i = N-1,\ldots,0$, from the relations above, we obtain

$$x(i+1) = [I + \Gamma(i)B^{-1}(i)\,\Gamma^T(i)S(i+1)]^{-1}\,\Phi(i)x(i)\,, \qquad x(0) \text{ specified.}$$

This is known as the *sweep method* for solving a linear two-point boundary-value problem. (For more on this method see Sections 6.10 and 6.11.)

Problem 2. Consider the problem of this section as a parameter optimization problem of Section 1.2 where x denotes the vector with component vectors $x(1),\ldots,x(N)$, u the vector with component vectors $u(0),\ldots,u(N-1)$, and f the vector with component vectors $x(1) - f^0$, $x(2) - f^1,\ldots,x(N) - f^{N-1}$. Show that the necessary conditions of Section 1.2 reduces to that of Equations (2.2.11)-(2.2.15).

2.3 Continuous systems; no terminal constraints, fixed terminal time

Optimal programming problems for continuous systems are problems in the *calculus of variations*. They may be considered as limiting cases of optimal programming problems for multistage systems in which the time increment between steps becomes small compared to times of interest. Actually, the reverse procedure is more common today; continuous systems are approximated by multistage systems for solution on digital computers. Consider the system described by the following nonlinear differential equations:

$$\dot{x} = f[x(t), u(t), t]\,; \qquad x(t_0) \text{ given}, \qquad t_0 \leq t \leq t_f, \tag{2.3.1}$$

where $x(t)$, an n-vector function, is determined by $u(t)$, an m-vector function. Consider a performance index (scalar) of the form

$$J = \varphi[x(t_f), t_f] + \int_{t_0}^{t_f} L[x(t), u(t), t]\, dt\,. \tag{2.3.2}$$

The problem is to find the functions $u(t)$ that minimize (or maximize) J. Adjoin the system differential equations (2.3.1) to J with multiplier functions $\lambda(t)$:

$$\bar{J} = \varphi[x(t_f), t_f] + \int_{t_0}^{t_f} [L[x(t), u(t), t] + \lambda^T(t)\{f[x(t), u(t), t] - \dot{x}\}]\, dt\,. \tag{2.3.3}$$

For convenience, define a scalar function H (the *Hamiltonian*), as follows:

$$H[x(t), u(t), \lambda(t), t] = L[x(t), u(t), t] + \lambda^T(t) f[x(t), u(t), t]\,. \tag{2.3.4}$$

Also, integrate the last term on the right side of (2.3.3) by parts, yielding

$$\bar{J} = \varphi[x(t_f), t_f] - \lambda^T(t_f)x(t_f) + \lambda^T(t_0)\, x(t_0) + \int_{t_0}^{t_f} \{H[x(t), u(t), t] + \dot{\lambda}^T(t)\, x(t)\}\, dt\,. \tag{2.3.5}$$

Now consider the *variation* in J due to variations in the control vector $u(t)$ for *fixed times* t_0 *and* t_f,

$$\delta\bar{J} = \left[\left(\frac{\partial\varphi}{\partial x} - \lambda^T\right)\delta x\right]_{t=t_f} + [\lambda^T\,\delta x]_{t=t_0} + \int_{t_0}^{t_f}\left[\left(\frac{\partial H}{\partial x} + \dot{\lambda}^T\right)\delta x + \frac{\partial H}{\partial u}\delta u\right] dt\,. \tag{2.3.6}$$

It would be tedious to determine the variations $\delta x(t)$ produced by a given $\delta u(t)$, so we *choose* the multiplier functions $\lambda(t)$ to cause the coefficients of δx in (2.3.6) to vanish:

$$\dot{\lambda}^T = -\frac{\partial H}{\partial x} = -\frac{\partial L}{\partial x} - \lambda^T\frac{\partial f}{\partial x}, \tag{2.3.7}$$

with boundary conditions

$$\lambda^T(t_f) = \frac{\partial\varphi}{\partial x(t_f)}\,. \tag{2.3.8}$$

Equation (2.3.6) then becomes

$$\delta \bar{J} = \lambda^T(t_o)\,\delta x(t_o) + \int_{t_o}^{t_f} \frac{\partial H}{\partial u}\,\delta u\,dt\,. \tag{2.3.9}$$

Thus, $\lambda^T(t_o)$ is the gradient of J with respect to variations in the initial conditions, while holding $u(t)$ constant and satisfying (2.3.1). The functions $\lambda(t)$ are also called *influence functions* on J of variations in $x(t)$ since t_o is arbitrary. The functions $\partial H/\partial u$ are called *impulse response functions* since each component of $\partial H/\partial u$ represents the variation in J due to a unit impulse (Dirac function) in the corresponding component of δu at time t, while holding $x(t_o)$ constant and satisfying (2.3.1).

For an extremum, δJ must be zero for arbitrary $\delta u(t)$; this can only happen if

$$\frac{\partial H}{\partial u} = 0\,, \qquad t_o \leq t \leq t_f\,. \tag{2.3.10}$$

Equations (2.3.7), (2.3.8), and (2.3.10) are known as the *Euler-Lagrange equations* in the calculus of variations.

In summary, to find a control vector function $u(t)$ that produces a stationary value of the performance index J, we must solve the following differential equations:

$$\dot{x} = f(x,u,t)\,, \tag{2.3.11}$$

$$\dot{\lambda} = -\left(\frac{\partial f}{\partial x}\right)^T \lambda - \left(\frac{\partial L}{\partial x}\right)^T, \tag{2.3.12}$$

where $u(t)$ is determined by

$$\frac{\partial H}{\partial u} = 0 \quad \text{or} \quad \left(\frac{\partial f}{\partial u}\right)^T \lambda + \left(\frac{\partial L}{\partial u}\right)^T = 0\,. \tag{2.3.13}$$

The boundary conditions for (2.3.11) and (2.3.12) are split; i.e., some are given for $t = t_o$, and some are given for $t = t_f$:

$$x(t_o) \quad \text{given}, \tag{2.3.14}$$

$$\lambda(t_f) = \left(\frac{\partial \varphi}{\partial x}\right)^T. \tag{2.3.15}$$

Thus, as in the multistage system optimal programming problems, we are faced with a *two-point boundary-value problem.*

A *first integral* of the boundary-value problem exists if L and f are *not* explicit functions of the time t, since we have

$$\begin{aligned}\dot{H} &= H_t + H_x\dot{x} + H_u\dot{u} + \dot{\lambda}^T f \\ &= H_t + H_u\dot{u} + (H_x + \dot{\lambda}^T)f \\ &= H_t + H_u\dot{u}\,.\end{aligned}$$

If L and f (hence, H) are not explicit functions of t and $u(t)$ is an optimal program (that is, $\partial H/\partial u = 0$), then we have

$$\dot{H} = 0 \quad \text{or} \quad H = \text{constant on the optimal trajectory}. \tag{2.3.16}$$

In order that J be a *local* minimum, not only must we have $\partial H/\partial u = 0$ but, in addition, the second-order expression for δJ, holding $\dot{x} - f = 0$, must be nonnegative for all values (infinitesimal) of δu; that is, we have

$$\delta J = \frac{1}{2}\left[\delta x^T \frac{\partial^2 \varphi}{\partial x^2}\delta x\right]_{t=t_f} + \frac{1}{2}\int_{t_0}^{t_f}[\delta x^T, \delta u^T]\begin{bmatrix} \dfrac{\partial^2 H}{\partial x^2}, & \dfrac{\partial^2 H}{\partial x\, \partial u} \\ \dfrac{\partial^2 H}{\partial u\, \partial x}, & \dfrac{\partial^2 H}{\partial u^2}\end{bmatrix}\begin{bmatrix}\delta x \\ \delta u\end{bmatrix} dt \geq 0, \tag{2.3.17}$$

where $\delta(\dot{x} - f) = 0$, or

$$\frac{d}{dt}(\delta x) = \frac{\partial f}{\partial x}\delta x + \frac{\partial f}{\partial u}\delta u, \qquad \delta x(t_0) = 0. \tag{2.3.18}$$

Equation (2.3.18) determines $\delta x(t)$ in terms of $\delta u(t)$, but in a complicated way. We will have more to say about this *second variation* in Chapter 6.

Example 1. *Hamilton's principle in mechanics.* The motion of a conservative system, from time t_0 to t_f is such that the integral

$$I = \int_{t_0}^{t_f} L(u,q)\, dt \tag{2.3.19}$$

has a stationary value, where

$$\begin{aligned} L &= T(u,q) - V(q) = \text{the Lagrangian of the system}, \\ T &= \text{kinetic energy of the system}, \\ V &= \text{potential energy of the system}, \\ q &= \text{generalized coordinate vector (state of system)}, \\ u &= \dot{q} = \text{generalized velocity vector}. \end{aligned} \tag{2.3.20}$$

The Hamiltonian is then

$$H = L + \lambda^T u\,.\dagger \tag{2.3.21}$$ †

Consequently, the Euler-Lagrange equations are

---

†In mechanics, H is defined as $-L + \lambda^T u$, and the vector λ is usually called p, where p is the generalized momentum vector.

$$\dot{\lambda}^T = -\frac{\partial H}{\partial q} = -\frac{\partial L}{\partial q}, \tag{2.3.22}$$

$$0 = \frac{\partial H}{\partial u} = \frac{\partial L}{\partial u} + \lambda^T. \tag{2.3.23}$$

Combining these last two vector equations, we have

$$\frac{d}{dt}\left(\frac{\partial L}{\partial \dot{q}}\right) - \frac{\partial L}{\partial q} = 0, \tag{2.3.24}$$

which are Lagrange's equations of motion for a conservative system.

If L is not an explicit function of time, a first integral of the motion is H = constant:

$$H = L - \frac{\partial L}{\partial u}u = T - V - \frac{\partial T}{\partial u}u = \text{const.} \tag{2.3.25}$$

Now T is a homogeneous quadratic form in u, so that we have

$$\frac{\partial T}{\partial u}u = 2T. \tag{2.3.26}$$

Hence, we have

$$-H = T + V = \text{const}; \tag{2.3.27}$$

that is, the kinetic plus potential energy is constant during the motion.

Example 2. *Variational principle for nonconservative mechanical systems.*† The motion of a nonconservative mechanical system from time t_o to t_f is such that

$$\delta \int_{t_0}^{t_f} T(u,q)\,dt + \int_{t_0}^{t_f} Q^T(q)\,\delta q\,dt = 0, \tag{2.3.28}$$

where

$$\delta \dot{q} = \delta u, \tag{2.3.29}$$

and $Q(q)$ is the generalized force vector. $Q(q)$ is defined by the fact that the *work* done on the system by these generalized forces is given by the (path-dependent) line integral

$$W = \int_{q_0}^{q_f} Q^T(q)\,dq. \tag{2.3.30}$$

The second term in (2.3.28) is the time integral of the *virtual work*; note that it is *not* $\delta \int_{t_0}^{t_f} W dt$, which prevents us from defining a Ham-

†See C. Lanczos, *The Variational Principle of Mechanics.* Toronto, Canada: University of Toronto Press, 1949, Chapter 5.

iltonian for nonconservative systems. However, we can adjoin the constraint (2.3.29) to Equation (2.3.28) with a Lagrange multiplier vector, as follows:

$$\int_{t_0}^{t_f} \left[\frac{\partial T}{\partial u} \delta u + \frac{\partial T}{\partial q} \delta q + Q^T \delta q + \lambda^T(\delta u - \delta \dot{q}) \right] dt = 0 . \tag{2.3.31}$$

Integrating the last term by parts, we have

$$\int_{t_0}^{t_f} \left[\left(\frac{\partial T}{\partial u} + \lambda^T \right) \delta u + \left(\frac{\partial T}{\partial q} + Q^T + \dot{\lambda}^T \right) \delta q \right] dt = 0 . \tag{2.3.32}$$

As usual, we choose $\lambda(t)$ to make the coefficient of δq vanish:

$$\dot{\lambda}^T = -\frac{\partial T}{\partial q} - Q^T . \tag{2.3.33}$$

Since u is arbitrary, the integral can vanish only if we have

$$\lambda^T = -\frac{\partial T}{\partial u} . \tag{2.3.34}$$

Combining (2.3.33) and (2.3.34) to eliminate λ, we obtain

$$\frac{d}{dt}\left(\frac{\partial T}{\partial \dot{q}} \right) - \frac{\partial T}{\partial q} = Q^T , \tag{2.3.35}$$

which are Lagrange's equations of motion for a nonconservative mechanical system.

Example 3. *Minimum-drag nose shape in hypersonic flow.** The pressure drag of a body of revolution at zero angle of attack in hypersonic flow is given quite accurately by the expression

$$D = -2\pi q \int_{x=0}^{\ell} C_p(\theta) r \, dr , \tag{2.3.36}$$

where

q = dynamic pressure,

x = axial distance from point of maximum radius,

$r = r(x)$ = radius of body,

$$\frac{dr}{dx} = -\tan\theta \text{ (see Figure 2.3.1)} \tag{2.3.37}$$

$$C_p = \begin{Bmatrix} 2\sin^2\theta \, ; \, \theta \geq 0 \\ 0 \quad ; \, \theta \leq 0 \end{Bmatrix} = \text{pressure coefficient (Newtonian approximation),†} \tag{2.3.38}$$

†This was the first problem ever solved in the calculus of variations; it was set up and solved in 1686 by Isaac Newton, whose model of aerodynamic forces happens to be very good at hypersonic speeds but *not* very good at subsonic speeds.

ℓ = length of body,

$r(0) = a$ = maximum radius of body,

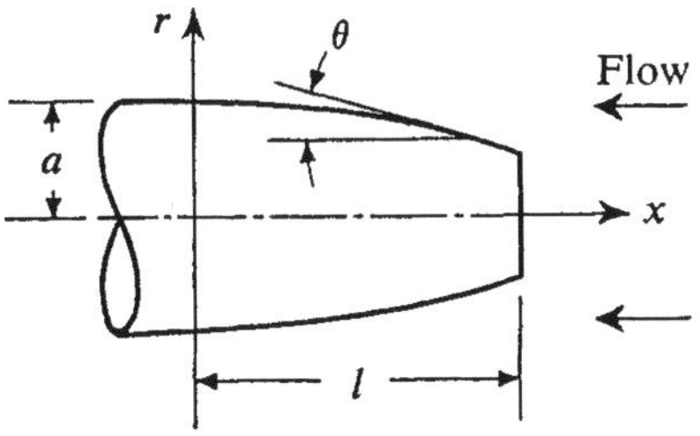

Figure 2.3.1. Nomenclature for analyzing minimum-drag nose shape.

The problem is to find $r(x)$ to minimize D for given values of q, ℓ, and a.

Let

$$-u = -\tan\theta = \frac{dr}{dx} \tag{2.3.39}$$

be the control variable, and allow for the possibility of a blunt tip by writing (2.3.36) in the form

$$\frac{D}{4\pi q} = \frac{1}{2}[r(\ell)]^2 + \int_0^\ell \frac{ru^3}{1+u^2}\,dx\,. \tag{2.3.40}$$

The Hamiltonian of the system is, therefore,

$$H = \frac{ru^3}{1+u^2} + \lambda(-u)\,. \tag{2.3.41}$$

The Euler-Lagrange equations are

$$\frac{d\lambda}{dx} = -\frac{\partial H}{\partial r} = -\frac{u^3}{1+u^2}\,, \tag{2.3.42}$$

$$0 = \frac{\partial H}{\partial u} = \frac{ru^2(3+u^2)}{(1+u^2)^2} - \lambda\,. \tag{2.3.43}$$

Now the first term on the right-hand side of (2.3.40) is a function of $r(\ell)$. According to (2.3.8), the optimal value of $r(\ell)$ is such that

$$\lambda(\ell) = r(\ell)\,. \tag{2.3.44}$$

Since $r(0) = a$ is specified, $\lambda(0)$ is not specified. Thus, the two boundary conditions for the second-order system of differential equations (2.3.37) and (2.3.42) are (2.3.44) and $r(0) = a$.

Instead of trying to solve (2.3.43) for u in terms of λ, substituting into (2.3.37) and (2.3.42) and integrating, we can take advantage of the fact that the Hamiltonian in (2.3.41) is not an explicit function of x, so that H = constant is an integral of the system. Eliminating λ between (2.3.41) and (2.3.43) gives

$$H = -\frac{2ru^3}{(1+u^2)^2} = \text{const.} \tag{2.3.45}$$

Eliminating $\lambda(\ell)$ between (2.3.43) and (2.3.44) yields

$$r(\ell)\left[1 - \frac{u^2(3+u^2)}{(1+u^2)^2}\right]_{x=\ell} = 0\,, \tag{2.3.46}$$

which is satisfied by $r(\ell) = 0$, or

$$u(\ell) = 1\,. \tag{2.3.47}$$

Using (2.3.47) in (2.3.45) at $x = \ell$, we find that

$$-H = \frac{r(\ell)}{2}\,. \tag{2.3.48}$$

Using (2.3.45) and (2.3.48), we have the radius of the body in terms of the slope u,

$$\frac{r}{r(\ell)} = \frac{(1+u^2)^2}{4u^3}\,. \tag{2.3.49}$$

From (2.3.37) and (2.3.39), we have

$$\frac{dx}{dr} = -\frac{1}{u}$$

or

$$\frac{\ell - x}{r(\ell)} = \int_1^u \frac{1}{u}\frac{d}{du}\frac{(1+u^2)^2}{4u^3}\,du\,. \tag{2.3.50}$$

Equation (2.3.50) can be integrated in terms of simple functions:

$$\frac{\ell - x}{r(\ell)} = \frac{1}{4}\left(\frac{3}{4u^4} + \frac{1}{u^2} - \frac{7}{4} - \log\frac{1}{u}\right). \tag{2.3.51}$$

Equations (2.3.49) and (2.3.51) are parametric equations for the optimum body shape. The tip radius $r(\ell)$ and the slope u_0 at $x = 0$ must be obtained by solution of the transcendental equations

$$\frac{a}{r(\ell)} = \frac{(1+u_0^2)^2}{4u_0^3}\,, \tag{2.3.52}$$

$$\frac{\ell}{r(\ell)} = \frac{1}{4}\left(\frac{3}{4u_0^4} + \frac{1}{u_0^2} - \frac{7}{4} - \log\frac{1}{u_0}\right). \tag{2.3.53}$$

Figure 2.3.2 shows some of these shapes for fixed a and several values of ℓ.

The minimum-drag coefficient is given by

$$C_D = \frac{D}{q\pi a^2} = \frac{u_0^2}{(1+u_0^2)^2}\left(3 + 10u_0^2 + 17u_0^4 + 2u_0^6 + 4u_0^4 \log\frac{1}{u_0}\right). \quad (2.3.54)$$

As $a/\ell \to 0$, it is easily shown that

$$\frac{r}{a} \to \left(\frac{\ell - x}{\ell}\right)^{3/4}, \quad (2.3.55)$$

$$C_D \to \frac{27}{16}\left(\frac{a}{\ell}\right)^2, \qquad \frac{a}{\ell} \to 0. \quad (2.3.56)$$

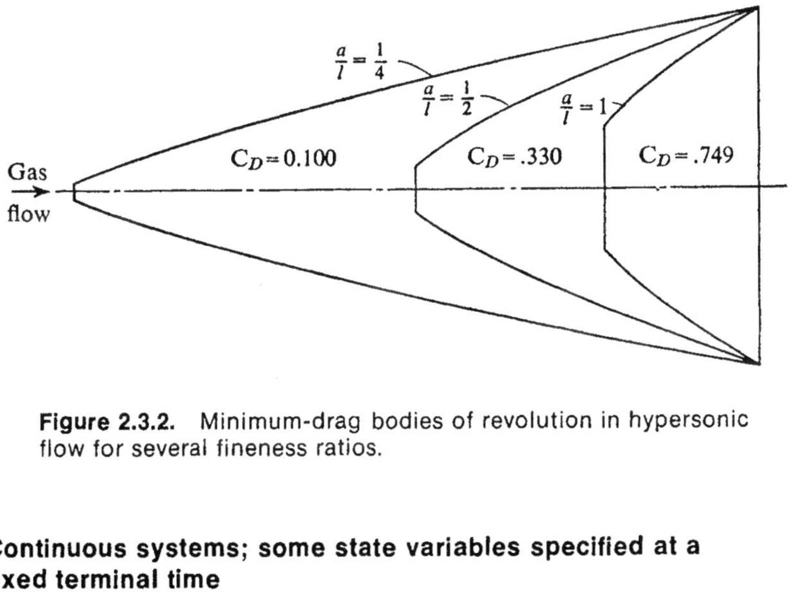

Figure 2.3.2. Minimum-drag bodies of revolution in hypersonic flow for several fineness ratios.

2.4 Continuous systems; some state variables specified at a fixed terminal time

Suppose that, in the optimization problem defined in Section 2.3, we wish to constrain *some* of the components of the state vector $x(t)$ to have prescribed values at $t = t_f$. Section 2.3 applies up to and including (2.3.7). Now, if x_i (the ith component of the vector x) is prescribed at $t = t_f$, it follows that admissible variations must produce $\delta x_i(t_f) = 0$ in (2.3.6). Thus, it is *not* necessary that $[(\partial\phi/\partial x_i) - \lambda_i^T]_{t=t_f} = 0$. Essentially, we have traded this latter boundary condition for another, namely, $x_i(t_f)$ given, so that the boundary-value problem (2.3.11)–(2.3.15) still has $2n$ boundary conditions.

Similarly, if x_k is *not* prescribed at $t = t_0$, it does *not* follow that $\delta x_k(t_0) = 0$; in fact, there will be an optimum value for $x_k(t_0)$ and it will be such that $\delta J = 0$ for arbitrary small variations of $x_k(t_o)$ around this value. For this to be the case, we choose

$$\lambda_k(t_o) = 0 \,, \tag{2.4.1}$$

which simply says that the influence of small changes in $x_k(t_o)$ on J is zero. Again we have simply traded one boundary condition, $x_k(t_o)$ given, for another, (2.4.1). Boundary conditions like (2.4.1) are sometimes called "natural boundary conditions."

However, the necessary condition (2.3.13), $\partial H/\partial u = 0$, needs additional justification for the problem with terminal constraints. In Section 2.3, we derived it under the assumption that $\delta u(t)$, $t_o < t < t_f$ is arbitrary. In the present case, $\delta u(t)$ is *not* completely arbitrary; the set of admissible $\delta u(t)$ is restricted by the constraints

$$\delta x_i(t_f) = 0 \,, \qquad i = 1, \ldots, q \,, \tag{2.4.2}$$

where we define "admissible" $\delta u(t)$, generally, as those $\delta u(t)$ which satisfy all constraints of the problem, for example, (2.4.2).

Now, it is still possible to determine influence functions for the performance index exactly as in Section 2.3. In this section we shall designate these influence functions with a superscript "J." However, since $x_i(t_f)$ for $i = 1, \ldots, q$ are specified, it is consistent to regard

$$\phi = \phi[x_{q+1}, \ldots, x_n]_{t=t_f} \,. \tag{2.4.3}$$

Thus (cf., Equations (2.3.7) through (2.3.9), we have (for $\delta x(t_o) = 0$)

$$\delta J = \int_{t_o}^{t_f} \left[\frac{\partial L}{\partial u} + (\lambda^{(J)})^T \frac{\partial f}{\partial u} \right] \delta u(t)\, dt \,, \tag{2.4.4}$$

where

$$\dot{\lambda}^{(J)} = -\left(\frac{\partial f}{\partial x}\right)^T \lambda^{(J)} - \left(\frac{\partial L}{\partial x}\right)^T \tag{2.4.5}$$

$$\lambda_j^{(J)}(t_f) = \begin{cases} 0\,; & j = 1, \ldots, q \\ \left.\dfrac{\partial \phi}{\partial x_j}\right|_{t=t_f}; & j = q+1, \ldots, n \,. \end{cases} \tag{2.4.6}$$

Suppose that, instead of $J = \phi[x(t_f)] + \int_{t_o}^{t_f} L(x,u,t)\, dt$, the performance index was $J = x_i(t_f)$; i.e., the ith component of the state vector at the final time. We could then determine influence functions for $x_i(t_f)$ by specializing the relations above; we would put $\phi = x_i(t_f)$ and $L(x,u,t) = 0$. We shall designate these influence functions with a superscript "i." Analogous to Equations (2.4.4), (2.4.5), and (2.4.6), we have

$$\delta x_i(t_f) = \int_{t_0}^{t_f} (\lambda^{(i)})^T \frac{\partial f}{\partial u} \delta u(t)\, dt\,, \tag{2.4.7}$$

where

$$\dot{\lambda}^{(i)} = -\left(\frac{\partial f}{\partial x}\right)^T \lambda^{(i)}\,, \tag{2.4.8}$$

$$\lambda_j^{(i)}(t_f) = \begin{cases} 0\,; & i \neq j\,, \\ 1\,; & i = j\,, \quad j = 1, \ldots, n \end{cases} \tag{2.4.9}$$

We could, in fact, determine q sets of such influence functions for $i = 1, \ldots, q$ (see Appendix A4).

We shall now construct a $\delta u(t)$ history that decreases J, i.e., produces $\delta J < 0$, *and* satisfies the q terminal constraints (2.4.2). Multiply each of the q equations in (2.4.7) by an undetermined constant, ν_i, and add the resulting equations to (2.4.4):

$$\delta J + \nu_i \delta x_i(t_f) = \int_{t_0}^{t_f} \left\{\frac{\partial L}{\partial u} + [\lambda^{(J)} + \nu_i \lambda^{(i)}]^T \frac{\partial f}{\partial u}\right\} \delta u\, dt\,.† \tag{2.4.10}$$

Now *choose*

$$\delta u = -k\left\{\left(\frac{\partial f}{\partial u}\right)^T [\lambda^{(J)} + \nu_i \lambda^{(i)}] + \left(\frac{\partial L}{\partial u}\right)^T\right\}, \tag{2.4.11}$$

where k is a positive scalar constant, and substitute this expression into (2.4.10), as follows:

$$\delta J + \nu_i \delta x_i(t_f) = -k \int_{t_0}^{t_f} \left\| \left(\frac{\partial f}{\partial u}\right)^T (\lambda^{(J)} + \nu_i \lambda^{(i)}) + \left(\frac{\partial L}{\partial u}\right)^T \right\|^2 dt < 0\,, \tag{2.4.12}$$

which is negative unless the integrand vanishes over the whole integration interval.

Next, we determine the ν_i's so as to satisfy the terminal constraints (2.4.2). Substituting (2.4.11) into (2.4.7), we have

$$0 = \delta x_i(t_f) = -k \int_{t_0}^{t_f} [\lambda^{(i)}]^T \frac{\partial f}{\partial u}\left[\left(\frac{\partial f}{\partial u}\right)^T (\lambda^{(J)} + \nu_j \lambda^{(j)}) + \left(\frac{\partial L}{\partial u}\right)^T\right] dt$$

$$0 = \int_{t_0}^{t_f} [\lambda^{(i)}]^T \frac{\partial f}{\partial u}\left[\left(\frac{\partial f}{\partial u}\right)^T \lambda^{(J)} + \left(\frac{\partial L}{\partial u}\right)^T\right] dt + \nu_j \int_{t_0}^{t_f} [\lambda^{(i)}]^T \frac{\partial f}{\partial u}\left(\frac{\partial f}{\partial u}\right)^T \lambda^{(j)}\, dt\,,$$

†Repeated indices indicate summation over the range of that index, for example:

$$\mu_i \delta x_i \equiv \sum_{i=1}^{q} \mu_i \delta x_i\,.$$

from which the appropriate choice of the ν_i's is

$$\nu = -Q^{-1}g, \tag{2.4.13}$$

where Q is a $(q \times q)$ matrix and g is a q-component vector:

$$Q_{ij} = \int_{t_0}^{t_f} (\lambda^{(i)})^T f_u f_u^T \lambda^{(j)} \, dt\,; \qquad i,j = 1, \ldots, q\,, \tag{2.4.14}$$

$$g_i = \int_{t_0}^{t_f} (\lambda^{(i)})^T \frac{\partial f}{\partial u} \left[\left(\frac{\partial f}{\partial u} \right)^T \lambda^{(J)} + \left(\frac{\partial L}{\partial u} \right)^T \right] dt\,; \qquad i = 1, \ldots, q\,. \tag{2.4.15}$$

The existence of the inverse of Q is the *controllability condition* (see Appendix B2 and Section 5.3). If Q^{-1} does not exist, it is not possible to control the system with $u(t)$ to satisfy one or more of the terminal conditions.

We have thus *constructed* a $\delta u(t)$ history that decreases the performance index *and* satisfies the terminal constraints (2.4.2); that is, $\delta u(t)$ is admissible and improving. From (2.4.12) the only case in which we cannot decrease the performance index is when

$$\frac{\partial L}{\partial u} + [\lambda^{(J)} + \nu_i \lambda^{(i)}]^T \frac{\partial f}{\partial u} = 0\,; \qquad t_o \le t \le t_f\,. \tag{2.4.16}$$

If (2.4.16) is satisfied, we have a *stationary solution* that satisfies the terminal constraints. Now, since the influence equations (2.4.5), (2.4.6), (2.4.8), and (2.4.9) are linear, the necessary condition (2.4.16) may be written as

$$\frac{\partial H}{\partial u} = 0\,, \tag{2.4.17}$$

where

$$H = L(x,u,t) + \lambda^T(t) f(x,u,t)\,, \tag{2.4.18}$$

and

$$\dot{\lambda}^T = -H_x\,, \; \lambda_j(t_f) = \begin{cases} \nu_j\,, & j = 1, \ldots, q\,, \\ \left. \dfrac{\partial \phi}{\partial x_j} \right|_{t=t_f}\,, & j = q+1, \ldots, n\,. \end{cases} \tag{2.4.19}$$

The development in this section represents the fundamental approach to the modern calculus of variations. By construction, we arrive at the equation

$$\delta J = \int_{t_0}^{t_f} H_u(t) \delta u(t) \, dt\,, \qquad \text{where} \qquad H_u(t) = \frac{\partial H}{\partial u}\,, \tag{2.4.20}$$

and the Hamiltonian is defined in terms of the multiplier functions $\lambda(t)$ and multipliers ν. *We then show that, unless we have* $H_u(t) \equiv 0$, *it is always possible (assuming controllability; that is, Q^{-1} exists) to choose ν such that $\delta u(t)$ as given by (2.4.11) is* ADMISSIBLE *and* IMPROVING.†

H_u may be interpreted as a function-space gradient of the performance index with respect to the control variable $u(t)$, while holding fixed the terminal values of x_i, $i = 1, \ldots, q$ and satisfying the system of differential equations.

Example. *Maximum velocity transfer to a rectilinear path.* Consider a particle of mass m, acted upon by a thrust force of magnitude ma. We assume planar motion and use an inertial coordinate system x,y to locate the particle; the velocity components of the particle are u,v. The thrust-direction angle $\beta(t)$ is the control variable for the system (see Figure 2.4.1). The equations of motion are

$$\dot{u} = a \cos \beta \,,$$

$$\dot{v} = a \sin \beta \,,$$

$$\dot{x} = u \,,$$

$$\dot{y} = v \,,$$

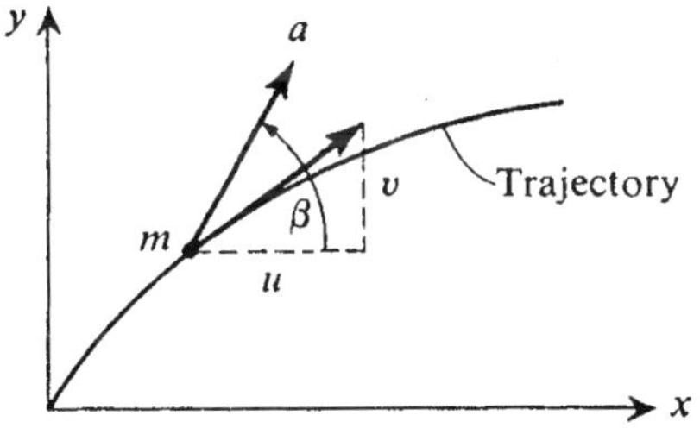

Figure 2.4.1. Nomenclature for planar motion with thrust acceleration = a.

where the thrust acceleration, a, is assumed to be a known function of time. The equations for the influence functions are particularly simple:

$$\dot{\lambda}_u = -\lambda_x \,, \qquad \dot{\lambda}_v = -\lambda_y \,, \qquad \dot{\lambda}_x = 0 \,, \qquad \dot{\lambda}_y = 0 \,.$$

These relations are easily integrated to yield

†For our purposes in this section, this "first-order" demonstration of the necessity of (2.4.17) is adequate. A more rigorous "second-order" demonstration is given in Section 6.3, where it is shown that the concept of "normality" is really what is needed, rather than "controllability."

$$\lambda_u = -c_1 t + c_3\,, \qquad \lambda_v = -c_2 t + c_4\,, \qquad \lambda_x = c_1\,, \qquad \lambda_y = c_2\,,$$

where c_1, c_2, c_3, c_4 are constants.

If we wish to extremalize a function of the end conditions only, then we have $L \equiv 0$, and the Hamiltonian of the system is

$$H = \lambda_u a \cos \beta + \lambda_v a \sin \beta + \lambda_x u + \lambda_y v\,,$$

which is constant for an optimal path if a is constant.

The optimality condition is

$$\frac{\partial H}{\partial \beta} = -\lambda_u \sin \beta + \lambda_v \cos \beta = 0\,.$$

Thus, the optimal control law is

$$\tan \beta = \frac{\lambda_v}{\lambda_u} = \frac{-c_2 t + c_4}{-c_1 t + c_3}\,,$$

which is often referred to as the "bilinear tangent law."

We wish to transfer the particle to a path parallel to the x-axis, a distance h away, in a given time T, arriving with the maximum value of $u(T)$. We do *not* care what the final x coordinate is (see Figure 2.4.2).

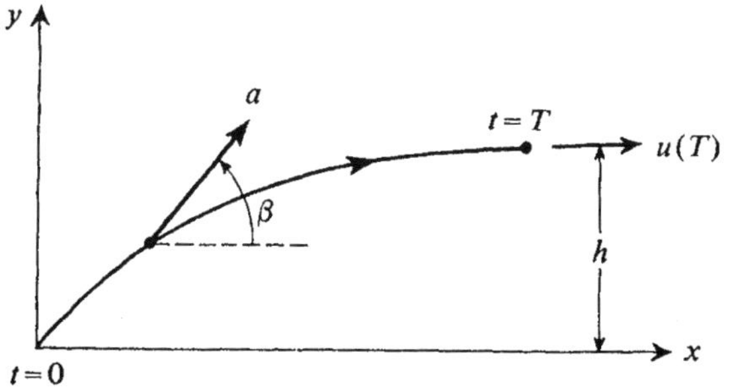

Figure 2.4.2. Nomenclature for transfer to a rectilinear path.

Thus, the boundary conditions for the problem are

$$\begin{aligned}
u(0) &= 0\,, & \lambda_u(T) &= 1\,;\\
v(0) &= 0\,, & v(T) &= 0\,, \lambda_v(T) = \nu_v\,;\\
x(0) &= 0\,, & \lambda_x(T) &= 0\,;\\
y(0) &= 0\,, & y(T) &= h\,, \lambda_y(T) = \nu_y\,;
\end{aligned}$$

where ν_v and ν_y are constants to be determined so that $v(T) = 0$, $y(T) = h$.

With $\lambda_x = 0$, it follows that $\lambda_u = 1$ throughout the flight, so the optimal control law becomes a "linear tangent law":

$$\tan\beta = \tan\beta_0 - ct\,, \qquad \text{where} \qquad \tan\beta_0 = \nu_v + \nu_y T\,, c = \nu_y\,.$$

For *constant-thrust acceleration*, a, the differential equations are readily integrated with the linear tangent law, using β as the independent variable instead of t, to obtain

$$u = \frac{a}{c}\log\frac{\tan\beta_0 + \sec\beta_0}{\tan\beta + \sec\beta}\,,$$

$$v = \frac{a}{c}(\sec\beta_0 - \sec\beta)\,,$$

$$x = \frac{a}{c^2}\left(\sec\beta_0 - \sec\beta - \tan\beta\log\frac{\tan\beta_0 + \sec\beta_0}{\tan\beta + \sec\beta}\right),$$

$$y = \frac{a}{2c^2}\left[(\tan\beta_0 - \tan\beta)\sec\beta_0 - (\sec\beta_0 - \sec\beta)\tan\beta - \log\frac{\tan\beta_0 + \sec\beta_0}{\tan\beta + \sec\beta}\right].$$

The constants β_0 and c (and, hence, ν_y and ν_v) are determined by the two final boundary conditions $v = 0$, $y = h$. These relations are *implicit* and may be put into the form

$$\frac{4h}{aT^2} = \frac{1}{\sin\beta_0} - \log\frac{\sec\beta_0 + \tan\beta_0}{\sec\beta_0 - \tan\beta_0}\Big/ 2\tan^2\beta_0\,,$$

$$c = \frac{2\tan\beta_0}{T} \Rightarrow \tan\beta = \tan\beta_0\left(1 - \frac{2t}{T}\right).$$

Clearly, the one dimensionless quantity, h/aT^2, determines β_0, which, in turn, determines c. The maximum velocity $u_{\max}$ and the final value of x are then determined from

$$\frac{u_{\max}}{aT} = \frac{2x_f}{aT^2} = \log\frac{\sec\beta_0 + \tan\beta_0}{\sec\beta_0 - \tan\beta_0}\Big/ 2\tan\beta_0\,.$$

These relations are shown on Figures 2.4.3 and 2.4.4. Note, also, that

$$\nu_v = -\tan\beta_0\,, \qquad \nu_y = \frac{2\tan\beta_0}{T}\,.$$

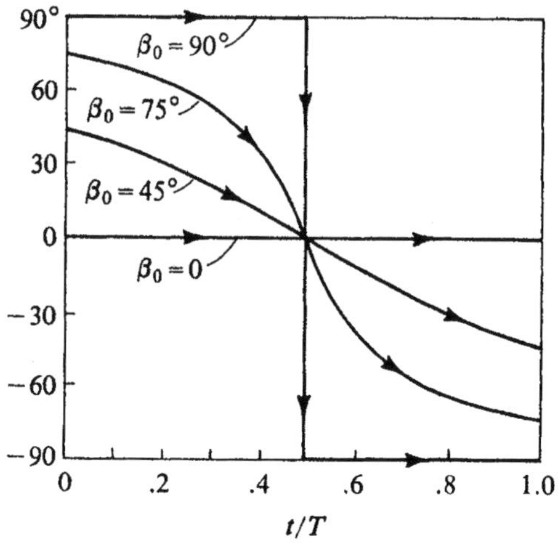

Figure 2.4.3. Thrust-angle programs for maximum velocity transfer to a rectilinear path.

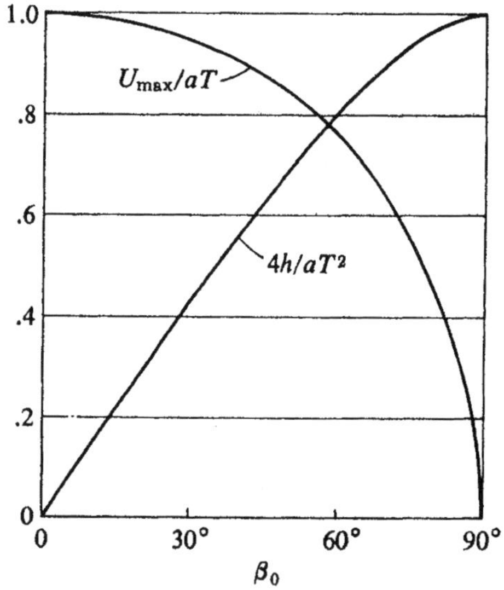

Figure 2.4.4. Maximum final velocity (U_{max}) vs. initial thrust angle (β_0), and β_0 as a function of $4h/aT^2$.

Problem 1. Consider an approximation to the optimal thrust-direction program in the example on pp. 59-62:

$$\beta = \begin{cases} \beta_1, & 0 < t < \dfrac{T}{2}, \\ -\beta_1, & \dfrac{T}{2} < t < T, \end{cases}$$

where β_1 is a constant. Note that this program gives $v(T) = 0$. Find β_1 so that $y(T) = h$ and determine $u(T)$, $x(T)$. Compare $u(T)$ with u_{max} in the example for a given h/aT^2.

ANSWER

$$\sin \beta_1 = \frac{4h}{aT^2}, \qquad u(T) = aT \cos \beta_1, \qquad x(T) = \frac{1}{2} aT^2 \cos \beta_1 .$$

Problem 2. *Airplane path in a wind to enclose maximum area in a given time.* An airplane has a fixed velocity V with respect to the air, and the wind velocity, u, is constant. *Find* the closed curve (as projected on the ground) the airplane should fly to enclose the maximum area in a given time T.

The equations of motion are

$$\dot{x} = V \cos \theta + u, \qquad \dot{y} = V \sin \theta,$$

where we have chosen the x-axis to be in the direction of the wind. If the airplane flies a closed curve, the area enclosed is given by

$$A = \oint y\, dx = \int_0^T y\dot{x}\, dt .$$

ANSWER. The closed curve is an ellipse of eccentricity $e = u/V$, minor axis parallel to the wind, and the maximum area enclosed is

$$A = \frac{V^2T^2}{4\pi}\left(1 - \frac{u^2}{V^2}\right)^{3/2} .$$

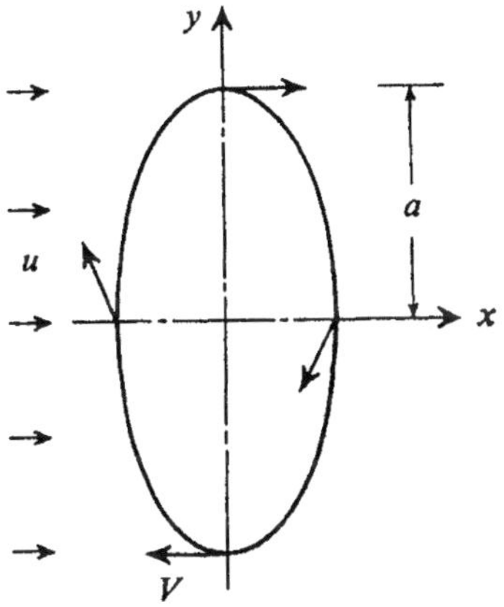

Figure 2.4.5. Airplane path in a wind to enclose maximum area in a given time.

Problem 3. *Minimum surface of revolution connecting two coaxial circular loops.* Given two coaxial circular loops, of radius a which are a distance 2ℓ apart, *find* the surface of revolution containing the two loops with minimum area. (This is the shape a soap film would take if stretched between two rings.) [HINT: Choose cylindrical coordinates r,x as shown in Figure 2.4.6. The annular element of area is

$$dA = 2\pi r\sqrt{(dr)^2 + (dx)^2}\,,$$

so the problem is to find $u(x)$ to minimize the integral.

$$A = 2\pi \int_{-\ell}^{\ell} r\sqrt{1 + u^2}\,dx\,,$$

where

$$\frac{dr}{dx} = u \qquad \text{and} \qquad r(\ell) = a\,, r(-\ell) = a\,.$$

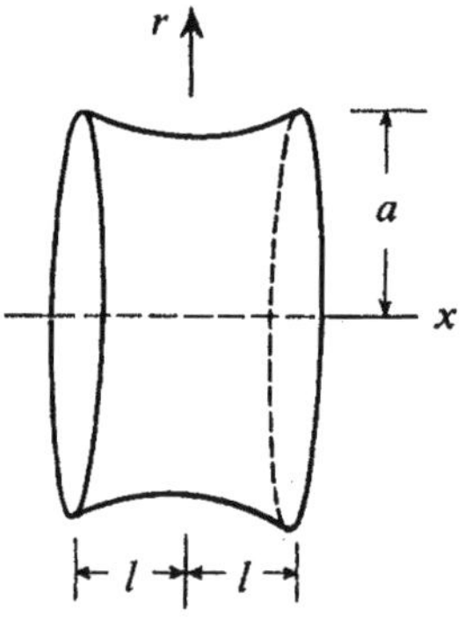

Figure 2.4.6. Minimum-area surface of revolution connecting two coaxial circular loops.

ANSWER. For $0 < \dfrac{\ell}{a} < .528$, the minimizing curve is

$$r = H\cosh\frac{x}{H}\,, \text{ where } \frac{H}{\ell} \text{ is determined by } \frac{a}{\ell}\frac{\ell}{H} = \cosh\frac{\ell}{H}\,.$$

This equation has *two* solutions for $0 < \ell/a < .663$ and no solution for $\ell/a > .663$. For $\ell/a > .528$, the minimizing curve is $r = 0$; i.e., two discs, each with area πa^2. The minimum area is given by

$$A_{\text{min}} = \begin{cases} 2\pi a^2\left(\tanh\dfrac{\ell}{H} + \dfrac{\ell}{H}\operatorname{sech}^2\dfrac{\ell}{H}\right), & 0 < \dfrac{\ell}{a} < .528\,, \\ 2\pi a^2\,, \quad \dfrac{\ell}{a} > .528\,. \end{cases}$$

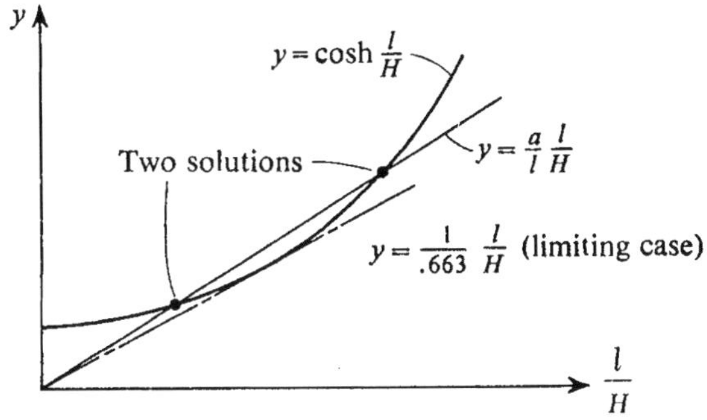

Figure 2.4.7. Solution of the minimum surface of revolution problem.

Problem 4. Find the minimum surface of revolution connecting two coaxial circular loops a distance ℓ apart, where one loop has radius a and the other loop has radius $b < a$. For each given value of b/a, show that a limiting value of $\ell/a < (\ell/a)_{\lim}$ exists beyond which the minimum surface is $r = 0$; that is, two flat discs within the circular loops.

2.5 Continuous systems with functions of the state variables prescribed at a fixed terminal time

In some problems we are interested in constraining *functions* of the terminal state to have prescribed values: that is, we have

$$\psi[x(t_f),t_f] = 0 \qquad (q \text{ equations}), \tag{2.5.1}$$

where ψ is a q-vector ($q \leq n-1$ if $L = 0$, $q \leq n$ if $L \neq 0$).

As in the previous section, we adjoin (2.5.1) to the performance index by a multiplier vector ν (a q-vector), also adjoining the system equations as in Section 2.3:

$$J = \phi[x(t_f),t_f] + \nu^T \psi[x(t_f),t_f] + \int_{t_0}^{t_f} \{L[x(t),u(t),t] + \lambda^T(f - \dot{x})\}\, dt\,. \tag{2.5.2}$$

If we define

$$\Phi = \phi + \nu^T \psi\,, \tag{2.5.3}$$

the development of Section 2.3 applies here without change. However, the final expressions for the necessary conditions for a stationary value of J satisfying (2.5.1) must be interpreted in a manner similar to that of Section 2.4; that is, we have a set of parameters, ν, which must be chosen to satisfy the q equations (2.5.1).† *In summary,* necessary conditions for J to have a stationary value are

†A controllability argument regarding the variations $\delta u(t)$ can be made, similar to the one made in Section 2.4, to justify (2.5.6).

$$\dot{x} = f(x,u,t) \qquad (n \text{ differential equations}), \tag{2.5.4}$$

$$\dot{\lambda} = -\left(\frac{\partial f}{\partial x}\right)^T \lambda - \left(\frac{\partial L}{\partial x}\right)^T \qquad (n \text{ differential equations}), \tag{2.5.5}$$

$$\left(\frac{\partial H}{\partial u}\right)^T = \left(\frac{\partial f}{\partial u}\right)^T \lambda + \left(\frac{\partial L}{\partial u}\right)^T = 0 \qquad (m \text{ algebraic equations}), \tag{2.5.6}$$

$$x_k(t_o) \text{ given or } \lambda_k(t_o) = 0 \; k = 1, \ldots, n \qquad (n \text{ boundary conditions}), \tag{2.5.7}$$

$$\lambda^T(t_f) = \left(\frac{\partial \phi}{\partial x} + \nu^T \frac{\partial \psi}{\partial x}\right)_{t=t_f} \qquad (n \text{ boundary conditions}), \tag{2.5.8}$$

$$\psi[x(t_f),t_f] = 0 \qquad (q \text{ side conditions}). \tag{2.5.9}$$

The stationarity conditions (2.5.6) determine the m-vector $u(t)$. The $2n$ differential equations (2.5.4) and (2.5.5), with the $2n$ boundary conditions (2.5.7) and (2.5.8), form a two-point boundary-value problem with q parameters ν to be found in (2.5.8) so that the q side conditions (2.5.9) are satisfied.

Example. *Maximum radius orbit transfer in a given time.* Given a constant-thrust rocket engine, T = thrust, operating for a given length of time, t_f, we wish to find the thrust-direction history, $\phi(t)$, to transfer a rocket vehicle from a given initial circular orbit to the largest possible circular orbit. The nomenclature is defined in Figure 2.5.1, below.

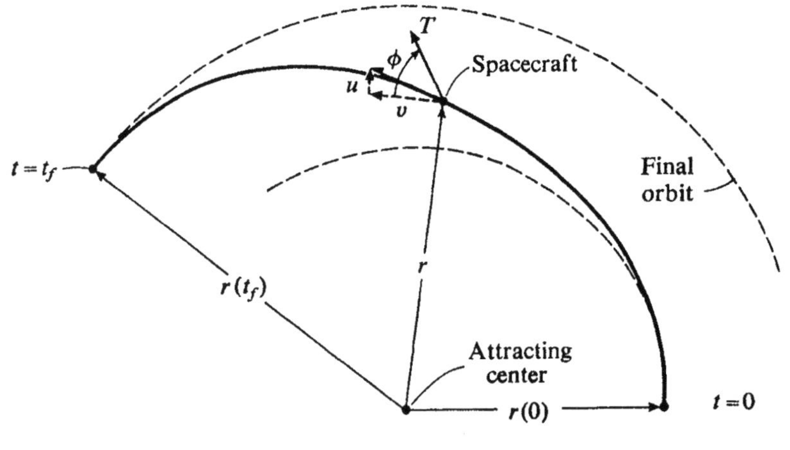

Figure 2.5.1. Maximum radius orbit transfer in a given time (or minimum time for a given final radius).

r = radial distance of spacecraft from attracting center
u = radial component of velocity
v = tangential component of velocity
m = mass of spacecraft, $\dot{m}$ = fuel consumption rate (constant)
ϕ = thrust direction angle
μ = gravitational constant of attracting center

Using this nomenclature, the problem may be stated as:
Find $\phi(t)$ to maximize $r(t_f)$ subject to

$$\dot{r} = u\,, \tag{2.5.10}$$

$$\dot{u} = \frac{v^2}{r} - \frac{\mu}{r^2} + \frac{T\sin\phi}{m_o - |\dot{m}|t}\,, \tag{2.5.11}$$

$$\dot{v} = -\frac{uv}{r} + \frac{T\cos\phi}{m_o - |\dot{m}|t}\,, \tag{2.5.12}$$

and

$$r(0) = r_o\,, \tag{2.5.13}$$

$$u(0) = 0\,, \tag{2.5.14}$$

$$v(0) = \sqrt{\frac{\mu}{r_o}}\,, \tag{2.5.15}$$

$$\psi_1 = u(t_f) = 0\,, \tag{2.5.16}$$

$$\psi_2 = v(t_f) - \sqrt{\frac{\mu}{r(t_f)}} = 0\,. \tag{2.5.17}$$

The Hamiltonian is, therefore,

$$H = \lambda_r u + \lambda_u\left(\frac{v^2}{r} - \frac{\mu}{r^2} + \frac{T\sin\phi}{m_o - |\dot{m}|t}\right) + \lambda_v\left(-\frac{uv}{r} + \frac{T\cos\phi}{m_o - |\dot{m}|t}\right)$$

and

$$\Phi = r(t_f) + \nu_1 u(t_f) + \nu_2\left[v(t_f) - \sqrt{\frac{\mu}{r(t_f)}}\right].$$

Thus, the necessary conditions (2.5.5), (2.5.6), and (2.5.9) become

$$\dot{\lambda}_r = -\lambda_u\left(-\frac{v^2}{r^2} + \frac{2\mu}{r^3}\right) - \lambda_v\left(\frac{uv}{r^2}\right), \tag{2.5.18}$$

$$\dot{\lambda}_u = -\lambda_r + \lambda_v\frac{v}{r}\,, \tag{2.5.19}$$

$$\dot{\lambda}_v = -\lambda_u\frac{2v}{r} + \lambda_v\frac{u}{r}\,, \tag{2.5.20}$$

$$0 = (\lambda_u \cos\phi - \lambda_v \sin\phi)\frac{T}{m_0 - |\dot{m}|t} \Rightarrow \tan\phi = \frac{\lambda_u}{\lambda_v}, \tag{2.5.21}$$

$$\lambda_r(t_f) = 1 + \frac{\nu_2\sqrt{\mu}}{2[r(t_f)]^{3/2}}, \tag{2.5.22}$$

$$\lambda_u(t_f) = \nu_1, \tag{2.5.23}$$

$$\lambda_v(t_f) = \nu_2. \tag{2.5.24}$$

The *six* differential equations (2.5.10), (2.5.11), (2.5.12), (2.5.18), (2.5.19), and (2.5.20) are to be solved subject to the *six* boundary conditions (2.5.13), (2.5.14), (2.5.15), (2.5.22), (2.5.23), and (2.5.24), with the choice of ν_1 and ν_2 available to satisfy the additional two boundary conditions (2.5.16) and (2.5.17). The control $\phi(t)$ is determined in terms of λ_u and λ_v from (2.5.21).

Minimum-time low-thrust orbit transfer;
thrust constant at 0.85,
initial spacecraft weight, 10,000 lb,
fuel consumption 12.9 lb per day

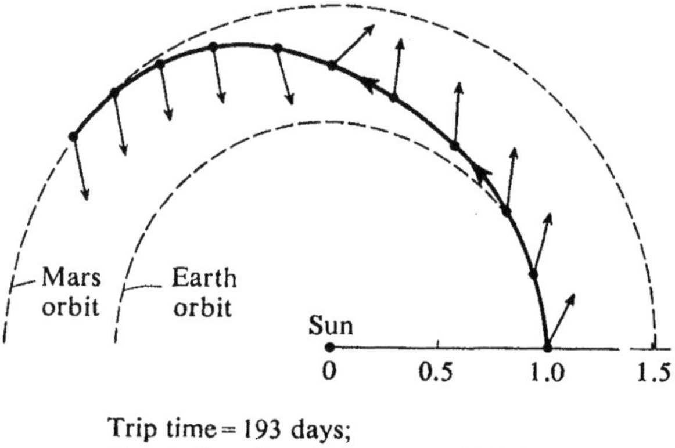

Trip time = 193 days;
thrust direction shown every 19.3 days

Figure 2.5.2. A particular minimum-time low-thrust orbit transfer path.

A numerical solution of this problem for

$$\frac{T/m_0}{\mu/r_0^2} = .1405, \qquad |\dot{m}|\sqrt{\mu/r_0}/T = 0.533, \qquad \frac{t_f}{\sqrt{r_0^3/\mu}} = 3.32$$

has been given by Kopp and McGill.† Interpreted for a 10,000-lb

†See A. V. Balakrishnan and L. W. Neustadt (eds.), *Computing Methods in Optimization Problems*. New York: Academic Press, 1964.

spacecraft moving out from the earth's orbit, the thrust would be 0.85 lb, the fuel consumption 12.9 lb/day, and the trip time 193 days. The optimal thrust direction and the resulting trajectory are shown in Figure 2.5.2. Note that the radial component of thrust is outward for the first half (roughly) of the flight, and inward for the second half.

2.6 Multistage systems; functions of the state variables specified at the terminal stage

Multistage systems, while of importance in their own right, have gained a special significance because of the use of digital computers to solve continuous problems. For numerical solution on a digital computer, the continuous optimization problems of Sections 2.3 to 2.5 must be converted to multistage optimization problems. Proper multistage formulation of such problems contributes significantly to the speed of convergence of iterative numerical solution procedures.

The following is a multistage version of the problems treated in Section 2.5. It differs from Section 2.2 only in the inclusion of terminal constraints. Find the sequence $u(0), \ldots, u(N-1)$ to minimize

$$J = \phi[x(N)] + \sum_{i=0}^{N-1} L^i[x(i),u(i)] \tag{2.6.1}$$

subject to the constraints

$$x(i+1) = f^i[x(i),u(i)], \tag{2.6.2}$$

$$\psi[x(N)] = 0, \tag{2.6.3}$$

where x is an n-vector, u is an m-vector, and ψ is a q-vector function, $q \leq n$.

As in Section 2.2, we adjoin Equation (2.6.2) to J with a multiplier sequence $\lambda(i)$ and, in addition, we adjoin Equation (2.6.3) with a set of q multipliers $(\nu_1, \ldots, \nu_q) \triangleq \nu^T$:

$$\bar{J} = \phi[x(N)] + \nu^T \psi[x(N)] + \sum_{i=0}^{N-1} \{L^i[x(i),u(i)] + \lambda^T(i+1)[f^i[x(i),u(i)] - x(i+1)]\}. \tag{2.6.4}$$

For convenience, define a scalar sequence H^i and a scalar function Φ, as follows:

$$H^i = L^i[x(i),u(i)] + \lambda^T(i+1) f^i[x(i),u(i)], \tag{2.6.5}$$

$$\Phi = \phi[x(N)] + \nu^T \psi[x(N)]. \tag{2.6.6}$$

Also, change indices of summation on the last term in (2.6.4), yielding

$$\bar{J} = \Phi[x(N)] - \lambda^T(N)\,x(N) + \sum_{i=1}^{N-1} [H^i - \lambda^T(i)\,x(i)] + H^o . \qquad (2.6.7)$$

Now consider differential changes in $\bar{J}$ due to differential changes in $u(i)$:

$$d\bar{J} = \left[\frac{\partial \Phi}{\partial x(N)} - \lambda^T(N)\right] dx(N) + \sum_{i=1}^{N-1} \left\{\left[\frac{\partial H^i}{\partial x(i)} - \lambda^T(i)\right] dx(i) + \frac{\partial H^i}{\partial u(i)}\, du(i)\right\}$$
$$+ \frac{\partial H^o}{\partial x(0)}\, dx(0) + \frac{\partial H^o}{\partial u(0)}\, du(0) . \qquad (2.6.8)$$

The coefficients multiplying $dx(i)(i = 0, \ldots, n)$ vanish if we *choose* the multiplier sequence $\lambda(i)$ so that we have

$$\lambda^T(i) - \frac{\partial H^i}{\partial x(i)} = 0 , \qquad (2.6.9)$$

or

$$\lambda^T(i) = \frac{\partial L^i}{\partial x(i)} + \lambda^T(i+1)\frac{\partial f^i}{\partial x(i)} , \qquad i = 0, \ldots, N-1 , \qquad (2.6.9a)$$

with boundary conditions

$$\lambda^T(N) = \frac{\partial \Phi}{\partial x(N)} , \qquad (2.6.10)$$

or

$$\lambda^T(N) = \frac{\partial \phi}{\partial x(N)} + \nu^T \frac{\partial \psi}{\partial x(N)} . \qquad (2.6.10a)$$

Equation (2.6.8) then becomes

$$d\bar{J} = \lambda^T(0)\, dx(0) + \sum_{i=0}^{N-1} \frac{\partial H^i}{\partial u(i)}\, du(i) . \qquad (2.6.11)$$

Thus $\partial H^i/\partial u(i)$ *is the gradient* of $\bar{J}$ with respect to $u(i)$ while holding $x(0)$ constant and satisfying (2.6.2), and $\lambda^T(0)$ is the gradient of $\bar{J}$ with respect to $x(0)$ while holding $u(i)$ constant and satisfying (2.6.2). If $x(0)$ is given, we have $dx(0) = 0$.

For a stationary value of $\bar{J}$, $d\bar{J}$ must be zero for admissible $du(i)$. If $u(i)$ is unconstrained and H^i is differentiable with respect to $u(i)$ and the problem is "normal", this can happen only if†

$$\frac{\partial H^i}{\partial u(i)} = 0 , \qquad (2.6.12)$$

†See Sections 5.3 and 6.3 for the argument concerning "normality" which relates to the existence of neighboring optimal paths.

or

$$\frac{\partial L^i}{\partial u(i)} + \lambda^T(i+1)\frac{\partial f^i}{\partial u(i)} = 0\,, \qquad i = 0,\ldots,N-1\,. \quad (2.6.12a)$$

In summary, to find a control-vector sequence $u(i)$ that produces a stationary value of the performance index J, we must solve the "two-point boundary-value problem" defined by (2.6.2), (2.6.3); (2.6.9), (2.6.10), and (2.6.12).

These equations constitute $(2n+m)\ N+n+p$ equations for as many unknowns: $x(0),\ldots,x(N)$ where x is an n-vector; $u(0),\ldots,u(N-1)$, where u is an m-vector; $\lambda(o), \lambda(1),\ldots,\lambda(N)$, where λ is an n-vector; and ν, a p-vector.

To solve (2.6.2) and (2.6.9a) together sequentially in the forward direction, using (2.6.12a) to determine $u(i)$, it is necessary to solve (2.6.9a) for $\lambda(i+1)$ in terms of $\lambda(i)$ and $x(i)$:

$$\lambda^T(i+1) = \left[\lambda^T(i) - \frac{\partial L^i}{\partial x(i)}\right]\left[\frac{\partial f^i}{\partial x(i)}\right]^{-1}. \qquad (2.6.13)$$

The inverse of $\partial f^i/\partial x(i)$ exists since it is, essentially, the linearized transition-matrix;† however, the computation of this inverse is time-consuming.‡ The alternative of sequential backward solution offers no improvement since (2.6.2), (2.6.9a), and (2.6.12a) would have to be viewed as a set of implicit equations for $x(i)$, $\lambda(i)$, and $u(i)$, given $x(i+1)$, $\lambda(i+1)$, $u(i+1)$.

2.7 Continuous systems; some state variables specified at an unspecified terminal time (including minimum-time problems)

This is almost the same set of problems as in Section 2.4, with the important difference that the terminal time, t_f, is not specified. It is convenient to regard t_f as a *control parameter* to be chosen in addition to the *control functions*, $u(t)$, so as to minimize the performance index and satisfy the constraints. We shall show that the same necessary conditions apply as in Section 2.4, but, in addition, the following condition must be satisfied by the optimal choice of the terminal time, t_f:

$$\left(\frac{\partial \phi}{\partial t} + \lambda^T f + L\right)_{t=t_f} = 0\,.$$

As in Section 2.3, we adjoin the system differential equations to the performance index, as follows:

†See Appendix A3.
‡The computation of the inverse is circumverted in the algorithm given in Section 7.7.

$$J = \phi[x(t_f),t_f] + \int_{t_0}^{t_f} [L(x,u,t) + \lambda^T(t)f(x,u,t) - \lambda^T\dot{x}]\, dt\,. \tag{2.7.1}$$

The *differential* of (2.7.1), taking into account differential changes in the terminal time, t_f, is

$$dJ = \left(\frac{\partial\phi}{\partial t} dt_f + \frac{\partial\phi}{\partial x} dx\right)_{t=t_f} + (L)_{t=t_f} dt_f + \int_{t_0}^{t_f} \left[\left(\frac{\partial L}{\partial x} + \lambda^T\frac{\partial f}{\partial x}\right)\delta x + \left(\frac{\partial L}{\partial u} + \lambda^T\frac{\partial f}{\partial u}\right)\delta u - \lambda^T\delta\dot{x}\right] dt\,. \tag{2.7.2}$$

Integrating (2.7.2) by parts and collecting terms gives

$$dJ = \left[\left(\frac{\partial\phi}{\partial t} + L\right) dt_f + \frac{\partial\phi}{\partial x} dx\right]_{t=t_f} - [\lambda^T\delta x]_{t=t_f} + [\lambda^T\delta x]_{t=t_0} + \int_{t_0}^{t_f} \left[\left(\frac{\partial L}{\partial x} + \lambda^T\frac{\partial f}{\partial x} + \dot{\lambda}^T\right)\delta x + \left(\frac{\partial L}{\partial u} + \lambda^T\frac{\partial f}{\partial u}\right)\delta u\right] dt\,. \tag{2.7.3}$$

Now δx, the variation in x, means "for time held fixed," so dx, the differential in x, may be written (see Figure 2.7.1)

$$dx(t_f) = \delta x(t_f) + \dot{x}(t_f)\, dt_f\,. \tag{2.7.4}$$

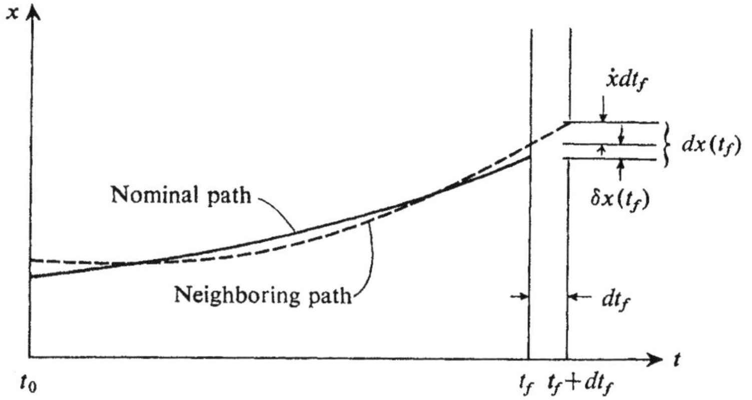

Figure 2.7.1. Relationship between $dx(t_f)$, $\delta x(t_f)$, and dt_f.

From (2.7.4), we have $\delta x(t_f) = dx(t_f) - \dot{x}(t_f)\, dt_f$; substituting this into (2.7.3) and collecting terms gives

$$dJ = \left[\left(\frac{\partial\phi}{\partial t} + L + \lambda^T\dot{x}\right) dt_f + \left(\frac{\partial\phi}{\partial x} - \lambda^T\right) dx\right]_{t=t_f} + \lambda^T(t_0)\,\delta x(t_0) + \int_{t_0}^{t_f} \left[\left(\frac{\partial L}{\partial x} + \lambda^T\frac{\partial f}{\partial x} + \dot{\lambda}^T\right)\delta x + \left(\frac{\partial L}{\partial u} + \lambda^T\frac{\partial f}{\partial u}\right)\delta u\right] dt\,. \tag{2.7.5}$$

Now, as in Section 2.4, we consider that

$$x_i(t_f)\,; \qquad i = 1,\ldots,q \text{ are specified,} \tag{2.7.6}$$

and, hence, it is consistent to consider ϕ to be a function only of the unspecified state variables:

$$\phi = \phi[x_j(t_f),t_f]\,; \qquad j = q+1,\ldots,n\,. \tag{2.7.7}$$

Next, we *choose* the functions $\lambda(t) \equiv \lambda^{(J)}(t)$ to make the coefficients of $\delta x(t)$, and $dx(t_f)$ vanish in (2.7.5)

$$\dot{\lambda}^{(J)} = -\left(\frac{\partial L}{\partial x}\right)^T - \left(\frac{\partial f}{\partial x}\right)^T \lambda^{(J)}\,, \tag{2.7.8}$$

$$\lambda_j^{(J)}(t_f) = \begin{cases} 0\,, & j = 1,\ldots,q\,, \\ \left(\dfrac{\partial \phi}{\partial x_j}\right)_{t=t_f}\,, & j = q+1,\ldots,n\,. \end{cases} \tag{2.7.9}$$

This choice of $\lambda(t)$ leaves (2.7.5) in the form

$$dJ = \left(\frac{\partial \phi}{\partial t} + L + f^T\lambda^{(J)}\right)_{t=t_f} dt_f + \int_{t_0}^{t_f} \left\{\frac{\partial L}{\partial u} + [\lambda^{(J)}]^T \frac{\partial f}{\partial u}\right\} \delta u\, dt\,, \tag{2.7.10}$$

where we have placed $\delta x(t_0) = 0$ since $x(t_0)$ is given.

Now, as in Section 2.4, let us consider the change in $x_i(t_f)$, $i = 1,\ldots,q$ for arbitrary $\delta u(t)$. Using the concept of influence (adjoint) functions (see Appendix A3) we have

$$dx_i\,(t_f) = [f_i]_{t=t_f}\, dt_f + \int_{t_0}^{t_f} [\lambda^{(i)}(t)]^T \frac{\partial f}{\partial u} \delta u\, dt\,, \tag{2.7.11}$$

where

$$\dot{\lambda}^{(i)} = -\left(\frac{\partial f}{\partial x}\right)^T \lambda^{(i)}\,, \tag{2.7.12}$$

$$\lambda_j^{(i)}(t_f) = \begin{cases} 1\,, & i = j\,, \\ 0\,, & i \neq j\,. \end{cases} \tag{2.7.13}$$

Note that Equation (2.7.11) may be regarded as a special case of (2.7.10) by replacing ϕ with x_i and placing $L = 0$.

We will now construct a $\delta u(t)$ history and select a value for dt_f that produces $dJ < 0$, and satisfies $dx_i(t_f) = 0, i = 1,\ldots,q$. Multiply each of the q equations (2.7.11) by an undetermined constant, ν_i, and add the resulting equations to (2.7.10)

$$\begin{aligned} dJ + \nu_i\, dx_i(t_f) = {} & \left\{\frac{\partial \phi}{\partial t} + L + [\lambda^{(J)}]^T f + \nu_i f_i\right\}_{t=t_f} dt_f \\ & + \int_{t_0}^{t_f} \left[\frac{\partial L}{\partial u} + (\lambda^{(J)} + \nu_i \lambda^{(i)})^T \frac{\partial f}{\partial u}\right] \delta u\, dt\,. \end{aligned} \tag{2.7.14}$$

Now *choose*

$$dt_f = -k_1\left\{\frac{\partial\phi}{\partial t} + L + [\lambda^{(J)}]^T f + \nu_i f_i\right\}_{t=t_f}, \tag{2.7.15}$$

$$\delta u = -k_2\left[\left(\frac{\partial L}{\partial u}\right)^T + \left(\frac{\partial f}{\partial u}\right)^T(\lambda^{(J)} + \nu_i\lambda^{(i)})\right], \tag{2.7.16}$$

where k_1 and k_2 are positive constants; substituting (2.7.15) and (2.7.16) into (2.7.14) yields

$$dJ + \nu_i\, dx_i(t_f) = -k_1\left\|\frac{\partial\phi}{\partial t} + L + (\lambda^{(J)})^T f + \nu_i f_i\right\|^2$$
$$-k_2\int_{t_0}^{t_f}\left\|\frac{\partial L}{\partial u} + (\lambda^{(J)} + \nu_i\lambda^{(i)})^T\frac{\partial f}{\partial u}\right\|^2 dt \le 0, \tag{2.7.17}$$

which is negative unless the squared terms are identically zero.

Next, we determine the ν_i's so as to satisfy the terminal constraints (2.7.11) with $dx_i(t_f) = 0$, $i = 1, \ldots, q$. Substitute (2.7.15) and (2.7.16) into (2.7.11):

$$0 = -k_1\{f_i[\phi_t + L + (\lambda^{(J)})^T f + \nu_j f_j]\}_{t=t_f}$$
$$-k_2\int_{t_0}^{t_f}(\lambda^{(i)})^T f_u[L_u^T + f_u^T(\lambda^{(J)} + \nu_j\lambda^{(j)})]\,dt, \tag{2.7.18}$$

or

$$0 = -k_1\{f_i[\phi_t + L + (\lambda^{(J)})^T f]\}_{t=t_f} - k_2\int_{t_0}^{t_f}(\lambda^{(i)})^T f_u[L_u^T + f_u^T\lambda^{(J)}]\,dt$$
$$-\left\{k_1(f_i f_j)_{t=t_f} + k_2\int_{t_0}^{t_f}(\lambda^{(i)})^T f_u f_u^T\lambda^{(j)}\,dt\right\}\nu_j,$$

from which the appropriate choice of ν is

$$\nu = -\left[Q + \frac{k_1}{k_2}S\right]^{-1}\left(g + \frac{k_1}{k_2}r\right), \tag{2.7.19}$$

where

$$Q_{ij} = \int_{t_0}^{t_f}(\lambda^{(i)})^T f_u f_u^T\lambda^{(j)}\,dt, \qquad S_{ij} = (f_i f_j)_{t=t_f},$$

$$g_i = \int_{t_0}^{t_f}(\lambda^{(i)})^T f_u[L_u^T + f_u^T\lambda^{(J)}]\,dt, \qquad r_i = \{f_i[\phi_t + L + (\lambda^{(J)})^T f]\}_{t=t_f}.$$

From (2.7.17), the only case in which we *cannot* decrease the performance index is when

$$(\phi_t + L + (\lambda^{(J)})^T f + \nu_j f_j)_{t=t_f} = 0, \tag{2.7.20}$$

$$L_u + (\lambda^{(J)} + \nu_i \lambda^{(i)})^T f_u = 0 \,; \qquad t_o < t < t_f \,. \tag{2.7.21}$$

If (2.7.20), (2.7.21) obtain, we have a ***stationary solution*** that satisfies the terminal constraints.

Using (2.7.20) in (2.7.18), we see that the ν_i's for a stationary solution are independent of k_1/k_2 and are given by

$$\nu = -Q^{-1} g \,. \tag{2.7.22}$$

Thus, as in the fixed-terminal-time case, the existence of the inverse of Q is the ***controllability condition.*** Since the influence equations are linear, the necessary conditions (2.7.20)–(2.7.21) may be written as

$$(\phi_t + H)_{t=t_f} = 0 \,, \tag{2.7.23}$$

$$\frac{\partial H}{\partial u} = 0 \,; \qquad t_o < t < t_f \,, \tag{2.7.24}$$

where

$$H = L + \lambda^T f \,, \tag{2.7.25}$$

$$\dot{\lambda}^T = -\frac{\partial H}{\partial x} = -\frac{\partial L}{\partial x} - \lambda^T \frac{\partial f}{\partial x} \,, \tag{2.7.26}$$

$$\lambda_j(t_f) = \begin{cases} \nu_j \,, & j = 1, \ldots, q \,, \\ \left(\dfrac{\partial \phi}{\partial x_j}\right)_{t=t_f} \,, & j = q+1, \ldots, n \,. \end{cases} \tag{2.7.27}$$

We may regard the ν_i's, $i = 1, \ldots, q$, as ***control parameters*** that control the terminal values of x_i, $i = 1, \ldots, q$, which must have the specified values for an admissible path. Similarly, t_f is a control parameter that controls the terminal value of $\phi_t + H$, which must vanish for a stationary path.

Conceptually, the problem of unspecified terminal time can always be approached as a series of optimization problems with fixed terminal time. In other words, we consider the terminal time t_f as an additional control parameter and solve a series of identical optimization problems, as in Section 2.4, with different values of t_f. The particular value of t_f that yields the minimal value of J for the series of optimization problems must be the solution to the problem with unspecified terminal time. Thus, we can expect that all necessary conditions derived in Section 2.4 hold. There must also be one additional condition that determines the optimal value of t_f, and this is (2.7.23).

Problem 1. Consider

$$\bar{J} = \Phi(x(t_f),t_f) + \int_0^{t_f} L(x,u,t)\, dt$$

and t_f as a control parameter. What is the variation of J due to a variation of t_f when all optimality conditions in Section 2.4 are to be satisfied? From this, derive the condition

$$\frac{\partial \Phi}{\partial t_f} = -H(t_f)$$

directly. [HINT:

$$d\bar{J} = \frac{\partial \Phi}{\partial x}\frac{dx}{dt}dt_f + \frac{\partial \Phi}{\partial t_f}dt_f + L dt_f .]$$

MINIMUM-TIME SOLUTIONS. In many problems, the performance index of interest is the *elapsed time* to transfer the system from its initial state to a specified state. In this case, we may place

$$\phi = 0, \qquad L = 1, \tag{2.7.28}$$

which implies that

$$J = t_f - t_0 . \tag{2.7.29}$$

The minimum-time control program is obtained, then, by solving the two-point boundary-value problem:

$$\dot{x} = f(x,u,t)\,; \qquad x(t_0) \text{ given}\dagger \ (n \text{ initial conditions}), \tag{2.7.30}$$

$$\begin{aligned} \dot{\lambda} = -(f_x)^T\lambda\,; \qquad & x_j(t_f) \text{ specified}; \qquad j = 1,\ldots,q\,; \\ \lambda_j(t_f) = 0\,, j = q+1,\ldots,n \ & (n \text{ terminal conditions}), \end{aligned} \tag{2.7.31}$$

$$0 = f_u^T\lambda \ (m \text{ optimality conditions}), \tag{2.7.32}$$

$$(\lambda^T f)_{t=t_f} = -1 . \tag{2.7.33}$$

Note that there are $2n$ boundary conditions for the $2n$ differential equations (2.7.30)–(2.7.31), m optimality conditions (2.7.32) for the m control variables, u, and one transversality condition (2.7.33) for the terminal time, t_f. The unspecified values of $\lambda_j(t_f)\,, j = 1,\ldots,q$, which we have called ν_j above, are part of the solution.

Note, also, that at least one state variable must be specified at $t = t_0$ and at $t = t_f$ or the minimum-time problem makes no sense.

†If $x_j(t_0)$ is not specified, we have $\lambda_j(t_0) = 0$.

Example 1. *Minimum-time paths through a region of position-dependent vector velocity (Zermelo's problem).*† A ship must travel through a region of strong currents. The magnitude and direction of the currents are known as functions of position:

$$u = u(x,y) \qquad \text{and} \qquad v = v(x,y)\,,$$

where (x,y) are rectangular coordinates and (u,v) are the velocity components of the current in the x and y directions, respectively. The magnitude of the ship's velocity relative to the water is V, a constant. The problem is to steer the ship in such a way as to minimize the time necessary to go from a point A to a point B.

The equations of motion are

$$\dot{x} = V\cos\theta + u(x,y)\,, \tag{2.7.34}$$

$$\dot{y} = V\sin\theta + v(x,y)\,, \tag{2.7.35}$$

where θ is the heading angle of the ship's axis relative to the (fixed) coordinate axes, and (x,y) represents the position of the ship.

The Hamiltonian of the system is

$$H = \lambda_x(V\cos\theta + u) + \lambda_y(V\sin\theta + v) + 1\,, \tag{2.7.36}$$

So the Euler-Lagrange equations are

$$\dot{\lambda}_x = -\frac{\partial H}{\partial x} = -\lambda_x\frac{\partial u}{\partial x} - \lambda_y\frac{\partial v}{\partial x}\,, \tag{2.7.37}$$

$$\dot{\lambda}_y = -\frac{\partial H}{\partial y} = -\lambda_x\frac{\partial u}{\partial y} - \lambda_y\frac{\partial v}{\partial y}\,, \tag{2.7.38}$$

$$0 = \frac{\partial H}{\partial \theta} = V(-\lambda_x\sin\theta + \lambda_y\cos\theta)\,, \Rightarrow \tan\theta = \frac{\lambda_y}{\lambda_x}\,. \tag{2.7.39}$$

Since the Hamiltonian (2.7.36) is not an explicit function of time, H = constant is an integral of the system. Furthermore, since we are minimizing time, this constant must be 0. We may solve (2.7.36) and (2.7.39) for λ_x and λ_y:

$$\lambda_x = \frac{-\cos\theta}{V + u\cos\theta + v\sin\theta}\,, \tag{2.7.40}$$

$$\lambda_y = \frac{-\sin\theta}{V + u\cos\theta + v\sin\theta}\,. \tag{2.7.41}$$

†For another derivation, using vector notation, see Section 3.2, Example 2, which treats the problem in three dimensions (e.g., for an aircraft in a region of strong winds).

We may now substitute (2.7.40) and (2.7.41) into either (2.7.37) or (2.7.38) (or demand consistency between $H_\theta = 0$, $\dot{H}_\theta = 0$) to obtain

$$\dot{\theta} = \sin^2\theta\frac{\partial v}{\partial x} + \sin\theta\cos\theta\left(\frac{\partial u}{\partial x} - \frac{\partial v}{\partial y}\right) - \cos^2\theta\frac{\partial u}{\partial y}. \tag{2.7.42}$$

This equation, solved simultaneously with (2.7.34) and (2.7.35), will give the desired minimum time paths; in order to go through a particular point B, starting at a point A, we must pick the correct value of θ_A.

Note that, if u and v are constant, (2.7.42) indicates that θ = const; that is, the minimum-time paths are straight lines.

ANALOG TO SNELL'S LAW. If we have $u = u(y)$, $v = v(y)$, then (2.7.37) becomes

$$\dot{\lambda}_x = 0 \Rightarrow \lambda_x = \text{const.} \tag{2.7.43}$$

Equation (2.7.40) is then

$$\frac{\cos\theta}{V + u(y)\cos\theta + v(y)\sin\theta} = \text{const}, \tag{2.7.44}$$

which is directly analogous to Snell's Law in optics since it (implicitly) gives the ship's heading, θ, in terms of the local current velocities.

Special case: linear variation of current velocity. If we have $u = -V(y/h)$, $v = 0$, and we wish to find the minimum-time path from a certain point x_0, y_0 to the origin (0,0) then we may use (2.7.44) to express the optimal heading angle, θ, in terms of the final heading angle, θ_f, and the present y coordinate, as follows:

$$\frac{\cos\theta}{V - V(y/h)\cos\theta} = \frac{\cos\theta_f}{V} = \text{constant},$$
$$\cos\theta = \frac{\cos\theta_f}{1 + (y/h)\cos\theta_f}. \tag{2.7.45}$$

It is convenient to use θ as the independent variable instead of t. From (2.7.45), we already have $y(\theta)$,

$$\frac{y}{h} = \sec\theta - \sec\theta_f. \tag{2.7.46}$$

Equation (2.7.42) becomes

$$\frac{dt}{d\theta} = \frac{h}{V}\sec^2\theta \Rightarrow \frac{V(t_f - t)}{h} = \tan\theta - \tan\theta_f, \tag{2.7.47}$$

where $t_f - t$ is the time to go to the origin.

Finally, (2.7.34), using (2.7.46) and (2.7.47), becomes

$$\frac{dx}{d\theta} = \frac{V\cos\theta + V(\sec\theta_f - \sec\theta)}{-(V/h)\cos^2\theta} = -h(\sec\theta + \sec\theta_f\sec^2\theta - \sec^3\theta)\,, \tag{2.7.48}$$

which may be integrated to give

$$\frac{x}{h} = \frac{1}{2}\left[\sec\theta_f(\tan\theta_f - \tan\theta) - \tan\theta\,(\sec\theta_f - \sec\theta) + \log\frac{\tan\theta_f + \sec\theta_f}{\tan\theta + \sec\theta}\right]. \tag{2.7.49}$$

Now, let us suppose that we want to find the minimum-time path from $x_o/h = 3.66$, $y_o/h = -1.86$ to the origin. Equations (2.7.46) and (2.7.49) are implicit equations for θ_o and θ_f, where θ_o is the value of θ at the initial point, as follows:

$$-1.86 = \sec\theta_o - \sec\theta_f\,, \tag{2.7.50}$$

$$3.66 = \tfrac{1}{2}[\sec\theta_f(\tan\theta_f - \tan\theta_o) - \tan\theta_o(\sec\theta_f - \sec\theta_o) + \sin h^{-1}(\tan\theta_f) - \sin h^{-1}(\tan\theta_o)]\,. \tag{2.7.51}$$

The solution to these equations is

$$\theta_o = 105^\circ\,, \qquad \theta_f = 240^\circ\,.$$

From (2.7.47), the time to go from the initial point to the origin is

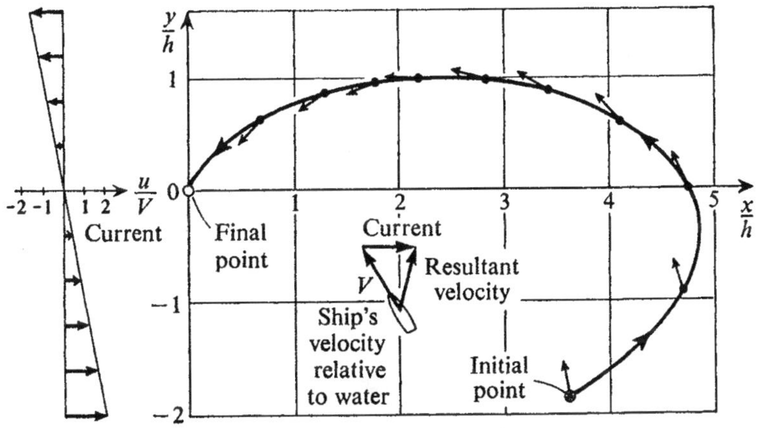

Figure 2.7.2. A minimum-time path through a region of linearly increasing current.

$$\frac{V(t_f - t_o)}{h} = 5.46 .$$

Figure 2.7.2 shows the optimal path with the ship's heading (the control function in this problem) indicated at several points along the path.

Problem 2. *Minimum-time paths through a region of position-dependent velocity magnitude.* A particle must travel through a region in which its instantaneous velocity magnitude is given as function of position, $V = V(x,y)$, where (x,y) are rectangular coordinates. We wish to find the minimum-time path from a point in the (x,y)-plane to the origin $(0,0)$. The equations of motion are

$$\dot{x} = V(x,y) \cos \theta , \qquad \dot{y} = V(x,y) \sin \theta ,$$

where θ is the angle between the x-axis and the direction of the path; in this case, θ is the "control variable."

Show that, along a minimum-time path, $\theta(t)$ must satisfy the differential equation

$$\dot{\theta} = \frac{\partial V}{\partial x} \sin \theta - \frac{\partial V}{\partial y} \cos \theta .$$

Note that $\theta(0)$ and the final time t_f are determined by requiring $x(t_f) = y(t_f) = 0$. If $V = \text{const}$, the minimum-time paths are straight lines $\dot{\theta} = 0)$.

Problem 3. *Fermat's principle* states that the paths of light rays are extremal time paths; they are usually minimum-time paths but they are occasionally only local minima, and, in some cases, they are merely stationary but not minimizing. The index of refraction, n, is defined as the ratio of the speed of light in a vacuum, c, to the local speed of light, $V(x,y)$:

$$n = \frac{c}{V(x,y)} .$$

Changing independent variables from t to distance along the path, s, where $ds = V\,dt$, show that:

$$\frac{d}{ds}\left(n \frac{dx}{ds}\right) = \frac{\partial n}{\partial x} , \qquad \frac{d}{ds}\left(n \frac{dy}{ds}\right) = \frac{\partial n}{\partial y} .$$

Problem 4. Consider a special case of Problem 2, in which the velocity is a function of only one coordinate, $V = V(y)$. Show that, in this case, an integral of the optimal path is

$$\frac{\cos \theta}{V(y)} = \text{const.}$$

This is known as *Snell's Law*, originating in optics.

Problem 5. Consider a special case of Problem 4, in which V is a linear function of y, as follows:

$$V = V_f\left(1 + \frac{y}{h}\right); \qquad V_f, h \text{ constants.}$$

Show that the minimum-time paths to the origin are *arcs of circles* whose centers lie on the line $y = -h$.

Problem 6. *The classical brachistochrone problem.*† A bead slides on a frictionless wire between two points A and B in a constant-gravity field. The bead has an initial velocity V_0 at point A. What is the shape of the wire that will produce a minimum-time path between the two points?

The two points and the gravity vector determine a vertical plane. Let the y-axis point downward and the origin be at point A (see Figure 2.7.3).

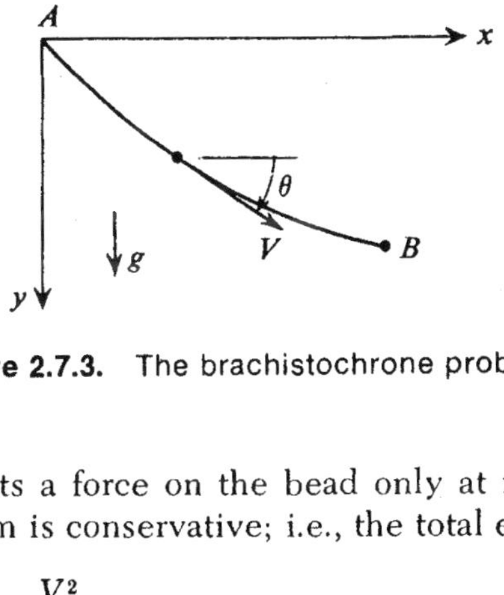

Figure 2.7.3. The brachistochrone problem.

Since the wire puts a force on the bead only at right angles to its velocity, the system is conservative; i.e., the total energy is constant:

$$\frac{V^2}{2} - gy = \frac{V_0^2}{2} \qquad \text{or} \qquad V = (V_0^2 + 2gy)^{1/2} = V(y).$$

The velocity components are, thus,

$$\dot{x} = V(y)\cos\theta, \qquad \dot{y} = V(y)\sin\theta.$$

The problem is to find $\theta(t)$ so as to minimize the time in going from A to B. Note that this is a special case of Problem 4; i.e., it is a *Fermat's* problem for a minimum-time path through a region of position-dependent velocity.

Show that the paths are *cycloids*, i.e., paths generated by a point

†Called the "brachistochrone" (Greek for "shortest time") problem by John Bernoulli, who proposed it in 1696.

on a circle rolling without slipping in a horizontal direction, and that $\dot{\theta} =$ constant.

Problem 7. (Courtesy T. N. Edelbaum). Find the path of minimum time connecting two points on the surface of the earth through a tunnel in the earth. The tunnel is assumed to be evacuated, gravity is the propelling force on the particle, and friction is negligible. Note that the gravitational force per unit mass inside the earth is directed radially toward the center of the earth and increases *linearly* with radius from zero at the center.

ANSWER. The paths are hypocycloids, i.e., curves generated by a point on a small circle rolling without slipping on the *inside* of the earth's surface.

Problem 8. *Thrust-direction programming with negligible external forces.* This is one of the simplest problems, of some practical interest, in optimal programming. As such, it is useful for fixing ideas.† Consider a particle of mass m, acted upon by a thrust force of magnitude ma. Assume planar motion and use an inertial coordinate system x,y to locate the particle; the velocity components of the particle are u,v. The thrust-direction angle $\beta(t)$ is the control variable for the system (see Figure 2.4.1). The equations of motion are

$$\dot{u} = a \cos\beta\,, \qquad \dot{v} = a \sin\beta\,, \qquad \dot{x} = u\,, \qquad \dot{y} = v\,,$$

where the thrust acceleration, a, is assumed to be a known function of time. If we wish to extremalize a function of the end conditions only *or* minimize the time, *show that* the optimal control law is

$$\tan\beta = \frac{-c_2 t + c_4}{-c_1 t + c_3}\,,$$

where c_1, c_2, c_3, c_4 are constants. This is often referred to as the "bilinear tangent law."

Problem 9. *Minimum-time orbit injection* ($g = 0$). We wish to transfer the particle of Problem 8 to a path a distance h away, arriving with a velocity of U parallel to the path in the least time; we do *not* care what the final x-coordinate is (see Figure 2.4.2).

Thus, the boundary conditions for the problem are

$$u(0) = 0\,, \quad u(T) = U\,, \quad v(0) = 0\,, \quad v(T) = 0\,, \quad x(0) = 0\,,$$
$$\lambda_x(T) = 0\,, \quad y(0) = 0\,, \quad y(T) = h\,,$$
$$(\lambda_u a \cos\beta + \lambda_v a \sin\beta)_{t=T} = -1\,.$$

†See Example, Section 2.4.

Since $x(T)$ is unspecified, we have $\lambda_x = c_1 = 0$, and the optimal control law becomes a "linear tangent law":

$$\tan\beta = \tan\beta_0 - ct, \qquad \text{where} \qquad \tan\beta_0 = \frac{c_4}{c_3}, c = \frac{c_2}{c_3} \text{ (see Problem 8).}$$

For *constant-thrust acceleration*, a, and using β as the independent variable instead of t, *show that*

$$u = \frac{a}{c}\log\frac{\tan\beta_0 + \sec\beta_0}{\tan\beta + \sec\beta}, \qquad v = \frac{a}{c}(\sec\beta_0 - \sec\beta),$$

$$x = \frac{a}{c^2}\left(\sec\beta_0 - \sec\beta - \tan\beta\log\frac{\tan\beta_0 + \sec\beta_0}{\tan\beta + \sec\beta}\right),$$

$$y = \frac{a}{2c^2}\left[(\tan\beta_0 - \tan\beta)\sec\beta_0 - (\sec\beta_0 - \sec\beta)\tan\beta - \log\frac{\tan\beta_0 + \sec\beta_0}{\tan\beta + \sec\beta}\right],$$

$$\lambda_u = -\frac{\cos\beta_0}{a}, \qquad \lambda_v = -\frac{\sin\beta_0}{a}\left(1 - 2\frac{t}{T}\right), \qquad \lambda_x = 0,$$

$$\lambda_y = -\frac{2\sin\beta_0}{aT},$$

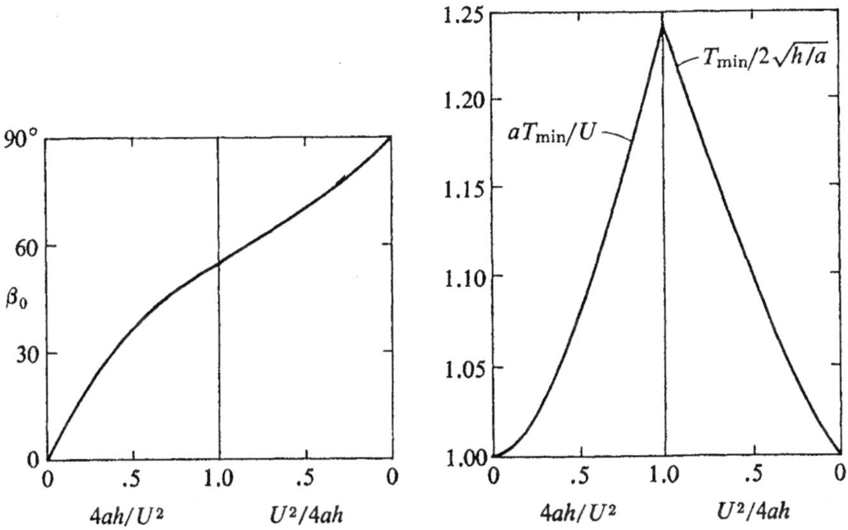

Figure 2.7.4. Initial thrust angle (β_0) and minimum time (T_{min}) as a function of $4ah/U^2$ for minimum-time transfer to a rectilinear path.

where the constants β_0 and c, as well as the final (minimum) time, T, are determined by the three final boundary conditions $v = 0$, $u = U$, $y = h$. *Show that* these relations may be put into the form

$$\frac{4ah}{U^2} = \frac{\tan\beta_0 \sec\beta_0 - \operatorname{logtan}[(\pi/4) + (1/2)\beta_0]}{\{\operatorname{logtan}[(\pi/4) + (1/2)\beta_0]\}^2},$$

$$\frac{aT}{U} = \frac{\tan\beta_0}{\operatorname{logtan}[(\pi/4) + (1/2)\beta_0]},$$

$$cT = 2\tan\beta_0 \Rightarrow \tan\beta = \left(1 - \frac{2t}{T}\right)\tan\beta_0 .$$

Clearly, the one dimensionless quantity, $4ah/U^2$ determines β_0 and, hence, also aT/U. These relationships are shown in Figure 2.7.4. Thrust-direction programs for various values of β_0 are shown in Figure 2.4.3.

Problem 10. *Minimum-time interception of a nonmaneuvering target* $(g = 0)$. Using the same equations of motion as in Problem 8, find the thrust-direction angle history, $\beta(t)$, to go to the origin, $x = y = 0$, in minimum time, starting from a given initial point, x_0, y_0, with initial velocity u_0, v_0. Assume constant-thrust acceleration a. Note that the final velocity is *not* specified, so this is an *interception* problem.

Problem 11. *Minimum-time rendezvous with a nonmaneuvering target* $(g = 0)$. This is the same as Problem 10 except that the final velocity is specified to be zero; that is, we have $u_f = v_f = 0$. This is a *rendezvous* problem. Note that the "bilinear tangent law" can be put into the form of a "linear tangent law":

$$\tan(\theta - \alpha) = \tan(\theta_f - \alpha) + c(T - t),$$

where α, θ_f, and c are parameters.

Problem 12. *Thrust-direction programming in a constant gravitational field.* If we align the y-axis in the opposite direction to the gravitational force, the only change from Problem 8 (without gravitational force) is in the vertical acceleration:

$$\dot{v} = a\sin\beta - g,$$

where g is the acceleration due to gravity. *Show that* the equations for the influence functions are unchanged, so that the "bilinear tangent law" is still the optimal law.

Problem 13. *Minimum-time orbit injection* (g = *constant*). *Show that* the only change from Problem 8 (with $g = 0$) is the addition of a term $-gt$ to the vertical velocity v, and a term $-\frac{1}{2}gt^2$ to the vertical height, y. For the case of constant-thrust acceleration a, there are three quantities to be determined: the initial thrust-direction angle, β_0; the final thrust-direction angle, β_f; and the minimum time, T. *Show that* the three equations available for determining them are

$$v_f = 0 = \frac{a}{c}(\sec\beta_0 - \sec\beta_f) - gT\,,$$

$$y_f = h = \frac{a}{2c^2}\left[(\tan\beta_0 - \tan\beta_f)\sec\beta_0 - (\sec\beta_0 - \sec\beta_f)\tan\beta_f - \log\frac{\tan\beta_0 + \sec\beta_0}{\tan\beta_f + \sec\beta_f}\right] - \frac{1}{2}gT^2\,,$$

$$u_f = U = \frac{a}{c}\log\frac{\tan\beta_0 + \sec\beta_0}{\tan\beta_f + \sec\beta_f}\,,$$

where

$$c = \frac{\tan\beta_0 - \tan\beta_f}{T}\,, \qquad \tan\beta = \tan\beta_0 - ct\,.$$

Eliminating c and T among these equations, derive the following two equations in the two unknowns β_0 and β_f:

$$\frac{a}{g} = \frac{\tan\beta_0 - \tan\beta_f}{\sec\beta_0 - \sec\beta_f}\,,$$

$$\frac{2ah}{U^2} = \left[\tan\beta_0\sec\beta_f - \tan\beta_f\sec\beta_0 - \log\frac{\tan\beta_0 + \sec\beta_0}{\tan\beta_f + \sec\beta_f}\right] \div \left[\log\frac{\tan\beta_0 + \sec\beta_0}{\tan\beta_f + \sec\beta_f}\right]^2\,,$$

with

$$\frac{aT}{U} = (\tan\beta_0 - \tan\beta_f)\Big/\left[\log\frac{\tan\beta_0 + \sec\beta_0}{\tan\beta_f + \sec\beta_f}\right].$$

Clearly, the quantities ah/U^2 and a/g determine β_0, β_f, aT/U, and cT.

Numerical Example. Figure 2.7.5 shows two example trajectories with $a/g = 3$, calculated for lunar takeoffs, using the constant g approximation. The moon's surface gravity is 5.3 ft sec^{-2}, and the radius of the moon is 938 nautical miles. Minimum-time ascent paths are shown for $h = 100$ nautical miles and $h = 50{,}000$ ft, the final velocity being circular-satellite velocity in the first case and slightly higher than circular-satellite velocity in the second case. The "characteristic velocity," ΔV_c, required is simply aT for $a =$ constant. For comparison, the impulsive injection into a 100 n mi orbit (Hohman transfer) requires a characteristic velocity of 5,780 ft sec^{-1} (5,640 at the moon's surface and 140 ft sec^{-1} at apolune). The velocity at the end of the 50,000-ft ascent is such that the spacecraft will coast to 100 n mi on the other side of the moon; there, an injection impulse of 464 ft sec^{-1} is required to put it into circular orbit at $h = 100$ n mi making the total $\Delta V_c = 6{,}584$ ft sec^{-1}. Note that the lunar surface is approximated as a parabola, which significantly extends the usefulness of the constant g approximation.

Obviously, the minimum-time "soft-landing" path with constant a (with down-range *not* specified) is the same as the minimum-time orbit-injection path run backwards.

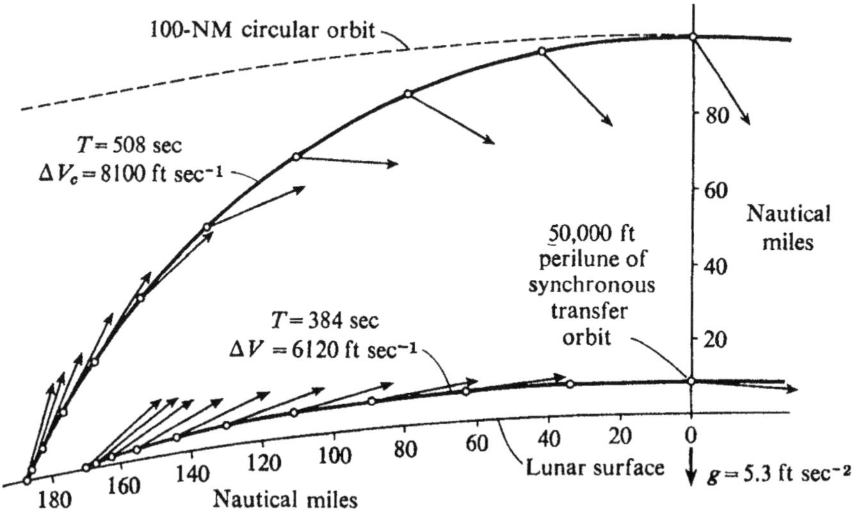

Figure 2.7.5. Minimum-time lunar take-offs (or landings) with constant-thrust acceleration = 15.9 ft sec^{-2} (thrust direction shown every tenth of total time).

Problem 14. (a) In the two-dimensional xt-plane, determine the extremal curve of stationary length which starts on the circle $x^2 + t^2 - 1 = 0$ and terminates on the line $t = T = 2$.

(b) Solve the same problem as (a) but consider that the termination is on the line $-x + t = 2\sqrt{2}$.

[NOTE: Parts (a) and (b) are *NOT* to be solved by inspection.]

2.8 Continuous systems; functions of the state variables specified at an unspecified terminal time, including minimum-time problems†

We again consider the performance index

$$J = \phi[x(t_f),t_f] + \int_{t_0}^{t_f} L[x(t),u(t),t]\, dt\,. \tag{2.8.1}$$

We adjoin the constraints

$$\psi[x(t_f),t_f] = 0 \qquad (\psi \text{ a } q\text{-vector function}) \tag{2.8.2}$$

and the system differential equations

$$\dot{x} = f[x(t),u(t),t], \qquad t_0 \text{ given} \tag{2.8.3}$$

to the performance index with Lagrange multipliers ν and $\lambda(t)$, as follows:

$$J = [\phi + \nu^T \psi]_{t=t_f} + \int_{t_0}^{t_f} \{L(x,u,t) + \lambda^T[f(x,u,t) - \dot{x}]\}\, dt\,. \tag{2.8.4}$$

The Hamiltonian is defined as

$$H = L(x,u,t) + \lambda^T(t) f(x,u,t)\,. \tag{2.8.5}$$

The *differential* of (2.8.4), taking into account differential changes in the terminal time, t_f, is

$$dJ = \left(\left(\frac{\partial \Phi}{\partial t} + L\right) dt + \frac{\partial \Phi}{\partial x}\, dx\right)_{t=t_f} + \int_{t_0}^{t_f} \left(\frac{\partial H}{\partial x}\, \delta x + \frac{\partial H}{\partial u}\, \delta u - \lambda^T \delta \dot{x}\right) dt - L|_{t=t_0}\, dt_0\,, \tag{2.8.6}$$

where

$$\Phi = \phi + \nu^T \psi\,. \tag{2.8.7}$$

Integrating by parts and using $\delta x = dx - \dot{x}\, dt$, we have

$$dJ = \left(\frac{\partial \Phi}{\partial t} + L + \lambda^T \dot{x}\right)_{t=t_f} dt_f + \left[\left(\frac{\partial \Phi}{\partial x} - \lambda^T\right) dx\right]_{t=t_f} + (\lambda^T \delta x)_{t=t_0} + \int_{t_0}^{t_f} \left[\left(\frac{\partial H}{\partial x} + \dot{\lambda}^T\right) \delta x + \frac{\partial H}{\partial u}\, \delta u\right] dt - L|_{t=t_0}\, dt_0\,. \tag{2.8.8}$$

We now *choose* the functions $\lambda(t)$ to make the coefficients of $\delta x(t)$, $dx(t_f)$, and dt_f vanish (if t_f is not prescribed):

†J. V. Breakwell, "The Optimization of Trajectories," *SIAM Journal*, Vol. 7, 1959.

$$\dot{\lambda}^T = -\frac{\partial H}{\partial x} = -\lambda^T \frac{\partial f}{\partial x} - \frac{\partial L}{\partial x}, \tag{2.8.9}$$

$$\lambda^T(t_f) = \left(\frac{\partial \Phi}{\partial x}\right)_{t=t_f} = \left(\frac{\partial \phi}{\partial x} + \nu^T \frac{\partial \psi}{\partial x}\right)_{t=t_f}, \tag{2.8.10}$$

$$\left(\frac{\partial \Phi}{\partial t} + L + \lambda^T \dot{x}\right)_{t=t_f} = \left(\frac{d\Phi}{dt} + L\right)_{t=t_f} = 0, \tag{2.8.11}$$

where

$$\frac{d\Phi}{dt} = \frac{\partial \Phi}{\partial t} + \frac{\partial \Phi}{\partial x}\dot{x}.$$

As a result of this choice of $\lambda(t)$, (2.8.8) is simplified to

$$dJ = \int_{t_0}^{t_f} \frac{\partial H}{\partial u} \delta u \, dt + \lambda^T(t_0)\, dx\,(t_0) - H(t_0)\, dt_0. \tag{2.8.12}$$

Clearly, as before, $\lambda^T(t_0)$ is the *influence* vector on J of changes in initial conditions $\delta x(t_0)$, while $\partial H/\partial u$ is a set of *impulse-response functions* indicating how J would change as a result of unit impulses in the controls at any point in the interval $t_0 \le t \le t_f$.

For a stationary value of J, clearly, we have

$$\frac{\partial H}{\partial u} = \lambda^T \frac{\partial f}{\partial u} + \frac{\partial L}{\partial u} = 0, \qquad t_0 \le t \le t_f, \tag{2.8.13}†$$

and if a component $x_k(t_0)$ is not specified, we have $\lambda_k(t_0) = 0$.

For *minimum time*, $t_f - t_0$, we may let $\phi[x(t_f)t_f] = 0$ and $L = 1$, so that condition (2.8.11) becomes

$$\left(\frac{d\Phi}{dt} + 1\right)_{t=t_f} = 0. \tag{2.8.14}$$

As in Section 2.6, the q constants ν must be determined to satisfy the terminal constraints (2.8.2). The condition (2.8.14) is the extra condition needed to determine the final time t_f.

In summary, a set of necessary conditions for J to have a stationary value is

$$\dot{x} = f(x,u,t) \tag{2.8.15}$$

$$\dot{\lambda} = -\left(\frac{\partial H}{\partial x}\right)^T = -\left(\frac{\partial f}{\partial x}\right)^T \lambda - \left(\frac{\partial L}{\partial x}\right)^T \tag{2.8.16}$$

---

†An argument regarding admissibility, similar to the one made in Section 2.7, must be made to justify (2.8.13).

$$0 = \left(\frac{\partial H}{\partial u}\right)^T = \left(\frac{\partial f}{\partial u}\right)^T \lambda + \left(\frac{\partial L}{\partial u}\right)^T \tag{2.8.17}$$

$$x_k(t_o) \text{ given,} \qquad \text{or} \qquad \lambda_k(t_o) = 0 \tag{2.8.18}$$

$$\lambda(t_f) = \left(\frac{\partial \phi}{\partial x} + \nu^T \frac{\partial \psi}{\partial x}\right)^T_{t=t_f} \tag{2.8.19}$$

$$\Omega = \left[\frac{\partial \phi}{\partial t} + \nu^T \frac{\partial \psi}{\partial t} + \left(\frac{\partial \phi}{\partial x} + \nu^T \frac{\partial \psi}{\partial x}\right) f + L\right]_{t=t_f} = 0 \tag{2.8.20}$$

$$\psi[x(t_f)t_f] = 0 \tag{2.8.21}$$

The optimality condition (2.8.17) determines the m-vector $u(t)$. The solution to the $2n$ differential equations (2.8.15) and (2.8.16) and the choice of the $q + 1$ parameters ν and t_f are determined by the $2n + 1 + q$ boundary conditions (2.8.18)–(2.8.21). Needless to say, this boundary-value problem is, in general, not very easy to solve.

Notice, however, that if we were to specify ν instead of ψ, and t_f instead of Ω, (2.8.18) and (2.8.19) provide $2n$ boundary conditions for a fixed-terminal-time, two-point boundary-value problem of order $2n$. By changing values of ν and t_f, it *may* be possible to bring ψ and Ω to zero at $t = t_f$ (see Chapter 7, Section 3).

Optimization problems for dynamic systems with path constraints

3

Introduction

In Chapter 2 we discussed optimization problems for nonlinear dynamic systems with *end-point constraints*; i.e., functions of the state variables were prescribed at the terminal point and the state variables were prescribed at the initial point. In this chapter we are going to consider problems with *path constraints*, i.e., constraints that apply at intermediate points or over the whole path, $t_o \leq t \leq t_f$, rather than just at the end points. We shall first consider equality constraints and then inequality constraints.

3.1 Integral constraints†

Consider the optimal programming problems of Sections 2.5 and 2.8 with the additional constraint that a certain integral must have a specified value along the optimal path, such as

$$x_{n+1}(t_f) = \int_{t_o}^{t_f} N(x,u,t)\, dt\,, \tag{3.1.1}$$

where N is a given scalar function.

One straightforward way to handle this problem is to add another system equation, namely,

$$\dot{x}_{n+1} = N(x,u,t)\,, \tag{3.1.2}$$

with boundary conditions

$$x_{n+1}(t_o) = 0 \qquad \text{and} \qquad x_{n+1}(t_f) \qquad \text{prescribed.} \tag{3.1.3}$$

†These are also called "isoperimetric constraints" in the calculus of variations; the reason is clear from Example 1, one of the first solved problems of this type.

Let μ be the influence function associated with x_{n+1}. The variational Hamiltonian of the augmented system is then

$$H = L + \lambda^T f + \mu N. \tag{3.1.4}$$

The Euler-Lagrange equations then become

$$\dot{\lambda}^T = -\frac{\partial H}{\partial x} = -\frac{\partial L}{\partial x} - \lambda^T \frac{\partial f}{\partial x} - \mu \frac{\partial N}{\partial x}, \tag{3.1.5}$$

$$0 = \frac{\partial H}{\partial u} = \frac{\partial L}{\partial u} + \lambda^T \frac{\partial f}{\partial u} + \mu \frac{\partial N}{\partial u}, \tag{3.1.6}$$

$$\dot{\mu} = -\frac{\partial H}{\partial x_{n+1}} = 0 \Rightarrow \mu = \text{constant}. \tag{3.1.7}$$

Here μ is the influence number of x_{n+1} on J, that is,

$$\mu = -\frac{\partial J}{\partial x_{n+1}}.$$

Now Equations (3.1.6) and (3.1.1) may be regarded as $m + 1$ equations for determining the m-component control vector $u(t)$ and the single constant μ. In effect, we have adjoined $N(x,u,t)$ to the variational Hamiltonian with a constant Lagrange multiplier μ.

Example 1. *Maximum area with given perimeter.* Given a rope of length P connected to each end of a straight line of length $2a < P$, find the shape of the rope necessary to enclose the maximum area between the rope and the straight line. Using the coordinate system shown in Figure 3.1.1, the problem is to find $\theta(x)$ to maximize

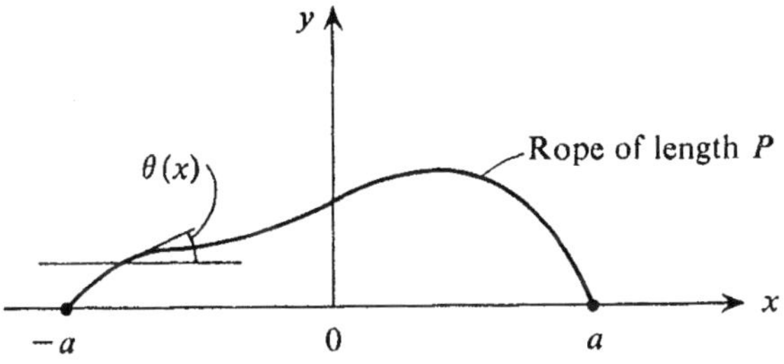

Figure 3.1.1. Maximum area with a given perimeter.

$$A = \int_{-a}^{a} y \, dx, \tag{3.1.8}$$

holding the perimeter (length of the rope) constant; that is, with

$$P = \int_{-a}^{a} \sec\theta \, dx \,, \tag{3.1.9}$$

where

$$\frac{dy}{dx} = \tan\theta \,.^{*} \tag{3.1.10}$$

The Hamiltonian of the system is, therefore,

$$H = y + \lambda \tan\theta + \mu \sec\theta \,, \tag{3.1.11}$$

and the Euler-Lagrange equations are

$$\frac{d\lambda}{dx} = -\frac{\partial H}{\partial y} = -1 \Rightarrow \lambda = -x + c \,, \qquad \text{where} \qquad c = \text{constant}, \tag{3.1.12}$$

$$0 = \frac{\partial H}{\partial \theta} = \mu \tan\theta \sec\theta + \lambda \sec^2\theta \Rightarrow \sin\theta = -\frac{\lambda}{\mu} . \tag{3.1.13}$$

Eliminating λ between (3.1.12) and (3.1.13) gives

$$x = \mu \sin\theta + c \,. \tag{3.1.14}$$

Since H is not an explicit function of x, H = constant is an integral of the optimal solution. Eliminating λ between (3.1.11) and (3.1.13) gives

$$y = -\mu \cos\theta + H \,, \qquad \text{where} \qquad H = \text{constant}. \tag{3.1.15}$$

The perimeter is given by substituting (3.1.14) into (3.1.9):

$$P = \int_{A}^{B} \sec\theta \frac{dx}{d\theta} \, d\theta = \mu \int_{A}^{B} d\theta = \mu(\theta_B - \theta_A) \,. \tag{3.1.16}$$

To evaluate the five unknowns c, H, μ, θ_A, and θ_B, we have the condition (3.1.16) and the four boundary conditions

$$x(\theta_A) = -a \,, \qquad x(\theta_B) = a \,, \qquad y(\theta_A) = 0 \,, \qquad y(\theta_B) = 0 \,. \tag{3.1.17}$$

The solution is easily obtained as

$$c = 0 \,, \qquad H = -\frac{P \cos\alpha}{2\alpha} \,, \qquad \mu = -\frac{P}{2\alpha} \,, \qquad \theta_A = \alpha \,, \qquad \theta_B = -\alpha \,, \tag{3.1.18}$$

where α is determined by the transcendental equation

$$\frac{\sin\alpha}{\alpha} = \frac{2a}{P} . \tag{3.1.19}$$

*This formulation assumes $-\pi/2 < \theta < \pi/2$, which is the case only if $P < \pi a$.

In summary, we have

$$x = -\frac{P}{2\alpha}\sin\theta\,, \qquad y = \frac{P}{2\alpha}(\cos\theta - \cos\alpha)\,;$$

$$x^2 + \left(y + \frac{P\cos\alpha}{2\alpha}\right)^2 = \frac{P^2}{4\alpha^2}\,, \tag{3.1.20}$$

which is the equation of a *circular arc* with center at $x = 0$ and $y = -(P\cos\alpha/2\alpha)$, and radius $P/2\alpha$.

Example 2. *Shape of a liquid drop on a horizontal surface.* A liquid drop on a horizontal surface assumes an axially symmetric shape that minimizes the sum of the liquid's potential energy in the earth's gravitational field plus the liquid's surface energy:

$$E = \int_{r=a}^{0} \gamma\pi r^2 z\frac{dz}{dr}\,dr + \int_{r=0}^{a} \sigma 2\pi r\frac{ds}{dr}\,dr\,, \tag{3.1.21}$$

where

γ = gravitational force per unit volume,

$z = z(r)$ is the equation of the surface (see Figure 3.1.2),

σ = surface energy per unit area (the surface tension),

$ds = \sqrt{(dz)^2 + (dr)^2}$ = differential element of arc length.

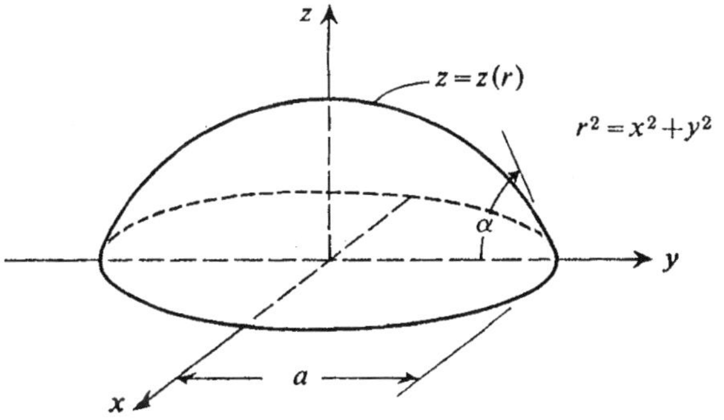

Figure 3.1.2. Shape of a liquid drop on a horizontal surface.

The volume of the liquid in the drop is fixed,

$$V = \int_{r=a}^{0} \pi r^2\frac{dz}{dr}\,dr\,, \tag{3.1.22}$$

and the contact angle α is given, where

$$\left(\frac{dz}{dr}\right)_{z=0} = -\tan\alpha . \tag{3.1.23}$$

The Hamiltonian for the problem is

$$H = 2\pi\sigma r\sqrt{1+u^2} - \pi\gamma r^2 z u + \lambda u + \mu\pi r^2 u , \tag{3.1.24}$$

where

$$\frac{dz}{dr} = u . \tag{3.1.25}$$

The Euler-Lagrange equations are, therefore,

$$\frac{d\lambda}{dr} = -\frac{\partial H}{\partial z} = \pi\gamma r^2 u , \tag{3.1.26}$$

$$0 = \frac{\partial H}{\partial u} = 2\pi\sigma r\frac{u}{\sqrt{1+u^2}} - \pi\gamma r^2 z + \lambda + \mu\pi r^2 . \tag{3.1.27}$$

Eliminating λ between these two equations and using (3.1.25), we obtain the nonlinear second-order differential equation for the droplet shape:

$$\frac{z''}{[1+(z')^2]^{3/2}} + \frac{z'}{r[1+(z')^2]^{1/2}} - \frac{\gamma}{\sigma}z = -\frac{\mu}{\sigma}, \tag{3.1.28}$$

which must be solved with the boundary conditions

$$z'(0) = 0 , \qquad z'(a) = -\tan\alpha , \qquad z(a) = 0 , \tag{3.1.29}$$

and μ is determined so that the volume equals the specified amount. Note that the extra boundary condition in (3.1.29) is used to determine the unknown radius a.

For $\alpha << 1$, (3.1.28) can be linearized to obtain

$$z'' + \frac{1}{r}z' - \frac{\gamma}{\sigma}z = -\frac{\mu}{\sigma}, \tag{3.1.30}$$

which has the solution

$$z = \frac{\mu}{\gamma} + AI_0\left(\sqrt{\frac{\gamma}{\sigma}}r\right) + BK_0\left(\sqrt{\frac{\gamma}{\sigma}}r\right), \tag{3.1.31}$$

where I_0 and K_0 are Bessel functions of order zero with imaginary argument, and μ, A, B are constants that must be chosen to satisfy (3.1.29) and the volume constraint (3.1.22).

Problem 1. *Maximum volume with given surface area.* Given an area of canvas, A, to build a tent, find the shape of the canvas necessary to cover a circular floor area of radius $a(\pi a^2 < A)$ with maximum volume inside the tent. Assume the shape is axially symmetric. (ANSWER. It is a spherical segment.)

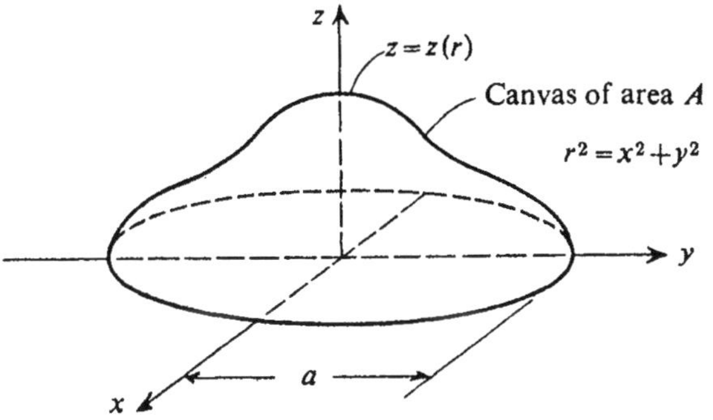

Figure 3.1.3. Maximum volume with a given surface area.

3.2 Control variable equality constraints

Again we consider the general optimal programming problems of Sections 2.5 and 2.8, this time with the control variable equality constraint

$$C(u,t) = 0\,, \tag{3.2.1}$$

where $u(t)$ is the m-component control vector, $m \geq 2$, and C is a scalar function. Clearly, we must have $m \geq 2$ for the problem to be interesting since for $m = 1$ the constraint (3.2.1) determines $u(t)$ completely and there is no optimization problem left. In the case in which we have $m \geq 2$, the effect of the constraint (3.2.1) is simply to reduce the freedom of choice of the control variables u. One possible method of handling this situation is to eliminate one of the control variables through (3.2.1) and require that the minimization condition be satisfied with respect to the rest of the control variables that are now free. All other conditions apply without change. Another possible method is to adjoin (3.2.1) to the variational Hamiltonian with a Lagrange multiplier $\mu(t)$, as follows:

$$H = L + \lambda^T f + \mu C\,. \tag{3.2.2}$$

The only change this brings about is in the optimality conditions,

$$0 = \frac{\partial H}{\partial u} = \frac{\partial L}{\partial u} + \lambda^T \frac{\partial f}{\partial u} + \mu \frac{\partial C}{\partial u}, \tag{3.2.3}$$

which, together with (3.2.1), represent $m + 1$ conditions for determining the m component vector $u(t)$ and the scalar function $\mu(t)$.

Example. *Minimum-time paths through a three-dimensional region of position-dependent vector velocity.*† An aircraft must travel through a region of strong winds. The magnitude and direction of the wind velocity are known as functions of position, $\mathbf{w} = \mathbf{w}(\mathbf{r})$, and the magnitude of the aircraft's velocity, relative to the air, is V, a constant. The problem is to program the aircraft's heading in such a way as to minimize the time to go from a point A to a point B. The total velocity relative to the ground is

$$\dot{\mathbf{r}} = V\hat{u} + \mathbf{w}, \tag{3.2.4}$$

where $\hat{u}$ is the direction of the aircraft's axis, a unit vector:

$$\hat{u} \cdot \hat{u} = 1. \tag{3.2.5}$$

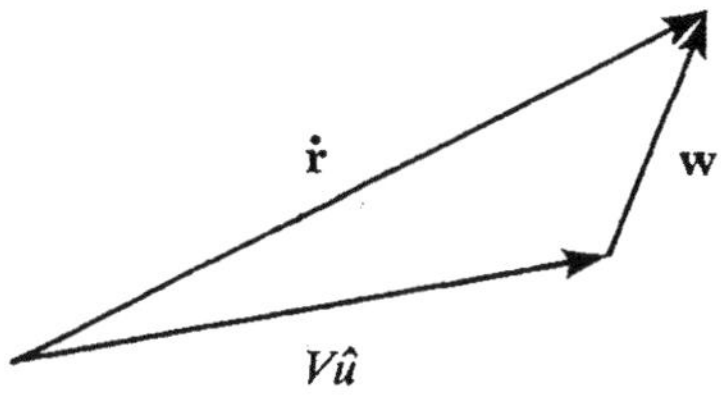

Figure 3.2.1. Vector addition of airplane velocity relative to air and wind velocity.

The Hamiltonian of the system is

$$H = \boldsymbol{\lambda} \cdot (V\hat{u} + \mathbf{w}) + \mu(1 - \hat{u} \cdot \hat{u}) + 1, \tag{3.2.6}$$

so the Euler-Lagrange equations are

$$\dot{\boldsymbol{\lambda}} = -\frac{\partial H}{\partial \mathbf{r}} = -\nabla(\boldsymbol{\lambda} \cdot \mathbf{w}), \tag{3.2.7}$$

$$0 = \frac{\partial H}{\partial \hat{u}} = V\boldsymbol{\lambda} - 2\mu\hat{u}. \tag{3.2.8}$$

To satisfy (3.2.5), the multiplier function $\mu(t)$ must be chosen as

†In two dimensions this is called the ship routing problem, or Zermelo's Problem; see Example in Section 2.7.

$$2\mu = \pm V|\boldsymbol{\lambda}| \Rightarrow \hat{u} = \pm \frac{\boldsymbol{\lambda}}{|\boldsymbol{\lambda}|} . \tag{3.2.9}$$

For the minimum-time problem, we must have $H = 0$; using (3.2.9) in (3.2.6), it is clear that the negative sign in (3.2.9) must be used; that is, we must have

$$\hat{u} = -\hat{\lambda} , \qquad \text{where} \qquad \hat{\lambda} = \frac{\boldsymbol{\lambda}}{|\boldsymbol{\lambda}|} . \tag{3.2.10}$$

In other words, the velocity vector should point in the opposite direction of the position influence vector, $\boldsymbol{\lambda}$.

If we use (3.2.10) in (3.2.4), the following differential equations define the desired minimum-time paths:

$$\dot{\mathbf{r}} = \mathbf{w} - V\hat{\lambda} , \tag{3.2.11}$$

$$\dot{\boldsymbol{\lambda}} = -\nabla(\boldsymbol{\lambda} \cdot \mathbf{w}) = -(\boldsymbol{\lambda} \cdot \nabla)\mathbf{w} - \boldsymbol{\lambda} \times (\nabla \times \mathbf{w}) . \tag{3.2.12}$$

The direction of $\boldsymbol{\lambda}$ at point A must be chosen so that the path goes through point B. The magnitude of $\boldsymbol{\lambda}$ must be chosen to make the Hamiltonian zero in order that $\boldsymbol{\lambda}$ be the influence vector on final time as indicated in (3.2.10). Since the Hamiltonian is independent of t, it is constant over the path; from (3.2.6) and (3.2.10), this condition yields

$$|\boldsymbol{\lambda}| = \frac{1}{V - \hat{\lambda} \cdot \mathbf{w}} , \tag{3.2.13}$$

which may be used as a check on the solution obtained from (3.2.11) and (3.2.12).

If the wind velocity field is irrotational; i.e., if $\nabla \times \mathbf{w} = 0$, then we obtain $\mathbf{w} = \nabla\phi$, where $\phi = \phi(\mathbf{r},t)$. In this case, the perturbation equations for $\mathbf{r}$ are identical, except for a minus sign, to the equations for $\boldsymbol{\lambda}$. The perturbation equations are said to be *self-adjoint* in this case.†

$$\frac{d}{dt}(\delta\mathbf{r}) = (\delta\mathbf{r} \cdot \nabla)\nabla\phi , \tag{3.2.14}$$

$$\frac{d}{dt}(\boldsymbol{\lambda}) = -(\boldsymbol{\lambda} \cdot \nabla)\nabla\phi . \tag{3.2.15}$$

Problem 1. *Minimum-time paths through a three-dimensional region of position-dependent velocity magnitude.*‡ A particle must travel through a region in which its instantaneous velocity magnitude is

†Note that the definition of "self-adjoint" for $\dot{x} = Fx$ is not equivalent to that for $(d^n x/dt^n) + a_1(d^{n-1}x/dt^{n-1}) + \cdots + a_n x = 0$, even though the definitions for "adjoint" are equivalent.

‡The corresponding problem in two dimensions is treated in Problem 2, Section 2.7.

given as a function of position, $V = V(\mathbf{r})$, where $\mathbf{r}$ is the position vector. We wish to find the minimum-time path from a point A to a point B in this three-dimensional space. Each individual path may be generated by

$$\dot{\mathbf{r}} = V(\mathbf{r})\hat{u}\,,$$

where $\hat{u}$ is a unit vector,

$$\hat{u} \cdot \hat{u} = 1\,.$$

Show that the minimum-time paths are generated by

$$\hat{u} = -\frac{\boldsymbol{\lambda}}{|\boldsymbol{\lambda}|}\,, \qquad |\boldsymbol{\lambda}(B)| = \frac{1}{V(B)}\,, \qquad \dot{\boldsymbol{\lambda}} = |\boldsymbol{\lambda}|\nabla V\,, \qquad \mathbf{r}(A),\mathbf{r}(B) \text{ given}$$

or

$$\frac{d}{dt}\left(\frac{1}{V^2}\frac{d\mathbf{r}}{dt}\right) = -\frac{1}{V}\nabla V\,.$$

Problem 2. *Minimum-time paths through a three-dimensional region of position-dependent forces.*† Consider the three-dimensional motion of a point mass acted upon by forces that depend on position $\mathbf{r}$ and not on velocity $\mathbf{v}$. (The inclusion of forces dependent on both $\mathbf{r}$ and $\mathbf{v}$ is a simple extension of this problem, see the reference.†) The equations of motion are

$$\dot{\mathbf{v}} = a(t)\hat{u} + \mathbf{F}(\mathbf{r},t)\,,$$

$$\dot{\mathbf{r}} = \mathbf{v}\,,$$

where $\hat{u}$, the control vector, is a unit vector:

$$\hat{u} \cdot \hat{u} = 1\,,$$

and the thrust acceleration $a(t)$ is presumed given. Here $\mathbf{F}(\mathbf{r},t)$ is the acceleration due to position-dependent forces; $\mathbf{F}$ may be an explicit function of the time due to motion of gravitating bodies.

Show that

$$\hat{u} = -\frac{\mathbf{p}}{|\mathbf{p}|}\,,$$

where

$$\dot{\boldsymbol{\lambda}} = -(\mathbf{p}\cdot\nabla)\mathbf{F} - \mathbf{p}\times(\nabla\times\mathbf{F})\,, \qquad \dot{\mathbf{p}} = -\boldsymbol{\lambda}\,;$$

$$[-a|\mathbf{p}| + \mathbf{p}\cdot\mathbf{F} + \boldsymbol{\lambda}\cdot\mathbf{v} + 1]_{t=t_f} = 0\,;$$

†This example is adapted from Burton D. Fried. *Space Technology*, H. S. Seifert (ed.), New York: Wiley, 1959, Chapter 4.

that is, $\boldsymbol{\lambda}$ = position influence vector, $\mathbf{p}$ = velocity influence vector.

Note that, for *conservative force fields*, we have $\nabla \times \mathbf{F} = 0 \Rightarrow \mathbf{F} = \nabla\phi$, where $\phi = \phi(\mathbf{r},t)$. In this case the perturbation equations (for $\delta\mathbf{v}$ and $\delta\mathbf{r}$) are identical, except for a minus sign, to the equations for $\boldsymbol{\lambda}$ and $\mathbf{p}$ respectively:

$$\frac{d}{dt}(\delta\mathbf{v}) = (\delta\mathbf{r}\cdot\nabla)\nabla\phi\,, \qquad \frac{d}{dt}(\delta\mathbf{r}) = \delta\mathbf{v}\,, \qquad \frac{d}{dt}(\boldsymbol{\lambda}) = -(\mathbf{p}\cdot\nabla)\nabla\phi\,,$$

$$\frac{d}{dt}(\mathbf{p}) = -\boldsymbol{\lambda}\,.$$

The perturbation equations are said to be self-adjoint in this case. The second-order equations for $\delta\mathbf{r}$ and $\mathbf{p}$, obtained by eliminating $\delta\mathbf{v}$ and $\boldsymbol{\lambda}$, are identical:

$$\frac{d^2}{dt^2}(\delta\mathbf{r}) = (\delta\mathbf{r}\cdot\nabla)\nabla\phi\,, \qquad \frac{d}{dt^2}(\mathbf{p}) = (\mathbf{p}\cdot\nabla)\nabla\phi\,.$$

This implies that, if a transition matrix is found for $\delta\mathbf{r},\delta\mathbf{v}$, it is then immediately known for $\mathbf{p}$ and $\boldsymbol{\lambda}$. This is useful on coasting arcs of space-vehicle trajectories.

Problem 3. Consider the linear dynamic system that is self-adjoint, that is,

$$\dot{x} = Fx + u \qquad \text{and} \qquad F = -F^T\,.$$

Determine $u(t)$ subject to the constraint $\|u\|^2 = 1$ such that $x(T) = 0$, where T is a minimum. Find the feedback solution; i.e., express $u(t)$ explicitly as a function of x and t. [HINT: See M. Athans, P. L. Falb, and R. T. Lacoss. "Time, Fuel, and Energy Optimal Control of Norm-Invariant Systems," *Trans. of IEEE on Automatic Control* AC-8, 3, July 1963.]

3.3 Equality constraints on functions of the control and state variables

Again we consider the general programming problem of Sections 2.5 and 2.7, this time with the equality constraints

$$C(x,u,t) = 0\,, \tag{3.3.1}$$

where $C_u \neq 0$ for any u. Following procedures identical to those in Section 3.2, we adjoin (3.3.1) to the variational Hamiltonian to form the new augmented Hamiltonian,

$$H = \lambda^T f + L + \mu C\,. \tag{3.3.2}$$

In addition to the change in the optimality condition,

$$0 = \frac{\partial H}{\partial u} = \lambda^T f_u + L_u + \mu C_u , \qquad (3.3.3)$$

we have the modified Euler equations,

$$\dot{\lambda}^T = -H_x = -\lambda^T f_x - L_x - \mu C_x . \qquad (3.3.4)$$

All other equations remain the same; Equations (3.3.1) and (3.3.3) are $1 + m$ equations for the $1 + m$ quantities μ and u. The new feature is the appearance of the term $\mu(\partial C/\partial x)$ in (3.3.4).

If $C(x,u,t) = 0$ had been a vector function with fewer components than u, Equations (3.3.2), (3.3.3), and (3.3.4) would remain applicable if we replace μC, μC_u, μC_x by $\mu^T C$, $\mu^T C_u$, and $\mu^T C_x$, respectively.

3.4 Equality constraints on functions of the state variables

If the constraint function has no explicit dependence on the control variables, an additional complexity occurs. Consider such a constraint:

$$S(x,t) = 0 . \qquad (3.4.1)$$

If this constraint is to apply for all $t_o \leq t \leq t_f$, its time derivative along the path must vanish; i.e., we must have

$$\frac{dS}{dt} = \frac{\partial S}{\partial t} + \frac{\partial S}{\partial x}\dot{x} = \frac{\partial S}{\partial t} + \frac{\partial S}{\partial x} f(x,u,t) = 0 . \qquad (3.4.2)$$

Now (3.4.2) may or may not have explicit dependence on u. If it does, (3.4.2) plays the role of a control-variable constraint of the type (3.3.1). However, we must either eliminate one component of x in terms of the remaining $n - 1$ components using (3.4.1) or add (3.4.1) as a boundary condition at $t = t_o$ or $t = t_f$.†

If (3.4.2) does not explicitly involve u, we may take another time derivative and substitute $\dot{x} = f(x,u,t)$; this procedure obviously can be repeated until, finally, some explicit dependence on u does occur. If this occurs on the qth time derivative, we shall call this a qth-*order state variable equality constraint.* In this case, the qth total time derivative of (3.4.1) plays the role of a control variable constraint of the type (3.3.1):

$$S^{(q)}(x,u,t) = 0 \qquad \text{where} \qquad S^{(q)} \triangleq \frac{d^q S}{dt^q} . \qquad (3.4.3)$$

In addition, we must either eliminate q components of x in terms of the remaining $n - q$ components, using the q relations

†The constraint (3.4.1) could be adjoined directly to the Hamiltonian function with the resulting necessary conditions differently stated but equivalent to those derived here. See Chang and Dreyfus.

$$\begin{bmatrix} S(x,t) \\ S^{(1)}(x,t) \\ \cdot \\ \cdot \\ \cdot \\ S^{(q-1)}(x,t) \end{bmatrix} = 0\,, \tag{3.4.4}$$

or add (3.4.4) as a set of boundary conditions at $t = t_o$ (or $t = t_f$).

As a computational alternative, it is possible to use an *integral penalty function* to satisfy the constraints approximately (see Section 1.9). In this case we may use the augmented performance index,

$$\bar{J} = J + K \int_0^T [S(x,t)]^2 \, dt\,,$$

where K is a large number. This procedure may produce numerical difficulties, as discussed in Section 1.9, and requires that $K \to \infty$ to satisfy (3.4.1).

3.5 Interior-point constraints

Suppose that, with the general optimal programming problem of Section 2.8, we have a set of interior boundary conditions,

$$N[x(t_1),t_1] = 0\,, \tag{3.5.1}$$

where t_1 is some intermediate time, $t_o < t_1 < t_f$, and N is a q-component vector function. We now have a three-point boundary-value problem instead of a two-point boundary-value problem.

Now (3.5.1) represents a set of terminal constraints for the part of the path from $t = t_o$ to $t = t_1$. If we let t_1- signify just before t_1 and t_1+ signify just after t_1, we may interpret the influence functions and the Hamiltonian at $t = t_1+$ as follows:

$$\lambda^T(t_1+) = \frac{\partial J}{\partial x(t_1)}\,, \tag{3.5.2}$$

$$H(t_1+) = -\frac{\partial J}{\partial t_1}\,. \tag{3.5.3}$$

It follows from (2.8.10) and (2.8.11) and Equations (3.5.2) and (3.5.3) that

$$\lambda^T(t_1-) = \lambda^T(t_1+) + \pi^T \frac{\partial N}{\partial x(t_1)}\,, \tag{3.5.4}$$

$$H(t_1-) = H(t_1+) - \pi^T \frac{\partial N}{\partial t_1}\,, \tag{3.5.5}$$

where π is a q-component vector of constant Lagrange multipliers, determined so that the q conditions (3.5.1) are satisfied. Equation (3.5.5) determines the time t_1. Note that (3.5.4) and (3.5.5) imply *discontinuities* in the influence functions λ and the Hamiltonian H at $t = t_1$. The state variables are continuous; i.e., we have $x(t_1-) = x(t_1+)$.

The extension to several interior points where boundary conditions must be met is straightforward. However, finding solutions to such problems is, in general, quite involved. The method of steepest descent may be used to solve such problems numerically.

Another approach to Equations (3.5.4) and (3.5.5), which is instructive, was given by Denham. The interior-point constraints, (3.5.1), are adjoined to the performance index by a set of multipliers π, in the same way that the terminal point constraints $\psi[x(t_f),t_f]$ were adjoined by multipliers ν in Section 2.5. The first variation of the augmented performance index is, then,

$$\delta\bar{J} = \delta(\Phi + \pi^T N) + \delta \int_{t_0}^{t_f} (H - \lambda^T \dot{x})\, dt\,. \tag{3.5.6}$$

Splitting the integral into $\int_{t_0}^{t_1-} + \int_{t_1+}^{t_f}$ and integrating by parts (allowing for possible discontinuities in λ at $t = t_1$), we have

$$\begin{aligned}\delta\bar{J} = {} & \frac{\partial \Phi}{\partial x}\delta x\Big|_{t=t_f} + \pi^T \frac{\partial N}{\partial t_1} dt_1 + \pi^T \frac{\partial N}{\partial x(t_1)}\, dx - \lambda^T \delta x\Big|_{t_1+}^{t_f} - \lambda^T \delta x\Big|_{t_0}^{t_1-} \\ & + (H - \lambda^T \dot{x})\Big|_{t=t_1-} dt_1 - (H - \lambda^T \dot{x})\Big|_{t=t_1+} dt_1 \\ & + \int_{t_0}^{t_f} \left[\left(\dot{\lambda}^T + \frac{\partial H}{\partial x}\right)\delta x + \frac{\partial H}{\partial u}\delta u\right] dt\,. \end{aligned} \tag{3.5.7}$$

We now make use of the relations

$$dx(t_1) = \begin{cases} \delta x(t_1-) + \dot{x}(t_1-)\, dt_1\,, \\ \delta x(t_1+) + \dot{x}(t_1+)\, dt_1\,, \end{cases} \tag{3.5.8}$$

which are illustrated (for x a scalar) in Figure 3.5.1.

Using (3.5.8) to eliminate $\delta x(t_1 -)$ and $\delta x(t_1 +)$ in (3.5.7), and regrouping terms, yields

$$\begin{aligned}\delta\bar{J} = {} & \left(\frac{\partial \Phi}{\partial x} - \lambda^T\right)\delta x\Big|_{t=t_f} + \lambda^T(t_1+) - \lambda^T(t_1-) + \pi^T \frac{\partial N}{\partial x(t_1)}\, dx(t_1) \\ & + H(t_1-) - H(t_1+) + \pi^T \frac{\partial N}{\partial t_1} dt_1 + \lambda^T \delta x\Big|_{t=t_0}\end{aligned}$$

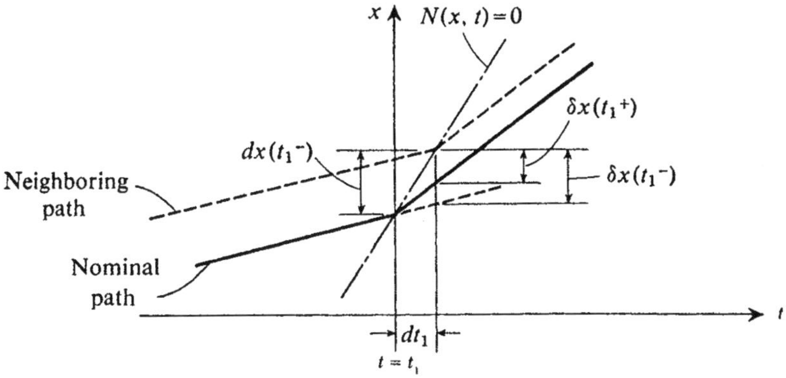

Figure 3.5.1. Relationships among the differential dx, the variations δx, and dt.

$$+\int_{t_0}^{t_f}\left[\left(\dot{\lambda}^T+\frac{\partial H}{\partial x}\right)\delta x+\frac{\partial H}{\partial u}\delta u\right]dt\,. \tag{3.5.9}$$

Now, let us choose $\lambda(t_1-)$ and $H(t_1-)$ to cause the coefficients of $dx(t_1)$ and dt_1 to vanish, yielding (3.5.4) and (3.5.5), and let us choose π to satisfy (3.5.1).

A controllability argument similar to the one in Section 2.7, regarding admissible variations, must be used here to justify placing the coefficient of $\delta u(t)$ equal to zero in (3.5.9) since $\delta u(t)$ is *not* arbitrary but must produce $dx(t_1)$ and dt_1 consistent with

$$dN=\frac{\partial N}{\partial t_1}dt_1+\frac{\partial N}{\partial x(t_1)}dx(t_1)=0\,. \tag{3.5.10}$$

Problem 1. *Interior quadratic penalty functions.* In guidance problems with noise, exact satisfaction of interior-point (or terminal-point) constraints is not possible. A useful alternative is to place a quadratic penalty on deviations from the desired interior-point constraints; i.e., place

$$J=\Phi[x(t_f),t_f]+\frac{1}{2}(N^TS_1N)|_{t=t_1}+\int_{t_0}^{t_f}(H-\lambda^T\dot{x})\,dt\,,$$

where S_1 is a positive definite $(q\times q)$-matrix, chosen by the optimizer.

Show that this technique requires the following jump in λ at $t=t_1$:

$$\lambda^T(t_1-)=\lambda^T(t_1+)+N^TS_1\frac{\partial N}{\partial x(t_1)}\,,\qquad H(t_1-)=H(t_1+)-N^TS_1\frac{\partial N}{\partial t_1}\,.$$

Problem 2. *Minimum effort interception of two targets* (Norbutas). Determine the acceleration program $a(t)$ that minimizes

$$J = \frac{1}{2} \int_0^{t_f} a^2 \, dt \,,$$

subject to the constraints

$$\dot{x} = v \,, \qquad \dot{v} = a$$

and

$$x(0) = 0 \,, v(0) = 0 \,, \qquad x(t_1) = x_1 \,, \qquad x(t_f) = 0 \,,$$

where x_1, t_1, and t_f are given with $0 < t_1 < t_f$.

Problem 3. *Minimum-time intercept passing through an intermediate point.* Determine the thrust direction program $\theta(t)$ that minimizes the time to go from $x = 0$, $y = 0$ to $x = x_f$, $y = 0$ subject to the constraints

$$\dot{u} = a \cos \theta \,, \qquad \dot{x} = u \,, \qquad \dot{v} = a \sin \theta \,, \qquad \dot{y} = v$$

and

$$x(0) = 0 \,, y(0) = 0 \,, \qquad u(0) = 0 \,, v(0) = 0 \,, \qquad x(t_1) = x_1 \,, \qquad y(t_1) = y_1 \,,$$
$$x(t_f) = x_f, \qquad y(t_f) = 0 \,,$$

where a, x_1, and x_f are given and $0 < t_1 < t_f$.

3.6 Discontinuities in the system equations at interior points

Suppose that one set of system equations,

$$\dot{x} = f^{(1)}(x,u,t) \,, \tag{3.6.1}$$

applies for $t < t_1$, where t_1 is specified by

$$\psi^{(1)}[x(t_1),t_1] = 0 \,, \tag{3.6.2}$$

and another set of system equations applies for $t > t_1$,† namely,

$$\dot{x} = f^{(2)}(x,u,t) \,. \tag{3.6.3}$$

Equation (3.6.2) is essentially an interior boundary condition as in Section 3.5. The necessary conditions (3.5.4)–(3.5.5) apply, with the obvious extension that we have

$$H^{(1)}(t_1-) = H^{(2)}(t_1+) - [\nu^{(1)}]^T \frac{\partial \psi^{(1)}}{\partial t_1} \,, \tag{3.6.4}$$

where

$$H^{(1)} = L + \lambda^T f^{(1)} \,, \qquad H^{(2)} = L + \lambda^T f^{(2)} \,.$$

†For example, an aircraft before and after takeoff or landing.

If we have $f^{(1)} = f^{(2)}$ and $\psi^{(1)}$ is scalar, $\nu^{(1)} = 0$ is the only solution, and $t = t_1$ is of no special significance.

Example. *Ship routing with discontinuous current.* Referring to Example 1, Section 2.7, suppose that the current is such that we have $v = 0$ and

$$u = \begin{cases} \epsilon V, & y > h, \\ 0, & y < h, \end{cases} \qquad \text{where} \qquad V = \text{speed of ship relative to water.}$$

We seek a minimum-time path from $x = y = 0$ to $x = ah$, $y = (1 + b)h$.

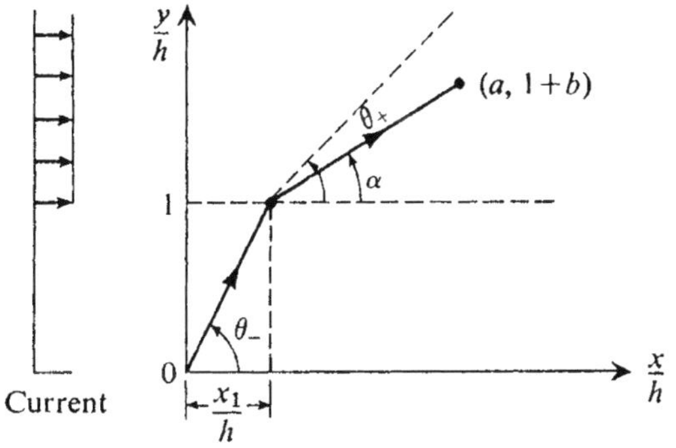

θ_+ is the control angle and α is the path angle.

Figure 3.6.1. Ship routing with discontinuous current.

Now λ_x and λ_y are constant in both regions ($y < h$, $y > h$), but a discontinuity occurs at $y - h = \psi^{(1)} = 0$. From (3.5.4), we obtain

$$\lambda_x(t_1-) = \lambda_x(t_1+) + 0, \qquad \lambda_y(t_1-) = \lambda_y(t_1+) + \nu^{(1)}.$$

Applying (3.6.4) above requires that we have

$$1 + \lambda_x(t_1-)V\cos\theta(t_1-) + \lambda_y(t_1-)V\sin\theta(t_1-) = 1$$
$$+ \lambda_x(t_1+)[V\cos\theta(t_1+) + \epsilon V] + \lambda_y(t_1+)V\sin\theta(t_1+) = 0.$$

From optimality we have

$$\tan\theta(t_1-) = \frac{\lambda_y(t_1-)}{\lambda_x(t_1-)}; \qquad \tan\theta(t_1+) = \frac{\lambda_y(t_1+)}{\lambda_x(t_1-)}.$$

From continuity of the path we have $x(t_1-) = x(t_1+) = x_1$ and $y(t_1-) = y(t_1+) = h$. The path, therefore, is made up of two connected straight lines, so we have

$$\tan\theta_- = \frac{h}{x_1}, \qquad \frac{\sin\theta_+}{\cos\theta_+ + \epsilon} = \frac{bh}{ah - x_1} = \tan\alpha$$

These equations in 8 unknowns are readily solved. Eliminating the λ's, $\nu^{(1)}$, and x_1 yields 2 equations in the 2 unknowns θ_- and θ_+ :

$$\sec\theta_- = \sec\theta_+ - \epsilon\,, \qquad \operatorname{ctn}\theta_- = a - b(\operatorname{ctn}\theta_+ + \epsilon\csc\theta_+)\,.$$

3.7 Discontinuities in the state variables at interior points

In some problems there will be discontinuities in the state variables as well as discontinuities of the system equations at interior points. Furthermore, the performance index and the constraints may be functions of the state and/or time at several discrete points rather than at only one point. A general class of such problems involves choosing $u(t)$ to minimize

$$J = \phi[x(t_o -), x(t_o +), \ldots, x(t_N -), x(t_N +); t_o, \ldots, t_N] + \sum_{i=1}^{N} \int_{t_{i-1}+}^{t_i -} L^{(i)}[x(t), u(t), t]\, dt \tag{3.7.1}$$

subject to constraints

$$\dot{x} = f^{(i)}(x,u,t)\,; \qquad t_{i-1} < t < t_i\,, \qquad i = 1, \ldots, N\,, \tag{3.7.2}$$

$$\psi^{(j)}[x(t_o -), x(t_o +), \ldots, x(t_N -), x(t_N +); t_o, \ldots, t_N] = 0\,, \quad j = 0, \ldots, N\,; \tag{3.7.3}$$

where $x(t_i -)$ signifies the state vector just before $t = t_i$, and $x(t_i +)$ signifies the state vector just after $t = t_i$.

Necessary conditions for a minimum are derived by adjoining (3.7.2) and (3.7.3) to (3.7.1) by Lagrange multiplier functions $\lambda(t)$ and constant Lagrange multipliers $\nu^{(i)}$, respectively:

$$\bar{J} = \phi + \sum_{j=0}^{N} [\nu^{(j)}]^T \psi^{(j)} + \sum_{i=1}^{N} \int_{t_{i-1}+}^{t_i -} [L^{(i)} + \lambda^T f^{(i)} - \lambda^T \dot{x}]\, dt\,. \tag{3.7.4}$$

As in previous sections, it is convenient to define

$$\Phi \triangleq \phi + \sum_{j=0}^{N} [\nu^{(j)}]^T \psi^{(j)}\,, \tag{3.7.5}$$

$$H^{(i)} \triangleq L^{(i)} + \lambda^T f^{(i)}\,; \qquad i = 1, \ldots, N\,. \tag{3.7.6}$$

The first variation of (3.7.4), using the usual integration by parts, is then

$$d\bar{J} = \sum_{i=0}^{N} \left[\frac{\partial \Phi}{\partial t_i} dt_i + \frac{\partial \Phi}{\partial x(t_i -)} dx(t_i -) + \frac{\partial \Phi}{\partial x(t_i +)} dx(t_i +) \right.$$

$$+ \sum_{i=1}^{N} \{[H^{(i)} - \lambda^T \dot{x}]_{t = t_i -} dt_i - [H^{(i)} - \lambda^T \dot{x}]_{t = t_{i-1}+} dt_{i-1}\}$$

$$+ \sum_{i=1}^{N} \left\{ \left[-\lambda^T \delta x \right]_{t_{i-1}+}^{t_i-} + \int_{t_{i-1}+}^{t_i-} \left[\left(\dot{\lambda}^T + \frac{\partial H^{(i)}}{\partial x} \right) \delta x + \frac{\partial H^{(i)}}{\partial u} \delta u \right] dt \right\}. \tag{3.7.7}$$

Eliminating $\delta x(t_i \pm)$ from (3.7.7) using the relations

$$dx(t_i \pm) = \delta x(t_i \pm) + \dot{x}(t_i \pm)\, dt_i \tag{3.7.8}$$

and regrouping terms, we have

$$\begin{aligned} d\bar{J} = {} & \sum_{i=0}^{N} \left[\frac{\partial \Phi}{\partial t_i} + H^{(i)}(t_i-) - H^{(i+1)}(t_i+) \right] dt_i \\ & + \sum_{i=1}^{N} \left[\frac{\partial \Phi}{\partial x(t_i-)} - \lambda^T(t_i-) \right] dx(t_i-) + \sum_{i=0}^{N-1} \left[\frac{\partial \Phi}{\partial x(t_i+)} + \lambda^T(t_i+) \right] dx(t_i+) \\ & + \sum_{i=1}^{N} \int_{t_{i-1}+}^{t_i-} \left[\left(\dot{\lambda}^T + \frac{\partial H^{(i)}}{\partial x} \right) \delta x + \frac{\partial H^{(i)}}{\partial u} \delta u \right] dt, \end{aligned} \tag{3.7.9}$$

where it is to be noted that $H^{(0)} = H^{(N+1)} \triangleq 0$.

Now, let us choose $\lambda(t)$ to satisfy

$$\dot{\lambda}^T = -\frac{\partial H^{(i)}}{\partial x}; \qquad t_{i-1}+ < t < t_i-, \qquad i = 1, \ldots, N, \tag{3.7.10}$$

$$\lambda^T(t_i-) = \frac{\partial \Phi}{\partial x(t_i-)}; \qquad i = 1, \ldots, N, \tag{3.7.11}$$

$$\lambda^T(t_i+) = -\frac{\partial \Phi}{\partial x(t_i+)}; \qquad i = 0, \ldots, N-1, \tag{3.7.12}$$

and let us choose t_i $(i = 0, \ldots, N)$ to satisfy

$$\frac{\partial \Phi}{\partial t_i} + H^{(i)}(t_i-) - H^{(i+1)}(t_i+) = 0, \qquad i = 0, \ldots, N, \tag{3.7.13}$$

where it is again noted that $H^{(0)} = H^{(N+1)} \triangleq 0$. If t_i is specified, the corresponding relation in (3.7.13) is, of course, *not* necessary since $dt_i = 0$ in (3.7.9). Similarly, if $x(t_0+)$ is specified, we have $dx(t_0+) = 0$ in (3.7.9), and Equation (3.7.12), with $i = 0$, is not necessary.

We must also choose $\nu^{(j)}$ to satisfy the constraints $\psi^{(j)} = 0$ in (3.7.13).

Again, a controllability argument must be made to justify setting

$$\frac{\partial H^{(i)}}{\partial u} = 0 \tag{3.7.14}$$

in (3.7.9) since $\delta u(t)$ is *not* arbitrary but must produce $dx(t_i-)$, $dx(t_i+)$, dt_i consistent with

$$d\psi^{(j)} = \sum_{i=0}^{N} \left[\frac{\partial \psi^{(j)}}{\partial t_i} dt_i + \frac{\partial \psi^{(j)}}{\partial x(t_i-)} dx(t_i-) + \frac{\partial \psi^{(j)}}{\partial x(t_i+)} dx(t_i+) \right] = 0, \qquad j = 0, \ldots, N. \tag{3.7.15}$$

Equations (3.7.10), (3.7.11), (3.7.12), and (3.7.14) are the Euler-Lagrange necessary conditions, and (3.7.13) is a set of necessary transversality conditions.

3.8 Inequality constraints on the control variables

Suppose that, instead of the equality constraint (3.2.1), we have the corresponding inequality constraint

$$C(u,t) \leq 0 . \tag{3.8.1}$$

If we define $H^{\circ} = \lambda^T f + L$, then, from Equation (2.3.9), we have

$$\delta J = \int_{t_0}^{t_f} H_u^{\circ}\, \delta u \, dt \triangleq \int_{t_0}^{t_f} \delta H^{\circ}(x,\lambda,u,t)\, dt , \tag{3.8.2}$$

where

$$\dot{\lambda}^T = -\lambda^T f_x - L_x , \qquad \lambda^T(t_f) = \phi_x \big|_{t=t_f} , \tag{3.8.3}$$

where we have assumed that the problem is fixed time and no terminal constraints. For $u(t)$ to be minimizing, we must have $\delta J \geq 0$ for all *admissible* $\delta u(t)$. This implies that we have $\delta H^{\circ} \geq 0$ for all t and all admissible $\delta u(t)$. Hence, at all points on $C = 0$ the optimal u has the property that

$$\delta H^{\circ} = H_u^{\circ}\, \delta u \geq 0 , \qquad \delta C = C_u\, \delta u \leq 0 . \tag{3.8.4}$$

Another way of stating (3.8.4) is to say that δH° must be nonimproving over the set of possible δu. Actually, a much stronger statement, "H° must be minimized over the set of all possible u," is true; this compact statement is due to McShane (1939) and Pontryagin (1962) and is known as the "Minimum Principle."† A rigorous justification of the statement, taking into account strong variations and terminal constraints is given in Pontryagin (1962). The argument above considers only the case without terminal constraints.

Another approach to Equations (3.8.4) has already been discussed in Section 1.7. If we define

$$H = \lambda^T f + L + \mu^T C , \tag{3.8.5}$$

the necessary condition on H is

$$H_u = L_u + \lambda^T f_u + \mu^T C_u = 0 , \tag{3.8.6}$$

which is the same as (3.2.3) with the additional requirement that

†In the Russian literature, it is called the "Maximum Principle" because of a different sign convention used in defining the variational Hamiltonian H°.

$$\mu\begin{cases} \geq 0\,, & C = 0\,, \\ = 0\,, & C < 0\,. \end{cases} \tag{3.8.7}$$

The positivity of the multiplier μ when we have $C = 0$ can be interpreted as the requirement that the gradient $H_u^* \triangleq \lambda^T f_u + L_u$ be such that improvement can only come by violating the constraints.

The decision to use H or H^* is a matter of taste. Translation from one to the other is easy and should not cause any confusion.

In solving a particular problem, constrained and unconstrained arcs must be pieced together to satisfy all the necessary conditions. At the junction points of constrained and unconstrained arcs, the control, u, may or may not be continuous; if it is discontinuous, the point is called a *corner*. (The nomenclature arises from the discontinuity in the time derivatives of some, or all, of the state variables.) Corners may occur at any point, but they are more likely to occur at junction points than in the middle of unconstrained arcs (see Section 3.12). There seem to be no a priori methods for determining the existence of corners. If $u(t)$ is continuous across a junction point, it follows from the continuity of λ, $\partial H/\partial u$, and H that $\mu(t)$ is continuous across the junction point.

Example. *Minimizing terminal norm subject to soft and hard constraints.*† We have

$$J = \frac{a^2}{2}\|x(T)\|^2 + \frac{1}{2}\int_0^T \|u\|^2\,dt\,; \tag{3.8.8}$$

$$\dot{x} = g(t)u\,; \qquad g(t) \text{ given function of time;} \tag{3.8.9}$$

$$|u(t)| \leq 1\,. \tag{3.8.10}$$

The Hamiltonian is

$$H = \tfrac{1}{2}\|u\|^2 + \lambda^T g u + \mu_1(u-1) + \mu_2(-u-1)\,, \tag{3.8.11}$$

$$\dot{\lambda}^T = -H_x = 0\,, \qquad \lambda(t) = \lambda(T) = a^2 x(T)\,, \tag{3.8.12}$$

$$H_u^* = u + a^2 g(t)x(T) \Rightarrow \begin{cases} > 0\,, & u = -1\,, \\ = 0\,, & -1 < u < 1\,, \\ < 0\,, & u = 1\,. \end{cases} \tag{3.8.13}$$

In view of the interpretation of H_u^*, we have

$$u_{\text{opt}} = -\text{Sat}[a^2 g(t)x(T)]\,, \tag{3.8.14}$$

where

†D. W. Tufts and D. A. Schnidman, "Optimum Waveform Subject to Both Energy and Scalar Value Constraints," *Proc. IEEE*, Vol. 52, Sept. 1964, pp. 1002–1007.

$$\text{Sat}(\alpha) = \begin{cases} \alpha\,, & |\alpha| < 1 \\ \text{sgn}\;\alpha, & |\alpha| \geq 1 \end{cases}$$

and $x(T)$ is computed from the implicit equations

$$x(T) = x_0 - \int_{t_0}^{T} g(t)\,\text{Sat}[a^2 g(t)x(T)]\,dt\,. \tag{3.8.15}$$

If solution of (3.8.15) leads to u_{opt} as shown in Figure 3.8.1,

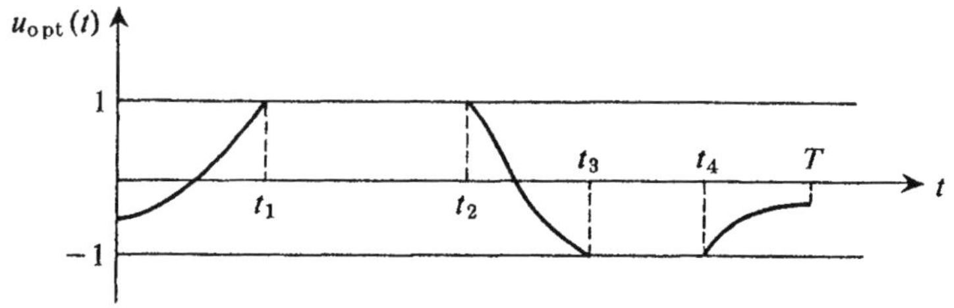

Figure 3.8.1. Typical optimal-control history subject to saturation and energy constraints.

then we have

$$\mu_1(t) = \begin{cases} -(1 + a^2 g(t)x(T))\,, & t_1 \leq t \leq t_2\,, \\ 0\,, & \text{other times}\,, \end{cases}$$

$$\mu_2(t) = \begin{cases} (-1 + a^2 g(t)x(T))\,, & t_3 \leq t \leq t_4\,, \\ 0\,, & \text{other times}\,, \end{cases}$$

and

$$\begin{aligned} 1 + a^2 g(t)x(T) &= 0 \qquad \text{at } t = t_1 \text{ and } t_2\,, \\ -1 + a^2 g(t)x(T) &= 0 \qquad \text{at } t = t_3 \text{ and } t_4\,. \end{aligned}$$

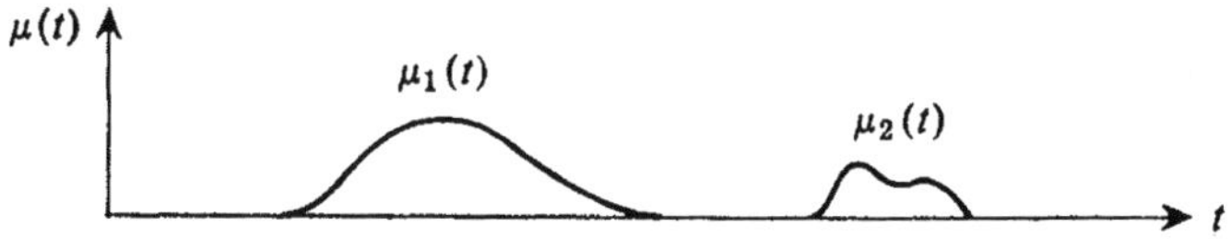

Figure 3.8.2. Typical multiplier histories for optimization problem with saturation and energy constraints.

3.9 Linear optimization problems; "bang-bang" control

An interesting special case of Section 3.8 is the case in which the performance index and the constraints are *linear* in the state and control variables. In general, no minimum will exist for such problems unless inequality constraints on the state and/or control variables are specified. If the inequality constraints are ***linear*** and are placed *only on the control variables*, it is reasonable to expect that the minimal solution, if it exists, will always require the control variables to be at one point or another of the ***boundary*** of the feasible control region. (See Section 1.8 on "linear programming" for the comparable situation in parameter optimization problems.) In general, one or more changes in control will occur during the time of operation of the system; thus the name "bang-bang" control, since the controls move suddenly from one point on the boundary of the feasible control region to another point on the boundary.

We shall consider only the ***minimum-time problem*** and limit our discussion to the case in which the control vector has only one component; that is, u is a scalar. The system equations are

$$\dot{x} = F(t)x + g(t)u \; ; \qquad x(0) = x^{o}(x \text{ an } n\text{-vector}) \,, \tag{3.9.1}$$

and the (scalar) control variable is bounded:

$$-1 \leq u(t) \leq 1 \,. \tag{3.9.2}$$

The problem is to find $u(t)$ to bring the system to

$$x(t_f) = 0 \tag{3.9.3}$$

in minimum time; that is, t_f is to be minimized. Since the problem is linear, Equation (3.9.3) is not as special as it might appear; transferring the system from any initial state to *any* final state may be put in this form by placing the origin of the state vector at the desired final state.†

Using the H° formulation of Section 3.8, we have, for this problem,

$$H^{\circ} = \lambda^T(Fx + gu) + 1 \,. \tag{3.9.4}$$

Clearly, to minimize H° with respect to u, taking into account (3.9.2), we have

$$u(t) = \begin{cases} 1 \,, & \lambda^T g < 0 \,, \\ -1 \,, & \lambda^T g > 0 \,. \end{cases} \tag{3.9.5}$$

The quantity $\lambda^T g$ is, for obvious reasons, called the "*switching function.*" The transversality condition (2.7.33) is, simply,

$$[\lambda^T(Fx + gu) + 1]_{t=t_f} = 0 \,. \tag{3.9.6}$$

The equations for the influence functions are

$$\dot{\lambda}^T = -\lambda^T F. \tag{3.9.7}$$

†If the final state lies on a *manifold* rather than being a point, the methods of Section 2.8, in addition to those of Section 3.8, may be used.

Equations (3.9.1) and (3.9.7) are to be integrated together with $u(t)$ determined by (3.9.5), boundary conditions $x(0) = x^0$, $x(t_f) = 0$; t_f is determined by (3.9.6).

Example. *Double integrator system.* We have

$$\dot{x}_1 = x_2\,, \qquad \dot{x}_2 = u\,, \qquad -1 \leq u(t) \leq 1\,.$$

For this very simple system, the influence equations are readily solved:

$$\begin{matrix} \dot{\lambda}_1 = 0\,, \\ \dot{\lambda}_2 = -\lambda_1 \end{matrix} \Rightarrow \begin{cases} \lambda_1 = \text{const}\,, \\ \lambda_2 = \lambda_2^f + \lambda_1(t_f - t)\,, \end{cases} \qquad \lambda_2^f = \text{const.}$$

Hence, the "switching function" is

$$\lambda^T g = \lambda_2(t)\,, \qquad \text{where} \qquad u = \begin{cases} +1\,, & \lambda_2 < 0\,, \\ -1\,, & \lambda_2 > 0\,. \end{cases}$$

The transversality condition (3.9.6), with $x_1(t_f) = x_2(t_f) = 0$, yields

$$\lambda_2(t_f)u(t_f) = -1\,,$$

which, together with the bang-bang control law, implies either

$$\lambda_2(t_f) = 1\,, \qquad u(t_f) = -1$$

or

$$\lambda_2(t_f) = -1\,, \qquad u(t_f) = +1\,.$$

Since the switching function is a linear function of time to go, it can change sign, at most, *only once.* Therefore, going *backward* in time from t_f, with $u = +1$ or $u = -1$, locates the switching curve in the state space:

$u = -1$	$u = +1$
$x_2 = t_f - t$	$x_2 = t - t_f$
$x_1 = -\dfrac{(t_f - t)^2}{2}$	$x_1 = \dfrac{(t - t_f)^2}{2}$
$\Rightarrow\ x_1 = -\dfrac{x_2^2}{2}$	$x_1 = +\dfrac{x_2^2}{2}$

This switching curve is made up of two parabolas in the x_1, x_2 space (see Figure 3.9.1).

If, by chance, the initial state lies on the switching curve, then we have $u = \pm 1$ accordingly as we have $x_1 \gtrless 0$. If, as in most cases, the

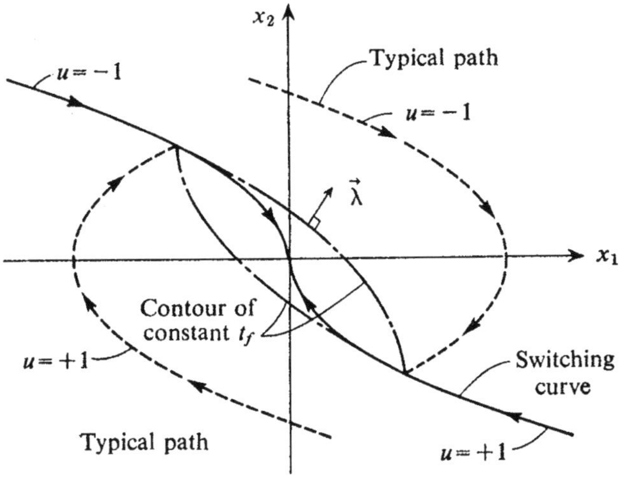

Figure 3.9.1. State space trajectories and switching curves for second-order "bang-bang" example.

initial state is *not* on the switching curve, then $u = \pm 1$ must be chosen to move the system *toward* the switching curve. By inspection, it is apparent that, *above the switching curve, we have* $u = -1$ *and below, we have* $u = +1$; typical paths (also parabolas) are shown in Figure 3.9.1. The control law may be written as

$$u = \begin{cases} +1, & x_2^2 \operatorname{sgn} x_2 < -2x_1 \quad \text{or} \quad x_2^2 \operatorname{sgn} x_2 = -2x_1, \quad x_1 > 0; \\ -1, & x_2^2 \operatorname{sgn} x_2 > -2x_1 \quad \text{or} \quad x_2^2 \operatorname{sgn} x_2 = -2x_1, \quad x_1 < 0. \end{cases}$$

The contours of constant time-to-go, t_f, are given by

$$(x_2 - t_f)^2 = 4[x_1 + (t_f^2/2)] \qquad \text{if} \qquad x_1 + \frac{1}{2} x_2|x_2| > 0;$$

$$(x_2 + t_f)^2 = 4[-x_1 + (t_f^2/2)] \qquad \text{if} \qquad x_1 + \frac{1}{2} x_2|x_2| < 0.$$

One such contour is shown in Figure 3.9.1. Note the *discontinuity in slope* of this contour at the switching curve. Since

$$\lambda_1 = \frac{\partial t_f}{\partial x_1}, \qquad \lambda_2 = \frac{\partial t_f}{\partial x_2},$$

it follows that $\lambda = \begin{bmatrix} \lambda_1 \\ \lambda_2 \end{bmatrix}$ is a vector normal to contours of constant t_f, except on the switching curve where the normal is undefined.

In a more general development, we let

$$\lambda(t_0) = \eta$$

and write

$$\lambda(t) = \Phi^T(t_0, t)\eta\,, \tag{3.9.8}$$

where $\Phi(t,\tau)$ is the fundamental matrix of (3.9.1). Substituting (3.9.8) into (3.9.5) and (3.9.1), we have

$$x(t_f) = 0 = \Phi(t_f, t_0)x_0 - \int_{t_0}^{t_f} \Phi(t_f, \tau)g(\tau) \operatorname{sgn}[g^T(\tau)\Phi^T(t_0, \tau)\eta]\, d\tau$$

or

$$x_0 = \int_{t_0}^{t_f} \Phi(t_0, \tau)g(\tau) \operatorname{sgn}[g(\tau)^T\Phi^T(t_0, \tau)\eta]\, d\tau\,. \tag{3.9.9}$$

The two-point boundary-value problem is essentially that of finding an "η" to satisfy Equation (3.9.9) for a given x_0. If we fix t_f and vary "η," the integral in (3.9.9) takes on those values of x_0 which can be brought to 0 in the time $t_f - t_0$ by a control which satisfies all the necessary conditions; i.e., we generate contours of constant $t_f - t_0$ as in Figure 3.9.1.

Numerically, Equation (3.9.9) suggests the following computational procedure:

(a) Guess an "η" and compute $\eta^T\omega \triangleq \eta^T \int_{t_0}^{t_f} \Phi(t_0, \tau)g \operatorname{sgn}[g^T\Phi^T(t_0, \tau)\eta]\, d\tau$ by integration until some t_f such that it is equal to $\eta^T x_0$. This t_f is, in general, less than the optimal t_f^o. Otherwise this implies that there exists a $u(t)^o$ and a t_f^o and η^o such that, for $t_f^o < t_f$,

$$\eta^T \int_{t_0}^{t_f^o} \Phi(t_0, \tau)g(\tau) \operatorname{sgn}[g^T(\tau)\Phi^T(t_0, \tau)\eta^o]\, d\tau$$
$$= \eta^T \int_{t_0}^{t_f^o} \Phi(t_0, \tau)g(\tau) \operatorname{sgn}[g^T(\tau)\Phi^T(t_0, \tau)\eta]\, d\tau\,, \tag{3.9.10}$$

which is a contradiction, since the right-hand side of (3.9.10) is increasing at the maximal rate.

(b) Change η so that ω is brought closer to x_0; for example, let

$$\delta\eta = \epsilon(x_0 - \omega)\,, \qquad \epsilon > 0\,.$$

(c) Repeat (a) until $t_f \to t_f^o$ from below.

The above scheme is essentially a successive approximation method in η (see Neustadt). The success of such a method is dependent on continuity of η with respect to x_0. Thus, we might expect computational difficulties for x_0 near or on a switching curve (see Figure 3.9.1).

If the system (a) with $u = 0$ is *unstable*, there will certainly be regions of the $x(0)$ space where the bounded control, $g(t)u$, is simply not large enough to overcome the terms, $F(t)x$; that is, it is impossible to bring the system to $x = 0$.

Similarly, if the system (a) is not *completely controllable*, it will, in general, be impossible to bring the system to $x = 0$.

Problem 1. Formulate the discrete version of the linear minimum-time problem (use scalar control for simplicity) and show that the assumptions on stability and controllability are directly concerned with the solubility of the linear equation $Au = b$ for some specified A and b (see Equation (3.9.9)).

Problem 2. For the discrete system

$$\begin{bmatrix} x_1(t+1) \\ x_2(t+1) \end{bmatrix} = \begin{bmatrix} 1 & 1 \\ 0 & 1 \end{bmatrix} \begin{bmatrix} x_1(t) \\ x_2(t) \end{bmatrix} + \begin{bmatrix} 1/2 \\ 1 \end{bmatrix} u(t),$$

$$|u(t)| \leq 1; \qquad t = 0,1,2,\ldots,$$

illustrate in two dimensions the region of $(x_1(0), x_2(0))$-space which can be brought to the origin in one, two, three, and four steps (see Problem 3).

Problem 3. The analogous continuous system is

$$\begin{bmatrix} \dot{x}_1 \\ \dot{x}_2 \end{bmatrix} = \begin{bmatrix} 0 & 1 \\ 0 & 0 \end{bmatrix} \begin{bmatrix} x_1 \\ x_2 \end{bmatrix} + \begin{bmatrix} 0 \\ 1 \end{bmatrix} u, \qquad |u(t)| \leq 1, \qquad t_0 = 0.$$

Calculate the regions in x_0-space for $t = 1,2,3$, and 4. Compare this with Problem 2.

Problem 4. Consider the problem of minimizing

$$J = \|x(t_f)\|^2$$

for the system

$$\dot{x} = Fx + Gu, \qquad x(0) = x_0, \qquad t_f \text{ given},$$

with constraints

$$|u(t)| \leq 1.$$

Show that the optimal control for $J_{\min} > 0$ is bang-bang. What is the analog of Figure 3.9.1 in this case?

Problem 5. *Undamped oscillator with bounded control*†

$$\dot{x}_1 = x_2, \qquad \dot{x}_2 = -x_1 + u; \qquad -1 \leq u(t) \leq 1.$$

†See Bushaw.

For this system, the general solution for the influence equation is easily found, as follows:

$$\dot{\lambda}_1 = \lambda_2 , \qquad \dot{\lambda}_2 = -\lambda_1 \Rightarrow \lambda_1 = a_1 \cos (T - t) + b_1 \sin (T - t) ,$$

$$\lambda_2 = a_1 \sin (T - t) - b_1 \cos (T - t) .$$

Hence, the "switching function" is

$$\lambda^T G = \lambda_2(t) ,$$

and the transversality condition (3.9.6) is

$$\lambda_2 u(T) + 1 = 0 ,$$

which, together with the constraint on $u(t)$, indicates that

$$\lambda_2(T) = \pm 1 , \qquad \text{which implies } b_1 = \pm 1 .$$

Sketch the switching curves and the contours of constant time-to-go in the state space ($x_1 x_2$-space).

Problem 6. A second-order system is governed by

$$\dot{x} = v , \qquad \dot{v} = u ,$$

where x,v are scalar state variables and u is a bounded scalar control variable:

$$-1 \leq u \leq 1 .$$

Given $x(0) = x_0$, $v(0) = v_0$, and t_f, the problem is to find $u(t)$ to minimize

$$J = \int_0^{t_f} |u(t)| \, dt$$

with specified terminal conditions

$$x(t_f) = 0 , \qquad v(t_f) = 0 .$$

Consider only the case $v_0 \geq 0$, $x_0 \geq -\frac{1}{2}v_0^2$, and assume $t_f > (t_f)_{\min}$, where $(t_f)_{\min}$ is the minimum time to bring the state from (x_0, v_0) to $(0,0)$ with $-1 \leq u \leq 1$.

Show that the control for this case is "bang-zero-bang"; that is

$$u = \begin{cases} -1 ; & 0 \leq t < t_1 , \\ 0 ; & t_1 \leq t < t_2 , \\ +1 ; & t_2 \leq t \leq t_f , \end{cases}$$

where the switching times are given by

$$\left.\begin{matrix} t_1 \\ \\ t_2 \end{matrix}\right\} = \frac{1}{2}\left[t_f + v_0 \mp \sqrt{(t_f - v_0)^2 - (4x_0 + 2v_0^2)}\right]$$

and

$$t_f \geq v_0 + \sqrt{4x_0 + 2v_0^2} = (t_f)_{\min} .$$

3.10 Inequality constraints on functions of the control and state variables

Instead of the equality constraint (3.3.1), suppose that we have the corresponding inequality constraint,

$$C(x,u,t) \leq 0 . \tag{3.10.1}$$

This problem is handled in the same way as the problem in Section 3.8. We define

$$H = L + \lambda^T f + \mu C , \tag{3.10.2}$$

where

$$\mu \begin{cases} > 0 , & C = 0 , \\ - & \\ = 0 , & C < 0 , \end{cases} \tag{3.10.3}$$

and the Euler-Lagrange equations become

$$\dot{\lambda}^T = -H_x = \begin{cases} -L_x - \lambda^T f_x - \mu C_x , & C = 0 , \\ \\ -L_x - \lambda^T f_x , & C < 0 . \end{cases} \tag{3.10.4}$$

Note the term μC_x in (3.10.4), which does not occur in the problems treated in Section 3.8. The condition determining $u(t)$ is

$$H_u \equiv L_u + \lambda^T f_u + \mu C_u = 0 . \tag{3.10.5}$$

For $C < 0$, we have $\mu = 0$, and (3.10.5) determines $u(t)$. For $C = 0$, Equations (3.10.1) and (3.10.5) together determine $u(t)$ and $\mu(t)$; $\mu(t)$ is needed for (3.10.4).

In solving a particular problem, constrained and unconstrained arcs must be pieced together to satisfy the necessary conditions. As in Section 3.8, the junction points may or may not be "corners," i.e., places where the control vector is discontinuous.

3.11 Inequality constraints on functions of the state variables

Instead of the state variable equality constraint (3.4.1), suppose that we have the corresponding inequality constraint,

$$S(x,t) \leq 0 . \tag{3.11.1}$$

For simplicity, we shall take S and u to be scalars.

As in Section 3.4, we take successive total time derivatives of (3.11.1) and substitute $f(x,u,t)$ for $\dot{x}$, until we obtain an expression that is explicitly dependent on u. If q time derivatives are required, we shall call (3.11.1) a *qth-order state variable inequality constraint.* Now $S^{(q)}(x,u,t)$, the qth total time derivative of S, plays the same role as $C(x,u,t)$ in Section 3.10. The Hamiltonian is

$$H = L + \lambda^T f + \mu S^{(q)} , \tag{3.11.2}$$

where

$$S^{(q)} = 0 \text{ on the constraint boundary,} \qquad S = 0 , \tag{3.11.3}$$

$$\mu = 0 \text{ off the constraint boundary,} \qquad S < 0 . \tag{3.11.4}$$

The Euler-Lagrange equations are identical to (3.10.4) and (3.10.5), with $S^{(q)}$ substituted for C.

A necessary condition for the influence function $\mu(t)$, as in Section 3.10, is

$$\mu(t) \geq 0 \quad \text{on} \quad S = 0 , \quad \text{if minimizing } J . \tag{3.11.5}$$

Since control of $S(x,t)$ is obtained only by changing its qth time derivative, no finite control will keep the system on the constraint boundary if the path entering onto the constraint boundary does not meet the following "tangency" constraints:

$$N(x,t) \triangleq \begin{bmatrix} S(x,t) \\ S^{(1)}(x,t) \\ \cdot \\ \cdot \\ \cdot \\ S^{(q-1)}(x,t) \end{bmatrix} = 0 . \tag{3.11.6}$$

Obviously, these same tangency constraints apply to the path leaving the constraint boundary.

Equations (3.11.6) form a set of interior boundary conditions such as we considered in Section 3.5. Consequently, the influence functions $\lambda(t)$ are, in general, discontinuous at junction points between constrained and unconstrained arcs.

For convenience, we may arbitrarily pick the entry point instead of the exit point as the place to satisfy these interior boundary conditions.† Thus the λ's and H are discontinuous at the entry point, $t = t_1$, and continuous at the exit point.

†A discussion of the lack of uniqueness of $\lambda(t)$ on $S = 0$ is given in Bryson, Denham, and Dreyfus (1963), Appendix B.

If we call the q quantities in (3.11.6) a vector $N(x,t)$, the "jump conditions" at the entry point are given directly by (3.5.4) and (3.5.5).

As in Section 3.8, the entry and exit points may or may not be "corners," i.e., places where the control vector is discontinuous.

Note that the control variable inequality constraint is a special case of the state variable inequality constraint in which $q = 0$. Since no N-vector exists in this case, there are no discontinuities in the λ's or H at $t = t_1$.

Other approaches to problems with state variable constraints involve (a) reducing the number of state variables from n to $n - q$ on constrained arcs or (b) adjoining $S(x,t)$ directly to the Hamiltonian instead of $S^{(q)}(x,u,t)$.†

Example 1. *A brachistochrone problem with a first-order state variable inequality constraint.* Given

$$\dot{x} = (2gy)^{1/2} \cos\gamma\,, \qquad \dot{y} = (2gy)^{1/2} \sin\gamma\,, \qquad x(0) = y(0) = 0\,,$$

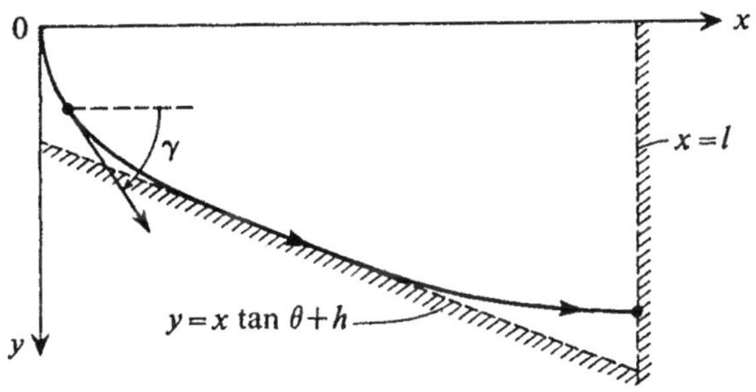

Figure 3.11.1. A brachistochrone problem with a state variable inequality constraint.

where x is horizontal distance, y is vertical distance (positive downward), g is the acceleration due to gravity, and γ is path angle to the horizontal (see Figure 3.11.1), find $\gamma(t)$ to minimize the time to reach $x = \ell$ with the constraint that $y < x \tan\theta + h$, with θ and h constant.

This is a problem with a first-order state variable inequality constraint, since $S = y - x\tan\theta - h \le 0$ does not contain the control variable γ, and $\dot{S} = (2gy)^{1/2} \sec\theta \sin(\gamma - \theta)$ does contain the control variable. On $S = 0$, $\dot{S} = 0$ implies that $\gamma = \theta$.

The solution to the *unconstrained problem,* $h/\ell > (2/\pi)\{1 - [(\pi/2) - \theta]\tan\theta\}$, is as follows:

†See J. L. Speyer and A. E. Bryson, Jr., "Optimal Programming Problems with a Bounded State Space," *AIAA Journal,* Vol. 6, 1968.

$$\gamma(t) = \frac{\pi}{2} - \omega t\,, \qquad \text{where} \qquad \omega = \left(\frac{\pi}{4}\frac{g}{\ell}\right)^{1/2};$$

$$\frac{x}{\ell} = \frac{2}{\pi}\left(\omega t - \frac{\sin 2\,\omega t}{2}\right), \quad \frac{y}{\ell} = \frac{2}{\pi}\sin^2 \omega t\,;$$

$$t_f = \left(\frac{\pi \ell}{g}\right)^{1/2} = \text{minimum final time}\,;$$

$$\lambda_x = -\omega/g\,, \qquad \lambda_y = -\frac{\omega}{g}\,\text{ctn}\,\omega t\,, \qquad \text{where} \qquad dt_f = (\lambda_x\,\delta x + \lambda_y\,\delta y)\,;$$

$$H = \lambda_x \dot{x} + \lambda_y \dot{y} + 1 = \text{variational Hamiltonian} = 0\,.$$

The solution to the *constrained problem,* $h/\ell < (2/\pi)\{1 - [(\pi/2) - \theta]\tan\theta\}$, is

$$\gamma(t) = \begin{cases} \dfrac{\pi}{2} - \omega_1 t\,, & 0 \le t \le t_1\,, \\ \theta\,, & t_1 \le t \le t_2\,, \\ \omega_2\,(t_f - t)\,, & t_2 \le t \le t_f\,, \end{cases}$$

where

$$\omega_1 = \left(\frac{g}{2}\,\frac{\theta - (\pi/2) + \text{ctn}\,\theta}{h\,\text{ctn}\,\theta}\right)^{1/2}, \qquad \omega_2 = \left(\frac{g}{2}\,\frac{\theta + \text{ctn}\,\theta}{\ell + h\,\text{ctn}\,\theta}\right)^{1/2},$$

$$t_1 = \frac{(\pi/2) - \theta}{\omega_1}\,, \qquad t_f - t_2 = \theta/\omega_2\,,$$

$$t_f = \left[\frac{2}{g}(\ell + h\,\text{ctn}\,\theta)\,(\theta + \text{ctn}\,\theta)\right]^{1/2} - \left[\frac{2h}{g}\,\text{ctn}\,\theta\left(\theta - \frac{\pi}{2} + \text{ctn}\,\theta\right)\right]^{1/2}$$

$$= \text{minimum final time},$$

$$\lambda_x(t_1-) - \lambda_x(t_1+) = -\mu_0 \tan\theta\,, \qquad \lambda_y(t_1-) - \lambda_y(t_1+) = \mu_0\,,$$

where $\mu_0 = (\text{ctn}\,\theta/g)\,(\omega_2 - \omega_1)$. Note that $\mu_0 \to 0$ and $t_1 \to t_2$ as $h/\ell \to (2/\pi)\,\{1 - [(\pi/2) - \theta]\,\tan\theta\}$:

$$H = \lambda_x \dot{x} + \lambda_y \dot{y} + 1 = 0\,; \qquad 0 \le t \le t_f\,.$$

Figure 3.11.2 shows the solutions for $\tan\theta = \frac{1}{2}$ for several values of h/ℓ .

Example 2. *A minimum-energy problem with a second-order state variable inequality constraint.*†

†This example was suggested by John V. Breakwell.

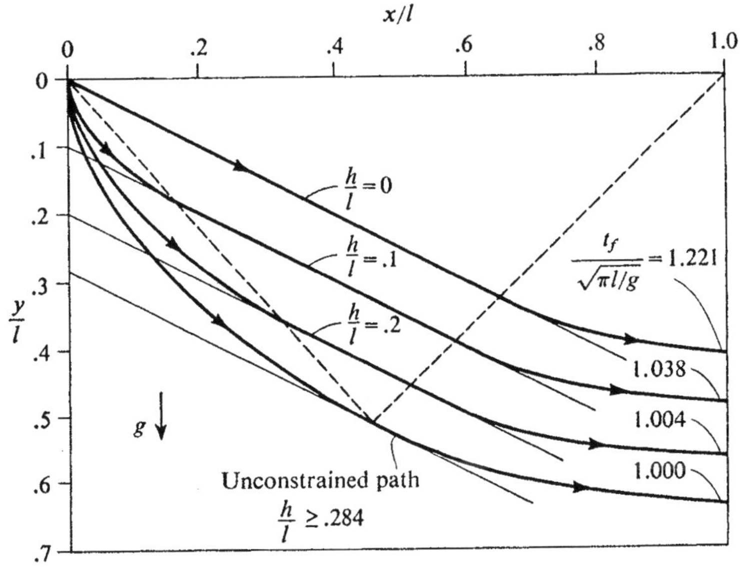

Figure 3.11.2. Constrained brachistochrone problem with $\tan\theta = \frac{1}{2}$ for several values h/ℓ.

Given

$$\dot{v} = a\,,$$

$$\dot{x} = v\,,$$

$$v(0) = -v(1) = 1\,,$$

$$x(0) = x(1) = 0\,,$$

find $a(t)$ in $0 \le t \le 1$ to minimize

$$J = \frac{1}{2}\int_0^1 a^2\,dt\,,$$

with the constraint that we have $x(t) \le \ell$. This is a problem with a second-order state variable inequality constraint, since $S = x - \ell$ and $\dot{S} = v$ do not explicitly contain the control variable $a(t)$, whereas $\ddot{S} = a(t)$ does explicitly contain the control variable.

The solution to the unconstrained problem, $(\ell \ge \frac{1}{4})$, is now obtained. $\dot{E} = \frac{1}{2}a^2$; $E(0) = 0$. Then $E(1)$ must be minimized. The Euler-Lagrange equations are

$$\dot{\lambda}_v = -\lambda_x\,, \qquad \dot{\lambda}_x = 0\,, \qquad \dot{\lambda}_E = 0\,, \qquad \lambda_v = -\lambda_x t + \text{const}\,; \qquad \lambda_x = \text{const}\,,$$

$$\lambda_E = \text{const} = 1\,; \qquad a = -\lambda_v\,.$$

The solution is easily obtained as

$$a = -2\,, \qquad v = 1 - 2t\,, \qquad x = t(1-t) \Rightarrow (x)_{\max} = \tfrac{1}{4}\,, \qquad \lambda_v = -a = 2\,,$$
$$\lambda_x = 0\,, \qquad J = 2\,, \qquad H = \lambda_x \dot{x} + \lambda_v \dot{v} + \lambda_E \dot{E} = -2\,.$$

The solution with constraint, $\frac{1}{6} \leq \ell \leq \frac{1}{4}$, is obtained as follows:†

$$a = \begin{cases} -8(1-3\ell) + 24(1-4\ell)t\,, & 0 \leq t \leq \frac{1}{2}\,, \\ -8(1-3\ell) + 24(1-4\ell)(1-t)\,, & \frac{1}{2} < t < 1\,; \end{cases}$$

$$v = \begin{cases} 1 - 8(1-3\ell)t + 12(1-4\ell)t^2\,, & 0 \leq t \leq \frac{1}{2}\,, \\ -1 + 8(1-3\ell)(1-t) - 12(1-4\ell)(1-t)^2\,, & \frac{1}{2} < t < 1\,, \end{cases}$$

$$x = \begin{cases} t - 4(1-3\ell)t^2 + 4(1-4\ell)t^3\,, & 0 \leq t \leq \frac{1}{2}\,, \\ 1 - t - 4(1-3\ell)(1-t)^2 + 4(1-4\ell)(1-t)^3\,, & \frac{1}{2} < t < 1\,; \end{cases}$$

$$\lambda_v = -a \rightarrow \lambda_v(\tfrac{1}{2}-) - \lambda_v(\tfrac{1}{2}+) = 0 (\ddot{S} = 0 \textit{ not} \text{ used here})\,;$$

$$\lambda_x = \begin{cases} 24(1-4\ell)\,, & 0 \leq t \leq \frac{1}{2}\,, \\ -24(1-4\ell)\,, & \frac{1}{2} \leq t \leq 1\,. \end{cases}$$

Note that

$$\lambda_x(\tfrac{1}{2}-) - \lambda_x(\tfrac{1}{2}+) = 48(1-4\ell)\,,$$
$$J = 2 + 6(1-4\ell)^2\,,$$
$$H = -8(1-6\ell)^2\,.$$

The solution with constraint, $0 < \ell \leq \frac{1}{6}$, is

$$a = \begin{cases} -\dfrac{2}{3\ell}\left(1 - \dfrac{t}{3\ell}\right), & 0 \leq t \leq 3\ell\,, \\ 0\,, & 3\ell \leq t \leq 1 - 3\ell\,, \\ -\dfrac{2}{3\ell}\left(1 - \dfrac{1-t}{3\ell}\right), & 1 - 3\ell \leq t \leq 1\,, \end{cases}$$

$$v = \begin{cases} \left(1 - \dfrac{t^2}{3\ell}\right), & 0 \leq t \leq 3\ell\,, \\ 0\,, & 3\ell \leq t \leq 1 - 3\ell\,, \\ -\left(1 - \dfrac{1-t}{3\ell}\right)^2, & 1 - 3\ell \leq t \leq 1\,, \end{cases}$$

$$x = \begin{cases} \ell\left[1 - \left(1 - \dfrac{t}{3\ell}\right)^3\right], & 0 \leq t \leq 3\ell\,, \\ \ell\,, & 3\ell \leq t \leq 1 - 3\ell\,, \\ \ell\left[1 - \left(1 - \dfrac{1-t}{3\ell}\right)^3\right], & 1 - 3\ell \leq t \leq 1\,; \end{cases}$$

$$\lambda_v = \begin{cases} \dfrac{2}{3\ell}\left(1 - \dfrac{t}{3\ell}\right), & 0 \leq t \leq 3\ell\,, \\ \dfrac{2}{3\ell}\left(1 - \dfrac{1-t}{3\ell}\right), & 3\ell \leq t \leq 1\,. \end{cases}$$

†By symmetry, the constrained problem can be viewed as two identical unconstrained problems with terminal conditions only.

Note that $\lambda_v(3\ell-)-\lambda_v(3\ell+)=(4/3\ell^2)(\frac{1}{6}-\ell)$:

$$\lambda_x=\begin{cases}2/9\ell^2, & 0\le t\le 3\ell,\\ -2/9\ell^2, & 3\ell\le t\le 1.\end{cases}$$

Note that $\lambda_x(3\ell-)-\lambda_x(3\ell+)=4/9\ell^2$:

$$J=4/9\ell, \qquad H=0.$$

Figure 3.11.3 shows the solutions for various values of ℓ. A most interesting feature of these solutions is the fact that the optimal path touches the constraint boundary at only one point for a finite range of values of the constraint parameter ($\frac{1}{6}\le\ell\le\frac{1}{4}$), and only one of the influence functions, λ_x, is discontinuous. For $0<\ell<\frac{1}{6}$, the path stays on the constraint boundary for a finite time, and both λ_v and λ_x are discontinuous. This behavior seems to be typical of a second-order state variable inequality constraint.

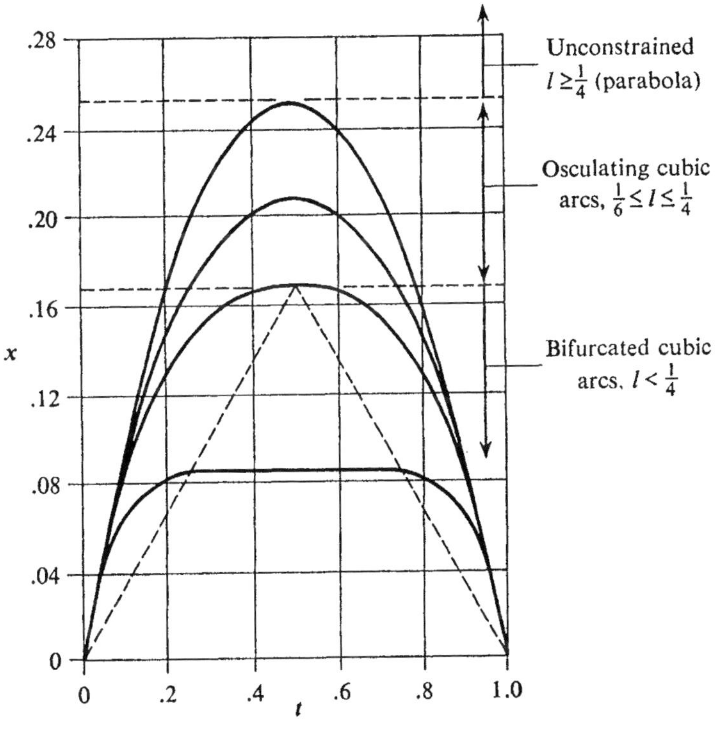

Figure 3.11.3. A constrained minimum-energy problem for several values of ℓ.

Problem. (Suggested by Walter F. Denham.) Let

$$\begin{aligned}\dot{x}_1&=x_2, & x_1(0)&>0\\ \dot{x}_2&=u; & x_2(0)&\end{aligned}\Bigg\}\text{specified;}\qquad \begin{aligned}x_1(t_f)&=0,\\ x_2(t_f)&=0,\end{aligned}$$

$$|u|\le 1, \qquad -x_2\le V>0.$$

Determine $u(t)$ to minimize t_f, assuming that initial conditions are such that solutions exist for the problem.

3.12 The separate computation of arcs in problems with state variable inequality constraints

Separate computation of the unconstrained arcs is possible in problems with state variable inequality constraints if they contain only one constrained arc in the middle, and the contribution of the constrained arc to the performance index depends only on the entry and exit values of one variable (t or one component of x). By definition, the contribution of the constrained arc to the performance index is

$$\Delta J(t_1,t_2) = \int_{t_1}^{t_2} L(x,u,t)\,dt\,. \tag{3.12.1}$$

If the integral in 3.12.1 is a function of its end points only

$$\Delta J(t_1,t_2) = g(x_1(t_2)) - g(x(t_1)) \tag{3.12.2}$$

or

$$\Delta J(t_1,t_2) = g(t_2) - g(t_1)\,, \tag{3.12.3}$$

the two unconstrained arcs in the intervals $t_0 \leq t \leq t_1$ and $t_2 \leq t \leq t_f$ may be determined independently of each other.

To show this, write the performance index as

$$J = \phi(x(t_f),t_f) + \int_{t_0}^{t_f} L(x,u,t)\,dt = J_1 + J_2\,, \tag{3.12.4}$$

where

$$J_1 = \int_{t_0}^{t_1} L(x,u,t)\,dt - g(x(t_1))\,, \tag{3.12.5}$$

$$J_2 = \phi(x(t_f),t_f) + \int_{t_2}^{t_f} L(x,u,t)\,dt + g(x(t_2))\,. \tag{3.12.6}$$

The performance indexes J_1 and J_2 and the interior point constraints, Equation (3.11.6), at the times t_1 and t_2, do not depend on each other, so the corresponding unconstrained arcs may be determined separately.

Example 1. *The bounded brachistochrone problem of Example 1, Section 3.11.* In this example, the final time, t_f, is to be minimized. We may write t_f as

$$t_f = \int_0^{\ell} \frac{dx}{\dot{x}} = t_1 + \int_{x(t_1)}^{x(t_2)} \frac{dx}{\dot{x}} + t_f - t_2\,.$$

On the constrained arc, we have $\gamma = \theta$, and $y = x \tan\theta + h$, hence we obtain

$$\int_{x(t_1)}^{x(t_2)} \frac{dx}{\cos\gamma\sqrt{2gy}} = \int_{x(t_1)}^{x(t_2)} \frac{dx}{\cos\theta\sqrt{2g(x\tan\theta + h)}}$$

$$= \sqrt{\frac{2}{g}}\csc\theta\,[\sqrt{x(t_2)\tan\theta + h} - \sqrt{x(t_1)\tan\theta + h}]$$

and, therefore,

$$J_1 = t_1 - \sqrt{2y(t_1)/g}\csc\theta\,, \qquad J_2 = t_f - t_2 + \sqrt{2y(t_2)/g}\csc\theta\,.$$

The solution must be tested, of course, to insure that $y(t_2) \leq y(t_1)$.

Example 2. *The minimum-fuel problem of Example 2, Section 3.11.* In this example the separation is immediately obvious: since $a = 0$ on the constraint boundary. Thus,

$$J_1 = \frac{1}{2}\int_0^{t_1} a^2\,dt\,, \qquad J_2 = \frac{1}{2}\int_{t_2}^{t_f} a^2\,dt\,.$$

Here, the solution must be tested to insure that $t_1 \leq t_2$.

Further discussion and a numerical example are given in Speyer, Mehra, and Bryson (1968).

3.13 Corner conditions

In optimal control problems, especially those involving inequality constraints on the control and/or state variables, we may encounter discontinuous changes in the control variable or, in other words, discontinuous changes in the slope of the state variable trajectory. These places of discontinuity are called *corners*. The purpose of this section is to summarize the conditions that must hold at these corners.

Let us denote by t^- the time just before the corner, and by t^+, immediately after. Then we have:

(i) for problems without path constraints and problems with control variable inequality constraints,

$$\lambda(t^-) = \lambda(t^+)\,, \tag{3.13.1}$$

$$H(t^-) = H(t^+)\,, \tag{3.13.2}$$

$$H_u(t^-) = H_u(t^+)\,; \tag{3.13.3}$$

(ii) for problems with state variable inequality constraints, at an entry point,

$$\lambda^T(t^-) = \lambda^T(t^+) + \pi^T N_x\,, \tag{3.13.4}$$

$$H(t^-) = H(t^+) - \pi^T N_t\,, \tag{3.13.5}$$

$$H_u(t^-) = H_u(t^+)\,, \tag{3.13.6}$$

where $N(x,t) = 0$ are the q constraints $S(x,t) = 0$, $\dot{S}(x,t) = 0, \ldots, S^{(q-1)}(x,t) = 0$.

At an exit point, the same conditions as in (3.13.1) through (3.13.3) hold.

A problem that demonstrates the existence of corners without necessarily encountering an inequality constraint is as follows:

Consider

$$J = \int_{t_0}^{t_f} \left[u^4(\tau) - (u(\tau) - a(\tau))^2\right] d\tau\,, \tag{3.13.7}$$

$$\dot{x} = u\,, \qquad x(t_0) = x_0\,, \qquad x(t_f) = x_{t_f}\,, \tag{3.13.8}$$

where $a(\tau)$ is a known time function. The Hamiltonian and the Euler equations are

$$\dot{\lambda} = 0 \Rightarrow \lambda = \text{constant}, \tag{3.13.9}$$

$$H = \lambda u - (u - a)^2 + u^4\,, \tag{3.13.10}$$

$$H_u = 0 = 4u^3 - 2u + (\lambda + 2a). \tag{3.13.11}$$

Equation (3.13.11) will, in general, have three solutions for u. It is entirely possible to choose $a(t)$ such that immediately before, during, and after a specific instant t the graphs of H vs. u appear as illustrated in Figures 3.13.1a, b, and c.

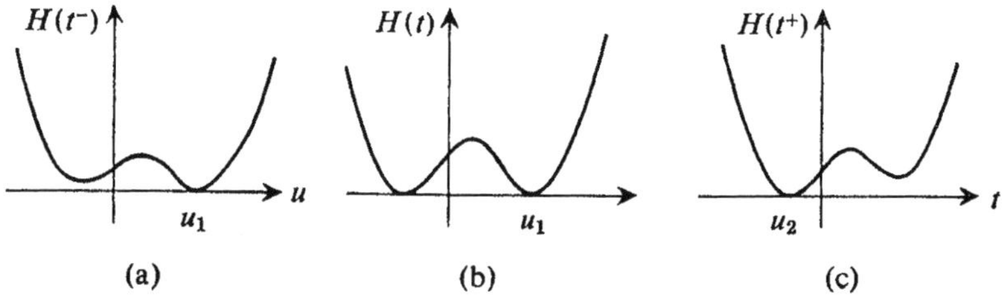

Figure 3.13.1. A possible behavior of H vs. u just before, at, and after a corner.

In this case, u will jump from the value u_1 to u_2 at t. A corner results.

Problem 1. Show that (3.13.1) through (3.13.3) reduce to the *Weierstrass-Erdmann Corner Conditions*:

$$F_{\dot{x}}\Big|_{t^-} = F_{\dot{x}}\Big|_{t^+}\,, \tag{3.13.7}$$

$$(F - \dot{x}F_{\dot{x}})_{t^-} = (F - \dot{x}F_{\dot{x}})_{t^+} \tag{3.13.8}$$

for the problem of minimizing $J = \int_0^T F(x,\dot{x},t)\, dt$.

Problem 2. Describe the situation in which an optimal trajectory leaves a constraint boundary with continuous control $u(t)$.

Optimal feedback control

4

4.1 The extremal field approach

Up to this point, we have studied problems of optimal programming which the control engineer would call optimal "open loop" control. We have aimed at obtaining the optimal control function†, $u(t)$, starting from one given initial state, $x(t_0)$, at the time t_0 and proceeding to a terminal hypersurface defined by the terminal constraints $\psi[x(t_f),t_f] = 0$.

Now, any point *directly on* the optimal path between $(x(t_0),t_0)$ and the terminal hypersurface is a possible "initial" point for the same optimal control functions; we may think of each point $(x(t),t)$ on this optimal path as being associated with the optimal control vector $u(t)$ at that point. However, if we wish to find the optimal control functions to take the system to the same terminal hypersurface from a point *not* directly on the optimal path from $(x(t_0),t_0)$, we must solve another optimal programming problem starting from this new point. It often happens in automatic control problems that we would like to know the optimal control functions $u(t)$ from a great many different initial points to a given terminal hypersurface, since we may not know where the system will start from or when it will start. To cover this situation we must calculate a *family* of optimal paths so that all of the possible initial points are on, or at least very close to, one of our calculated optimal paths. In the literature on the calculus of variations, such a family is called a *field of extremals.*

In general, only one optimal path to the terminal hypersurface will pass through a given point $(x(t),t)$, and a unique optimal control vector $u^\circ(t)$ is then associated with each point.‡ Hence, we may write

$$u^\circ = u^\circ(x,t)\,. \tag{4.1.1}$$

†To be more precise, we have considered some first-order necessary conditions that *may* determine the optimal $u(t)$; in Chapter 6 we will discuss *sufficient* conditions for $u(t)$ to be optimal.

‡In Chapter 6 we will consider what happens in the unusual circumstance in which more than one extremal passes through a given point.

This is the *optimal feedback control law*; i.e., the control vector is given as a function of the present state $x(t)$ and the present time t.† In Figure 4.1.1, a sketch of a system with only one state variable x, is shown, giving the optimal paths and a set of contours of constant values of u°. Note that the terminal "hypersurface," $\psi(x,t) = 0$, is in this case simply a curve in the x,t-plane.

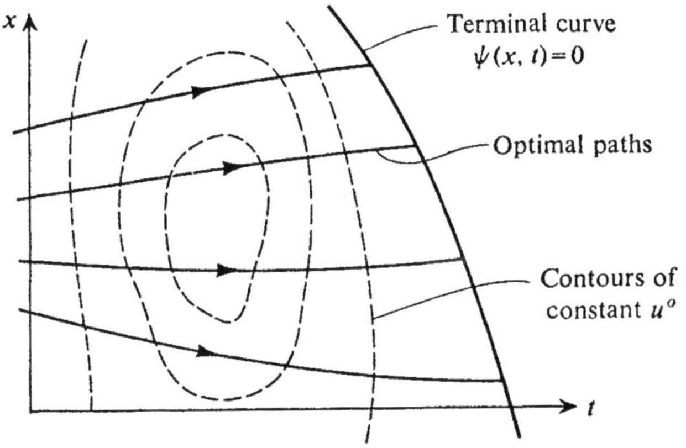

Figure 4.1.1. A family of optimal paths and contours of constant optimal control u^{o}.

Associated with starting from a point (x,t) and proceeding optimally to the terminal hypersurface, there is a unique optimal value of the performance index, J°. We may therefore regard J° as a function of the initial point, that is,

$$J^{\circ} = J^{\circ}(x,t) \,. \tag{4.1.2}$$

This is the *optimal return function.* For the one-dimensional case, contours of constant J° may be sketched on the x,t-plane as in Figure 4.1.2.

The contours of constant J° are like "wave fronts," and the optimal paths are like "rays." However, in general, the rays are not orthogonal to the wave fronts; in optics, they are orthogonal because $\mathbf{r}$ is proportional to $\boldsymbol{\lambda}$ (see Problem 2, Section 2.7, and Problem 1, Section 3.2).

One aspect of the classical *Hamilton-Jacobi theory* is concerned with finding the partial differential equation satisfied by the optimal return function $J^{\circ}(x,t)$. There is also a (vector) partial differential equation satisfied by the optimal control functions $u^{\circ}(x,t)$. Bellman has generalized the Hamilton-Jacobi theory to include multistage

†Such a feedback control scheme is often called an *explicit guidance scheme*.

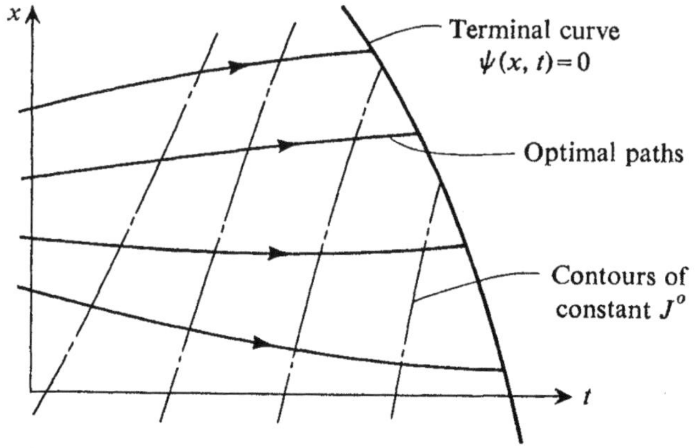

Figure 4.1.2. A family of optimal paths and contours of constant optimal return function J^o.

systems and combinatorial problems and he calls this overall theory *dynamic programming*. This approach is discussed in the next section of this chapter.

Stationary systems. If the system, the performance index, and the constraints are not explicit functions of the time, and if the terminal time is unspecified, the optimal control functions, the optimal return function, and the time-to-go are not explicit functions of time; i.e., we have

$$u^\circ = u^\circ(x)\,, \tag{4.1.3}$$

$$J^\circ = J^\circ(x)\,, \tag{4.1.4}$$

$$T^\circ = T^\circ(x)\,. \tag{4.1.5}$$

We will call such systems *stationary systems*.†

Example. A simple stationary system was discussed in Example 1 of Section 2.7, where a minimum-time path was found from a given initial point to the origin for a particle moving through a region with linear variation in current velocity. Now, it is quite simple to generate a field of extremals for this problem, using the parametric equations developed there:

$$\frac{y}{h} = \sec\theta - \sec\theta_f\,, \tag{4.1.6}$$

†For minimum-time problems in which it is "time-to-go" that is being minimized, the optimal return function is identical with the time-to-go.

$$\frac{x}{h} = \frac{1}{2}\left[\sec\theta_f(\tan\theta_f - \tan\theta) - \tan\theta(\sec\theta_f - \sec\theta) + \log\frac{\tan\theta_f + \sec\theta_f}{\tan\theta + \sec\theta}\right], \tag{4.1.7}$$

$$\frac{V(t_f - t)}{h} = \tan\theta - \tan\theta_f. \tag{4.1.8}$$

Extremal paths are generated by picking a constant value of θ_f and plotting x,y varying with θ in (4.1.6), (4.1.7). Contours of constant performance index $[V(t_f - t)/h]$ and contours of constant control angle θ are then easily obtained as "cross-plots"; this is done in Figures 4.1.4 and 4.1.3, respectively.

Problem. Make plots similar to Figures 4.1.3 and 4.1.4 for the minimum-time problem of Section 2.7, Problem 5.

4.2 Dynamic programming; the partial differential equation for the optimal return function

Consider the general control problem discussed in Section 2.7 for an arbitrary initial point (x,t). The performance index is

$$J = \varphi[x(t_f),t_f] + \int_t^{t_f} L[x(\tau),u(\tau),\tau]\, d\tau, \tag{4.2.1}$$

and the system equations are

$$\dot{x} = f(x,u,t), \tag{4.2.2}$$

with terminal boundary conditions

$$\psi[x(t_f),t_f] = 0. \tag{4.2.3}$$

The optimal return function, defined in (4.1.2), is given symbolically by

$$J^\circ(x,t) = \min_{u(t)} \{\varphi[x(t_f),t_f] + \int_t^{t_f} L[x,u,\tau]\, d\tau\}, \tag{4.2.4}$$

with the boundary condition

$$J^\circ(x,t) = \varphi(x,t) \text{ on the hypersurface } \psi(x,t) = 0. \tag{4.2.5}$$

Let us assume that $J^\circ(x,t)$ exists, is continuous, and possesses continuous first and second partial derivatives at all points of interest in the x,t-space.

Now suppose that the system starts at (x,t) and proceeds for a short

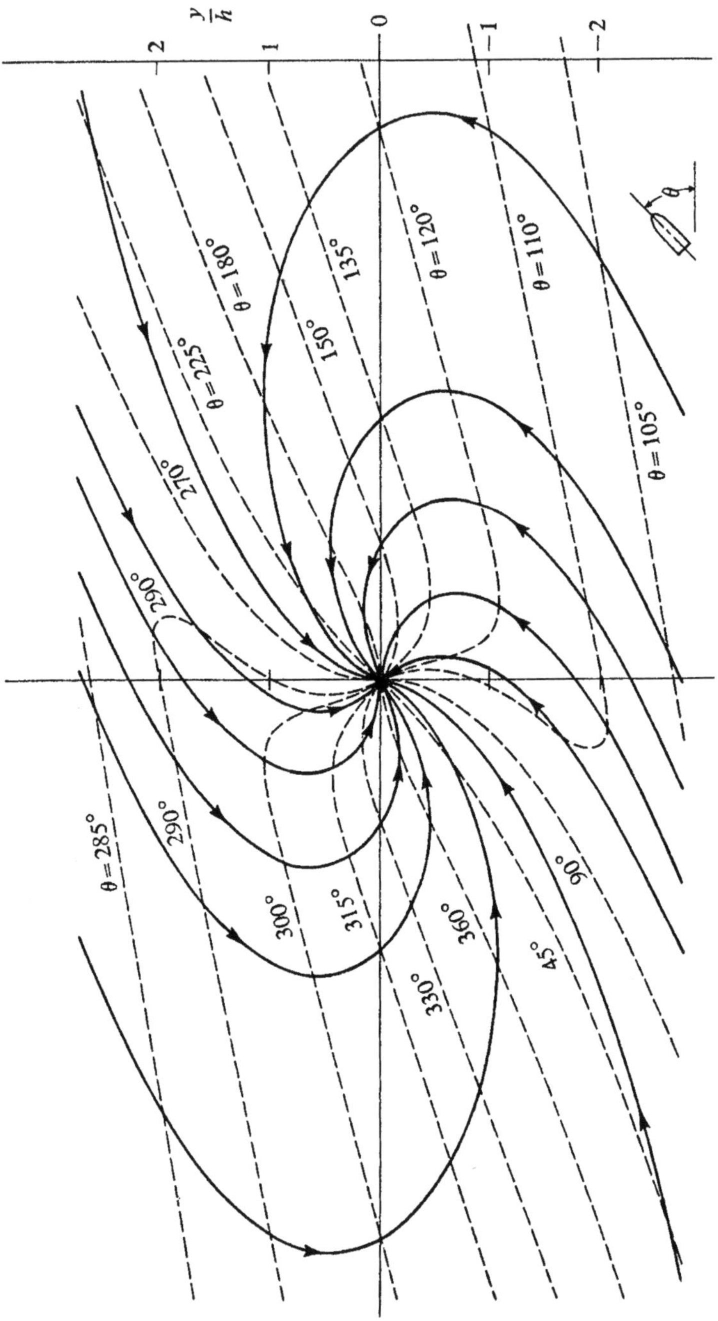

Figure 4.1.3. Minimum-time ship paths with linear variation in current and contours of *constant heading angle*.

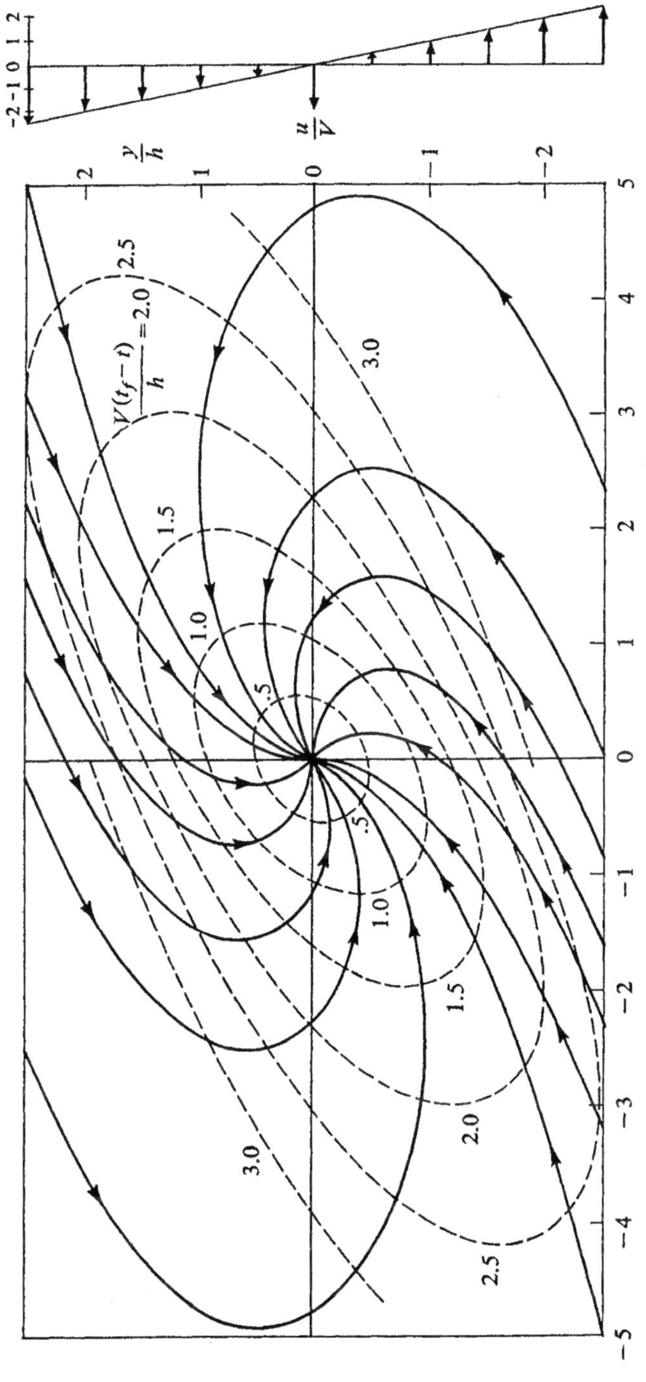

Figure 4.1.4. Minimum-time ship paths with linear variation in current and contours of *constant time-to-go*.

time Δt using a control $u(t)$, which is not the optimal control. In view of (4.2.2), the system will proceed to the new point,

$$[x + f(x,u,t)\Delta t, t + \Delta t]. \tag{4.2.6}$$

Now, supposing optimal control is used from this point onwards, the return function is then given to first order by

$$J^\circ(x + f(x,u,t)\Delta t, t + \Delta t) + L(x,u,t)\Delta t \triangleq J^1(x,t). \tag{4.2.7}$$

Now, clearly, since *nonoptimal* control was used in the interval t to $(t + \Delta t)$, we have

$$J^\circ(x,t) \leq J^1(x,t). \tag{4.2.8}$$

The equality sign in (4.2.8) will obtain only if we choose $u(t)$ in the interval t to $t + \Delta t$ so as to minimize the right-hand side of (4.2.8):

$$J^\circ(x,t) = \min_u \{J^\circ[x + f(x,u,t)\Delta t, t + \Delta t] + L(x,u,t)\Delta t\}. \tag{4.2.9}$$

In view of our continuity and differentiability assumptions about J°, we may expand the right-hand side of (4.2.9) in a Taylor series about x,t, as follows:

$$J^\circ(x,t) = \min_u \left\{J^\circ(x,t) + \frac{\partial J^\circ}{\partial x} f(x,u,t)\Delta t + \frac{\partial J^\circ}{\partial t}\Delta t + L(x,u,t)\Delta t\right\}. \tag{4.2.10}$$

Since J° and, hence $\partial J^\circ/\partial t$ do not explicitly depend on u, (4.2.10) may be written by passing to the limit as $\Delta t \to 0$:*

$$-\frac{\partial J^\circ}{\partial t} = \min_u \left[L(x,u,t) + \frac{\partial J^\circ}{\partial x} f(x,u,t)\right]. \tag{4.2.11}$$

In Section 2.3 we showed that the Lagrange multipliers $\lambda(t)$ were influence functions; i.e., small changes in the initial conditions dx, and small changes in the initial time dt, produce small changes in the performance index dJ according to

$$dJ^\circ = \lambda^T(t)\, dx - H(t)\, dt, \tag{4.2.12}$$

where

$$H(x,\lambda,u,t) = L(x,u,t) + \lambda^T f(x,u,t). \tag{4.2.13}$$

Clearly, (4.2.12) implies that

$$\lambda^T \equiv \frac{\partial J^\circ}{\partial x} \text{ (on the optimal trajectory)}, \qquad H \equiv -\frac{\partial J^\circ}{\partial t}. \tag{4.2.14}$$

*For a justification of interchanging the minimization and limit operation, see Berkovitz and Dreyfus.

Using (4.3.13), Equation (4.2.11) may be written as

$$-\frac{\partial J^\circ}{\partial t} = H^\circ\left(x, \frac{\partial J^\circ}{\partial x}, t\right), \tag{4.2.15}$$

where

$$H^\circ\left(x, \frac{\partial J^\circ}{\partial x}, t\right) = \min_u H\left(x, \frac{\partial J^\circ}{\partial x}, u, t\right). \tag{4.2.16}$$

Equation (4.2.15) or (4.2.11) is called the *Hamilton-Jacobi-Bellman Equation*; it is a first-order nonlinear partial differential equation and it must be solved with the boundary condition (4.2.5).

Equation (4.2.16) states that u° is the value of u that minimizes globally the Hamiltonian $H(x, \partial J^\circ/\partial x, u, t)$, holding $x, \partial J^\circ/\partial x$, and t constant; this is another statement of the *Minimum Principle*.† If there are no bounds on x and u, it follows from our differentiability assumptions and (4.2.16) that u° must be such that we have

$$\frac{\partial H}{\partial u} \equiv \frac{\partial L}{\partial u} + \frac{\partial J^\circ}{\partial x}\frac{\partial f}{\partial u} = 0, \tag{4.2.17}$$

$$\frac{\partial^2 H}{\partial u^2} \geq 0 \tag{4.2.18}$$

for all $t \leq t_f$; that is, every component of the vector $\partial H/\partial u$ must vanish and $\partial^2 H/\partial u^2$ must be a nonnegative-definite matrix. Equation (4.2.18) is known as the *Legendre-Clebsch condition* in the calculus of variations.‡

One of the most effective ways to solve the nonlinear partial differential equations (4.2.15) is by the "method of characteristics,"§ which amounts to finding a "field of extremals" as in Section 4.1, using the calculus of variations.

†In the USSR and much of the classic literature, H is defined with the opposite sign, so that, to minimize J, one must maximize H. Thus it is called the Pontryagin Maximum Principle in the USSR. To be precise, the minimum condition derived this way is part of a sufficient condition (i.e., if we can solve (4.2.15) and (4.2.16), then u from (4.2.16) is the optimal control), while the Pontryagin Maximum Principle is derived directly as a necessary condition in the spirit of Chapters 2 and 3. For more connection between dynamic programming and variational calculus, see Section 6.3 and Dreyfus (1965).

‡Note that (4.2.17) and (4.2.18) are local versions of the condition (4.2.16), which is a *global* result in the control space. (See Section 6.9 for another discussion of (4.2.16) via the *Weierstrass necessary condition*.)

§R. Courant and D. Hilbert, *Methoden der mathematischen Physik*. Berlin: Springer, 1937, Vol. II, Chapter 2.

The great drawback of dynamic programming is, as Bellman himself calls it, the "curse of dimensionality." Even recording the solution to a moderately complicated problem involves an enormous amount of storage. If we want only one optimal path from a known initial point, it is wasteful and tedious to find a whole field of extremals; if we need feedback, a perturbation feedback scheme is often quite adequate (see Chapter 6, Neighboring Extremals and the Second Variation).

Derivation of the Euler-Lagrange equations from the Hamilton-Jacobi equation. Consider a particular optimal path and its associated optimal control function. Then we have

$$\frac{d\lambda^T}{dt} \equiv \frac{d}{dt}\left(\frac{\partial J^\circ}{\partial x}\right) = \frac{\partial^2 J^\circ}{\partial x^2}\dot{x} + \frac{\partial^2 J^\circ}{\partial x \partial t}. \tag{4.2.19}$$

Partial differentiation of (4.2.15) with respect to x, considering $u^\circ = u^\circ(x,t)$, gives

$$\frac{\partial^2 J^\circ}{\partial x \partial t} + \frac{\partial L}{\partial x} + \frac{\partial L}{\partial u}\frac{\partial u^\circ}{\partial x} + f^T \frac{\partial^2 J^\circ}{\partial x^2} + \frac{\partial J^\circ}{\partial x}\left(\frac{\partial f}{\partial x} + \frac{\partial f}{\partial u}\frac{\partial u^\circ}{\partial x}\right) = 0. \tag{4.2.20}$$

Now the coefficient of $\partial u^\circ/\partial x$ in (4.2.20) vanishes on an optimal path according to (4.2.17).† Using (4.2.20) in (4.2.19), then, we obtain

$$\frac{d\lambda^T}{dt} = -\lambda^T \frac{\partial f}{\partial x} - \frac{\partial L}{\partial x}, \tag{4.2.21}$$

which, along with (4.2.17), are the Euler-Lagrange equations.

Furthermore, the fact that J° is equal to ϕ on $\psi = 0$ implies that there exists a vector ν such that

$$\left.\frac{\partial J^\circ}{\partial x}\right|_{t_f} = \left(\frac{\partial \phi}{\partial x} + \nu^T \frac{\partial \psi}{\partial x}\right)_{t=t_f} \triangleq \lambda^T(t_f). \tag{4.2.22}$$

In words, the change in cost due to admissible change in state (that is, $d\psi = 0$) is given by a linear combination of the gradient of ϕ with respect to state and the gradients of ψ (constraints) with respect to state (see Section 1.2).

Combinatorial problems. Dynamic programming is especially useful in solving multistage optimization problems in which there are only a small number of possible choices of the control at each

†When there are inequality constraints on the control variable, it can be shown (e.g., by defining a modified Hamiltonian as in Chapter 3 or see Kalman (1963) or Dreyfus (1965)) that the term $(L_u + J_x^\circ f_u)u_x^\circ$ still vanishes.

stage and in which no derivative information is available. Consider a simple example with only two possible choices of control at each stage. We would like to find the path from A to B in Figure 4.2.1,

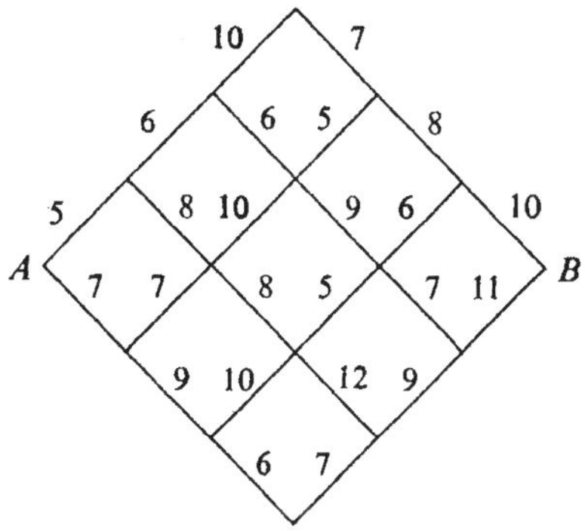

Figure 4.2.1. A minimum-time combinatorial problem. Numbers are times to travel legs of grid.

traveling only to the right (either up-right or down-right at each corner), such that the sum of the numbers on the segments along this path is a minimum. If we consider these numbers to be travel times, we are looking for the minimum-time path.

There are 20 possible routes from A to B, traveling only to the right. It would be tedious to try out all 20 routes. Instead of starting at A and trying different routes to B, we work *backward* from B to find the minimum-time routes to B from each of the 15 corners on the grid, as indicated on Figure 4.2.2.

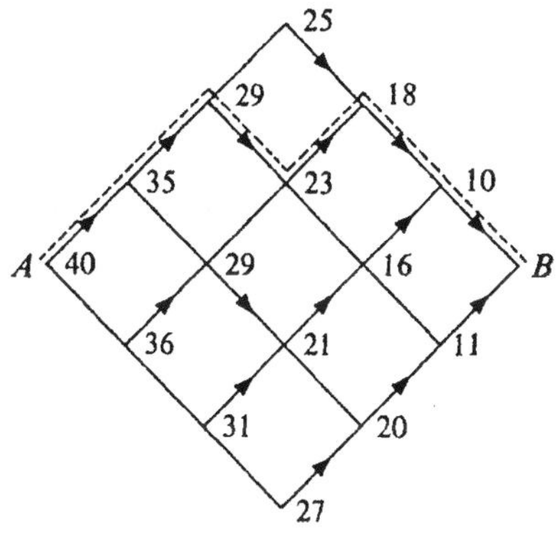

Figure 4.2.2. Dynamic programming solution to the problem of Figure 4.2.1.

The first step backward can be taken either up or down. In Figure 4.2.3 we have placed the time figures (10 and 11) near these two corners, and we have indicated by a small arrow the direction from these corners to the end point, B.

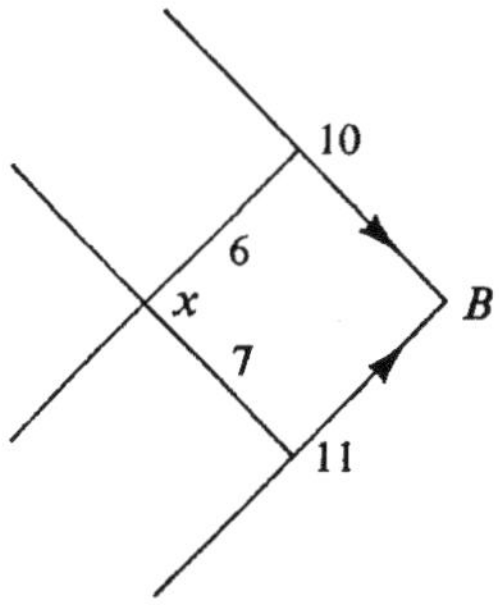

Figure 4.2.3. First step in dynamic programming solution to the problem of Figure 4.2.1.

Now let us examine the minimum time from the corner marked with an "x" in Figure 4.2.3 to the terminal point, B. Two paths are possible: one through the "10" corner, which would take 16 units of time (6 + 10); and the other through the "11" corner, which would take 18 units of time (7 + 11). Clearly, the faster path passes through the "10" corner, so we replace the "x" by a "16" and draw an arrow pointing from that corner to the "10" corner.

This same procedure is repeated over and over for new corners lying to the left of corners that have already been labeled by their shortest times to B. Finally, we have the minimum-time values for every corner, shown in Figure 4.2.2, *and* the better direction to follow in leaving each corner.

The shortest-time path can now be traced from A to B by moving always in the direction indicated by the arrows; this route is shown on Figure 4.2.2 by a dotted line and takes 40 units of time. Note that we had to find only 15 numbers using this algorithm instead of computing the time for each of the 20 possible paths. The savings in computation are more impressive as the number of segments on a side increases:

Segments on a side	3	4	5	6	7	n
Possible routes	20	70	252	724	2632	$(2n)!/n!n!$
Computations	15	24	35	48	63	$(n + 1)^2 - 1$

We also have a useful by-product, namely the minimum-time routes from each of the corners to B.

Problem 1. Solve the "minimum distance in the plane" problem by the dynamic programming method, and identify the optimal return function and the feedback control function.

Problem 2. In the five-by-five grid shown in Figure 4.2.4, find the minimum-time path from A to B moving only to the right. There are 70 possible routes, but you need to calculate only 24 numbers.

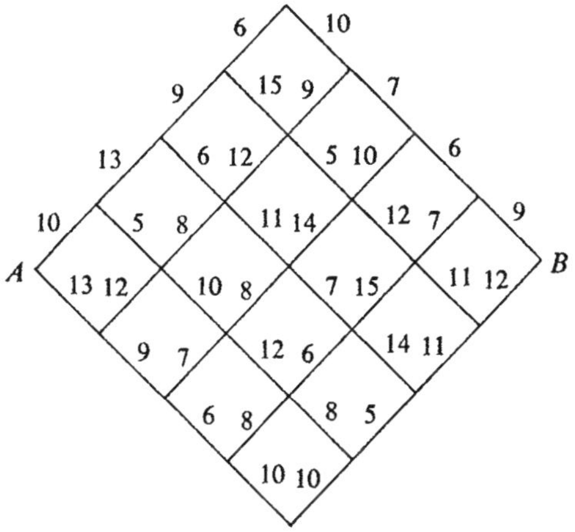

Figure 4.2.4. Grid for Problem 2. Numbers are times to travel legs of grid.

Problem 3. Find the *maximum*-time path from A to B for the grid of Figure 4.2.4, still moving only to the right.

Problem 4. Some of the airlines flying the North Atlantic now use computers to find minimum-time jet-flight paths, taking into account the strong winds that usually occur at jet cruising altitudes and the restrictions on possible routes imposed by air traffic control. Savings on the order of 15 minutes on the nominal 7-hour flight are obtained this way. A grid of "checkpoints" is selected and all routes must consist of generally east-west straight-line segments between checkpoints. A simplified version of such a grid is shown in Figure 4.2.5. Imagine that point A is New York and point M is London. Checkpoints B through L are points above the ocean, located by giving their longitude and latitude. Using wind data collected by "weather ships," the flight planner computes the time to fly each segment. In practice, there are many more checkpoints than are shown in Figure 4.2.5, so a computer solution is essential. For the times (in minutes) given on the figure, use the dynamic programming algorithm to find the minimum-time route.

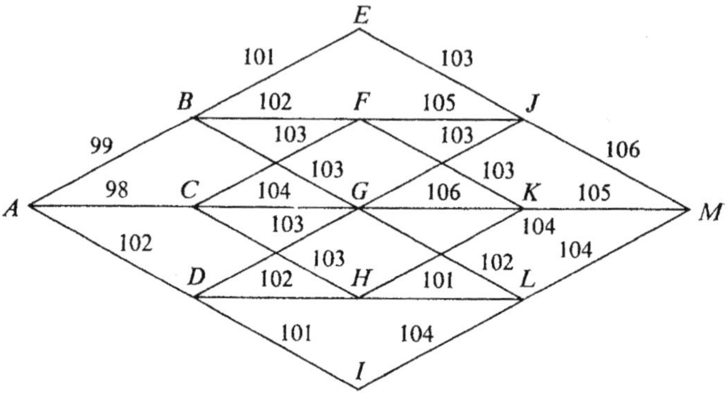

Figure 4.2.5. Grid for Problem 4. Numbers are times to travel legs of grid.

Problem 5. The dynamic programming algorithm developed in the example can also be used on irregular patterns with more than two choices at some corners, as in Figure 4.2.6. Find the minimum-time path from corner 12 to corner 1, moving only to the right.

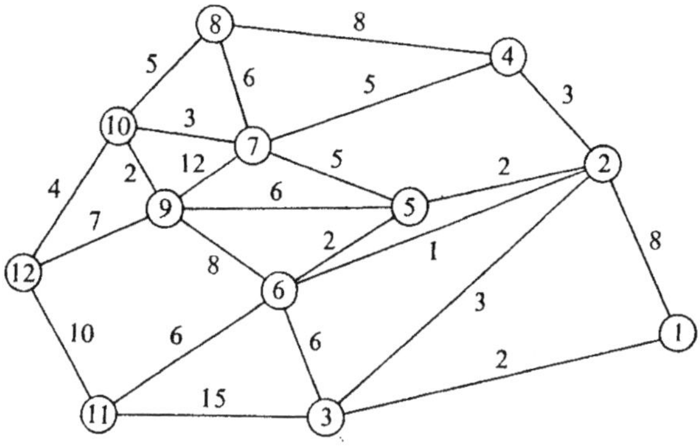

Figure 4.2.6. Grid for Problem 5. Numbers are times to travel legs of grid.

Problem 6. Consider the following routing problem, as illustrated in Figure 4.2.7.

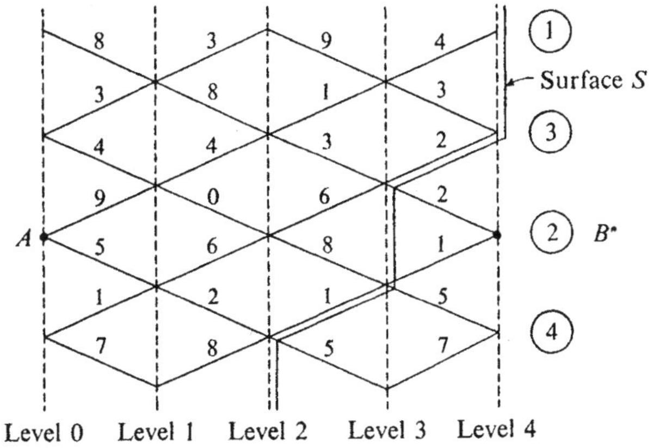

Figure 4.2.7. Grid for Problem 6. Numbers are costs to travel legs of grid.

We wish to proceed along a path from level 0 successively to level 4, with the cost for each leg of the path given by the number associated with that particular segment. The total cost is measured by the sum of the cost of each segment and the terminal cost associated with each position of level 4.

(a) Find the minimal-cost path from point A to level 4.
(b) Find the minimal-cost path from point A to point B.
(c) Find the minimal-cost path from level 0 to level 4.
(d) Find the minimal-cost path from point A to the surface S, assuming 0 terminal costs on S.
(e)† Solve (a) with the added consideration that, each time we change directions at a given level, one additional unit of cost is added to the cost of the segment following. What is different about this problem?
(f) What is the analog of $J^\circ(x,t)$, $u^\circ = k(x,t)$, $x,u,t,f(x,u.t)$, $\phi(x(t_f),t_f)$, $L(x,u,t)$, and $\psi(x(t_f),t_f)$ in this problem?

Problem 7. Find the minimum time and the minimum-time path from point A to point B, moving only to the right. The time to travel between points is shown on each segment of Figure 4.2.8.

4.3 Reducing the dimension of the state space by use of dimensionless variables

One difficulty in using fields of extremals for optimal feedback control is the large amount of memory storage required for even

†Problem suggested by S. Dreyfus.

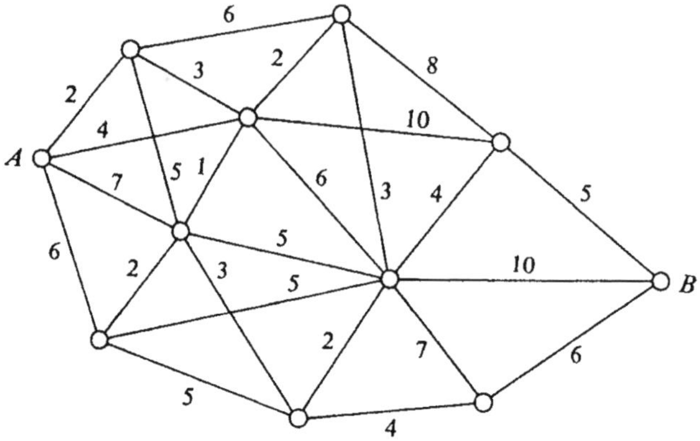

Figure 4.2.8. Grid for Problem 7. Numbers are times to travel legs of grid.

moderately complicated systems. In some problems it is possible to reduce the amount of memory storage by noticing that the control depends only on certain dimensionless groupings of the state variables. This is best illustrated by examples.

Example 1. *A brachistochrone problem.* Find the feedback law $\gamma = \gamma(V,x,y)$ to yield minimum-time paths from an arbitrary state V,x,y to the wall $x = 0$, where

$$\dot{V} = g \sin \gamma , \tag{4.3.1}$$

$$\dot{x} = -V \cos \gamma , \tag{4.3.2}$$

$$\dot{y} = V \sin \gamma , \tag{4.3.3}$$

and V = velocity, g = gravitational force per unit mass.

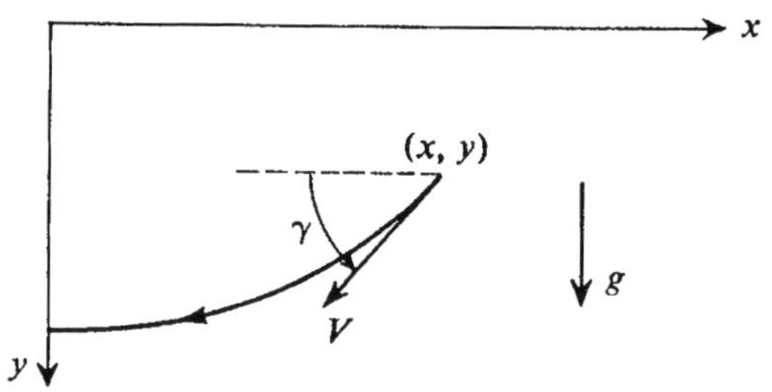

Figure 4.3.1. Nomenclature for a brachistochrone problem.

The first thing we notice is that final y is *not* specified and y does *not* enter into the equation for $\dot{V}$ and $\dot{x}$. Hence, the feedback law is independent of y:

$$\gamma = \gamma(V,x)\,. \tag{4.3.4}$$

Now, γ is dimensionless (an angle), so we *must* be able to find dimensionless groupings of the variables and the parameter g on the right-hand side of Equation (4.3.4). In fact, there is only *one* such dimensionless grouping, gx/V^2 ; hence, we know that

$$\gamma = \text{function of } \frac{gx}{V^2}. \tag{4.3.5}$$

Similarly, the minimum time-to-go, $t_f - t$, and the change in y, $y - y_f$, where y_f is the final value of y, are given by

$$\frac{g(t_f - t)}{V} = \text{function of } \frac{gx}{V^2}, \tag{4.3.6}$$

$$\frac{y - y_f}{x} = \text{function of } \frac{gx}{V^2}. \tag{4.3.7}$$

Problem 1.† Using the results of Problem 6, Section 2.7, *show that* the functions (4.3.5), (4.3.6), and (4.3.7) of Example 1 are given *implicitly* by

$$\gamma = \cos^{-1}\left(\frac{V}{V_f}\right) \qquad (V_f \text{ is the final velocity}),$$

$$\frac{g(t_f - t)}{V} = \frac{(\pi/2) - \sin^{-1}(V/V_f)}{(V/V_f)},$$

$$\frac{y_f - y}{x} = \frac{V^2}{2gx}\left(\frac{V_f^2}{V^2} - 1\right),$$

where (V/V_f) is the solution to the transcendental equation

$$\frac{2gx}{V^2} = \frac{V_f}{V}\sqrt{1 - V^2/V_f^2} + (V_f/V)^2\left(\frac{\pi}{2} - \sin^{-1}\frac{V}{V_f}\right).$$

Plot a graph of γ, $[g(t_f - t)/V]$, and $(y - y_f/x)$ vs. $(2gx/V^2)$.

Example 2. *Thrust-direction programming for rendezvous* ($g = 0$, a = constant). A spacecraft wishes to rendezvous in minimum time with another (nonmaneuvering) spacecraft, using constant-thrust acceleration, a, and controllable thrust direction, β. Using a coordinate system with origin at the target spacecraft (see Figure 4.3.2), the

†(See J. L. Speyer, Tech. Report 492, Div. Engineering and Appl. Physics, Harvard University, December 1965.)

problem is to find the feedback law, $\beta = \beta(V,\gamma,r,\sigma)$, that takes the pursuing spacecraft to $r = V = 0$ in minimum time, where

$$\dot{V} = a \cos(\gamma + \beta)\,, \tag{4.3.8}$$

$$V(\dot{\gamma} - \dot{\sigma}) = -a \sin(\gamma + \beta)\,, \tag{4.3.9}$$

$$\dot{r} = -V \cos\gamma\,, \tag{4.3.10}$$

$$r\dot{\sigma} = V \sin\gamma\,. \tag{4.3.11}$$

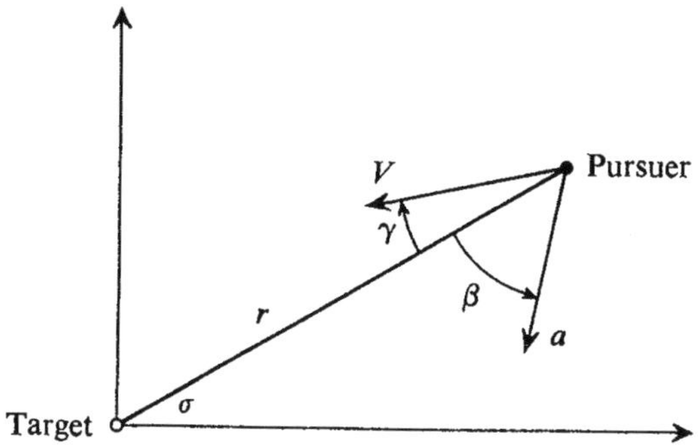

Figure 4.3.2. Nomenclature for minimum-time rendezvous problem.

The first thing we notice is that final σ is *not* specified, and σ does *not* enter into the equations of motion for $\dot{V}$, $\dot{\gamma}$, and $\dot{r}$, so the feedback law must be independent of σ. Next we recognize that β is dimensionless (an angle), so we must be able to find dimensionless groupings of V, r, and the parameter a. (Note that γ is already dimensionless.) In fact, again, there is only one such dimensionless grouping, ar/V^2. Hence, we know that

$$\beta = \text{function of } \frac{ar}{V^2} \text{ and } \gamma\,. \tag{4.3.12}$$

Similarly, the minimum time to go, $t_f - t$, and the change in σ, $\sigma_f - \sigma$ (where σ_f = final value of σ) are given by

$$\frac{a(t_f - t)}{V} = \text{function of } \frac{ar}{V^2} \text{ and } \gamma\,, \tag{4.3.13}$$

$$\sigma_f - \sigma = \text{function of } \frac{ar}{V^2} \text{ and } \gamma\,. \tag{4.3.14}$$

This problem was solved by Bryson,† and the results are shown in Figures 4.3.3 and 4.3.4. The analysis is not difficult but it is somewhat tedious.

†A. E. Bryson. "Nonlinear Feedback Solution for Minimum Time Rendezvous with Constant Thrust Acceleration," 16th Int. Astro. Congress, Athens, Greece, Sept. 1965 (also Report 478, Div. Engineering and Appl. Physics, Harvard University, July 15, 1965).

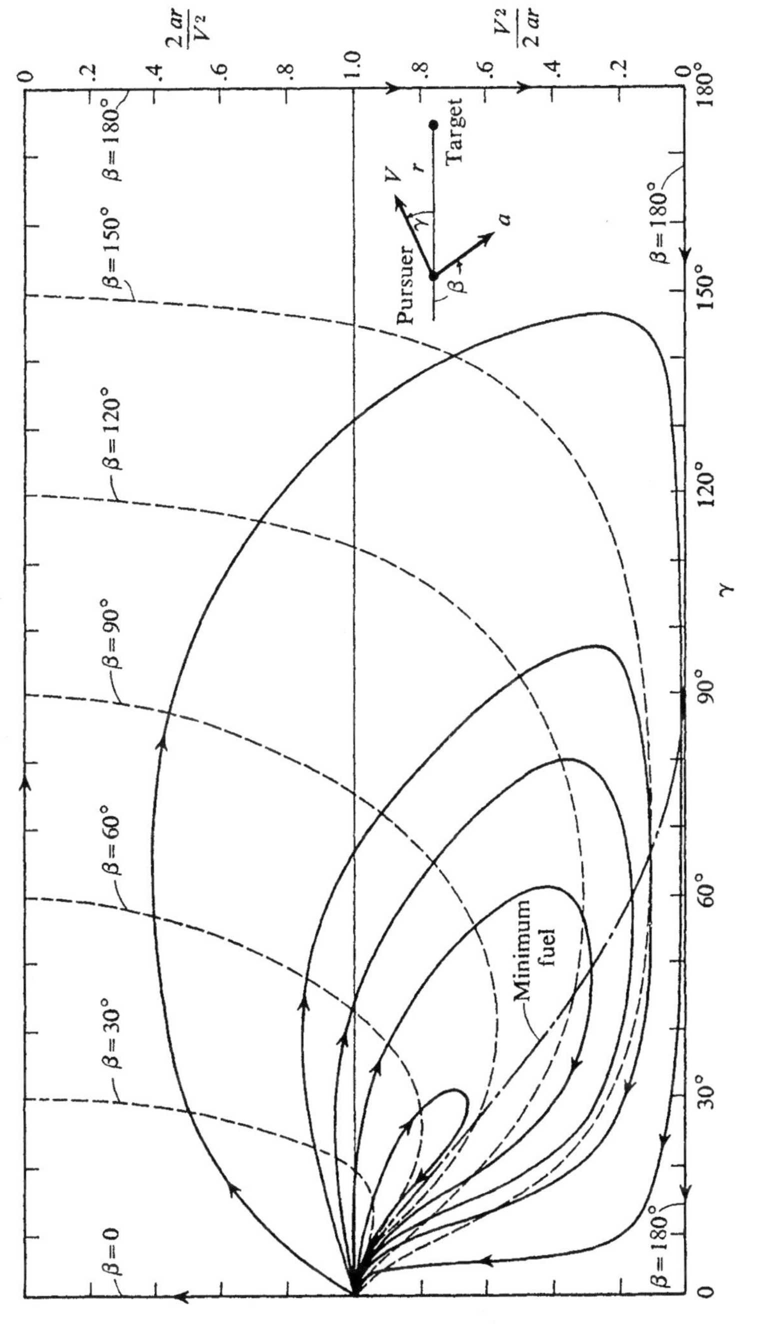

Figure 4.3.3. Minimum-time rendezvous paths and contours of constant thrust direction β.

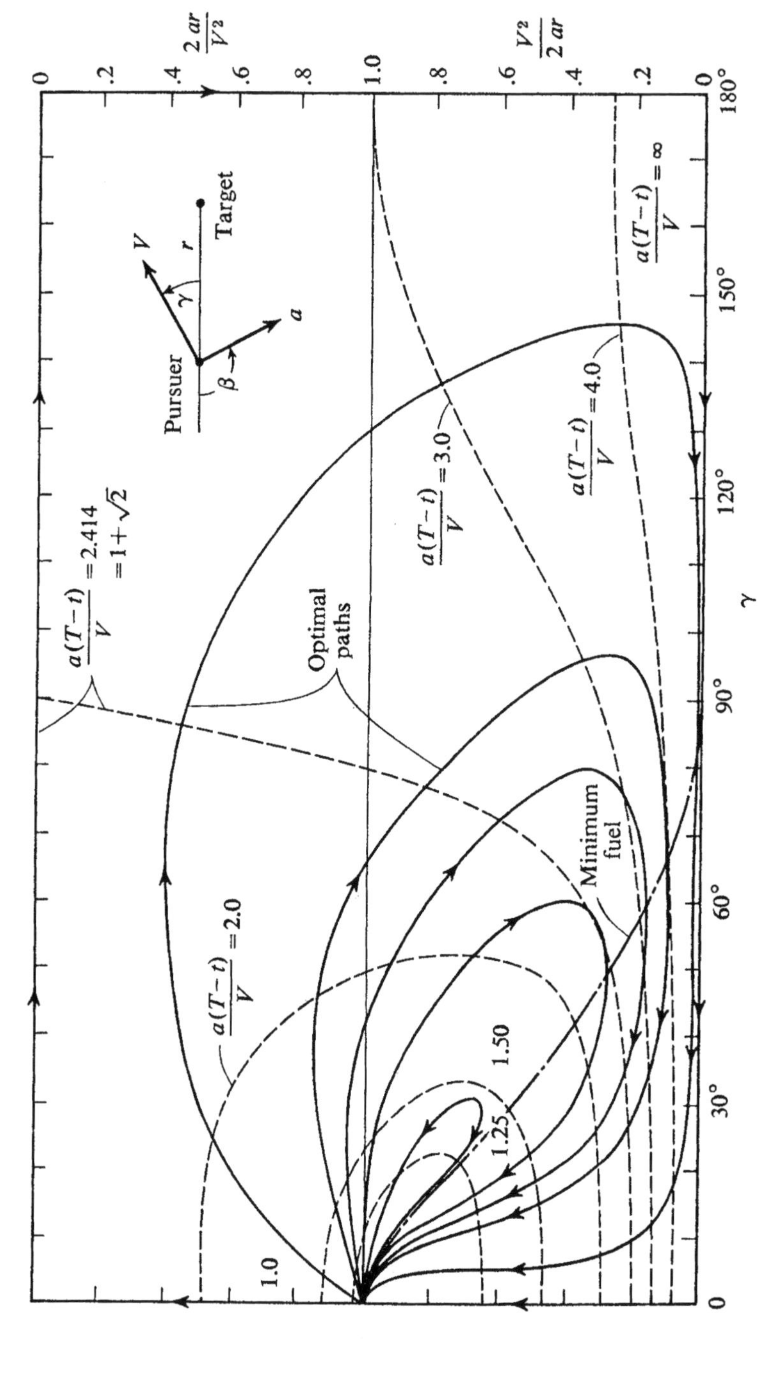

Figure 4.3.4. Minimum-time rendezvous paths and contours of constant time-to-go, $T-t$.

Problem 2. With reference to Example 1, consider the brachistochrone problem with terminal conditions $x = y = 0$. (Reverse the direction of the y-axis.) Show that the feedback law is of the form

$$\gamma = \gamma\left(\frac{2gx}{V^2}, \frac{x}{y}\right),$$

and find the implicit equations necessary to calculate this function and the dimensionless time to go, $[g(t_f - t)/V]$.

Problem 3. With reference to Example 2, show that the feedback law for *minimum-time interception* of a target at $r = 0$, with $g = 0$, $a =$ constant, may be put into the form

$$\beta = \beta\left(\frac{ar}{V^2}, \gamma\right),$$

and find the implicit equations necessary to calculate this function and the dimensionless time to go, $[a(t_f - t)]/V$. (See G. Smuck, M.Sc. Thesis, M.I.T., June, 1966.)

Problem 4.† Show that the feedback law for *minimum-time injection into orbit*, with $g =$ constant, $a =$ constant, may be put into the form

$$\beta = \beta\left(\frac{\sqrt{2ay}}{u_f - u}, \frac{v}{u_f - u}\right),$$

where

$u_f =$ final (orbital) velocity,

$y_f = 0 =$ final (orbital) altitude,

$v_f = 0 =$ final vertical velocity.

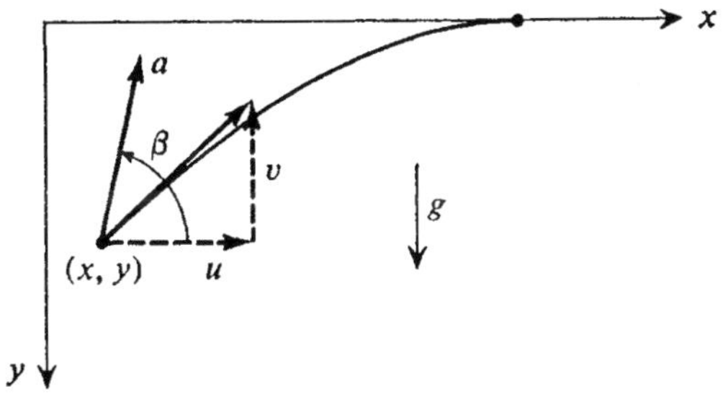

Figure 4.3.5. Nomenclature for minimum-time injection into orbit problem.

†(See D. Winfield, Tech. Report 507, Div. Engineering and Appl. Physics, Harvard University, July 1966.)

Linear systems with quadratic criteria: linear feedback

5

5.1 Terminal controllers and regulators; introduction

Only rarely is it feasible to solve the Hamilton-Jacobi-Bellman partial differential equation for a nonlinear system of any practical significance, and, hence, the development of "exact" explicit feedback guidance and control schemes for nonlinear systems is usually out of reach. However, as mentioned in Chapter 4, *perturbation guidance* schemes, which involve only those extremal paths in the immediate neighborhood of a nominal extremal path, appear, in many cases, to be feasible. We shall show in Chapter 6 that the perturbation guidance approach leads to a discussion of *time-varying linear systems with quadratic performance criteria.*

Many systems that we wish to control are already adequately described by linear dynamic models. In this case, it is possible to synthesize very satisfactory linear feedback controllers by the proper choice of quadratic performance criteria and quadratic constraints.

We shall distinguish between terminal controllers and regulators. A *terminal controller* is designed to bring a system close to desired conditions at a terminal time (which may or may not be specified) while exhibiting acceptable behavior on the way. A *regulator* is designed to keep a stationary system within an acceptable deviation from a reference condition using acceptable amounts of control.

5.2 Terminal controllers; quadratic penalty function on terminal error

Given a time-varying linear system of the form

$$\dot{x} = F(t)x + G(t)u\,, \qquad x = n\text{-component state vector}\,, \tag{5.2.1}$$

$u = m$-component control vector,

we desire to bring it from an initial state $x(t_0)$ to a terminal state,

$$x(t_f) \cong 0\,, \qquad t_f = \text{terminal time}\,, \tag{5.2.2}$$

using "acceptable" levels of control $u(t)$ and not exceeding "acceptable" levels of the state on the way.

One way to do this is to minimize a performance index made up of a quadratic form in the terminal state plus an integral of quadratic forms in the state and the control:

$$J = \frac{1}{2}(x^T S_f x)_{t=t_f} + \frac{1}{2}\int_{t_0}^{t_f} (x^T A x + u^T B u)\,dt\,, \tag{5.2.3}$$

where S_f and $A(t)$ are positive semidefinite matrices and B is a positive definite matrix.† An appropriate choice of these matrices must be made to obtain "acceptable" levels of $x(t_f)$, $x(t)$, and $u(t)$; for example one might choose these matrices to be diagonal with

$1/(S_f)_{ii}$ = maximum acceptable value of $[x_i(t_f)]^2$

$1/A_{ii} = (t_f - t_o)$ X maximum acceptable value of $[x_i(t)]^2$

$1/B_{ii} = (t_f - t_o)$ X maximum acceptable value of $[u_i(t)]^2$

Using the methods of Chapter 2 (the methods of Chapter 4 can also be used here; see last part of this section), the control $u(t)$ that minimizes (5.2.3) will be obtained by solving (5.2.1) simultaneously with the Euler-Lagrange equations,

$$\dot{\lambda}^T = -\frac{\partial H}{\partial x}\,; \qquad \lambda(t_f) = S_f x(t_f)\,; \tag{5.2.4}$$

$$0 = \frac{\partial H}{\partial u}\,, \tag{5.2.5}$$

where

$$H = \tfrac{1}{2}x^T A x + \tfrac{1}{2}u^T B u + \lambda^T(Fx + Gu)\,. \tag{5.2.6}$$

Performing the differentiations indicated in (5.2.4) and (5.2.5), we have

$$\dot{\lambda} = -Ax - F^T\lambda\,, \tag{5.2.4$'$}$$

$$0 = Bu + G^T\lambda \Rightarrow u = -B^{-1}G^T\lambda\,. \tag{5.2.5$'$}$$

Substituting (5.2.5′) into (5.2.1) and repeating (5.2.4′), we have a linear two-point boundary-value problem:

$$\begin{bmatrix} \dot{x} \\ \dot{\lambda} \end{bmatrix} = \begin{bmatrix} F\,, & -GB^{-1}G^T \\ -A\,, & -F^T \end{bmatrix} \begin{bmatrix} x \\ \lambda \end{bmatrix}; \qquad \begin{matrix} x(t_0)\ \text{given}, & (5.2.7) \\ \lambda(t_f) = S_f x(t_f)\,. & (5.2.8) \end{matrix}$$

†A more general expression is considered in Problem 4 of this section.

Since the problem is linear and the differential equations and the terminal boundary conditions are homogeneous, it is clear that both $x(t)$ and $\lambda(t)$ are proportional to $x(t_0)$.

We shall now discuss *two* ways of solving this boundary-value problem.

Solution by transition matrix. One way to solve (5.2.7), (5.2.8), is by ***linear superposition.*** We determine a set of n linearly-independent solutions to the $2n$ differential equations,

$$x^{(i)}(t) \qquad \text{and} \qquad \lambda^{(i)}(t)\,, \qquad i = 1, \ldots, n\,,$$

each solution satisfying the terminal boundary condition; that is, we have

$$\lambda^{(i)}(t_f) = S_f x^{(i)}(t_f)\,.$$

A convenient way to do this is to determine ***unit solutions***, where

$$x_j^{(i)}(t_f) = \begin{cases} 1\,, i = j\,, \\ 0\,, i \neq j\,; \end{cases} \qquad \lambda_j^{(i)}(t_f) = (S_f)_{ji}.$$

Let these n solutions ($i = 1, \ldots, n$) be arranged in columns, giving rise to two ($n \times n$) ***transition matrices***, $X(t)$ and $\Lambda(t)$, where

$$X_{ji}(t) = x_j^{(i)}(t), \qquad \Lambda_{ji}(t) = \lambda_j^{(i)}(t)\,, \tag{5.2.9}$$

and

$$X(t_f) = I = \text{unit matrix}, \qquad \Lambda(t_f) = S_f\,. \tag{5.2.10}$$

Then, by superposition, if we knew $x(t_f)$, we could write the solution as

$$x(t) = X(t)x(t_f)\,, \tag{5.2.11}$$

$$\lambda(t) = \Lambda(t)x(t_f)\,. \tag{5.2.12}$$

Now, at $t = t_0$, we can invert Equation (5.2.11) to find $x(t_f)$ in terms of the given data, $x(t_0)$:

$$x(t_f) = [X(t_0)]^{-1}x(t_0)\,. \tag{5.2.13}$$

Substituting (5.2.13) into (5.2.11) and (5.2.12), we have.

$$x(t) = X(t)\,[X(t_0)]^{-1}x(t_0)\,, \tag{5.2.14}$$

$$\lambda(t) = \Lambda(t)\,[X(t_0)]^{-1}x(t_0)\,. \tag{5.2.15}$$

Substituting (5.2.15) into (5.2.5′) gives

$$u(t) = -C(t,t_0)x(t_0)\,, \tag{5.2.16}$$

where the gain matrix $C(t,t_0)$ is given by

$$C(t,t_0) = [B(t)]^{-1}G^T(t)\,\Lambda(t)\,[X(t_0)]^{-1}. \qquad (5.2.17)$$

Now (5.2.16) may be regarded as a *sampled-data feedback law*, where t_0 is the most recent sample time.

If a continuous determination of the state, x, is made, the most recent sample time is the current time, that is, $t_0 \equiv t$, and (5.2.16) becomes a *continuous feedback law*,

$$u(t) = -C(t)x(t), \qquad (5.2.18)$$

where the time-varying gain matrix $C(t)$ is given by

$$C(t) = [B(t)]^{-1}\,G^T(t)\,\Lambda(t)[X(t)]^{-1}. \qquad (5.2.19)$$

The solution for $\lambda(t)$ in this case may be written, from (5.2.15), as

$$\lambda(t) = S(t)x(t), \qquad (5.2.20)$$

where

$$S(t) = \Lambda(t)\,[X(t)]^{-1}. \qquad (5.2.21)$$

Solution by the sweep method. In some problems, particularly those involving dissipation, the numerical determination of the unit solution matrices $X(t)$ and $\Lambda(t)$ of Equations (5.2.11) and (5.2.12) may become quite difficult because of the different growth rates of the unit solutions; i.e., accuracy will be lost because of the change in size of elements of $X(t)$ and $\Lambda(t)$ by different orders of magnitude over the time interval t_0 to t_f. In this case, as well as in other cases, it may be useful to use the sweep method of solution.† (See Chapter 7 and Kalman (Reference 9), 1966).

The idea of the sweep method is contained in Equations (5.2.20) and (5.2.21). Instead of determining the unit solution matrices $X(t)$ and $\Lambda(t)$, we determine $S(t) \equiv \Lambda(t)\,(X(t))^{-1}$ directly. We may think of this as generating a boundary condition for (5.2.7) and (5.2.8) *equivalent* to the terminal condition $\lambda(t_f) = S_f\,x(t_f)$, but at earlier times; in effect, the coefficients of the terminal condition are "swept" backward to the initial time. Then, since $x(t_0)$ is known, $\lambda(t_0)$ may be computed from $\lambda(t_0) = S(t_0)x(t_0)$, and (5.2.7) and (5.2.8) may be integrated forward once as an initial-value problem.

Substituting (5.2.20) into (5.2.8) yields

$$\dot{S}x + S\dot{x} = -Ax - F^T Sx. \qquad (5.2.22)$$

Next, substituting $\dot{x}$ from (5.2.7) into (5.2.22) and using (5.2.20) again, we have

†See I. M. Gelfand and S. V. Fomin, *Calculus of Variations.* Englewood Cliffs, N.J.: Prentice-Hall, 1963, Chapter 6.

$$(\dot{S} + SF + F^TS - SGB^{-1}G^TS + A)\,x = 0\,. \tag{5.2.23}$$

Since $x(t) \neq 0$, Equation (5.2.23) requires that

$$\dot{S} = -SF - F^TS + SGB^{-1}G^TS - A\,. \tag{5.2.24}$$

From the terminal boundary condition in Equation (5.2.8) it is clear that

$$S(t_f) = S_f\,. \tag{5.2.25}$$

Equation (5.2.24) is a *matrix Riccati equation* since it contains a quadratic term in S.

Since S_f is symmetric and Equation (5.2.24) is symmetric, it is clear that S is a symmetric matrix.

Equation (5.2.24) must be integrated (or "swept") backward from the terminal time $t = t_f$ to the initial time $t = t_0$; it is then possible to determine $\lambda(t_0)$ from (5.2.20), as follows:

$$\lambda(t_0) = S(t_0)x(t_0)\,, \tag{5.2.26}$$

which may be regarded as the equivalent of the terminal boundary condition in (5.2.8) at an earlier time. If desired, the solution to (5.2.7) and (5.2.8) can then be found by *forward integration* since $\lambda(t_0)$ and $x(t_0)$ are both known.

However, we are often interested mainly in the continuous feedback law for terminal control, which is determined from (5.2.5′) and (5.2.20) once $S(t)$ has been found, as follows:

$$\begin{aligned} u(t) &= -C(t)x(t)\,, \\ C(t) &= [B(t)]^{-1}G^T(t)\,S(t)\,. \end{aligned} \tag{5.2.27}$$

Dynamic programming interpretation. Still a third way to approach the solution of the linear-quadratic problem (5.2.1) and (5.2.3) is by the Hamilton-Jacobi-Bellman partial differential equation, which in this case becomes

$$-\frac{\partial J^\circ}{\partial t} = \min_u \left\{ \frac{\partial J^\circ}{\partial x}(Fx + Gu) + \frac{1}{2}(x^TAx + u^TBu) \right\}, \tag{5.2.28}$$

with the terminal boundary condition

$$J^\circ(x,t_f) = \tfrac{1}{2}\,x^TS_fx\,. \tag{5.2.29}$$

The minimization of the Hamiltonian over u in Equation (5.2.28) is, of course, the same as in Equation (5.2.6) and leads to Equation (5.2.5) where $\lambda^T \equiv \partial J^\circ/\partial x$ on an optimal trajectory; substituting this into (5.2.28), one obtains

$$-\frac{\partial J^\circ}{\partial t} = \frac{\partial J^\circ}{\partial x}Fx + \frac{1}{2}x^TAx - \frac{1}{2}\left(\frac{\partial J^\circ}{\partial x}\right)GB^{-1}G^T\left(\frac{\partial J^\circ}{\partial x}\right)^T. \tag{5.2.30}$$

This is a nonlinear partial differential equation of first order. It happens to have a product solution of the form

$$J^{\circ} = \tfrac{1}{2} x^T S(t) x \,. \tag{5.2.31}$$

Substituting (5.2.31) into (5.2.30), we have

$$0 = \tfrac{1}{2} x^T [\dot{S} + SF + F^T S - SGB^{-1}G^T S + A] x \,, \tag{5.2.32}$$

which, since it must hold for all x, implies Equation (5.2.24).

Thus the "sweep method" and dynamic programming are one and the same thing for linear-quadratic problems. Equation (5.2.31) gives us another interpretation of the meaning of $S(t)$; that is, $\frac{1}{2} x^T S(t)x$ is the optimal return function (minimum value of J starting at time t with state vector x).

Example 1. *A simple first-order linear system with quadratic criteria. Given*

$$\dot{x} = u \text{ with } x(t_o)\,,\, t_o\,, \text{and } t_f \text{ specified; } x, u \text{ scalar variables,}$$

$$J = \frac{1}{2} c[x(t_f)]^2 + \frac{1}{2}\int_{t_0}^{t_f} u^2\, dt\,; \qquad c \text{ scalar constant}.$$

Find $u(t;t_o)$ to minimize J.

SOLUTION.

$$H = \tfrac{1}{2} u^2 + \lambda u \qquad (\lambda \text{ a scalar}),$$

$$\dot{\lambda} = -\frac{\partial H}{\partial x} = 0 \Rightarrow \lambda = \text{const},$$

$$\frac{\partial H}{\partial u} = u + \lambda = 0 \Rightarrow u = -\lambda \qquad \text{for optimality},$$

$$\lambda(t_f) = cx(t_f) \qquad \text{(boundary condition)},$$

$$x(t) = -[cx(t_f)](t - t_o) + x(t_o) \Rightarrow x(t_f) = \frac{x(t_o)}{1 + c(t_f - t_o)}\,,$$

$$u(t;t_o) = -\frac{1}{(1/c) + t_f - t_o} x(t_o) \qquad \text{(a sampled-data feedback control law).}$$

If $t_o = t$, we have

$$u(t) = -\frac{1}{1/c + t_f - t} x(t) \qquad \text{(a continuous-data feedback control law)}$$

Note that $x(t_f) \to 0$ as $c \to \infty$.

Example 2. *A simple intercept (or rendezvous) problem.*†
Given

$$\dot{v} = a(t), \qquad \dot{y} = v,$$

$$J = \frac{1}{2}\left[[v,y]\begin{bmatrix} c_1 & 0 \\ 0 & c_2 \end{bmatrix}\begin{bmatrix} v \\ y \end{bmatrix}\right]_{t=t_f} + \frac{1}{2}\int_t^{t_f} a^2\, dt.$$

$$c_1, c_2, t_f \text{ constants.}$$

Find $a(v,y,t)$ to minimize J.

SOLUTION.

$$a(t) = -\Lambda_v(t)\, v(t) - \Lambda_y(t)\, y(t);$$

$$\Lambda_v = \frac{1/c_2 + 1/c_1\,(t_f - t)^2 + 1/3\,(t_f - t)^3}{D(t_f - t)},$$

$$\Lambda_y = \frac{1/c_1\,(t_f - t) + 1/2\,(t_f - t)^2}{D(t_f - t)},$$

where

$$D(t_f - t) = \left[\frac{1}{c_2} + \frac{1}{3}(t_f - t)^3\right]\left[\frac{1}{c_1} + t_f - t\right] - \frac{1}{4}(t_f - t)^4.$$

COMMENTS. If $c_2 \to 0$, then $y(t_f)$ is *not* controlled, and we have

$$\Lambda_v = \frac{1}{(1/c_1) + t_f - t}, \qquad \Lambda_y = 0,$$

$$\Rightarrow a = -\frac{v(t)}{[(1/c_1) + t_f - t]},$$

which is "velocity-to-be-gained" control.

If we have $c_1 \to 0$, $v(t_f)$ is *not* controlled and

$$\Lambda_v = \frac{(t_f - t)^2}{1/c_2 + \frac{1}{3}(t_f - t)^3}, \qquad \Lambda_y = \frac{t_f - t}{1/c_2 + \frac{1}{3}(t_f - t)^3}.$$

If $c_1 \to 0$ and $c_2 \to \infty$, then $y(t_f) \to 0$, and

$$a(t) = -3\left[\frac{v(t)}{t_f - t} + \frac{y(t)}{(t_f - t)^2}\right].$$

Let V = closing velocity along the line of sight, and let

$$\sigma \cong \frac{y(t)}{V(t_f - t)} = \text{the line-of-sight angle.}$$

Then

$$a = -3V\dot{\sigma},$$

†Additional discussion of this problem is given on pages 280 and 424.

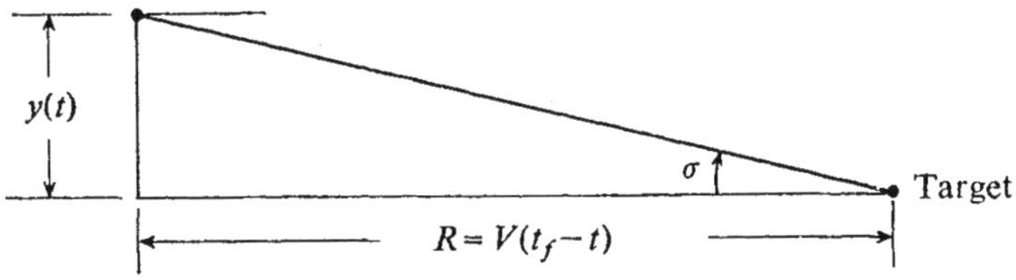

Figure 5.2.1. Nomenclature for intercept and rendezvous example.

which is "proportional navigation" leading to a perfect *intercept*, $y(t_f) = 0$.

If both c_1 and $c_2 \to \infty$, then we have $v(t_f) \to 0$, $y(t_f) \to 0$, and

$$a(t) = -\frac{4v(t)}{t_f - t} - \frac{6y(t)}{(t_f - t)^2} \qquad \text{or} \qquad a = -V\left(4\dot{\sigma} + \frac{2\sigma}{t_f - t}\right),$$

which is a modified form of proportional navigation leading to a "perfect" *rendezvous*, $y(t_f) = v(t_f) = 0$.

Note that if $v(t_f)$ and $y(t_f)$ in J are replaced by $v(t_f) - v_d$ and $y(t_f) - y_d$, the feedback law is

$$a(t) = -\Lambda_v(t)[v(t) - v_d] - \Lambda_y(t)[y(t) - y_d].$$

Also note that v, y, and a may be replaced by ***three-dimensional vectors*** **v**, **r**, and **a**, and the three-dimensional intercept (or rendezvous) solutions are obtained. In this case, only the two components perpendicular to the line of sight can be given the angle σ interpretation.

Problem 1. For one more method of solving the general linear-quadratic control problem, consider the following: Let $u = Kx$, where $K(t)$ now plays the role of control variables to be determined. Use the variational techniques of Chapter 2 to derive the results of this section. A useful identity to know

$$\mathrm{Tr}(ABC) = \mathrm{Tr}(CAB) = \mathrm{Tr}(BCA), \quad \frac{\partial \mathrm{Tr}(ABC)}{\partial C} = AB.$$

Problem 2. *First-order system. Given* the first-order linear system with quadratic criterion

$$\dot{x} = -ax + bu; \qquad x(t_0) \text{ given},$$

$$J = \frac{1}{2}c[x(t_f)]^2 + \frac{1}{2}\int_{t_0}^{t_f} [u(t)]^2\, dt,$$

where x, u are scalar variables and a, b, c are scalar constants, *show that* the sampled-data and continuous feedback control laws for minimum J are, respectively,

$$u(t,t_o) = -b\frac{\exp[-a(t_f - t_o) - a(t_f - t)]}{(1/c) + (b^2/2a)\{1 - \exp[-2a(t_f - t_o)]\}}x(t_o),$$

$$u(t) = -b\frac{\exp[-2a(t_f - t)]}{(1/c) + (b^2/2a)\{1 - \exp[-2a(t_f - t)]\}}x(t).$$

Find expressions for $x(t)$ and $\lambda(t)$ for the sampled-data case and for $S(t)$ in the continuous case.
Show also that $x(t_f) \to 0$ as $c \to \infty$.

Problem 3. *Second-order system. Given*

$$\dot{x}_2 = -\omega^2 x_1 + u,$$

$$\dot{x}_1 = x_2; \qquad x_1(t_o), x_2(t_o) \text{ given};$$

$$J = \frac{1}{2}c(x_1(t_f))^2 + \frac{1}{2}\int_{t_0}^{t_f} u^2\, dt.$$

Show that the sampled-data and continuous feedback control laws that minimize J are, respectively,

$$u(t;t_o) = -\frac{[4\omega^2\cos\omega(t_f - t_o)]\,x_1(t_o) + [4\omega\sin\omega(t_f - t_o)]\,x_2(t_o)}{4\omega^3/c + 2\omega(t_f - t_o) - \sin 2\omega(t_f - t_o)}\sin\omega(t_f - t),$$

$$u(t) = -\frac{[4\omega^2\cos\omega(t_f - t)]\,x_1(t) + [4\omega\sin\omega(t_f - t)]\,x_2(t)}{4\omega^3/c + 2\omega(t_f - t) - \sin 2\omega(t_f - t)}\sin\omega(t_f - t),$$

and that the terminal value of x_1 is

$$x_1(t_f) = \frac{x_1(t_o)\cos\omega(t_f - t_o) + x(t_o)\sin\omega(t_f - t_o)/\omega}{1 + (c/4\omega^3)[2\omega(t_f - t_o) - \sin 2\omega(t_f - t_o)]}.$$

Note that, for $c \to \infty$, $x_1(t_f) \to 0$ and the feedback gains $\to \infty$ as $t \to t_f$. Note that the continuous-data feedback gains, while negative near $t = t_f$, alternate in sign for increasing values of $\omega(t_f - t)$.

Problem 4. *More general quadratic performance index. Given* the linear system and quadratic criterion

$$\dot{x} = F(t)x + G(t)u$$

$$J = \frac{1}{2}(x^T S_f x)_{t=t_f} + \frac{1}{2}\int_{t_0}^{t_f} (x^T, u^T)\begin{bmatrix} A(t), & N(t) \\ N^T(t), & B(t) \end{bmatrix}\begin{bmatrix} x \\ u \end{bmatrix} dt.$$

Show that the continuous feedback control law that minimizes J is

$$u(t) = -B^{-1}(N^T + G^T S)\,x(t)\,,$$

where

$$\begin{aligned}\dot{S} &= -S(F - GB^{-1}N^T) - (F - GB^{-1}N^T)^T S + S\,GB^{-1}G^T S - (A - NB^{-1}N^T)\\ &= -SF - F^T S + (SG + N)\,B^{-1}(N^T + G^T S) - A\,; \qquad S(t_f) = S_f\end{aligned}$$

and, hence, it is equivalent to the problem $\dot{x} = (F - GB^{-1}N^T)x + Gu$ with

$$J = \frac{1}{2}(x^T S_f x)_{t=t_f} + \frac{1}{2}\int_{t_0}^{t_f} (x^T, u^T)\begin{bmatrix} A - NB^{-1}N^T & 0 \\ 0 & B \end{bmatrix}\begin{bmatrix} x \\ u \end{bmatrix} dt\,.$$

Problem 5. *Symplectic character of transition matrices for optimum linear-quadratic system.* For the problem of Equations (5.2.7) and (5.2.8), consider the partitioned transition matrix $\Phi(t,t_0)$, where

$$\begin{bmatrix} x(t) \\ \lambda(t) \end{bmatrix} = \begin{bmatrix} \Phi_{xx}(t,t_0)\,, \Phi_{x\lambda}(t,t_0) \\ \Phi_{\lambda x}(t,t_0)\,, \Phi_{\lambda\lambda}(t,t_0) \end{bmatrix}\begin{bmatrix} x(t_0) \\ \lambda(t_0) \end{bmatrix}.$$

Show that $\Phi(t,t_0)$ is symplectic, that is, $\Phi^T(t,t_0)J\Phi(t,t_0) = J$, where $J = \begin{bmatrix} 0\,, I \\ -I\,, 0 \end{bmatrix}$, and this implies that

$$\Phi^{-1}(t,t_0) = \begin{bmatrix} \Phi^T_{\lambda\lambda}(t,t_0)\,, -\Phi^T_{x\lambda}(t,t_0) \\ -\Phi^T_{\lambda x}(t,t_0)\,, \quad \Phi^T_{xx}(t,t_0) \end{bmatrix}.$$

Note that this is useful in numerical solutions.

Problem 6. A simple first-order system is governed by

$$\dot{x} = u \qquad \text{for} \qquad 0 \le t \le t_f\,,$$

where both x and u are *scalars*. A continuous, but not very accurate, measurement of $x(t)$ is made over the interval; call it $z(t)$. We believe that the initial condition $x(0)$ was equal to zero and that $u(t)$ was equal to zero, but we are not certain of either.

To estimate $x(t)$ for $0 \le t \le t_f$, we decide to make a "least square fit" in the sense that we want the $x(t)$ that minimizes the following quadratic form:

$$J = \frac{1}{2}\frac{[x(0)]^2}{p} + \frac{1}{2}\int_0^{t_f}\left[\frac{u^2}{q} + \frac{(z-x)^2}{r}\right] dt\,,$$

where p, q, and r are (scalar) constants chosen so as to express our *relative* confidence in our estimates of $x(0)$, $u(t)$, and the measurement $z(t)$, respectively.

Develop a procedure for finding the $x(t)$ that minimizes J, given $z(t)$, p, q, r, and t_f (see Chapter 13).

5.3 Terminal controllers; zero terminal error and controllability

Suppose that we wish to design a terminal controller that would bring certain components of $x(t_f)$ *exactly* to zero instead of approximately to zero.†

$$x_i(t_f) = 0\,, \qquad i = 1, \ldots, q\,, \qquad \text{where } q \le n\,. \tag{5.3.1}$$

We can do this by using Equations (5.2.7) and (5.2.8) with

$$S_{ij}(t_f) = \begin{cases} 0\,, & i \ne j\,, \\ \infty\,, & i = j\,, \quad i \le q\,, \\ 0\,, & i = j\,, \quad i > q\,.‡ \end{cases}$$

However, this approach runs into difficulties (infinite boundary conditions) in attempting to integrate Equation (5.2.24) in Section 5.2. These difficulties can be circumvented (see Problem 1) but, instead, we might reformulate the problem of Section 5.2 as follows: Find $u(t)$ to minimize

$$J = \frac{1}{2}\int_{t_o}^{t_f} (x^T A x + u^T B u)\, dt \tag{5.3.2}$$

subject to the terminal constraints (5.3.1) and the constraints

$$\dot{x} = F(t)x + G(t)u\,, \tag{5.3.3}$$

$$x(t_o) \text{ given.} \tag{5.3.4}$$

This problem is a special case of the problems treated in Sections 2.4 and 2.5.

The constraints (5.3.1) may be adjoined to (5.3.2) by multipliers $(\nu_1, \ldots, \nu_q) = \nu^T$:

$$\bar{J} = \sum_{i=1}^{q} \nu_i x_i(t_f) + \frac{1}{2}\int_{t_o}^{t_f} (x^T A x + u^T B u)\, dt\,. \tag{5.3.5}$$

The Euler-Lagrange equations for this problem are easily found to be

$$\dot{\lambda} = -Ax - F^T\lambda\,; \qquad \lambda_j(t_f) = \begin{cases} \nu_j\,; & j = 1, \ldots, q\,, \\ 0\,; & j = q+1, \ldots, n\,; \end{cases} \tag{5.3.6}$$

$$u = -B^{-1}G^T\lambda\,. \tag{5.3.7}$$

Substituting (5.3.7) into (5.3.3), we have the two-point boundary-value problem

---

†A more general terminal constraint is considered in Problem 1 of this section.

‡By this we mean that $x^T S_f x = \begin{cases} \infty, & x \ne 0\,, \\ 0\,, & x = 0\,. \end{cases}$

$$\begin{bmatrix} \dot{x} \\ \dot{\lambda} \end{bmatrix} = \begin{bmatrix} F, & -GB^{-1}G^T \\ -A, & -F^T \end{bmatrix} \begin{bmatrix} x \\ \lambda \end{bmatrix}; \qquad (5.3.8)\ (5.3.9)$$

$$x(t_0) \text{ given,} \qquad (5.3.10)$$

$$\begin{aligned} x_i(t_f) &= 0\,, \qquad i = 1,\ldots,q\,; \\ \lambda_i(t_f) &= 0\,, \qquad i = q+1,\ldots,n\,. \end{aligned} \qquad (5.3.11)$$

Solution by transition matrix. The two-point boundary-value problem (5.3.8) through (5.3.11) may be solved by finding a set of n unit solutions to Equations (5.3.8) and (5.3.9), where all unit solutions satisfy (5.3.11) and the ith unit solution (denoted $x^{(i)}(t),\lambda^{(i)}(t)$) satisfies

$$\lambda_j^{(i)}(t_f) = \begin{cases} 1\,, & i = j\,, \\ 0\,, & i \neq j\,, \end{cases} \qquad i = 1,\ldots,q\,; \qquad (5.3.12)$$

$$x_j^{(i)}(t_f) = \begin{cases} 1\,, & i = j\,, \\ 0\,, & i \neq j\,, \end{cases} \qquad i = q+1,\ldots,n\,. \qquad (5.3.13)$$

The general solution may then be written as

$$x(t) = X(t)\mu\,, \qquad (5.3.14)$$

$$\lambda(t) = \Lambda(t)\mu\,,\dagger \qquad (5.3.15)$$

where

$$X_{ji} = x_j^{(i)}(t)\,, \qquad (5.3.16)$$

$$\Lambda_{ji} = \lambda_j^{(i)}(t)\,; \qquad i,j = 1,\ldots,n\,; \qquad (5.3.17)$$

$$\mu^T = [\nu_1,\ldots,\nu_q, x_{q+1}(t_f),\ldots,x_n(t_f)]\,. \qquad (5.3.18)$$

Clearly, at $t = t_0$, if $X(t_0)$ is nonsingular, we can solve (5.3.14) for μ in terms of $x(t_0)$:

$$\mu = [X(t_0)]^{-1}x(t_0)\,. \qquad (5.3.19)$$

This yields $\lambda(t)$ and, hence, $u(t)$ by substituting (5.3.19) into (5.3.15) and the resulting expression into (5.3.7):

$$u(t) = -C(t,t_0)x(t_0)\,, \qquad (5.3.20)$$

$$C(t,t_0) = B^{-1}G^T\,\Lambda(t)[X(t_0)]^{-1}\,. \qquad (5.3.21)$$

Equations (5.3.20) and (5.3.21) appear exactly like (5.2.16) and (5.2.17) of Section 5.2, but the boundary conditions on the unit solutions $\Lambda(t)$ and $X(t)$ are *different.*

If we go to the limit of continuous data so that $t_0 \to t$, we run into some practical difficulties, since $X(t_f)$ is singular, meaning that the feedback gains $C(t) \to \infty$ as $t \to t_f$. This is understandable since

†Note that these are *not* the same unit solutions $X(t)$ and $\Lambda(t)$ as in Section 5.2.

we are insisting that we have zero terminal error, Equation (5.3.1). As we shall see in later chapters, $C(t) \to \infty$ as $t \to t_f$ is *not* acceptable when there is noise (uncertainty) in the data or in the process, since, in general, it would require infinite values of the control $u(t)$ as $t \to t_f$.

Solution by the sweep method. The two-point boundary value problem (5.3.8) through (5.3.11) may also be solved by the sweep method, which, as pointed out in Section 5.2, usually has advantages for numerical solution over the transition matrix method. However, we must extend the sweep method, as presented in Section 5.2, to handle the linear terminal constraints (5.3.11). To do this, it is convenient to replace the n boundary conditions (5.3.11) by the n boundary conditions

$$\lambda_i(t_f) = \begin{cases} \nu_i ; & i = 1, \ldots, q , \\ 0 ; & i = q+1, \ldots, n , \end{cases} \tag{5.3.22}$$

and then to postulate the specified boundary values $[x_1 \ldots x_q]_{t=t_f}$ as linear functions of $x(t_o)$ and $(\nu_1, \ldots, \nu_q)$, as follows:

$$\boxed{\psi = U(t_o)x(t_o) + Q(t_o)\nu ,} \tag{5.3.23}$$

where

$$\psi^T = (x_1, \ldots, x_q)_{t=t_f} , \tag{5.3.24}$$

$$\nu^T = (\lambda_1, \ldots, \lambda_q)_{t=t_f} . \tag{5.3.25}$$

From the linearity of (5.3.8) through (5.3.11), it is clear that $\lambda(t_o)$ is a linear function of $x(t_o)$ and ψ or, equivalently, of $x(t_o)$ and ν:

$$\boxed{\lambda(t_o) = S(t_o)\,x(t_o) + R(t_o)\,\nu .} \tag{5.3.26}$$

Now any time, $t \leq t_f$ is a possible "initial time," so we may write (5.3.23) and (5.3.26) as†

$$\lambda(t) = S(t)\,x(t) + R(t)\,\nu , \tag{5.3.27}$$

$$\psi = U(t)\,x(t) + Q(t)\,\nu . \tag{5.3.28}$$

Since these relations must be valid at $t = t_f$, it is clear that we must have

†This method is not always foolproof, since S,R may not exist while the solution to the problem exists. See Section 6.3 and Example 2.

$$S(t_f) = 0\,, \tag{5.3.29}$$

$$U_{ji}(t_f) = R_{ij}(t_f) = \left(\frac{\partial \psi_j}{\partial x_i}\right)_{t=t_f} = \begin{cases} 1\,, & i = j\,, \quad i = 1, \ldots, n\,, \\ 0\,, & i \neq j\,, \quad j = 1, \ldots, q\,; \end{cases} \tag{5.3.30}$$

$$Q(t_f) = 0\,. \tag{5.3.31}$$

Now, as in Section 5.2, we substitute (5.3.27) into (5.3.9), treating ν as a *constant* vector:

$$\dot{S}x + S\dot{x} + \dot{R}\nu = -Ax - F^T(Sx + R\nu)\,. \tag{5.3.32}$$

Substituting $\dot{x}$ from (5.3.8) into (5.3.32) and again using (5.3.27) to eliminate λ yields

$$\dot{S}x + S(Fx - GB^{-1}G^T(Sx + R\nu)) + \dot{R}\nu = -(A + F^TS)x - F^TR\nu\,. \tag{5.3.33}$$

However, (5.3.33) must be true for any x and ν, so the coefficients of x and ν must vanish:

$$\boxed{\dot{S} + SF + F^TS + A - SGB^{-1}G^TS = 0\,; \qquad S(t_f) = 0\,,} \tag{5.3.34}$$

$$\boxed{\dot{R} + (F^T - SGB^{-1}G^T)\,R = 0\,; \qquad R^T(t_f) = \left(\frac{\partial \psi}{\partial x}\right)_{t=t_f}} \tag{5.3.35}$$

Next, differentiate (5.3.28) with respect to time, treating ψ and ν as constant vectors:

$$0 = \dot{U}x + U\dot{x} + \dot{Q}\nu\,. \tag{5.3.36}$$

Substitute $\dot{x}$ from (5.3.8) into (5.3.36) and use (5.3.27) to eliminate λ:

$$0 = \dot{U}x + U(Fx - GB^{-1}G^T(Sx + R\nu)) + \dot{Q}\nu\,. \tag{5.3.37}$$

However, (5.3.37) must be true for any x and ν, so the coefficients of x and ν must vanish:

$$\dot{U} + U(F - GB^{-1}G^TS) = 0\,, \tag{5.3.38}$$

$$\dot{Q} - UGB^{-1}G^TR = 0\,. \tag{5.3.39}$$

Examination of (5.3.35) and (5.3.38) and the boundary conditions (5.3.30) shows that

$$\boxed{U(t) \equiv R^T(t)\,.} \tag{5.3.40}$$

Hence, (5.3.39) may be written as

$$\boxed{\dot{Q} = R^T G B^{-1} G^T R, \qquad Q(t_f) = 0.} \tag{5.3.41}$$

Equation (5.3.34) is the same matrix Riccati equation obtained in Section 5.2, but the boundary condition is *different*. Equation (5.3.35) is a *linear* matrix differential equation with coefficients depending on S, and (5.3.41) is simply a quadrature. All three matrix equations (5.3.34), (5.3.35), and (5.3.41) may be integrated *backwards* to yield $S(t)$, $R(t)$, and $Q(t)$. Note that $Q \leq 0$ since $\dot{Q} \geq 0$ and $Q(t_f) = 0$.

At some particular initial time $t = t_o$, if $Q(t_o)$ is nonsingular, (5.3.28) may be solved for ν:

$$\boxed{\nu = [Q(t_o)]^{-1} [\psi - R^T(t_o)\, x(t_o)] .} \tag{5.3.42}$$

If $Q(t_o)$ is singular, the optimization problem (5.3.1) through (5.3.4) is said to be *abnormal*; no neighboring minimum solutions exist in this case.

If the problem is not abnormal, values of ν from (5.3.42) may be substituted into (5.3.27) to find $\lambda(t_o)$:

$$\lambda(t_o) = (S - RQ^{-1} R^T)_{t=t_o} x(t_o) + (RQ^{-1})_{t=t_o} \psi . \tag{5.3.43}$$

Having $\lambda(t_o)$ from (5.3.43), we could, if we wished, integrate (5.3.8) and (5.3.9) forward as an initial-value problem. Alternatively, if we stored $S(t)$ and $R(t)$ when solving (5.3.34) and (5.3.35), we need only integrate (5.3.8) forward, using (5.3.27) to evaluate $\lambda(t)$, which, in turn, requires ν from (5.3.42).

However, we are usually more interested in *optimal feedback laws* as in (5.3.20). Thus, we could evaluate ν, as in (5.3.42), at several intermediate times to produce a sampled-data feedback law, or we could evaluate ν continuously to produce a continuous feedback law. From (5.2.5′) and (5.3.43), with $t_o = t$, we have

$$\boxed{u(t) = -C(t)\,x(t) - D(t)\,\psi ,} \tag{5.3.44}$$

where

$$\boxed{\begin{aligned} C &= B^{-1}G^T(S - RQ^{-1}R^T) , && (5.3.45)\\ D &= B^{-1}G^T R Q^{-1} . && (5.3.46)\end{aligned}}$$

In contrast to (5.3.20), this feedback law shows *explicit dependence on the specified terminal values of* $[x_1(t_f), \ldots, x_q(t_f)] = \psi^T$. If $\psi = 0$, it is the same as the continuous version of (5.3.20), i.e., where t_o is replaced by t. Referring to (5.3.22), $\nu = 0$ corresponds to the case of *no* terminal constraints. From (5.3.23) and (5.3.40), the resulting value of ψ for $\nu = 0$ is

$$\hat{\psi} = R^T(t_o)x(t_o)$$

that is, $R^T(t_o)x(t_o)$ is the predicted value of ψ if J is minimized with no terminal constraints. Using this interpretation, (5.3.44) can be written as

$$u(t) = -B^{-1}G^T Sx(t) - B^{-1}G^T RQ^{-1}(\psi - \hat{\psi}) . \qquad (5.3.44a)$$

Minimum integral square control. A special case of interest is when $A = 0$ in the performance index (5.3.2). We are then simply minimizing the integral of a quadratic form in the control:

$$J = \frac{1}{2}\int_{t_0}^{t_f} (u^T Bu)\, dt , \qquad (5.3.47)$$

still subject to the constraints (5.3.1), (5.3.3), and (5.3.4). The Riccati equation (5.3.34), with $A = 0$, has the simple solution

$$\boxed{S(t) = 0 .} \qquad (5.3.48)$$

This implies that $u = 0$ is the solution to the problem with no terminal constraints (that is, $J = 0$).

Equations (5.3.35) and (5.3.41) become, simply,

$$\dot{R} + F^T R = 0 , \qquad R^T(t_f) = \left(\frac{\partial \psi}{\partial x}\right)_{t=t_f} , \qquad (5.3.49)$$

$$Q(t) = -\int_t^{t_f} (R^T GB^{-1}G^T R)\, dt . \qquad (5.3.50)$$

The continuous feedback control can then be written as

$$u(t) = -D(t)[\psi - R^T(t)\, x(t)] , \qquad (5.3.51)$$

where

$$D = B^{-1}G^T RQ^{-1} . \qquad (5.3.52)$$

Note that the *predicted value of* ψ, *if no control is used at all from* t *to* t_f, is given by

$$\hat{\psi} = R^T(t)x(t) , \qquad (5.3.53)$$

so that (5.3.51) may also be written as

$$\boxed{u(t) = -D(t)[\psi - \hat{\psi}(t)]\,.} \tag{5.3.54}$$

Controllability (see also Appendix B). It is straightforward to show that the minimum value of the performance index (5.3.47) is

$$J_{\min} = \tfrac{1}{2}\,(\hat{\psi} - \psi)^T\,(-Q^{-1})\,(\hat{\psi} - \psi)\big|_{t=t_0}\,, \tag{5.3.55}$$

where $\hat{\psi}$ is given by (5.3.53).

Clearly, if $Q(t_0)$ is singular, $J_{\min} = \infty$, and, from (5.3.51), $u(t_0) = \infty$. *The system is said to be uncontrollable in the reduced sense if $Q(t)$ is singular for any t between t_0 and t_f.*

A system that is controllable for $q = n$ is called *completely controllable* since all components of the terminal state may then be controlled.

For a *stationary system* (F, G constant matrices), the criterion for complete controllability may be expressed as

$$\text{Rank}(G, FG, F^2G, \ldots, F^{n-1}G) = n\,. \tag{5.3.56}$$

If the matrix F has distinct eigenvalues (real or complex) and the linear transformation $y = Tx$ is such that $D = TFT^{-1}$ is a diagonal matrix, the system equations become

$$\dot{y} = Dy + TGu\,. \tag{5.3.57}$$

Clearly, if TG has any zero rows, the corresponding component of y is not affected by any of the control components and the system is *not* completely controllable.

Another way a system may become uncontrollable is for the D-matrix in (5.3.57) to have identical diagonal terms. Consider

$$\dot{y}_1 = \lambda y_1 + u\,, \qquad \dot{y}_2 = \lambda y_2 + 2u\,;$$

the linear combination $2y_1 - y_2$ cannot be affected by the control u.

These conclusions about controllability are *not* affected by the choice of B so long as it is positive definite; one could choose B to be the identity matrix. Thus...

Example. Two pendulums, coupled by a spring, are to be controlled by two equal and opposite forces u which are applied to the pendulum bobs as shown. The equations of motion are:

$$m\ell^2\,\ddot{\theta}_1 = -ka^2(\theta_1 - \theta_2) - mg\ell\,\theta_1 - u\,,$$
$$m\ell^2\,\ddot{\theta}_2 = -ka^2(\theta_2 - \theta_1) - mg\ell\,\theta_2 + u\,.$$

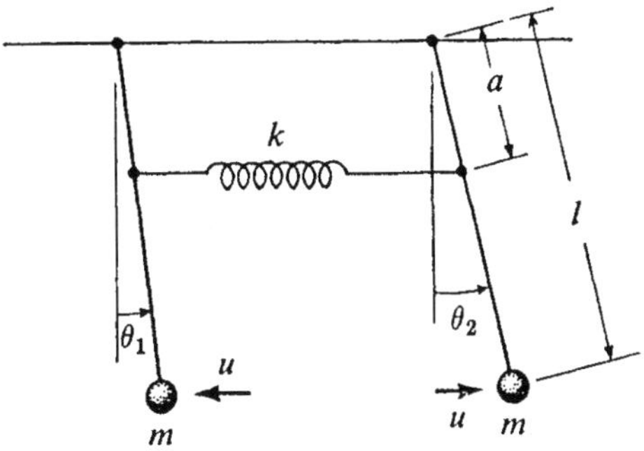

Figure 5.3.1. Example of uncontrollable system.

The diagonalization of this system can be done by inspection; simply add and subtract the two equations

$$\ddot{y}_1 = -\frac{g}{\ell} y_1 , \qquad m\ell^2 \ddot{y}_2 = -(2ka^2 + mg\ell)\, y_2 - 2u ,$$

where

$$y_1 = \theta_1 + \theta_2 , \qquad y_2 = \theta_1 - \theta_2 .$$

Clearly, there is no control over the symmetric mode of vibration $y_1 \neq 0$, so the system is *not* completely controllable.

Note that, if one of the forces were eliminated (or made unequal to the other force), the system would be completely controllable.

Problem 1. *More general performance index and terminal constraint.* Find $u(t)$ to minimize

$$J = \frac{1}{2} \int_{t_0}^{t_f} (x^T, u^T) \begin{bmatrix} A(t) & , N(t) \\ N^T(t) & , B(t) \end{bmatrix} \begin{bmatrix} x \\ u \end{bmatrix} dt ,$$

with the constraints

$\dot{x} = F(t)x + G(t)u , \qquad x(t_0)$ given;

$\psi = M\, x(t_f) , \qquad$ where $\quad \psi$, a q-vector, and M, a $(q \times n)$-matrix, are given, $\quad q \leq n$.

ANSWER.

$$u = -B^{-1}[N^T + G^T(S - RQ^{-1}R^T)]x - B^{-1}G^T RQ^{-1} \psi$$

or

$$u = -B^{-1}(N^T + G^T S)x - B^{-1} G^T RQ^{-1} (\psi - \hat{\psi}) ,$$

where

$$\dot{S} = -SF - F^T S - A + (SG + N) B^{-1} (N^T + G^T S), \qquad S(t_f) = 0;$$
$$\dot{R} = -(F^T - (SG + N) B^{-1} G^T) R, \qquad R(t_f) = M^T;$$
$$\dot{Q} = R^T G B^{-1} G^T R, \qquad Q(t_f) = 0;$$

and

$$J_{\min} = \tfrac{1}{2}[x^T S x + (\hat{\psi} - \psi)^T (-Q^{-1})(\hat{\psi} - \psi)]_{t=t_0},$$

where $\hat{\psi} = R^T(t_0) x(t_0)$ is the predicted value of ψ if J is minimized with no terminal constraints.

Problem 2. Show that the Riccati equation (5.2.24) with $A = 0$, may be written as a *linear* differential equation for S^{-1}:

$$\frac{d}{dt}(S^{-1}) = FS^{-1} + S^{-1}F^T - GB^{-1}G^T.$$

Even for $S^{-1}(t_f) = 0$ (that is, $S_f \to \infty$), this matrix equation may be integrated backwards. Show that the solution may be expressed in terms of the $R(t)$-matrix (when we have $q = n$) defined by Equations (5.3.49) as

$$[S(t)]^{-1} = [R^T(t)]^{-1} [S_f^{-1} + \int_t^{t_f} R^T G B^{-1} G^T R \, dt] [R(t)]^{-1}$$

(see Kalman, Ho, and Narendra, 1963; Appendix B2).

Problem 3. Let F and G be constant matrices. Show that the condition $S^{-1}(t) > 0$ is equivalent to

$$\text{Rank}(G, FG, F^2G, \ldots, F^{n-1}G) = n.$$

[HINT: Differentiate $\Phi(t,\tau)$ and apply the Cayley-Hamilton theorem.]

Problem 4. Show that the predicted terminal state, defined in Equation (5.3.53), is changed by the use of control according to

$$\dot{\hat{\psi}} = R^T G u,$$

where R is given by Equation (5.3.49) and $\hat{\psi}(t_0) = R^T(t_0) x(t_0)$.

Problem 5. The simple second-order system

$$\ddot{x} + x = u$$

has a control $u(t)$, where both x and u are scalar quantities. Find $u(t)$ to bring the system from $x(0) = x_0$, $\dot{x}(0) = v_0$ to $x(t_f) = \dot{x}(t_f) = 0$ with a minimum value of

$$J = \frac{1}{2}\int_0^{t_f} u^2\, dt \,.$$

Useful fact: A particular solution of $\ddot{x} + x = A \sin t + B \cos t$ is

$$x_p = \tfrac{1}{2} Bt \sin t - \tfrac{1}{2} At \cos t \,.$$

ANSWER.

$$u = -\frac{2}{t_f^2 - \sin^2 t_f}\,[\sin (t_f - t) \sin t_f - t_f \sin t$$

$$- \cos (t_f - t) \sin t_f + t_f \cos t] \begin{bmatrix} x_0 \\ v_0 \end{bmatrix} .$$

5.4 Regulators and stability

A *regulator* is a feedback controller designed to keep a stationary system within an acceptable deviation from a reference condition using acceptable amounts of control. The disturbances to the system are often unpredictable, i.e., random. In later chapters we will treat the statistical behavior of regulated systems in the presence of random forcing functions. In this section we will consider only deterministic initial disturbances, that is, $x(t_0) \neq 0$, and no forcing functions.†

For a stationary system (F, G constant matrices), let the matrices A and B in the performance index of (5.2.3) be constant, and consider the case $t_f - t_0 \to \infty$. Examining the Riccati equation for the matrix $S(t)$ in (5.2.24), it is *possible* that a steady, finite solution exists as

$$\dot{S} = 0 = -SF - F^T S + SGB^{-1} G^T S - A \Rightarrow S(t) \to S^o \text{ as } t_f - t \to \infty \,. \qquad (5.4.1)$$

If so, the feedback gain matrix of (5.2.27) is constant:

$$C \to C^o = B^{-1} G^T S^o \,. \qquad (5.4.2)$$

Furthermore, from (5.2.31), the optimal return function is given by

$$J^o{}_{\text{min}} = \tfrac{1}{2} x^T(t_0)\, S^o\, x(t_0) \qquad \text{as} \qquad t_f - t_0 \to \infty \,, \qquad (5.4.3)$$

which is independent of time. Thus if a steady-state *finite* solution S^o, of (5.4.1) exists *and is positive definite*, clearly $x(t)$ and $u(t)$ are *bounded* (never become infinite), and

$$u(t) = -C^o x(t) \qquad (5.4.4)$$

is an *asymptotically stable regulator* (see Problem 1).

In general, solution of the quadratic Equations (5.4.1), will produce more than one value for the steady state S. The extraneous roots can

†Since t_0 is arbitrary, the method is, of course, applicable to any disturbance in state that occurs during the control interval by whatever agent it is produced.

usually be eliminated by the requirement that $S > 0$. However, another approach is to integrate the differential Equations (5.4.1) backwards with boundary conditions $S_f = 0$ until $\dot{S} \approx 0$. In this case Kalman (1961) has shown that $S(t) \to S^o$. This is a valuable technique for *synthesizing regulators.*

Example 1. *Regulator for a first-order system. Given* the first-order system

$$\dot{x} = -\frac{1}{\tau}x + u\,,$$

synthesize a regulator that will keep x near zero.

SOLUTION. If we aim to keep x^2 below x_m^2 = const, using u^2 below u_m^2 = constant, we might try the performance index

$$J = \frac{1}{2}\int_{t_0}^{t_f}(ax^2 + bu^2)\,dt\,, \qquad \text{where} \qquad a = \frac{1}{x_m^2}\,, \quad b = \frac{1}{u_m^2}\,.$$

The corresponding Riccati equation is

$$\dot{S} = +\frac{2}{\tau}S - a + \frac{S^2}{b}\,.$$

In steady state, we have

$$S^2 + \frac{2b}{\tau}S - ab = 0 \Rightarrow S = -(b/\tau) \pm \sqrt{(b^2/\tau^2) + ab}\,.$$

Since S must be positive, only the "+" is acceptable. So we have

$$\boxed{S^o = \sqrt{\frac{b^2}{\tau^2} + ab} - \frac{b}{\tau} \Rightarrow u(t) = -\left[\sqrt{\frac{1}{\tau^2} + \frac{a}{b}} - \frac{1}{\tau}\right]x(t)\,.}$$

Example 2. *Regulator for a third-order system (a roll attitude regulator for a missile).* We wish to design a feedback controller for a missile, using hydraulic-powered ailerons that will keep the roll attitude ϕ close to zero, while staying within the physical limits of aileron deflection δ and aileron deflection rate $\dot{\delta}$. A sketch of the system (Figure 5.4.1) and the equations of motion are given below:

$$\dot{\delta} = u\,, \qquad \dot{\omega} = -\frac{1}{\tau}\omega + \frac{Q}{\tau}\delta + (\text{noise}), \qquad \dot{\phi} = \omega\,,$$

where

τ = roll-time constant, $\quad Q$ = aileron effectiveness,

u = command signal to aileron actuators,

ω = roll angular velocity.

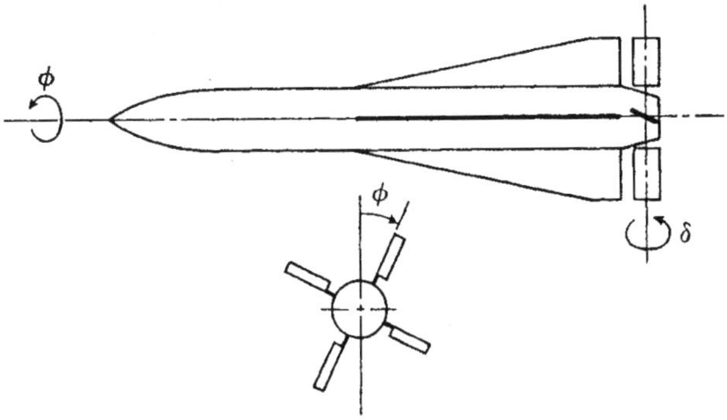

Figure 5.4.1. Nomenclature for controlling roll altitude of a missile.

Using quadratic synthesis, we shall minimize the performance index

$$J = \lim_{t_f - t_o \to \infty} \frac{1}{2} \int_{t_o}^{t_f} \left(\frac{\phi^2}{\phi_o^2} + \frac{\delta^2}{\delta_o^2} + \frac{u^2}{u_o^2} \right) dt \,,$$

where

ϕ_o = maximum desired value of ϕ,

δ_o = maximum available value of δ,

u_o = maximum available value of u.

The steady-state matrix Riccati equation is

$$0 = -SF - F^T S + SGB^{-1}G^T S - A \,,$$

where

$$F = \begin{bmatrix} 0 & 0 & 0 \\ \dfrac{Q}{\tau} & -\dfrac{1}{\tau} & 0 \\ 0 & 1 & 0 \end{bmatrix}, \qquad G = \begin{bmatrix} 1 \\ 0 \\ 0 \end{bmatrix}, \qquad S = \begin{bmatrix} S_{11} & S_{12} & S_{13} \\ S_{12} & S_{22} & S_{23} \\ S_{13} & S_{23} & S_{33} \end{bmatrix},$$

$$B = \frac{1}{u_o^2}, \qquad A = \begin{bmatrix} \dfrac{1}{\delta_o^2} & 0 & 0 \\ 0 & 0 & 0 \\ 0 & 0 & \dfrac{1}{\phi_o^2} \end{bmatrix},$$

and the control law is given by

$$u = -B^{-1}G^T S x \equiv -u_o^2[S_{11}, S_{12}, S_{13}] \begin{bmatrix} \delta \\ \omega \\ \phi \end{bmatrix}.$$

Substitution into the Riccati equation yields the six equations

$$0 = -2S_{12}\frac{Q}{\tau} - \frac{1}{\delta_o^2} + u_o^2 S_{11}^2, \qquad 0 = \frac{S_{12}}{\tau} - S_{13} - \frac{Q}{\tau}S_{22} + u_o^2 S_{11} S_{12},$$

$$0 = \frac{2S_{22}}{\tau} - 2S_{23} + u_o^2 S_{12}^2, \qquad 0 = -S_{23}\frac{Q}{\tau} + u_o^2 S_{11} S_{13},$$

$$0 = \frac{S_{23}}{\tau} - S_{33} + u_o^2 S_{12} S_{13}, \qquad 0 = -\frac{1}{\phi_o^2} + u_o^2 S_{13}^2.$$

By straightforward manipulation, these equations may be reduced to solving a single *quartic* equation for $\sigma \equiv u_o S_{11}$:

$$\sigma^4 + \frac{4}{u_o\tau}\sigma^3 + \frac{4}{u_o^2\tau^2}\left(1 - \frac{u_o^2\tau^2}{2\delta_o^2}\right)\sigma^2 - \frac{4}{u_o\tau}\left(\frac{2Q}{\phi_o u_o} + \frac{1}{\delta_o^2}\right)\sigma + \frac{1}{\delta_o^4} - \frac{4}{\delta_o^2 u_o^2 \tau^2} - \frac{8Q}{\phi_o u_o^3 \tau^2} = 0,$$

where

$$S_{11} = \frac{\sigma}{u_o}, \qquad S_{12} = \frac{\tau}{2Q}\left(\sigma^2 - \frac{1}{\delta_o^2}\right), \qquad S_{13} = \frac{1}{u_o\phi_o}, \qquad S_{23} = \frac{\tau\sigma}{Q\phi_o},$$

$$S_{22} = \frac{\tau^2}{2}(S_{23} - u_o^2 S_{12}^2), \qquad S_{33} = \frac{\sigma}{Q\phi_o} + \frac{u_o}{\phi_o}S_{12}.$$

We are, of course, interested only in positive real values of σ.

Numerical case. Suppose that $\tau = 1$ sec, $Q = 10$ sec^{-1}, $u_o = \pi$ rad sec^{-1}, $\delta_o = \pi/12$ rad, $\phi_o = \pi/180$ rad. The quartic equation then becomes

$$\sigma^4 + 1.272\sigma^3 - 28.3\sigma^2 - 482\sigma + 59.4 = 0,$$

which has only two positive real roots, $\underline{\sigma = 8.55 \text{ and } \sigma = .12}$. Using the smaller root makes $S_{33} < 0$, so the larger root is the one we want. This gives

$$C_1 = u_o\sigma = \underline{26.9}\ \text{sec}^{-1}, \qquad C_2 = \frac{u_o^2\tau}{2Q}\left(\sigma^2 - \frac{1}{\delta_o^2}\right) = \underline{28.9},$$

$$C_3 = \frac{u_o}{\phi_o} = \underline{180}\ \text{sec}^{-1}.$$

Problem 1. Consider the general problem of this section with the additional assumption that (F,G) represents a controllable system. Show that in this case with $A > 0$, it is possible to establish an upper bound to the integral

$$J = \min_{\substack{u(t),\\ t_o \le t \le t_f}} \int_{t_0}^{t_f} \left(\tfrac{1}{2}\|x\|_A^2 + \tfrac{1}{2}\|u\|_B^2\right) dt$$

and that the limit of J as $(t_f - t_0) \to \infty$ converges. Furthermore, show that the optimal cost $J = \frac{1}{2}\|x(t_0)\|_{S_0}^2$ is a Lyapunov function (see Appendix B4), which, in turn, implies that the controlled system $\dot{x} = (F - GB^{-1}G^T S_0)x$ is asymptotically stable.

It is worthwhile to point out that optimality does not necessarily imply stability. Certain criterion functions may, in fact, force the control to destabilize the system. Thus, an important problem in the *qualitative* theory of optimal control is to investigate the relationship between the criterion and the properties of the controlled system. The development of this section can be further extended to yield results along this line (Kalman, 1960, 1961, and 1964).

Problem 2. Take a controllable, stationary, and stable linear system

$$\dot{x} = Fx + Gu$$

with the constraint $\|u\|^2 \le 1$. Let $\frac{1}{2}\|x\|_S^2$ be a Lyapunov function for the above equation with $u = 0$. Now choose the control law to minimize the derivative of the Lyapunov function. Show that this control law is also optimal for *some* criterion of the the type in Problem 1.

Problem 3. *Longitudinal autopilot to maintain small vertical acceleration.* The longitudinal perturbations of an airplane in horizontal cruising flight are reasonably well described by the second-order system

$$\dot{\alpha} = -\frac{1}{T}\alpha + q\,, \qquad \dot{q} = -\omega_0^2(\alpha - Q\delta)\,,$$

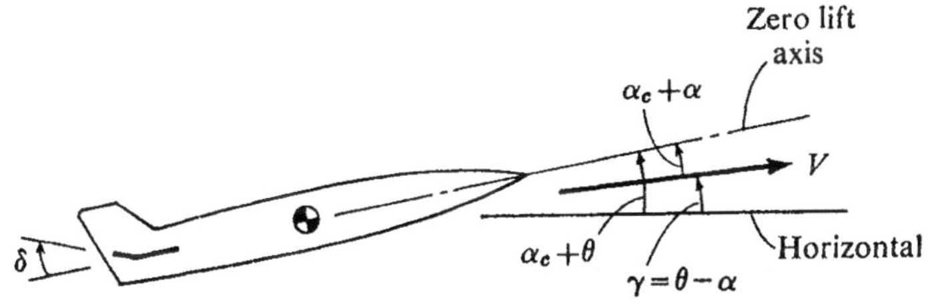

Figure 5.4.2. Nomenclature for controlling longitudinal motions of an aircraft.

where

α = perturbation from cruise angle of attack, α_c,

$q = \dot{\theta}$, and θ is the perturbation from cruise pitch angle of the zero-lift axis, α_c,

$\omega_o = \left(-\dfrac{M_\alpha}{I}\right)^{1/2}$ = undamped pitch natural frequency,

$Q = \dfrac{-M_\delta}{M_\alpha}$ = elevator effectiveness,

$\tau = \dfrac{mV}{L_\alpha}$ = lifting time constant, δ = elevator deflection.

(a) Determine the steady-state regulator gains C_1 and C_2, where

$$\delta = -C_1\alpha - C_2 q,$$

to minimize

$$J = \lim_{t_f - t_o \to \infty} \int_{t_o}^{t_f} \left[\frac{\delta^2}{\delta_o^2} + \frac{\alpha^2}{\alpha_o^2} + \frac{q^2}{q_o^2}\right] dt.$$

(b) For the controlled system, *plot* contours of constant damping ratio ζ on a graph of $Q\delta_o/\alpha_o$.

Note that such an autopilot will *not* maintain horizontal flight (i.e., zero vertical velocity or $\gamma = 0$).

Problem 4. *Longitudinal autopilot to maintain small vertical velocity.* With reference to Problem 3, add the equation

$$\dot{\theta} = q$$

to the perturbation equations to produce a *third-order system.*

Set up the equations to determine the steady-state regulator gains C_1, C_2, and C_3, where

$$\delta = -C_1\alpha - C_2 q - C_3\theta,$$

to minimize

$$J = \lim_{t_f - t_o \to \infty} \int_{t_o}^{t_f} \left[\frac{\delta^2}{\delta_o^2} + \frac{(\theta - \alpha)^2}{\gamma_o^2}\right] dt.$$

Note that such an autopilot will maintain nearly horizontal flight but will *not* maintain constant altitude.

Problem 5. *Longitudinal autopilot to maintain constant altitude.* With reference to Problem 4, add the equation

$$\dot{h} = V(\theta - \alpha)$$

to the perturbation equations to produce a *fourth-order system*, where h = perturbation from desired altitude.

Set up the equations to determine the steady-state regulator gains C_1, C_2, C_3, and C_4, where

$$\delta = -C_1\alpha - C_2 q - C_3\theta - C_4 h ,$$

to minimize

$$J = \lim_{t_f - t_o \to \infty} \int_{t_o}^{t_f} \left(\frac{\delta^2}{\delta_o^2} + \frac{h^2}{h_o^2} \right) dt .$$

Problem 6. *Lateral autopilot to maintain heading and roll attitude.* The roll, yaw, sideslip motions of an airplane (the "lateral" motions) are coupled together but are almost negligibly coupled to the pitch, plunge motions (the "longitudinal" motions). The perturbation equations of lateral motion (a fifth-order system) are

$$\dot{\beta} + r = \frac{Y_\beta}{mV}\beta + \frac{g}{V}\phi ,$$

$$\dot{r} + \frac{I_{xz}}{I_{zz}}\dot{p} = \frac{n_\beta}{I_{zz}}\beta + \frac{n_r}{I_{zz}} r + \frac{n_p}{I_{zz}} p + \frac{n_{\delta_r}}{I_{zz}} \delta_r ,$$

$$\dot{p} + \frac{I_{xz}}{I_{xx}}\dot{r} = \frac{\ell_p}{I_{xx}} p + \frac{\ell_\beta}{I_{xx}}\beta + \frac{\ell_r}{I_{xx}} r + \frac{\ell_{\delta_a}}{I_{xx}}\delta_a ,$$

$$\dot{\phi} = p ,$$

$$\dot{\psi} = r ,$$

where

β = sideslip angle, ψ = yaw angle, r = yaw angular velocity,
ϕ = roll angle, p = roll angular velocity,
δ_r = rudder deflection, δ_a = aileron deflection (see Figure 5.4.3).

Set up the equations to determine the ten steady-state regulator gains $C_{r_1}, \ldots, C_{r_5}$ and $C_{a_1}, \ldots, C_{a_5}$, where

$$\begin{bmatrix} \delta_r \\ \delta_a \end{bmatrix} = -\begin{bmatrix} C_{r_1}, C_{r_2}, C_{r_3}, C_{r_4}, C_{r_5} \\ C_{a_1}, C_{a_2}, C_{a_3}, C_{a_4}, C_{a_5} \end{bmatrix} \begin{bmatrix} \beta \\ r \\ p \\ \phi \\ \psi \end{bmatrix},$$

to minimize

$$J = \lim_{t_f - t_o \to \infty} \frac{1}{2} \int_{t_o}^{t_f} \left[\frac{\delta_a^2}{\delta_{a_o}^2} + \frac{\delta_r^2}{\delta_{r_o}^2} + \frac{(\beta + \psi)^2}{\epsilon_o^2} + \frac{\phi^2}{\phi_o^2} \right] dt .$$

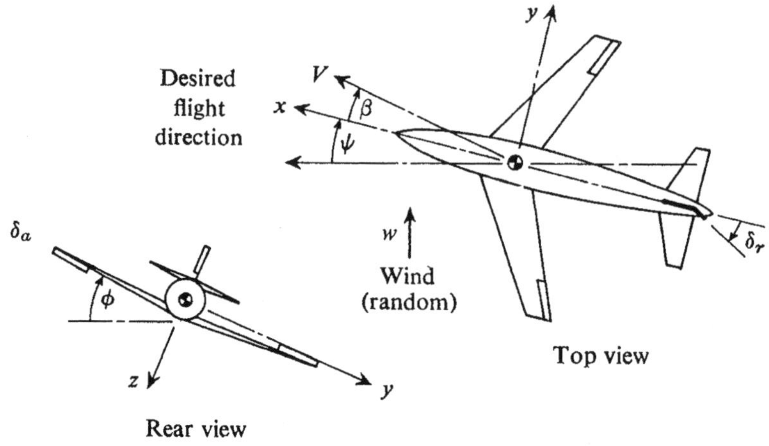

Figure 5.4.3. Nomenclature for controlling lateral motions of an aircraft.

For an airplane weighing 100,000 lb, flying at 30,000 ft altitude at 500 mph, typical values of the coefficients are

$$\frac{Y_\beta}{mV} = -.0297\ \text{sec}^{-1}, \quad \frac{n_{\delta_r}}{I_{zz}} = +.379\ \text{sec}^{-2}, \quad \frac{\ell_{\delta_a}}{I_{xx}} = +1.580\ \text{sec}^{-1},$$

$$\frac{\ell_\beta}{I_{xx}} = -1.17\ \text{sec}^{-2}, \quad \frac{\ell_p}{I_{xx}} = -.790\ \text{sec}^{-1}, \quad \frac{\ell_r}{I_{xx}} = .129\ \text{sec}^{-1},$$

$$\frac{n_\beta}{I_{zz}} = .379\ \text{sec}^{-2}, \quad \frac{n_p}{I_{zz}} = -.0125\ \text{sec}^{-1}, \quad \frac{n_r}{I_{zz}} = -.0096\ \text{sec}^{-1},$$

$$\frac{g}{V} = .0438\ \text{sec}^{-1}, \quad \frac{I_{xz}}{I_{zz}} = -.0423, \quad \frac{I_{xz}}{I_{xx}} = -.106.$$

Using these numerical values and

$$\delta_{r_0} = \delta_{a_0} = \epsilon_0 = \phi_0 = 1,$$

use a computer program (e.g., the RIAS Automatic Synthesis Program (ASP)) to find the feedback gains on a digital computer. Determine the eigenvalues and eigenvectors (mode shapes) for the controlled and the uncontrolled systems (see Chapter 14, Section 14.2, Problem 1).

ANSWER. Numerical computations by R. K. Mehra gave

$$\begin{bmatrix}\delta_r\\ \delta_a\end{bmatrix} = -\begin{bmatrix}.317, 1.01, .069, .076, .551\\ .177, .388, .737, 1.03, .834\end{bmatrix}\begin{bmatrix}\beta\\ r\\ p\\ \phi\\ \psi\end{bmatrix}$$

Problem 7. *Linear feedback to follow a desired output.* The output of a time-varying linear system,

$$\dot{x} = F(t)x + G(t)u\,,$$

is a linear combination of the components of the state

$$y = M(t)x\,,$$

where x is an n-component state vector, u is an m-component control vector, and y is a p-component output vector.

Using an integral quadratic penalty function, show how to determine $C(t)$ and $w(t)$ in a control law of the form

$$u = -C(t)x + w(t)$$

that will cause the system to follow, rather closely, a *desired* output $y(t)$ over the interval $t_o \le t \le t_f$, with reasonable amounts of control.

ANSWER. Choose

$$J = \frac{1}{2}\int_{t_o}^{t_f} [(y - Mx)^T A(y - Mx) + u^T Bu]\, dt\,.$$

Then we have

$$C = B^{-1}G^TS\,, \qquad w = -B^{-1}G^Tg\,,$$

where

$$\dot{S} = -SF - F^TS + SGB^{-1}G^TS - M^TAM\,, \qquad S(t_f) = 0\,;$$

$$\dot{g} = -(F^T - SGB^{-1}G^T)g + M^TAy\,, \qquad g(t_f) = 0\,.$$

Problem 8. *An inhomogeneous linear-quadratic optimization problem* (V. Garber). Find $u(t)$ to minimize

$$J = \frac{1}{2}(x^TS_fx)_{t=t_f} + \frac{1}{2}\int_{t_o}^{t_f} (x^TAx + u^TBu)\, dt\,,$$

subject to the constraints

$$\dot{x} = F(t)x + G(t)u + C(t)\,, \qquad x(t_o)\text{ given.}$$

ANSWER. We have

$$u(t) = -B^{-1}G^T(Sx + k)\,,$$

where

$$\dot{S} = -SF - F^TS + SGB^{-1}G^TS - A\,; \qquad S(t_f) = S_f\,;$$
$$\dot{k} = (SGB^{-1}G^T - F^T)k - SC\,, \qquad k(t_f) = 0\,.$$

Problem 9. *Solution of a general linear two-point boundary-value problem* (G. B. Rybicki and P. D. Usher). Consider the linear two-point boundary-value problem

$$\dot{x} = F(t)x + w(t)\,, \qquad Ax(t_0) = a\,, \qquad Bx(t_f) = b\,,$$

where x is an n-vector, a is an $(n - k)$-vector, b is a k-vector, and $F(t)$, $w(t)$, A, B, a, and b are given.

(a) Show that a forward sweep solution may be obtained by letting

$$Ax(t) = S(t)\,Bx(t) + m(t)\,,$$

where

$$\dot{S} = C_1S - SC_4 - SC_3S + C_2\,, \qquad S(t_0) = 0\,;$$
$$\dot{m} = (C_1 - SC_3)m + (A - SB)w\,, \qquad m(t_0) = a\,;$$

and

$$\begin{bmatrix} C_1 & C_2 \\ C_3 & C_4 \end{bmatrix} = \begin{bmatrix} A \\ B \end{bmatrix} F(t) \begin{bmatrix} A \\ B \end{bmatrix}^{-1}$$

Note that S is an $(n - k \times k)$-matrix, while m is an $(n - k)$-vector.

Integrating the differential equations for S and m *forward* generates a family of solutions that satisfies the initial conditions. At $t = t_f$, we have

$$Ax(t_f) = S(t_f)b + m(t_f)\,, \qquad Bx(t_f) = b\,,$$

which are n equations for $x(t_f)$. Having $x(t_f)$, the original equations may be integrated *backwards* to obtain the desired solution.

(b) Obviously a similar procedure could be used for a backward sweep solution by letting $Bx(t) = QAx(t) + n(t)$.

Neighboring extremals and the second variation

6

6.1 Neighboring extremal paths (final time specified)

Let us suppose that we have determined a control vector function $u(t)$ that meets all the first-order necessary conditions for the Bolza problem of Section 2.5. That is, we have

$$\dot{x} = f(x,u,t)\,, \tag{6.1.1}$$

$$\dot{\lambda}^T = -\frac{\partial H}{\partial x}\,, \tag{6.1.2}$$

$$0 = \frac{\partial H}{\partial u}\,, \tag{6.1.3}$$

$$x(t_o) \text{ specified}, \quad t_f \text{ specified}, \tag{6.1.4}$$

$$\lambda^T(t_f) = \left(\frac{\partial \phi}{\partial x} + \nu^T \frac{\partial \psi}{\partial x}\right)_{t=t_f}\,, \tag{6.1.5}$$

$$\psi[x(t_f)] = 0\,, \tag{6.1.6}$$

where the performance index is

$$J = \phi[x(t_f)] + \int_{t_o}^{t_f} L[x(t),u(t),t]\,dt\,. \tag{6.1.7}$$

The augmented performance index is

$$\bar{J} = \phi[x(t_f)] + \nu^T\psi[x(t_f)] + \int_{t_o}^{t_f} [H(x,u,\lambda,t) - \lambda^T\dot{x}]\,dt\,, \tag{6.1.8}$$

where

$$H = L + \lambda^T f\,. \tag{6.1.9}$$

Next, let us consider small perturbations from this extremal path produced by small perturbations in the initial state $\delta x(t_o)$ and in the terminal conditions $\delta\psi$. We expect that such perturbations will give rise to perturbations $\delta x(t)$, $\delta\lambda(t)$, $\delta u(t)$, $d\nu$ governed by linearizing (6.1.1) through (6.1.6) around the extremal path:

$$\delta\dot{x} = f_x \delta x + f_u \delta u\,, \tag{6.1.10}$$

$$\delta\dot{\lambda} = -H_{xx}\delta x - f_x^T \delta\lambda - H_{xu}\,\delta u\,; \qquad \left(H_{xu} \triangleq \frac{\partial}{\partial u}(H_x)^T\right), \tag{6.1.11}$$

$$0 = H_{ux}\,\delta x + f_u^T \delta\lambda + H_{uu}\,\delta u\,, \tag{6.1.12}$$

$$\delta x(t_0) \text{ specified,} \tag{6.1.13}$$

$$\delta\lambda(t_f) = [(\phi_{xx} + (\nu^T\psi_x)_x)\,\delta x + \psi_x^T d\nu]_{t=t_f}\,, \tag{6.1.14}$$

$$\delta\psi = [\psi_x\,\delta x]_{t=t_f}\,. \tag{6.1.15}$$

Alternatively, we may consider an expansion of the performance criterion and constraints to second order (since all first-order terms vanish about a trajectory that satisfies (6.1.1) through (6.1.9)). As pointed out in Section 1.3, this may be accomplished by expanding the augmented criterion to *second* order and all the constraints to *first* order. Thus we have

$$\delta^2\bar{J} = \frac{1}{2}[\delta x^T(\phi_{xx} + (\nu^T\psi_x)_x)\,\delta x]_{t=t_f}$$

$$+ \frac{1}{2}\int_{t_0}^{t_f} [\delta x^T\ \delta u^T] \begin{bmatrix} H_{xx} & H_{xu} \\ H_{ux} & H_{uu} \end{bmatrix} \begin{bmatrix} \delta x \\ \delta u \end{bmatrix} dt\,, \tag{6.1.16}$$

subject to

$$\delta\dot{x} = f_x\,\delta x + f_u\,\delta u\,, \tag{6.1.17}$$

$$\delta x(t_0) \text{ specified,} \tag{6.1.18}$$

$$\delta\psi = (\psi_x\,\delta x)_{t=t_f}\,, \qquad \text{where} \qquad \delta\psi \text{ is specified.} \tag{6.1.19}$$

Since we are interested in a neighboring extremal path, we must determine $\delta u(t)$ such that $\delta^2\bar{J}$ is minimized subject to (6.1.17) through (6.1.19). This problem is of the linear-quadratic type that we have solved in detail in Chapter 5. Defining the multipliers as $\delta\lambda$ and $d\nu$ (with obvious forethought), we find the associated two-point boundary-value problem to be given precisely by Equations (6.1.10) through (6.1.15).

Equations (6.1.10) through (6.1.15) represent a *linear* two-point boundary-value problem since the coefficients are evaluated on the extremal path. Provided that $H_{uu}(t)$ is nonsingular for $t_0 \leq t \leq t_f$, we may solve (6.1.12) for $\delta u(t)$ in terms of $\delta\lambda(t)$ and $\delta x(t)$:

$$\delta u(t) = -H_{uu}^{-1}(H_{ux}\delta x + f_u^T \delta\lambda)\,. \tag{6.1.20}$$

Substituting (6.1.20) into (6.1.10) and (6.1.11), we have

$$\delta\dot{x} = A(t)\,\delta x - B(t)\,\delta\lambda\,, \tag{6.1.21}$$

$$\delta\dot{\lambda} = -C(t)\,\delta x - A^T(t)\,\delta\lambda\,, \tag{6.1.22}$$

where

$$A(t) = f_x - f_u H_{uu}^{-1} H_{ux}\,, \tag{6.1.23}$$

$$B(t) = f_u H_{uu}^{-1} f_u^T\,, \tag{6.1.24}$$

$$C(t) = H_{xx} - H_{xu} H_{uu}^{-1} H_{ux}\,. \tag{6.1.25}$$

We may regard the "forcing" perturbations in this two-point boundary-value problem as $\delta x(t_o)$ and $d\nu$ (instead of $\delta x(t_o)$ and $\delta\psi$); we must then find the correct $d\nu$ to yield the desired $\delta\psi$.

6.2 Determination of neighboring extremal paths by the backward sweep method

The sweep method for linear-quadratic problems was introduced in Section 5.2. The extension to problems containing cross-product terms in x and u in the integrand of the performance index was considered in Problem 4 of Section 5.2. The extension to handle linear terminal constraints was given in Section 5.3. A slight further extension is needed to handle nonlinear terminal constraints of the form (6.1.6).† As in Section 5.3, we seek solutions of (6.1.22) and (6.1.19) in the form

$$\delta\lambda(t) = S(t)\,\delta x(t) + R(t)\,d\nu\,, \tag{6.2.1}$$

$$\delta\psi = R^T(t)\,\delta x(t) + Q(t)\,d\nu\,, \tag{6.2.2}$$

where $d\nu$ and $\delta\psi$ are constant infinitesimal vectors and $S(t)$, $R(t)$, and $Q(t)$ are matrix functions. Clearly these matrices must be such that (6.1.14) and (6.1.15) are satisfied, i.e., that

$$S(t_f) = [\phi_{xx} + (\nu^T\psi_x)_x]_{t=t_f}\,, \tag{6.2.3}$$

$$R(t_f) = [\psi_x^T]_{t=t_f}\,, \tag{6.2.4}$$

$$Q(t_f) = 0\,. \tag{6.2.5}$$

Now differentiate (6.2.1) and (6.2.2) with respect to time, treating $d\nu$ and $\delta\psi$ as constants:

$$\delta\dot{\lambda} = \dot{S}\,\delta x + S\,\delta\dot{x} + \dot{R}\,d\nu\,, \tag{6.2.6}$$

$$0 = \dot{R}^T\,\delta x + R^T\,\delta\dot{x} + \dot{Q}\,d\nu\,. \tag{6.2.7}$$

†See also Problem 1 of Section 5.3.

Using (6.2.1) in (6.1.21) gives

$$\delta\dot{x} = (A - BS)\,\delta x - BR\,d\nu\,. \tag{6.2.8}$$

Equating (6.2.6) to (6.1.22) and using (6.2.1) and (6.2.8) to eliminate $\delta\dot{x}$ and $\delta\lambda$, we have

$$(-C - A^TS - SA + SBS - \dot{S})\,\delta x - [(A^T - SB)R + \dot{R}]\,d\nu = 0\,. \tag{6.2.9}$$

In a similar manner, substituting (6.2.8) into (6.2.7), we obtain

$$[\dot{R}^T + R^T(A - BS)]\,\delta x + (-R^TBR + \dot{Q})\,d\nu = 0\,. \tag{6.2.10}$$

If we view (6.2.9) and (6.2.10) as identities, valid for arbitrary δx and $d\nu$, it follows that the coefficients of δx and $d\nu$ must vanish:

$$\dot{S} = -SA - A^TS + SBS - C\,, \quad \text{or} \quad \dot{S} = -Sf_x - f_x^TS - H_{xx} + (Sf_u + H_{ux}^T)H_{uu}^{-1}(H_{ux} + f_u^TS)\,, \tag{6.2.11}$$

$$\dot{R} = -(A^T - SB)R\,, \tag{6.2.12}$$

$$\dot{Q} = R^TBR\,. \tag{6.2.13}$$

The boundary conditions for these matrix differential equations are given in (6.2.3), (6.2.4), and (6.2.5). If these differential equations are integrated backward from $t = t_f$, the relations (6.2.1) and (6.2.2) represent boundary conditions *equivalent* to the terminal boundary conditions (6.1.14) and (6.1.15) *at earlier times*; thus, we are "sweeping" the terminal boundary conditions backward to earlier times.

Having integrated (6.2.11) through (6.2.13) backward all the way to the initial time $t = t_0$, we may solve (6.2.2) at $t = t_0$ for the required $d\nu$ to produce the desired $\delta\psi$:

$$d\nu = Q^{-1}(t_0)\,[\delta\psi - R^T(t_0)\,\delta x(t_0)]\,. \tag{6.2.14}$$

Note that the existence of $d\nu$ for all $\delta\psi$ depends on the *nonsingularity* of $Q(t_0)$. (cf. Section 5.3.)

Having these values of $d\nu$, we could substitute them into (6.2.1) at $t = t_0$ to find $\delta\lambda(t_0)$:

$$\delta\lambda(t_0) = [S(t_0) - R(t_0)\,Q^{-1}(t_0)\,R^T(t_0)]\,\delta x(t_0) + R(t_0)\,Q^{-1}(t_0)\,\delta\psi\,. \tag{6.2.15}$$

Then we could find $\delta x(t)$ and $\delta\lambda(t)$ by integrating (6.1.21) and (6.1.22) forward as an initial-value problem, using desired $\delta x(t_0)$ and the required $\delta\lambda(t_0)$ from (6.2.15).

Alternatively, if we stored $S(t)$ and $R(t)$ on the backward sweep, we could simply integrate (6.2.8) forward, using $d\nu$ from (6.2.14). This would yield $\delta x(t)$ and we could obtain $\delta\lambda(t)$ from (6.2.1).

Still another alternative, if we stored $S(t)$ and $R(t)$ on the backward sweep, would be to regard (6.1.20) as a *linear feedback law*. Substituting (6.2.1) into (6.1.20) gives

$$\delta u(t) = -H_{uu}^{-1}\,[(H_{ux} + f_u^T S)\,\delta x + f_u^T R\,d\nu]\,, \tag{6.2.16}$$

which could be used with (6.1.10) and (6.2.14) to determine $\delta x(t)$.

Now, $d\nu$ in (6.2.14) was evaluated at $t = t_0$. Thinking in terms of feedback laws, as in (6.2.16), we see that we could evaluate $d\nu$ at several intermediate times in the manner of a sampled-data feedback law *or* we could evaluate it continuously in the manner of a continuous-data feedback law. If we evaluate $d\nu$ continuously, (6.2.16), using $d\nu$ from (6.2.14) with t_0 replaced by t, becomes

$$\begin{aligned}\delta u(t) &= -H_{uu}^{-1}\,\{[H_{ux} + f_u^T(S - RQ^{-1}R^T)]\,\delta x + f_u^T RQ^{-1}\,\delta\psi\}\\ &\equiv -\Lambda_1(t)\,\delta x - \Lambda_2(t)\,\delta\psi\,.\end{aligned} \tag{6.2.17}$$

This is a continuous linear feedback law that will produce desired small changes in the terminal conditions $\delta\psi$, while minimizing J of (6.1.7); hence, we call it a *neighboring optimum feedback law*.

Problem 1. Consider the dynamic system

$$\dot{x} = (F + GK)x\,, \qquad x(0) = x_0\,,$$

where x is a scalar variable, K is a scalar control variable, and we have the criterion

$$J = \frac{1}{2}\,x^2(T)S_T + \frac{1}{2}\int_0^T x^2(t)\,(Q + KRK)\,dt\,,$$

where F, G, S_T, R, Q are given scalars. Write down the necessary conditions for the first variations to vanish and solve the resultant two-point boundary-value problem. Note that this problem is derived from the usual linear-quadratic problem with the added assumption that $u = Kx$.

6.3 Sufficient conditions for a local minimum

In this section we shall show that neighboring stationary paths exist (in a weak sense; that is, δx and δu are small) if

$$H_{uu}(t) > 0 \quad \text{for} \quad t_0 \le t \le t_f\,, \tag{6.3.1}$$

$$Q(t) < 0 \quad \text{for} \quad t_0 \le t < t_f\,, \tag{6.3.2}$$

$$S(t) - R(t)Q^{-1}(t)R^T(t) \quad \text{finite for} \quad t_0 \le t < t_f\,. \tag{6.3.3}$$

In the calculus of variations these conditions are called (i) *the convexity condition* (or strengthened Legendre-Clebsch condition),

(ii) the *normality condition*, and (iii) the condition that *no conjugate points* exist on the path (Jacobi condition).

Equations (6.3.1), (6.3.3) and the necessary conditions given by Equations (6.1.1) through (6.1.9) in Section 6.1 form a set of *sufficient conditions* for the trajectory to be a *local minimum*. To see this, let us consider the expansion of $\delta^2\bar{J}$ in (6.1.16) once again:

$$\delta^2\bar{J} = \frac{1}{2}[\delta x^T(\phi_{xx} + \nu^T\psi_{xx})\,\delta x]_{t=t_f} + \frac{1}{2}\int_{t_0}^{t_f}[\delta x^T\;\delta u^T]\begin{bmatrix} H_{xx} & H_{xu} \\ H_{ux} & H_{uu}\end{bmatrix}\begin{bmatrix}\delta x \\ \delta u\end{bmatrix} dt\,. \tag{6.3.4}$$

If we can show that $\delta^2\bar{J} > 0$ for all $\delta u(t) \neq 0$, we have essentially established the minimality of the trajectory in question.†

Now add to (6.3.4) the identically zero quantity‡

$$0 = [(\delta x^T\psi_x^T - \delta\psi^T)\,d\nu]_{t=t_f} + \int_{t_0}^{t_f}\{d\nu^T R^T(f_x\,\delta x + f_u\,\delta u - \delta\dot{x}) + \tfrac{1}{2}\,\delta x^T S(f_x\delta x + f_u\,\delta u - \delta\dot{x})\}\,dt\,,$$

where $d\nu$, R, and S are quantities yet to be specified. Integrating the terms $d\nu^T R^T\,\delta\dot{x}$ and $\delta x^T S\,\delta\dot{x}$ by parts, we obtain

$$\begin{aligned}\delta^2\bar{J} = {} & \frac{1}{2}[\delta x^T(\phi_{xx} + \nu^T\psi_{xx} - S)\,\delta x]_{t=t_f} + d\nu^T[(\psi_x - R^T)\,\delta x - \delta\psi]_{t=t_f} \\ & + \frac{1}{2}[\delta x^T S\,\delta x]_{t=t_0} + [\delta x^T R\,d\nu]_{t=t_0} + \frac{1}{2}\int_{t_0}^{t_f}\Big\{2\,d\nu^T(\dot{R}^T + R^T f_x)\,\delta x \\ & + 2\,d\nu^T R^T f_u\,\delta u + \delta x^T\dot{S}\,\delta x + 2(\delta x^T f_x + \delta u^T f_u)\,S\,\delta x \\ & + [\delta x^T\;\delta u^T]\begin{bmatrix} H_{xx} & H_{xu} \\ H_{ux} & H_{uu}\end{bmatrix}\begin{bmatrix}\delta x \\ \delta u\end{bmatrix}\Big\}\,dt\,. \end{aligned} \tag{6.3.5}$$

The integral in (6.3.5), when expanded, becomes

$$\begin{aligned}\frac{1}{2}\int_{t_0}^{t_f} & [\delta x^T(\dot{S} + Sf_x + f_x^T S + H_{xx})\,\delta x + \delta u^T H_{uu}\,\delta u + \delta x^T(\dot{R} + f_x^T R)\,d\nu \\ & + d\nu^T(\dot{R}^T + R^T f_x)\,\delta x + \delta x^T(H_{xu} + Sf_u)\,\delta u + \delta u^T(H_{ux} + f_u^T S)\,\delta x \\ & + d\nu^T R^T f_u\,\delta u + \delta u^T f_u^T R\,d\nu]\,dt\,.\end{aligned}$$

Now let us *choose*

$$\dot{S} + Sf_x + f_x^T S + H_{xx} = (H_{xu} + Sf_u)\,H_{uu}^{-1}\,(H_{ux} + f_u^T S)\,,$$

$$S(t_f) = (\phi_{xx} + (\nu^T\psi_x)_x)_{t=t_f}, \tag{6.3.6}$$

---

†This is called the "accessory minimum problem" in the classical literature.

‡This proof is due to S. R. McReynolds, Ph.D Thesis, Harvard University, 1966.

$$\dot{R} + f_x^T R - (H_{xu} + S f_u^T) H_{uu}^{-1} f_u^T R = 0\,, \qquad R(t_f) = \psi_x^T \qquad (6.3.7)$$

and define

$$\dot{Q} = R^T f_u H_{uu}^{-1} f_u^T R\,, \qquad Q(t_f) = 0\,; \qquad (6.3.8)$$

that is, S, R, and Q satisfy Equations (6.2.11), (6.2.12), and (6.2.13) of Section 6.2. Then the integral of $\delta^2 \bar{J}$ in (6.3.4) can be written as a *perfect square*:

$$\delta^2 \bar{J} = \tfrac{1}{2}(\delta x^T S\, \delta x)_{t=t_o} + (\delta x^T R\, d\nu)_{t=t_o} - (\delta\psi^T d\nu) + \tfrac{1}{2} d\nu^T Q(t_o)\, d\nu$$

$$+ \frac{1}{2}\int_{t_o}^{t_f} \|(H_{uu})^{-1}[(H_{ux} + f_u^T S)\,\delta x + f_u^T R\, d\nu] + \delta u\|^2_{H_{uu}}\, dt\,. \qquad (6.3.9)$$

Now, choosing $d\nu$ to satisfy Equation (6.2.14) or (6.2.2), we finally reduce this to

$$\delta^2 \bar{J} = [\tfrac{1}{2}\delta x^T (S - RQ^{-1}R^T)\,\delta x + \delta\psi^T Q^{-1} R\,\delta x - \tfrac{1}{2}\delta\psi^T Q^{-1}\delta\psi]_{t=t_o}$$

$$+ \frac{1}{2}\int_{t_o}^{t_f} \|(H_{uu})^{-1}[(H_{ux} + f_u^T(S - RQ^{-1}R^T))\,\delta x + f_u^T R Q^{-1}\,\delta\psi] + \delta u\|^2_{H_{uu}}\, dt\,. \qquad (6.3.10)$$

If we compare two trajectories with the same initial and terminal conditions, $\delta x(t_o) = 0$ and $\delta\psi = 0$, then it is clear that $\delta^2\bar{J} > 0$ unless $\delta u(t)$ is chosen so as to make the integrand in (6.3.10) vanish:

$$\delta u(t) = -H_{uu}^{-1}(H_{ux} + f_u^T(S - RQ^{-1}R^T))\,\delta x \qquad \text{for all } t\,. \qquad (6.3.11)$$

For $\delta x(t_o) = 0$ and $\delta\psi = 0$, (6.3.11) indicates that $\delta u(t) = 0$ provided that (6.3.1), (6.3.2), and (6.3.3) are satisfied. This means that $\delta^2\bar{J} > 0$ for any nonzero $\delta u(t)$, when $\delta x(t_o) = 0$, $\delta\psi = 0$; that is, $u(t)$ produces a local minimum.

Furthermore, if the feedback law, Equation (6.2.17) is used when $\delta x(t_o) \neq 0$, $\delta\psi \neq 0$, the change in the performance index to second order as given by Equation (6.3.10) is,

$$\delta\bar{J} = \lambda^T(t_o)\delta x(t_o) - \nu^T \delta\psi$$

$$+ \frac{1}{2}[\delta x^T(t_o)\,,\, \delta\psi^T] \begin{bmatrix} S - RQ^{-1}R^T, & R^T Q^{-1} \\ Q^{-1}R & , -Q^{-1} \end{bmatrix}_{t=t_o} \begin{bmatrix} \delta x(t_o) \\ \delta\psi \end{bmatrix},$$

showing that the partial derivatives of $J^o[x(t_o), \psi]$ are

$$\frac{\partial J^o}{\partial x} = \lambda^T\,, \qquad \frac{\partial J^o}{\partial \psi} = -\nu^T\,, \qquad \frac{\partial^2 J^o}{\partial x^2} = S - RQ^{-1}R^T\,,$$

$$\frac{\partial^2 J^o}{\partial x\, \partial\psi} = R^T Q^{-1}\,, \qquad \frac{\partial^2 J^o}{\partial \psi^2} = -Q^{-1}\,.$$

It can also be shown that a second-order necessary condition for a minimum is the weakened version of (6.3.1) (see also Chapter 4, Section 2):

$$H_{uu}(t) \geq 0 \qquad \text{for} \qquad t_o \leq t \leq t_f. \qquad (6.3.12)\dagger$$

This convexity condition is easily understood from Chapter 4, where it was shown that $u(t)$ is determined by minimizing the Hamiltonian H with respect to u, holding x, λ, and t fixed. For a smooth H with no constraints on u, this requires that

$$H_u = 0\,, \qquad H_{uu} \geq 0\,.$$

An interpretation of the normality condition (6.3.2) is obtained from (6.2.14). Small changes $\delta\psi$ can be produced by small changes $d\nu$ only if $Q(t)$ is nonsingular over $t_o \leq t < t_f$. If $H_{uu} > 0$, it follows from (6.3.8) that $\dot{Q} \geq 0$. Since $Q(t_f) = 0$, it then follows that $Q(t) \leq 0$.

If $S - RQ^{-1}R^T \to \infty$ at $t = t'$ where $t_o \leq t' < t_f$, it is necessary that certain linear combinations of $\delta x(t')$ be zero; i.e., the set of possible perturbations is restricted to a dimension less than n, the number of state variables. This means that $\partial^2 J^o/\partial x^2 \to \infty$ at $t = t'$. In fact, the paths will not be minimizing if continued for $t < t'$. (See Examples 1 and 2 and Problems 1 and 2 of this section.) Note that $S \to \infty$ does not necessarily imply that $S - RQ^{-1}R^T \to \infty$ (see Example 2).

Example 1. *Shortest path between a point and a great circle on a sphere.* To find the shortest path between a point 0 and a great circle, choose a coordinate system with origin at 0 and let the great circle be the meridian $\phi = \phi_1$ (θ is latitude, ϕ is longitude (see Figure 6.3.1).

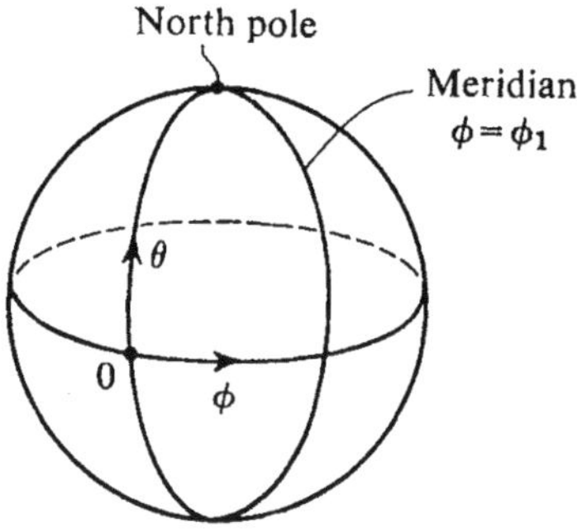

Figure 6.3.1. Nomenclature for shortest path between a point and a great circle on a sphere.

†Classically, this is called a first-order necessary condition; see M. Hestenes, *Calculus of Variations and Optimal Control Theory.* New York: Wiley, 1967, p. 252.

The element of distance, ds, on the surface of the sphere is

$$ds = [r^2(d\theta)^2 + r^2 \cos^2 \theta (d\phi)^2]^{1/2} ,$$

where r = radius of the sphere. Thus, the problem is to find $u(\phi)$ to minimize

$$J = \int_0^{\phi_1} (u^2 + \cos^2 \theta)^{1/2} \, d\phi ,$$

where

$$\frac{d\theta}{d\phi} = u \qquad \text{and} \qquad \theta(0) = 0 .$$

It is straightforward to show that $u = 0$, $\theta = 0$ satisfies the first-order necessary conditions, so let us consider paths neighboring to this one. Expanding the performance index to second order yields

$$\delta J = J - \phi_1 \cong \frac{1}{2} \int_0^{\phi_1} (u^2 - \theta^2) \, d\phi .$$

The Hamiltonian for this *accessory minimum problem* is, therefore,

$$H = \tfrac{1}{2}(u^2 - \theta^2) + \lambda u ,$$

and the Euler-Lagrange equations are

$$\frac{d\lambda}{d\phi} = -\frac{\partial H}{\partial \theta} = \theta , \qquad \lambda(\phi_1) = 0 ,$$

$$0 = \frac{\partial H}{\partial u} = u + \lambda .$$

Eliminating λ and u, with $d\theta/d\phi = u$, we have

$$\frac{d^2\theta}{d\phi^2} + \theta = 0 , \qquad \theta(0) = 0 , \qquad \left(\frac{d\theta}{d\phi}\right)_{\phi = \phi_1} = 0 .$$

Now $\theta = A \sin \phi$ satisfies the differential equation and $\theta(0) = 0$, but we must also have

$$\left(\frac{d\theta}{d\phi}\right)_{\phi = \phi_1} = A \cos \phi_1 = 0 ,$$

which is satisfied *only* by $A = 0$ if $\phi_1 < \pi/2$, but is satisfied by *any* A if $\phi_1 = \pi/2$. For $\phi_1 = \pi/2$, note that

$$\delta J = \frac{1}{2} \int_0^{\pi/2} (A^2 \cos^2 \phi - A^2 \sin^2 \phi) \, d\phi = 0 .$$

Point 0 is said to be a focal point or a conjugate point to $\phi_1 = \pi/2$ for this problem.

An alternate method is to use the *Riccati equation approach*. Here, we have $F = 0$, $G = 1$, $H_{\theta\theta} = -1$, $H_{\theta u} = 0$, $H_{uu} = 1$, so the Riccati equation is, simply,

$$\frac{dS}{d\phi} = S^2 + 1, \qquad S(\varphi_1) = 0.$$

The solution is easily obtained as

$$S = -\tan(\phi_1 - \phi).$$

Clearly, we see that $S \to \infty$ when $\phi_1 - \phi \to \pi/2$; that is, a conjugate point occurs at $\phi_1 - \phi = \pi/2$. The neighboring optimum feedback law is

$$\delta u = [\tan(\phi_1 - \phi)]\, \delta\theta.$$

Note that the feedback gain is positive for $0 \le \phi_1 - \phi \le \pi/2$.

Example 2. *Shortest path between two points on a sphere.* This is nearly the same problem as Example 1 except for the terminal boundary condition. If we place the other point at $\theta = 0$, $\phi = \phi_1$, then we have

$$\theta(\phi_1) = 0 \qquad (\text{instead of } \lambda(\phi_1) = 0).$$

Again, we find that $\theta = A \sin\phi$ satisfies the Euler-Lagrange equations for the accessory minimum problem and $\theta(0) = 0$, but we must also have

$$\theta(\phi_1) = A \sin\phi_1 = 0,$$

which is satisfied *only* by $A = 0$ if $\phi_1 < \pi$ but is satisfied by *any* A if $\phi_1 = \pi$. Point 0 is a conjugate point to $\theta = 0$, $\phi = \pi$ and we have

$$\delta J = \frac{1}{2}\int_0^{\pi} (A^2 \cos^2\phi - A^2 \sin^2\phi)\, d\phi = 0;$$

that is, there is an infinite number of paths between the two points with equal performance index (other great circles).

For $\phi_1 > \pi$, $u = 0$, $\theta = 0$ is *not even a local minimum*, even though it satisfies the first-order necessary conditions. We can show this by trying the neighboring path,

$$\theta = A \sin \pi \frac{\phi}{\phi_1},$$

which yields

$$\delta J = \frac{1}{2}\int_0^{\phi_1} \left(\frac{A^2\pi^2}{\phi_1^2}\cos^2\frac{\pi\phi}{\phi_1} - A^2 \sin^2\frac{\pi\phi}{\phi_1}\right) = -\frac{A^2}{4\phi_1}(\phi_1^2 - \pi^2),$$

which can be made as much less than zero as we wish by increasing A.

The *Riccati equation approach* here involves the auxiliary quantities R and Q since we have a terminal boundary condition:

$$\frac{dS}{d\phi} = S^2 + 1\,, \qquad S(\phi_1) = 0\,; \qquad \frac{dR}{d\phi} = SR\,, \qquad R(\phi_1) = 1\,;$$

$$\frac{dQ}{d\phi} = R^2\,, \qquad Q(\phi_1) = 0\,.$$

The solutions are easily obtained for $0 \leq \phi_1 - \phi < \pi/2$:

$$S = -\tan(\phi_1 - \phi)\,, \qquad R = \sec(\phi_1 - \phi)\,, \qquad Q = -\tan(\phi_1 - \phi)\,,$$

implying that $S - RQ^{-1}R^T = \text{ctn}(\phi_1 - \phi)$ and $R^TQ^{-1} = -\csc(\phi_1 - \phi)$. The neighboring optimum feedback law is therefore

$$\delta u = -[\text{ctn}(\phi_1 - \phi)]\delta\theta + [\csc(\phi_1 - \phi)]\delta\theta_f\,.$$

Although S, R, and Q do not even exist for $\phi_1 - \phi \geq \pi/2$, $S - RQ^{-1}R^T$ does exist for $0 \leq \phi_1 - \phi < \pi$. Clearly, we have $S - RQ^{-1}R^T \to \infty$ as $\phi_1 - \phi \to \pi$; that is, a conjugate point occurs at $\phi_1 - \phi = \pi$. *Note* that $S \to \infty$ at $\phi_1 - \phi = \pi/2$, but $S - RQ^{-1}R^T$ does *not* $\to \infty$ at $\phi_1 - \phi = \pi/2$. Note also that the same result would have been obtained without the auxiliary quantities R and Q if we had used $S(\phi_1) = \infty$ instead of $S(\phi_1) = 0$ with the Riccati equation.

Example 3. *Minimum-time paths with velocity magnitude* $V(y) = V_0\sqrt{1 + (y^2/h^2)}$.

From Problems 2 through 5, Section 2.7, the minimum-time paths from the origin are given by

$$\dot{x} = V_0\sqrt{1 + (y^2/h^2)}\cos\theta\,, \qquad x(0) = 0\,,$$

$$\dot{y} = V_0\sqrt{1 + (y^2/h^2)}\sin\theta\,, \qquad y(0) = 0\,,$$

$$\dot{\theta} = -\frac{V_0}{h^2}\frac{y}{\sqrt{1 + (y^2/h^2)}}\cos\theta\,, \qquad \theta(0) = \theta_0\,,$$

and a first integral of this set is given by Snell's Law,

$$\frac{\cos\theta}{V_0\sqrt{1 + (y^2/h^2)}} = \frac{\cos\theta_0}{V_0} = \text{const.}$$

Using the first integral, we can express y/h in terms of θ:

$$\frac{y}{h} = \frac{\sqrt{\cos^2\theta - \cos^2\theta_0}}{\cos\theta_0}\,.$$

Using above in Equations for $\dot{x}$ and $\dot{\theta}$ we obtain

$$\frac{dx}{d\theta} = \frac{h\cos^2\theta \sec\theta_0}{\sqrt{\cos^2\theta - \cos^2\theta_0}},$$

$$\frac{dt}{d\theta} = -\frac{h}{V_0\sqrt{\cos^2\theta - \cos^2\theta_0}}.$$

These latter equations can be integrated in terms of standard elliptic integrals:

$$\frac{x}{h} = \sec\theta_0\left[E\left(\theta_0,\frac{\pi}{2}\right) - E(\theta_0,\phi)\right],$$

$$\frac{V_0 t}{h} = F\left(\theta_0,\frac{\pi}{2}\right) - F(\theta_0,\phi),$$

where

$$\sin\phi = \frac{\sin\theta}{\sin\theta_0},$$

$$F(\theta_0,\phi) = \int_0^\phi \frac{d\alpha}{\sqrt{1-\sin^2\theta_0\sin^2\alpha}} = \text{incomplete elliptic integral of the first kind,}$$

$$E(\theta_0,\phi) = \int_0^\phi \sqrt{1-\sin^2\theta_0\sin^2\alpha}\, d\alpha = \text{incomplete elliptic integral of the second kind.}$$

These integrals are tabulated, for example, in Jahnke and Emde, pp. 61–72. We may express y/h in terms of ϕ as follows:

$$\frac{y}{h} = \tan\theta_0 \cos\phi.$$

Figure 6.3.2 shows some of the minimum-time paths (the "rays") and some of the contours of constant $V_0 t/h$ (the "wavefronts"). Note the *conjugate point* at $x/h = \pi$, $y = 0$; $y = 0$ is a minimum-time path for $0 < x_f/h < \pi$ but *not* for $x_f/h > \pi$.

Note that the contours of constant $V_0 t/h$ on the figure develop an infinite curvature at the conjugate point (that is, $\partial^2 J/\partial y^2 \to \infty$). Furthermore, on $y = 0$ beyond the conjugate point $(x/h > \pi)$, the contours of constant $V_0 t/h$ have a discontinuity in slope going from $y < 0$ to $y > 0$. (See also Problem 3.)

Problem 1. Let $Q^{-1}(t)$ exist for $t_0 \le t < t_f$ but let $S - RQ^{-1}R^T \to \infty$ at $t = t_1$ (i.e., a conjugate point at t_1). Show that:

(i) It is possible to find a distinct neighboring trajectory from t_1 to t_f that satisfies

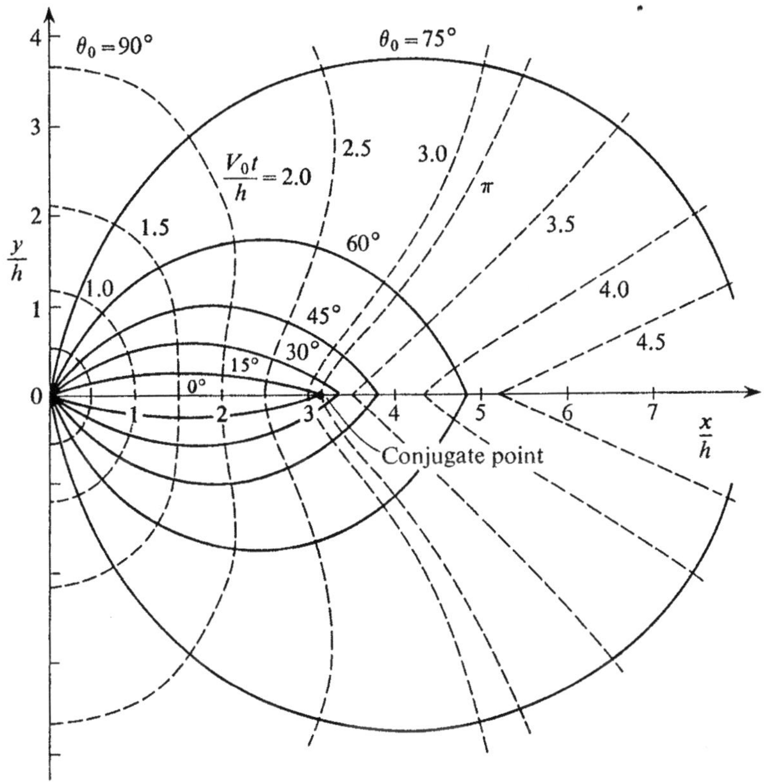

Figure 6.3.2. Minimum-time paths for velocity magnitude $V = V_0\sqrt{1 + (y/h)^2}$.

$$\delta\psi = \psi_x\,\delta x = 0\,, \qquad \delta\dot{x} = f_x\,\delta x + f_u\,\delta u\,,$$
$$\delta\dot{\lambda} = -f_x^T\,\delta\lambda - H_{xx}\,\delta x - H_{xu}\,\delta u\,,$$
$$\delta u = -H_{uu}^{-1}\,[H_{ux} + f_u^T\,(S - RQ^{-1}R^T)]\,\delta x\,,$$

and

$$\delta x(t_1) = 0\,.$$

This is shown pictorially in Figure 6.3.3.

(ii) The integral

$$\frac{1}{2}\int_{t_1}^{t_f} [\delta x^T \delta u^T] \begin{bmatrix} H_{xx} & H_{xu} \\ H_{ux} & H_{uu} \end{bmatrix} \begin{bmatrix} \delta x \\ \delta u \end{bmatrix} dt + \frac{1}{2}\,(\delta x^T S\,\delta x)_{t=t_f} = 0$$

on the conjugate path. Hence, cost of $A\,B\,C$ = cost of $A\,B\,D$.

(iii) From (ii), argue that the cost of going from A to D without

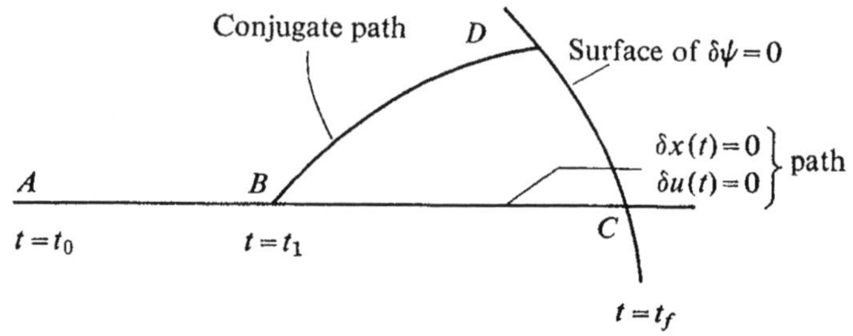

Figure 6.3.3. Conjugate point and path.

passing through B is less than the cost of ABD. Hence, nonexistence of conjugate point in $t_0 < t < t_f$ is *necessary* for minimality. [HINT: Consider the path $A\ E\ B\ F\ D$ (see Figure 6.3.4), where $E \to B$, $F \to B$.]

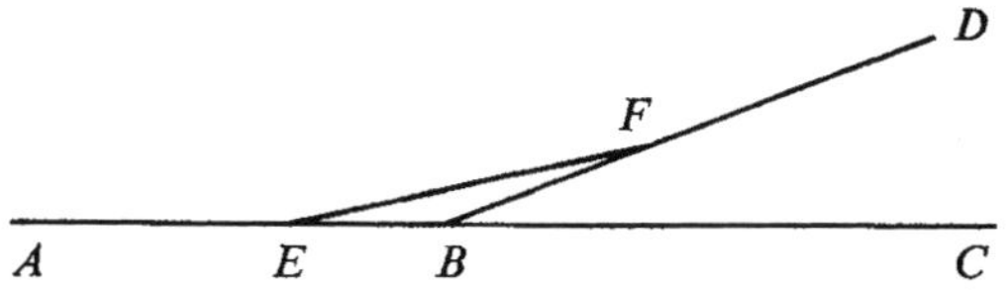

Figure 6.3.4. Conjugate path and nonoptimality.

Problem 2. What part of the dynamic programming derivation in Section 4.2, Chapter 4, breaks down when a conjugate point exists at t_1? What part of the transition matrix solution method of Section 5.3 breaks down?

Problem 3. Example 3 in Section 6.3 may also be posed as minimizing

$$J = \frac{1}{V_0} \int_0^{x_f} \left[\frac{1 + u^2}{1 + (y^2/h^2)} \right]^{1/2} dx ,$$

subject to

$$\frac{dy}{dx} = u , \qquad y(0) = 0 , \qquad y(x_f) = 0 .$$

For $y << h$ and $u << 1$, the performance index may be approximated as

$$J \cong \frac{x_f}{V_0} + \frac{1}{2V_0} \int_0^{x_f} \left(u^2 - \frac{y^2}{h^2} \right) dx .$$

Show that a conjugate point exists at $x = \pi h$, $y = 0$ for this approximation to the original problem.

Problem 4. For example 2 in Section 6.3, show that the minimum-distance paths (great circles) through $\theta = \phi = 0$ are given by the one-parameter family

$$\tan \theta = \tan \theta_m \sin \phi \, ,$$

where

$$u = \pm \cos \theta \left(\frac{\cos^2\theta}{\cos^2\theta_m} - 1 \right)^{1/2} ,$$

θ_m = maximum value of θ on the great circle.

Problem 5. Find the minimum-time paths from $x = 1$, $y = 0$ in a medium in which the velocity magnitude is given by $V = 1 + x^2 + y^2$. In particular, show that $x = -1$, $y = 0$ is a conjugate point to $x = 1$, $y = 0$. [HINT: Use polar coordinates, $x = r\cos\theta$, $y = r\sin\theta$.]

Problem 6. *Buckling of a column.* The equilibrium-deflected shape of a column is such that it minimizes

$$J = \int_0^\ell \left[\frac{EI}{2} u^2 - P(1 - \cos\theta) \right] ds \, ,$$

where

$\dfrac{d\theta}{ds} = u$ = curvature,

P = load on column,

E = modulus of elasticity,

I = second moment of cross-sectional area about neutral axis,

$\tan\theta$ = slope of neutral axis,

ℓ = length of column,

s = distance along neutral axis of column.

The column in Figure 6.3.5 has $\theta(0) = \theta(\ell) = 0$.

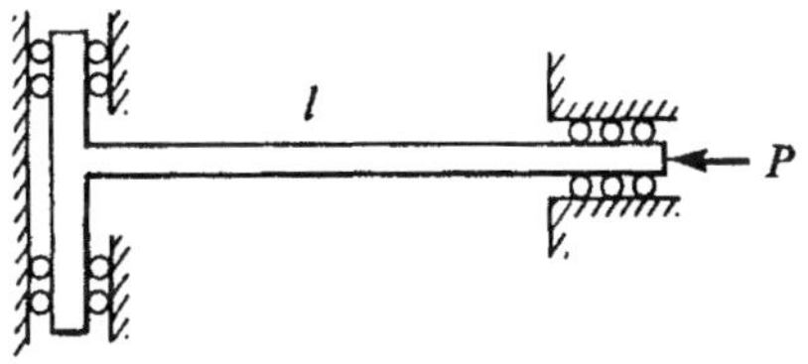

Figure 6.3.5. Buckling of a column.

Find the shape of such a column for EI and P constant, with ℓ as an increasing parameter. Note that, if y = deflection of neutral axis from straight, then

$$\frac{dy}{ds} = \sin\theta .$$

In particular, show that $u = \theta = y = 0$ is the only solution until ℓ reaches a certain length; for lengths greater than this, $\theta \neq 0$ is minimizing.

Problem 7. Inside a sphere of radius R, the velocity magnitude is given by $V = V_0/\sqrt{2 - (r^2/R^2)}$. Show that a plane wave moving through a medium with speed V_0 and then impinging on the sphere will be focused down to a point on the opposite surface of the sphere (see Figure 6.3.6).
This is called a "Luneberg Lens," useful in optics and radar applications.

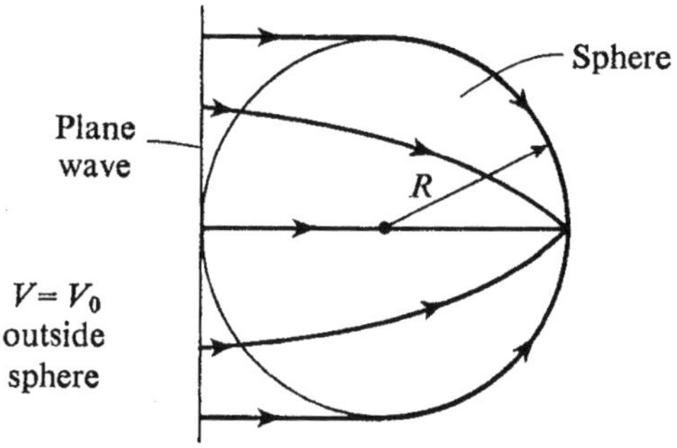

Figure 6.3.6. Luneberg Lens.

Problem 8. (a) Show that the Sturm-Liouville problem,

$$\frac{d}{dt}\left[r(t)\frac{dx}{dt}\right] + q(t)x = 0 , \qquad x(0) = 0 , \qquad ax(t_f) + r(t_f)\dot{x}(t_f) = 0$$

(all quantities are scalars), may be interpreted as finding the minimum of

$$J = \frac{1}{2}a[x(t_f)]^2 + \frac{1}{2}\int_0^{t_f}[r(t)u^2 - q(t)x^2]\,dt ,$$

where $\dot{x} = u$, $x(0) = 0$.

(b) For r and q positive constants and $a = 0$, show that $u = 0$ is the only solution if $t_f < \pi/2[\sqrt{r/q}\,]$ and that an infinite number of solutions, all yielding $J = 0$, exist if $t_f = \pi/2[\sqrt{r/q}\,]$.

Problem 9. Verify the expression for δJ immediately following Equation (6.3.11).

6.4 Perturbation feedback control (final time specified)

For systems with three or more state variables, the computation, and even the storage, of nonlinear optimum feedback laws,† as in Chapter 4, becomes increasingly plagued by the quantity of numbers to handle. For practical purposes, we are forced to consider perturbation feedback control, i.e., control in the vicinity of a nominal path.

If an optimum nominal path is used, the feedback gains determined in Section 6.2 yield neighboring optimum paths. In Section 6.1 it was shown that this type of feedback control is identical to the linear feedback control discussed in Chapter 5, where the weighting factors in the quadratic performance index are the *second* partial derivatives of the variational Hamiltonian, Equation (6.1.16), and the linear system equations are the linear perturbation equations around the nominal optimum path, Equation (6.1.17).‡ Figure 6.4.1 shows a block

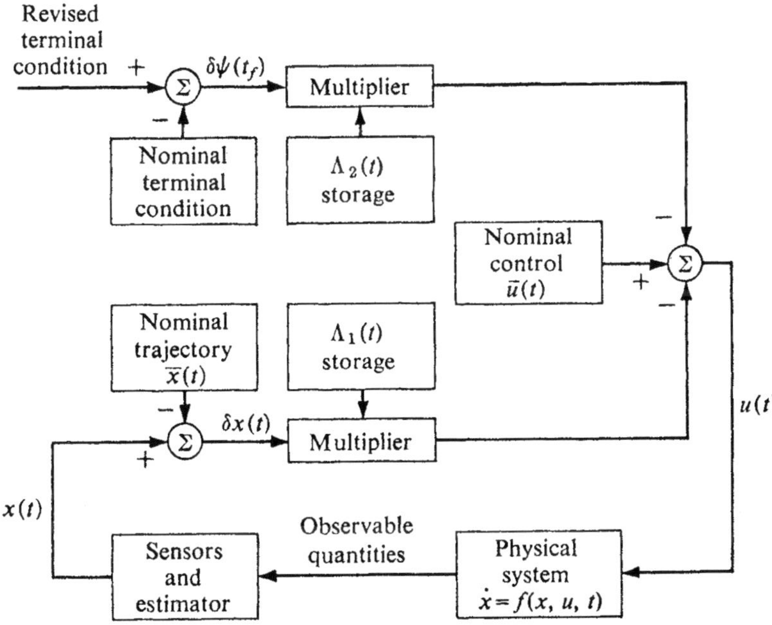

Figure 6.4.1. Neighboring optimum feedback control.

---

†These are also called "explicit guidance laws," "closed-loop feedback laws," or "dynamic programming solutions."

‡This interpretation partially answers the criticism that quadratic criteria are often arbitrary and artificial.

diagram of the neighboring optimum feedback control scheme based on Equation (6.2.17).

Example. *Perturbation guidance for injection into orbit with maximum velocity.* Using the approximation of constant gravitational force per unit mass, we consider the problem of thrust-direction control to place a rocket vehicle at a given altitude at a given time with zero vertical velocity and maximum horizontal velocity (see Figure 6.4.2 for nomenclature). The problem is "reciprocal" to Problem 13 of

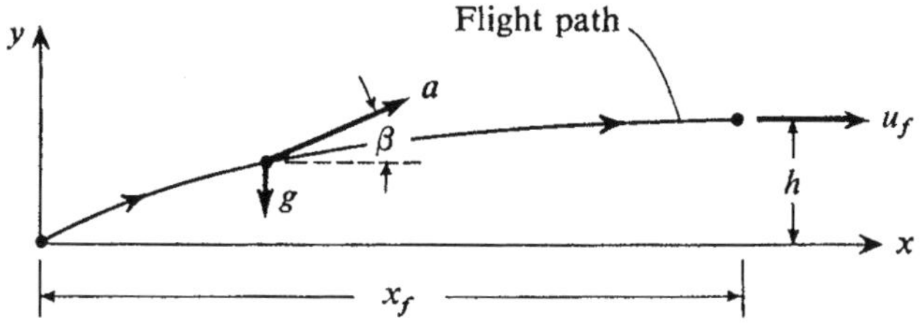

Figure 6.4.2. Nomenclature for injection into orbit.

Section 2.7, where time was minimized for a given final horizontal velocity; the nominal is determined by solving the first two equations of that problem for β_o and β_f, assuming final time, T, as given and a as constant. Then the thrust-direction history $\beta(t)$ is given by

$$\tan\beta = \tan\beta_o + (\tan\beta_f - \tan\beta_o)\frac{t}{T}.$$

The trajectory is then given by

$$\dot{v} = a\sin\beta - g, \qquad v(0) = 0; \qquad \dot{y} = v, \qquad y(0) = 0,$$

and the performance index by

$$J = a\int_0^T \cos\beta\, dt.$$

The neighboring extremals are given by (6.2.11) through (6.2.13) with boundary conditions (6.2.3) through (6.2.5). In this case, we have

$$\psi_1 = v, \qquad \psi_2 = y - h; \qquad H = a\cos\beta + \lambda_v(a\sin\beta - g) + \lambda_y v;$$
$$\lambda_v = \nu_1 + \nu_2(T - t) = \tan\beta, \qquad \lambda_y = \nu_2;$$
$$H_{xx} = 0, \qquad H_{\beta x} = 0, \qquad H_{\beta\beta} = -a\cos\beta - \lambda_v a\sin\beta \equiv -a\sec\beta;$$
$$f_\beta = \begin{bmatrix} a\cos\beta \\ 0 \end{bmatrix}, \qquad f_x = \begin{bmatrix} 0 & 0 \\ 1 & 0 \end{bmatrix} = A, \qquad C = 0,$$
$$B = -a\cos^3\beta \begin{bmatrix} 1 & 0 \\ 0 & 0 \end{bmatrix}, \qquad S = 0.$$

Hence, we have

$$\left.\begin{array}{ll}\dot{R}_{11} = -R_{21}\,; & R_{11}(T) = 1 \\ \dot{R}_{12} = -R_{22}\,; & R_{12}(T) = 0 \\ \dot{R}_{21} = 0 \quad ; & R_{21}(T) = 0 \\ \dot{R}_{22} = 0 \quad ; & R_{22}(T) = 1\end{array}\right\} \Rightarrow \left\{\begin{array}{l} R_{11} = 1\,, \\ R_{12} = T - t\,, \\ R_{21} = 0\,, \\ R_{22} = 1\,;\end{array}\right.$$

$$\dot{Q}_{11} = -a\cos^3\beta\,; \qquad Q_{11}(T) = 0\,,$$

$$\dot{Q}_{12} = -a(T-t)\cos^3\beta\,; \qquad Q_{12}(T) = 0\,,$$

$$\dot{Q}_{22} = -a(T-t)^2\cos^3\beta\,; \qquad Q_{22}(T) = 0\,.$$

However, we have $\dot{\beta} = -\nu_2\cos^2\beta$ and $(\tan\beta - \nu_1)/\nu_2 = T - t$, hence, it follows that

$$dQ_{11} = \frac{a}{\nu_2}\cos\beta\,d\beta\,, \qquad Q_{11} = \frac{a}{\nu_2}(\sin\beta - \sin\beta_f)\,,$$

$$dQ_{12} = \frac{a}{\nu_2^2}(\tan\beta - \nu_1)\cos\beta\,d\beta\,, \qquad Q_{12} = \frac{a}{\nu_2^2}\frac{1-\cos(\beta-\beta_f)}{\cos\beta_f}\,,$$

$$dQ_{22} = \frac{a}{\nu_2^3}(\tan\beta - \nu_1)^2\cos\beta\,d\beta\,, \qquad Q_{22} = \frac{a}{\nu_2^3}\left[\frac{\sin\beta - \sin\beta_f}{\cos^2\beta_f} + \log\frac{\tan\beta + \sec\beta}{\tan\beta_f + \sec\beta_f}\right];$$

$$Q^{-1} = \frac{\begin{bmatrix} Q_{22}\,, & -Q_{12} \\ -Q_{12}\,, & Q_{11}\end{bmatrix}}{D}\,, \qquad \text{where} \qquad D = Q_{11}Q_{22} - Q_{12}^2\,,$$

$$RQ^{-1} = \begin{bmatrix} 1\,, T-t \\ 0,\,1\end{bmatrix}\frac{\begin{bmatrix} Q_{22}\,, & -Q_{12} \\ -Q_{12}\,, & Q_{11}\end{bmatrix}}{D}$$

$$= \begin{bmatrix} Q_{22} - (T-t)Q_{12}\,, & (T-t)Q_{11} - Q_{12} \\ -Q_{12} & , \quad Q_{11}\end{bmatrix}\frac{1}{D}\,.$$

Finally, the neighboring optimum feedback law is given by

$$\delta\beta = \frac{\cos^2\beta}{D}[Q_{22} - (T-t)Q_{12}\,, (T-t)Q_{11} - Q_{12}]\begin{bmatrix}\delta v_f - \delta v \\ \delta y_f - [(T-t)\,\delta v + \delta y]\end{bmatrix}.$$

Figure 6.4.3 shows a maximum velocity path for $a/g = 3$, and $(2h/aT^2) = .258$, and Figure 6.4.4 shows the neighboring optimum feedback gains associated with the path for $\delta v_f = \delta y_f = 0$. Also needed would be the nominal time histories $v^{\circ}(t)$, $y^{\circ}(t)$, and $\beta^{\circ}(t)$.

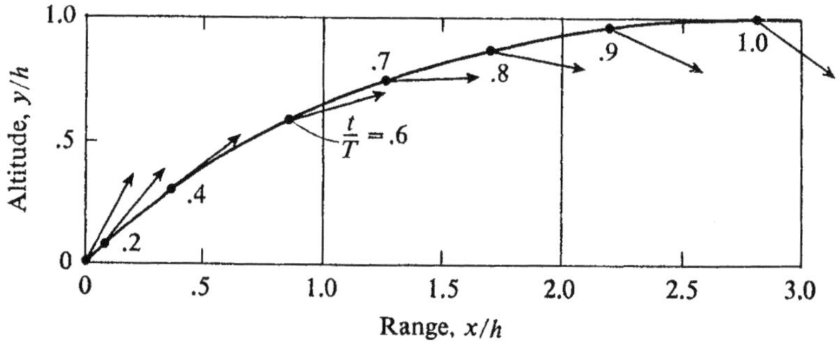

Figure 6.4.3. Maximum final horizontal velocity path for $a/g = 3, 2h/aT^2 = .258$.

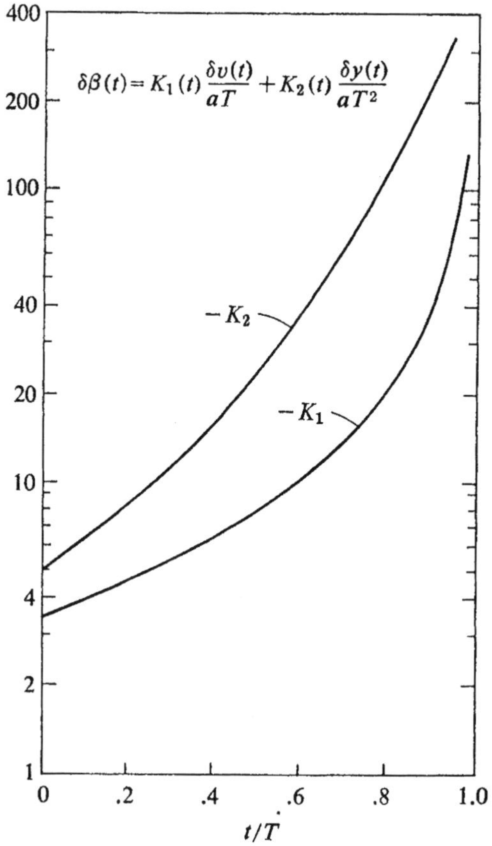

Figure 6.4.4. Time history of feedback gains for $a/g = 3$, $2h/aT^2 = .258$.

Problem. *Minimum effort control (final time specified).* A nominal path $u(t) \Rightarrow x(t)$ satisfies end conditions $\psi[x(t_f)] = 0$ at a specified final time t_f. Find the perturbation control law that meets slightly different end conditions $\psi[x(t_f)] = d\psi$ and minimizes the perturbation effort

$$E = \frac{1}{2}\int_t^{t_f} (\delta u)^T B\, \delta u\, dt\,,$$

where

$$B(t) = \text{a positive definite matrix.}$$

ANSWER.

$$\delta u(t) = -\,B^{-1} f_u^T R Q^{-1}\,(d\psi - R^T\,\delta x)\,,$$

where

$$\dot{R} = -f_x^T R\,, \qquad R(t_f) = \left(\frac{\partial \psi}{\partial x}\right)^T_{t=t_f}; \qquad \dot{Q} = R^T f_u B^{-1} f_u^T R\,, \qquad Q(t_f) = 0\,.$$

6.5 Neighboring extremal paths with final time unspecified

If the final time, t_f, is determined *implicitly*, rather than explicitly, by the terminal constraints, as in Section 2.7, an extension of the methods presented in Sections 6.1 through 6.4 is necessary. The nominal optimum solution must satisfy the additional necessary condition (2.7.23):

$$\Omega(x,u,\nu,t)\big|_{t=t_f} \triangleq \left(\frac{d\Phi}{dt} + L\right)_{t=t_f} = 0\,, \tag{6.5.1}$$

where

$$\Phi = \phi(x,t) + \nu^T \psi(x,t)\,, \qquad \frac{d\Phi}{dt} = \frac{\partial \Phi}{\partial t} + \frac{\partial \Phi}{\partial x}\dot{x}\,.$$

This scalar equation determines the additional unknown parameter, t_f.

Perturbation of the necessary conditions (6.1.5), (6.1.6), and (6.5.1) must take into account the perturbation in the final time, dt_f:

$$d\lambda(t_f) = \left[\frac{\partial}{\partial x}\left(\frac{\partial \Phi}{\partial x}\right)^T dx + \left(\frac{\partial \psi}{\partial x}\right)^T d\nu + \frac{\partial}{\partial t}\left(\frac{\partial \Phi}{\partial x}\right)^T dt_f\right]_{t=t_f}, \tag{6.5.2}$$

$$d\psi = \left[\frac{\partial \psi}{\partial x} dx + \frac{\partial \psi}{\partial t} dt_f\right]_{t=t_f}, \tag{6.5.3}$$

$$0 = \left[\frac{\partial \Omega}{\partial x} dx + d\nu^T \frac{d\psi}{dt} + \frac{\partial \Omega}{\partial t} dt_f\right]_{t=t_f} . \dagger \tag{6.5.4}$$

Now, for computation we need $\delta\lambda(t_f)$ rather than $d\lambda(t_f)$, and δx rather than dx. Hence, we substitute

$$d\lambda(t_f) = \delta\lambda(t_f) + \dot{\lambda}(t_f)\, dt_f, \tag{6.5.5}$$

$$dx(t_f) = \delta x(t_f) + \dot{x}(t_f)\, dt_f, \tag{6.5.6}$$

into (6.5.2) to obtain

$$\delta\lambda(t_f) = \frac{\partial^2 \Phi}{\partial x^2} \delta x + \left(\frac{\partial \psi}{\partial x}\right)^T d\nu + \left(\frac{d}{dt}\left(\frac{\partial \Phi}{\partial x}\right)^T - \dot{\lambda}\right) dt_f. \tag{6.5.7}$$

Next, by using (6.1.2) and (6.1.5), we can establish that

$$\frac{d}{dt}\left(\frac{\partial \Phi}{\partial x}\right)^T - \dot{\lambda} \equiv \left(\frac{\partial \Omega}{\partial x}\right)^T. \tag{6.5.8}$$

Substituting (6.5.8) into (6.5.7), and (6.5.6) into (6.5.3) and (6.5.4), we have, finally,

$$\begin{bmatrix} \delta\lambda(t_f) \\ d\psi \\ 0 \end{bmatrix} = \begin{bmatrix} \frac{\partial^2 \Phi}{\partial x^2}, & \left(\frac{\partial \psi}{\partial x}\right)^T, & \left(\frac{\partial \Omega}{\partial x}\right)^T \\ \frac{\partial \psi}{\partial x}, & 0, & \frac{d\psi}{dt} \\ \frac{\partial \Omega}{\partial x}, & \left(\frac{d\psi}{dt}\right)^T, & \frac{d\Omega}{dt} \end{bmatrix}_{t=t_f} \begin{bmatrix} \delta x(t_f) \\ d\nu \\ dt_f \end{bmatrix}, \qquad \begin{matrix} (6.5.9) \\ (6.5.10) \\ (6.5.11) \end{matrix}$$

where

$$\frac{d\Omega}{dt} = \frac{\partial \Omega}{\partial t} + \frac{\partial \Omega}{\partial x} f, \qquad \frac{d\psi}{dt} = \frac{\partial \psi}{\partial t} + \frac{\partial \psi}{\partial x} f.$$

Equations (6.1.10) through (6.1.13) plus (6.5.9) through (6.5.11) represent a linear two-point boundary-value problem for a neighboring extremal with small changes in initial conditions, $\delta x(t_0)$, and/or small changes in the terminal conditions, $d\psi$. These changes, $\delta x(t_0)$ and $d\psi$ will, in general, produce small changes $\delta x(t_f)$, $d\nu$, and dt_f. The problem may be solved, in principle, by linear superposition, as in Section 5.3. However, an extension of the sweep method, presented in the next section, appears to offer greater stability for numerical computation.

Another derivation of the boundary conditions (6.5.9) through

†Note that $(\partial\Omega/\partial u) \equiv (\partial H/\partial u) = 0$.

(6.5.11) is possible through the careful expansion of the augmented performance criterion as described below Equation (6.1.15) for fixed final time (see Breakwell and Ho, 1965).

A great improvement in the effectiveness of perturbation control is made if the feedback gains are indexed by *time-to-go* rather than clock time. This requires that an estimate of time to go be made during operation of the system; this can be done by using Equation (6.6.15) of the next section. Further discussion and numerical examples are presented in Speyer and Bryson (1968).

6.6 Determination of neighboring extremal paths by the backward sweep method with final time unspecified

The extension of the sweep method (beyond Section 6.2) needed to handle unspecified final time starts by observing the symmetry of the coefficient matrix in (6.5.9–6.5.11). This motivates the substitution

$$\begin{bmatrix} \delta\lambda(t) \\ d\psi \\ d\Omega \end{bmatrix} = \begin{bmatrix} S(t) & , R(t) & , m(t) \\ R^T(t) & , Q(t) & , n(t) \\ m^T(t) & , n^T(t) & , \alpha(t) \end{bmatrix} \begin{bmatrix} \delta x(t) \\ d\nu \\ dt_f \end{bmatrix} . \qquad \begin{matrix} (6.6.1) \\ (6.6.2) \\ (6.6.3) \end{matrix}$$

Next, differentiate (6.6.1) through (6.6.3) with respect to time, using the fact that $d\psi$, $d\nu$, and dt_f are constants, and $d\Omega = 0$:

$$\begin{bmatrix} \delta\dot{\lambda} \\ 0 \\ 0 \end{bmatrix} = \begin{bmatrix} \dot{S} & , \dot{R} & , \dot{m} \\ \dot{R}^T & , \dot{Q} & , \dot{n} \\ \dot{m}^T & , \dot{n}^T & \dot{\alpha} \end{bmatrix} \begin{bmatrix} \delta x \\ d\nu \\ dt_f \end{bmatrix} + \begin{bmatrix} S \\ R^T \\ m^T \end{bmatrix} [\delta\dot{x}] . \qquad \begin{matrix} (6.6.4) \\ (6.6.5) \\ (6.6.6) \end{matrix}$$

The perturbation equations are still as given by (6.1.21) through (6.1.25). Substituting the expressions for $\delta\dot{x}$ and $\delta\lambda$ from (6.1.21) and (6.1.22) into (6.6.4) through (6.6.6) and using (6.6.1) to eliminate $\delta\lambda$, we have

$$\begin{bmatrix} 0 \\ 0 \\ 0 \end{bmatrix} = \begin{bmatrix} \dot{S} + SA + A^TS - SBS + C & , \dot{R} + (A^T - SB)R & , \dot{m} + (A^T - SB)m \\ \dot{R}^T + R^T(A - BS) , & \dot{Q} - R^TBR , & \dot{n} - R^TBm \\ \dot{m}^T + m^T(A - BS) , & \dot{n}^T - m^TBR , & \dot{\alpha} - m^TBm \end{bmatrix} \cdot \begin{bmatrix} \delta x(t) \\ d\nu \\ dt_f \end{bmatrix} . \qquad (6.6.7)$$

If equations (6.6.7) are to be identities, true for all $\delta x(t)$, $d\nu$, dt_f, and if (6.5.9) through (6.5.11) are to be true at $t = t_f$, we must have

$$\dot{S} = -SA - A^T S + SBS - C\,, \qquad S(t_f) = \left(\frac{\partial^2 \Phi}{\partial x^2}\right)_{t=t_f}; \tag{6.6.8}$$

$$\dot{R} = -(A^T - SB)R\,, \qquad R(t_f) = \left(\frac{\partial \psi}{\partial x}\right)^T_{t=t_f}; \tag{6.6.9}$$

$$\dot{Q} = R^T BR\,, \qquad Q(t_f) = 0\,; \tag{6.6.10}$$

$$\dot{m} = -(A^T - SB)m\,, \qquad m(t_f) = \left(\frac{\partial \Omega}{\partial x}\right)^T_{t=t_f}; \tag{6.6.11}$$

$$\dot{n} = R^T Bm\,, \qquad n(t_f) = \left(\frac{d\psi}{dt}\right)_{t=t_f} \tag{6.6.12}$$

$$\dot{\alpha} = m^T Bm\,, \qquad \alpha(t_f) = \left(\frac{d\Omega}{dt}\right)_{t=t_f}. \tag{6.6.13}$$

Equation (6.6.8) is a matrix Riccati equation, whereas (6.6.9) and (6.6.11) are linear matrix equations, and (6.6.10), (6.6.12), and (6.6.13) are simply quadratures. Note that (6.6.8) through (6.6.10) are identical to (6.2.11) through (6.2.13).

If these equations are integrated backward from t_f to t_o, we can use (6.6.2) and (6.6.3), evaluated at $t = t_o$, to determine $d\nu$ and dt_f in terms of $\delta x(t_o)$ and $d\psi$, as follows:

$$d\nu = [\bar{Q}^{-1}(d\psi - \bar{R}^T \delta x)]_{t=t_o}, \tag{6.6.14}$$

$$dt_f = -\left[\left(\frac{m^T}{\alpha} - \frac{n^T}{\alpha}\bar{Q}^{-1}\bar{R}^T\right)\delta x + \frac{n^T}{\alpha}\bar{Q}^{-1}\,d\psi\right]_{t=t_o}, \tag{6.6.15}$$

where

$$\bar{Q} = Q - \frac{nn^T}{\alpha}, \tag{6.6.16}$$

$$\bar{R} = R - \frac{mn^T}{\alpha}. \tag{6.6.17}$$

Since we have $d\nu$ and dt_f from (6.6.14) and (6.6.15), $\delta\lambda(t_o)$ can be determined from (6.6.1):

$$\delta\lambda(t_o) = [(\bar{S} - \bar{R}\bar{Q}^{-1}\bar{R}^T)\,\delta x + \bar{R}\bar{Q}^{-1}\,d\psi]_{t=t_o}, \tag{6.6.18}$$

where

$$\bar{S} = S - \frac{mm^T}{\alpha}. \tag{6.6.19}$$

Now, with $\delta x(t_o)$ and $\delta\lambda(t_o)$ from (6.6.18), the perturbation equations

(6.1.21) and (6.1.22) may be integrated forward *once* to determine the neighboring optimum solution.

If $(d\Omega/dt)_{t=t_f} \neq 0$, (6.5.11) may be solved for dt_f in terms of $\delta x(t_f)$ and $d\nu$.

$$dt_f = \left\{ \left(\frac{d\Omega}{dt}\right)^{-1} \left[d\Omega - \frac{\partial \Omega}{\partial x}\,\delta x - \left(\frac{d\psi}{dt}\right)^T d\nu \right] \right\}_{t=t_f}. \tag{6.6.20}$$

This may be used in (6.5.9) and (6.5.10) to obtain

$$\begin{bmatrix} \delta\lambda(t_f) \\ \\ d\psi \end{bmatrix} = \begin{bmatrix} \dfrac{\partial^2 \Phi}{\partial x^2} - \left(\dfrac{\partial \Omega}{\partial x}\right)^T \left(\dfrac{d\Omega}{dt}\right)^{-1} \dfrac{\partial \Omega}{\partial x}, & \left(\dfrac{\partial \psi}{\partial x}\right)^T - \left(\dfrac{\partial \Omega}{\partial x}\right)^T \left(\dfrac{d\Omega}{dt}\right)^{-1} \left(\dfrac{d\psi}{dt}\right)^T \\ \\ \dfrac{\partial \psi}{\partial x} - \dfrac{d\psi}{dt}\left(\dfrac{d\Omega}{dt}\right)^{-1} \dfrac{\partial \Omega}{\partial x}, & -\dfrac{d\psi}{dt}\left(\dfrac{d\Omega}{dt}\right)^{-1} \left(\dfrac{d\psi}{dt}\right)^T \end{bmatrix}_{t=t_f} \begin{bmatrix} \delta x(t_f) \\ \\ d\nu \end{bmatrix}. \quad \begin{matrix} (6.6.21) \\ \\ (6.6.22) \end{matrix}$$

A simpler backward sweep is then possible:

$$\begin{bmatrix} \delta\lambda(t) \\ \\ d\psi \end{bmatrix} = \begin{bmatrix} \bar{S}(t)\ , \bar{R}(t) \\ \\ \bar{R}^T(t), \bar{Q}(t) \end{bmatrix} \begin{bmatrix} \delta x(t) \\ \\ d\nu \end{bmatrix}. \quad \begin{matrix} (6.6.23) \\ \\ (6.6.24) \end{matrix}$$

It is easily verified that $\bar{S}$, $\bar{R}$, $\bar{Q}$, satisfy the same differential equations as S, R, Q in (6.6.8) through (6.6.10); the difference lies in the terminal boundary conditions that are obtained from (6.6.21) and (6.6.22).

6.7 Sufficient conditions for a local minimum with final time unspecified

The existence of neighboring minimal paths when the final time is unspecified depends on three conditions (as in Section 6.3) with fixed final time:

$$H_{uu}(t) > 0 \qquad \text{for} \qquad t_0 \leq t \leq t_f, \tag{6.7.1}$$

$$\bar{Q}(t) < 0, \alpha(t) > 0 \qquad \text{for} \qquad t_0 \leq t < t_f, \tag{6.7.2}$$

and

$$\bar{S}(t) - \bar{R}(t)[\bar{Q}(t)]^{-1}\bar{R}^T(t) \text{ finite} \qquad \text{for} \qquad t_o \leq t < t_f, \qquad (6.7.3)$$

where $\bar{Q}$, $\bar{R}$, $\bar{S}$ are as defined in (6.6.16), (6.6.17), and (6.6.19). These are, again, the *convexity condition*, the *normality condition*, and the *no-conjugate-point condition*, respectively.

Sufficient conditions for a weak local minimum are the first-order necessary conditions (6.1.1) through (6.1.6), (6.5.1), *and* the second-order conditions (6.7.1), (6.7.2), and (6.7.3). Second-order necessary conditions for a minimum are weakened versions of (6.7.1) and (6.7.3):

$$H_{uu}(t) \geq 0, \qquad t_o \leq t \leq t_f, \qquad (6.7.4)†$$

$$\bar{S} - \bar{R}\bar{Q}^{-1}\bar{R}^T \text{ finite}, \qquad t_o < t < t_f. \qquad (6.7.5)$$

$$Q(t) \leqslant 0, \alpha(t) \leqslant 0 \qquad (6.7.6)$$

6.8 Perturbation feedback control with final time unspecified

The perturbation feedback control law for a neighboring optimum path is obtained by substituting (6.6.18) into (6.1.20):

$$\delta u(t) = -H_{uu}^{-1}[(H_{ux} + f_u^T(\bar{S} - \bar{R}\bar{Q}^{-1}\bar{R}^T))\,\delta x + f_u^T\bar{R}\bar{Q}^{-1}\,d\psi], \quad (6.8.1)$$

which is of the same form as (6.2.17). In addition, it will often prove desirable to predict the change in the final time from (6.6.15):

$$dt_f = -\left(\frac{m^T}{\alpha} - \frac{n^T}{\alpha}\bar{Q}^{-1}\bar{R}^T\right)\delta x - \frac{n^T}{\alpha}\bar{Q}^{-1}\,d\psi. \qquad (6.8.2)$$

If the problem has no explicit dependence on time, then time to go, $t_f - t$, is really the important time, and the gains in (6.8.1) should be "keyed" to $t_f - t$ rather than t. A first-order correction of t_f using (6.8.2) should, in most cases, be adequate for changing the time of applying gains in (6.8.1). This latter condition avoids the problem of "running out of gains" if actual $t_f >$ nominal t_f.

Example. *Perturbation guidance for atmospheric entry.*‡ Figure 6.8.1 shows an example of a perturbation guidance problem. The problem was to design guidance logic for an entry glider to guide it to horizontal flight ($\gamma = 0$) and a specified altitude ($h = 250{,}000$ ft in this case), while minimizing the energy loss due to aerodynamic drag (i.e., maximizing final velocity). First the *nominal* optimum path was computed from the nominal entry conditions ($V_o = 36{,}000$ ft sec^{-1}, $\gamma_o = -7.5°$, $h_o = 400{,}000$ ft), and this nominal path ($V_{\text{nom}}(t)$, $\gamma_{\text{nom}}(t)$, $h_{\text{nom}}(t)$), with its associated control history $\alpha_{\text{nom}}(t)$, was

†See remark associated with Equation (6.3.12).

‡Numerical computation for this example was done by J. Speyer based on data in Reference 2 of Chapter 6.

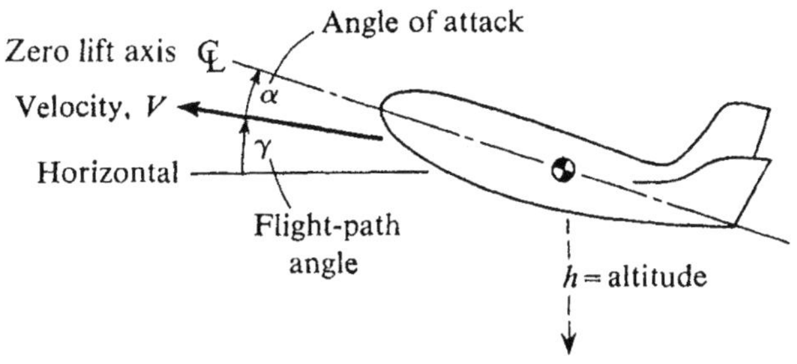

$$\alpha = \alpha_{nom} - C_1(V - V_{nom}) - C_2(\gamma - \gamma_{nom}) - C_3(h - h_{nom}),$$

7 numbers at 24 sample times $\cong$ 1700 bits

(nominal flight time = 240 secs)

Figure 6.8.1. Perturbation guidance of an entry glider to horizontal flight at a specified altitude.

stored at 10-second intervals along the 240 sec nominal path. Next, the perturbation feedback gains $C_1(t)$, $C_2(t)$, and $C_3(t)$ were computed and stored. A simulation was then made on the digital computer to see how well the guidance scheme behaved in the presence of initial condition deviations; Figures 6.8.2 and 6.8.3 show velocity-versus-altitude plots of the closed-loop trajectories for deviations of $\pm .6°$ in entry angle and ± 1000 ft sec^{-1} in entry velocity, respectively.

The guidance scheme behaved very well and, in fact, came very close to guiding the vehicle on the appropriate neighboring optimal path.

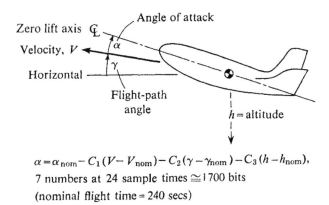

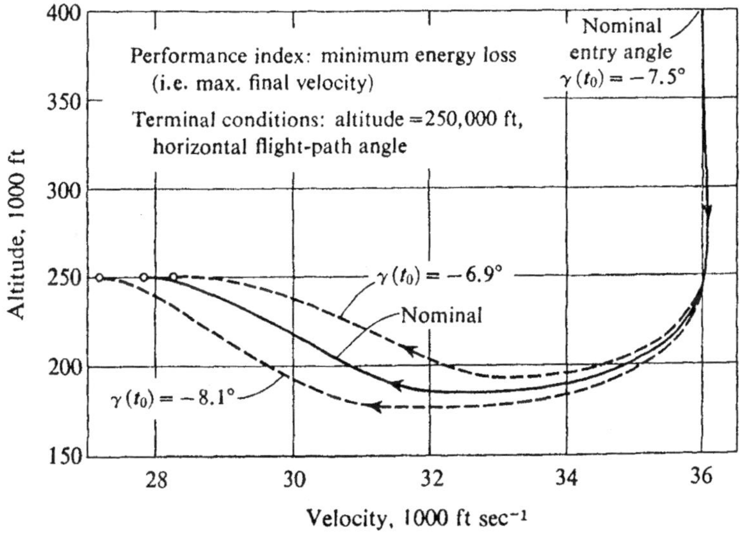

Figure 6.8.2. Entry angle perturbations.

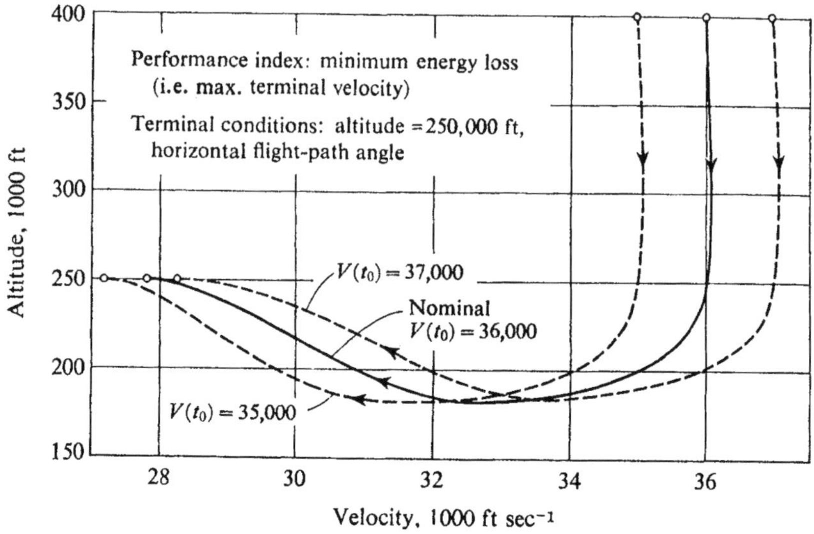

Figure 6.8.3. Entry velocity perturbations.

Problem 1. *Minimum-effort control with final time unspecified.* A nominal path $u(t) \Rightarrow x(t)$ satisfies end conditions $\psi[x(t_f),t_f] = 0$ at the nominal final time t_f. Find the perturbation control law that will cause the system to meet slightly different end conditions, $\psi[x(t_f),t_f] = d\psi$, with minimum perturbation effort. We have

$$E = \frac{1}{2} \int_t^{t_f} (\delta u)^T B \, \delta u \, dt ,$$

where

$$B(t) = \text{positive definite matrix.}$$

ANSWER.

$$\delta u(t) = -B^{-1} f_u^T \bar{R} \bar{Q}^{-1} [d\psi - \bar{R}^T \delta x] - B^{-1} f_u^T \frac{m m^T}{\alpha} \delta x ,$$

where

$$\dot{R} = -f_x^T R , \qquad R(t_f) = \left(\frac{\partial \psi}{\partial x} \right)^T_{t=t_f} ;$$

$$\dot{Q} = R^T f_u B^{-1} f_u^T R , \qquad Q(t_f) = 0 ;$$

$$\dot{m} = -f_x^T m , \qquad m(t_f) = \left(\frac{\partial \Omega}{\partial x} \right)_{t=t_f} ;$$

$$\dot{n} = R^T f_u B^{-1} f_u^T m\,, \qquad n(t_f) = \left(\frac{d\psi}{dt}\right)_{t=t_f}\,;$$

$$\dot{\alpha} = m^T f_u B^{-1} f_u^T m\,, \qquad \alpha(t_f) = \left(\frac{d\Omega}{dt}\right)_{t=t_f}\,;$$

and

$$\Omega = \nu^T \frac{d\psi}{dt}\,, \qquad \overline{R} = R - \frac{mn^T}{\alpha}\,, \qquad \overline{Q} = Q - \frac{nn^T}{\alpha}\,.$$

Problem 2. *Perturbation guidance for minimum time orbit injections.* Find the neighboring optimum feedback law for the problem given in Section 2.7, Problem 13. For hints, see Example, Section 6.4.

6.9 Sufficient conditions for a strong minimum

In our discussions so far, we have used

$$\delta J = \int_{t_0}^{t_f} H_u\, \delta u\, dt + 0(\|\delta u\|^2, \|\delta x\|^2)\,, \tag{6.9.1}$$

where the variations δu are "weak" variations, i.e., sufficiently small that higher-order terms $\|\delta x\|^2$, $\|\delta u\|^2$ are negligible. However, this may not be valid when we consider arbitrary variations in u that only assure negligible $\|\delta x\|^2$. This is the "strong" variation.† It will be desirable to consider an exact formula for δJ directly comparing two controls $u^1(t)$ and $u^o(t)$. To do this, let us define

$J^1(x,t)$ = Cost of starting the system in state x at a time t using the control $u^1(x,t)$,

$J^o(x,t)$ = Cost of starting the system in state x at time t using the control $u^o(x,t)$, which is optimal.

We shall assume that both u^1 and u^o are admissible controls. Now, by the usual dynamic programming arguments,‡ we immediately get the partial differential equations for J^1 and J^o as

$$J_t^1 + J_x^1 f(x,u^1,t) + L(x,u^1,t) = 0\,; \qquad J^1(x(t_f),t_f) = \phi(x(t_f),t_f)\,, \tag{6.9.2}$$

$$J_t^o + J_x^o f(x,u^o,t) + L(x,u^o,t) = 0\,; \qquad J^o(x(t_f),t_f) = \phi(x(t_f),t_f)\,, \tag{6.9.3}$$

†More precisely, strong variations only bound the magnitude of δx, while weak variations bound both the magnitude of δx and the magnitude of the derivative of δx.

‡The existence of solutions of (6.9.2) and (6.9.3) is tantamount to the assumption that a field of extremals exists for the optimization problem and that no conjugate points exist in the interval $t_0 - t_1$ (see Section 6.3, Problem 2).

where we have assumed for convenience that $\phi(x(t_f),t_f)$ includes a penalty function to reflect the presence of terminal constraints $\psi = 0$ if needed. Now let $\delta J \equiv J^1 - J^o$, and subtract (6.9.3) from (6.9.2):

$$J_t^1 - J_t^o + J_x^1 f^1 - J_x^o f^o + L^1 - L^o = 0\,,$$

$$J_t^1 - J_t^o + (J_x^1 - J_x^o) f^1 + J_x^o(f^1 - f^o) + (L^1 - L^o) = 0\,,$$

$$\frac{d}{dt}\delta J = -[J_x^o(f^1 - f^o) + (L^1 - L^o)] = -\delta H(x,J_x^o,u^1,u^o,t)\,, \tag{6.9.4}$$

where the time derivative is taken along path (1). Integrating both sides of (6.9.4) and noting that $\delta J[x(t_f),t_f] = 0$, we get

$$\delta J(x,t) = \int_{\text{path (1)}} \delta H(x,J_x^o,u^1,u^o,t)\,dt\,. \tag{6.9.5}$$†

Thus, a sufficient condition for a strong minimum (in addition to other conditions, implicitly assumed in connection with (6.9.2) and (6.9.3) is

$$\delta H = H(x,J_x^o,u^1,t) - H(x,J_x^o,u^o,t) > 0\,, \qquad \text{for all } t \text{ and } u^1 \neq u^o. \tag{6.9.6}$$

This is the so-called strengthened *Weierstrass condition*; that is, u^o minimizes the Hamiltonian along the optimal trajectory. Furthermore, using the mean value theorem, we can write

$$\delta H = (J_x^o f_{u^o} + L_{u^o})(u^1 - u^o) + \tfrac{1}{2}\|u^1 - u^o\|^2_{H_{uu}|_\theta} \tag{6.9.7}$$

where H_{uu} is evaluated at some point $u^o < \theta < u^1$. Since we have $H_u = 0$ on an optimal trajectory, we may replace (6.9.7) by

$$H_{uu} > 0 \text{ for all } x\,,\, u \text{ in the neighborhood of } u^o\,,\, x^o\,. \tag{6.9.8}$$

In other words, the strengthened convexity (Legendre-Clebsch) condition holds not only *on* the optimal trajectory but also *in a neighborhood* around the optimal trajectory. Furthermore, it turns out that (6.9.6) or (6.9.8), with the strict inequality sign replaced by "≥," is also necessary for the optimality of the control $u^o(t)$ (as was shown in Chapter 4 via the dynamic programming method).

To summarize, we have the following conditions for the optimality of a variational problem (assuming the problem is normal; see Section 6.3, Equation (6.3.2)):

NECESSARY CONDITION I. Euler Equations:

$$\dot{\lambda}^T = -H_x\,, \qquad H_u = 0\,,$$

and associated transversality conditions.

†Idea for this derivation due to Dr. R. E. Kalman.

NECESSARY CONDITION II. Legendre-Clebsch condition:

$$H_{uu} \geq 0 .$$

NECESSARY CONDITION III. Weierstrass condition:

$$\delta H(x, J_x^o, u^1, u^o, t) \geq 0 .$$

NECESSARY CONDITION IV. Nonexistence of conjugate point in (t_o, t_f).

Conditions I, II, and IV are necessary for weak minimum, while III is necessary for strong minimum. By strengthening II and III to strict inequalities and closing the interval in IV, that is, $[t_o, t_f]$, we get sufficient conditions.

Example. *Problem with a weak but not a strong minimum.* (From S. E. Dreyfus, *Dynamic Programming and the Calculus of Variations.* New York: Academic Press, 1965.) Consider

$$J = \int_0^1 u^3\, dt , \qquad \dot{x} = u , \qquad x(0) = 0 , \qquad x(1) = 1 .$$

The first-order necessary condition yields

$$H = \lambda u + u^3 , \qquad \dot{\lambda} = 0 , \qquad \lambda(1) = \nu = \lambda(t) , \qquad H_u = \lambda + 3u^2 ,$$

$$u = \sqrt{-\nu/3} ,$$

implying that the straight line, $x = t$, is an extremal. Furthermore, we have $H_{uu}|_{u=1} > 0$, and the conjugate-point condition can be shown to be satisfied also. Then we have a weak local minimum. On the other hand, on the optimal trajectory, we have $\delta H = u^3 - 1 - 3(u - 1) = (u - 1)^2 (u + 2)$, which is negative for $u < 2$ (or H is not at an absolute minimum). Thus the Weierstrass condition is violated. A curve that yields a smaller value of J than 1 but involves a discontinuity in the slope of x (i.e., strong variation) is shown in Figure 6.9.1.

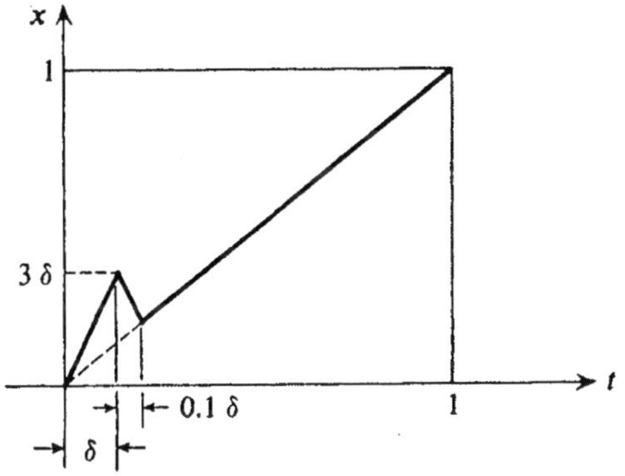

Figure 6.9.1. Problem with a weak, not a strong, minimum.

This example also illustrates the necessity of the Weierstrass condition. For another example illustrating the need for considering the minimum of H instead of merely verifying $H_{uu} > 0$, see the example in Section 3.12.

6.10 A multistage version of the backward sweep†

A backward-sweep method is presented in this section for determining neighboring extremal solutions for multistage problems of the type considered in Section 2.6. We assume that a nominal solution is available that satisfies all of the first-order necessary conditions (2.6.2), (2.6.3), (2.6.9), (2.6.10), and (2.6.12) of Section 2.6, which are linearized about the nominal solution:

$$dx(i+1) = f_x^i\,dx(i) + f_u^i\,du(i)\,, \qquad i = 0,\ldots,N-1\,, \tag{6.10.1}$$

$$dx(0) \text{ specified}, \tag{6.10.2}$$

$$d\psi(N) = \psi_x\,dx(N) \text{ specified}, \tag{6.10.3}$$

$$d\lambda(i) = H_{xx}^i\,dx(i) + (f_x^i)^T\,d\lambda(i+1) + H_{xu}^i\,du(i)\,, \qquad i = 0,\ldots,N-1\,, \tag{6.10.4}$$

$$d\lambda(N) = \Phi_{xx}\,dx(N) + \psi_x^T\,d\nu\,, \tag{6.10.5}$$

$$0 = H_{uu}^i\,du(i) + H_{ux}^i\,dx(i) + (f_u^i)^T\,d\lambda(i+1)\,, \qquad i = 0,\ldots,N-1\,, \tag{6.10.6}$$

$$f_x^i = \frac{\partial f^i}{\partial x(i)}\,, \qquad H_{xu}^i = \frac{\partial^2 H^i}{\partial x(i)\,\partial u(i)}\,, \qquad \text{etc.}$$

Equations (6.10.1) through (6.10.6) constitute a linear two-point boundary-value problem for $dx(i)$, $d\lambda(i)$, $du(i)$, and $d\nu$, since the coefficients are evaluated on the nominal solution. In principle, one can solve it by the discrete analog of the transition matrix method used in Section 5.2, but again, this may present formidable numerical difficulties if the system is dissipative.

The discrete analog of the sweep method used in Section 6.2 generates a sequence of relations equivalent to (6.10.3) and (6.10.5) of the form

$$d\lambda(i) = S(i)\,dx(i) + R(i)\,d\nu\,, \tag{6.10.7}$$

$$d\psi = R^T(i)\,dx(i) + Q(i)\,d\nu\,. \tag{6.10.8}$$

---

†This section is based on the Harvard University. Ph.D. Thesis of S. R. McReynolds (1966).

Assuming that (6.10.7) and (6.10.8) are known for $i = k + 1$, Equations (6.10.1) through (6.10.6) can be used to obtain equivalent relations for $i = k$ as follows:

STEP 1. Substitute (6.10.1), with $i = k$, into (6.10.7) and 6.10.8, with $i = k + 1$:

$$d\lambda(k+1) = S(k+1)\,[f_x^k\,dx(k) + f_u^k\,du(k)] + R(k+1)\,d\nu\,, \quad (6.10.9)$$

$$d\psi = R^T(k+1)\,[f_x^k\,dx(k) + f_u^k\,du(k)] + Q(k+1)\,d\nu\,. \quad (6.10.10)$$

STEP 2. Place $k = i$ in (6.10.9) and substitute the expression for $d\lambda(i + 1)$ into (6.10.4) and (6.10.6):

$$d\lambda(i) = [H_{xx}^i + (f_x^i)^T S(i+1) f_x^i]\,dx(i) + [H_{xu}^i + (f_x^i)^T S(i+1) f_u^i]\,du(i) + (f_x^i)^T R(i+1)\,d\nu\,, \quad (6.10.11)$$

$$0 = [H_{uu}^i + (f_u^i)^T S(i+1) f_u^i]\,du(i) + [H_{ux}^i + (f_u^i)^T S(i+1) f_x^i]\,dx(i) + (f_u^i)^T R(i+1)\,d\nu\,. \quad (6.10.12)$$

STEP 3. Assuming that the coefficient of $du(i)$ in (6.10.12) is nonsingular, Equation (6.10.12) may be solved for $du(i)$:

$$du(i) = -[Z_{uu}(i)]^{-1}\,[Z_{ux}(i)\,dx(i) + Z_{u\nu}(i)\,d\nu]\,, \quad (6.10.13)$$

where

$$Z_{uu}(i) = H_{uu}^i + (f_u^i)^T S(i+1) f_u^i\,, \quad (6.10.14)$$

$$Z_{ux}(i) = H_{ux}^i + (f_u^i)^T S(i+1) f_x^i = [Z_{xu}(i)]^T\,, \quad (6.10.15)$$

$$Z_{u\nu}(i) = (f_u^i)^T R(i+1) = [Z_{\nu u}(i)]^T\,. \quad (6.10.16)$$

STEP 4. Equation (6.10.13) is now used to eliminate $du(i)$ from (6.10.11) and (6.10.10):

$$d\lambda(i) = \{Z_{xx}(i) - Z_{xu}(i)\,[Z_{uu}(i)]^{-1}\,Z_{ux}(i)\}\,dx(i) + \{Z_{x\nu}(i) - Z_{xu}(i)\,[Z_{uu}(i)]^{-1}\,Z_{u\nu}(i)\}\,d\nu\,, \quad (6.10.17)$$

$$d\psi = \{Z_{\nu x}(i) - Z_{\nu u}(i)\,[Z_{uu}(i)]^{-1}\,Z_{ux}(i)\}\,dx(i) + \{Q(i+1) - Z_{\nu u}(i)\,[Z_{uu}(i)]^{-1}\,Z_{u\nu}(i)\}\,d\nu\,, \quad (6.10.18)$$

where

$$Z_{xx}(i) = H_{xx}^i + (f_x^i)^T S(i+1) f_x^i\,, \quad (6.10.19)$$

$$Z_{x\nu}(i) = (f_x^i)^T R(i+1) = [Z_{\nu x}(i)]^T\,. \quad (6.10.20)$$

STEP 5. For (6.10.17) and (6.10.18) to be equivalent to (6.10.7) and (6.10.8), the coefficients must be equal:

$$S(i) = (f_x^i)^T S(i+1) f_x^i + H_{xx}^{(i)} - [H_{ux}^i + (f_u^i)^T S(i+1) f_x^i]^T [H_{uu}^i + (f_u^i)^T S(i+1) f_u^i]^{-1} [H_{ux}^i + (f_u^i)^T S(i+1) f_x^i], \tag{6.10.21}$$

$$R(i) = \{f_x^i - f_u^i [H_{uu}^i + (f_u^i)^T S(i+1) f_u^i]^{-1} [H_{ux}^i + (f_u^i)^T S(i+1) f_x^i]\}^T R(i+1), \tag{6.10.22}$$

$$Q(i) = Q(i+1) - [R(i+1)^T f_u^i [H_{uu}^i + (f_u^i)^T S(i+1) f_u^i]^{-1} (f_u^i)^T R(i+1). \tag{6.10.23}$$

These are the discrete analogs of (6.2.11) through (6.2.13). For another version, see Problem in this section.

The relations (6.10.21) through (6.10.23) may be regarded as recursive relations, and $S(i)$, $R(i)$, and $Q(i)$ may be calculated sequentially, in descending order, from $i = N-1$ to $i = 0$, starting with boundary conditions obtained from (6.10.3) and (6.10.5):

$$S(N) = \Phi_{xx}(N), \tag{6.10.24}$$

$$R(N) = \{\psi_x[x(N)]\}^T, \tag{6.10.25}$$

$$Q(N) = 0. \tag{6.10.26}$$

Improved values of ν and $u(i)$ can then be obtained by adding $d\nu$ and $du(i)$ to the values on the previous nominal, where, from (6.10.8),

$$d\nu = [Q(0)]^{-1} [d\psi - R^T(0)\, dx(0)], \tag{6.10.27}$$

and $du(i)$ is given in (6.10.13), using (6.10.27). These improved values of ν and $u(i)$ can be used to produce a new nominal that will come closer to satisfying the boundary conditions on $x(0)$ and $\psi[x(t_f)]$.

As in Section 6.2, improved values of ν and $u(i)$ could be found at every stage in the manner of a sampled-data *feedback law*:

$$d\nu = [Q(i)]^{-1} [d\psi - R^T(i)\, dx(i)], \tag{6.10.28}$$

$$du(i) = -[Z_{uu}^i]^{-1} [\{H_{ux}^i + (f_u^i)^T [S(i+1) f_x^i - R(i+1) Q^{-1}(i) R^T(i)]\}\, dx(i) + (f_u^i)^T R(i+1) Q^{-1}(i)\, d\psi]. \tag{6.10.29}$$

Equation (6.10.29) is the *neighboring optimum feedback law*, which will produce desired changes in the terminal conditions, $d\psi$, given the present deviation from the nominal optimum path, $dx(i)$, while minimizing J.

Problem. For the case where $S(i)$ is nonsingular, use the matrix inversion lemma (see Section 1.3, Problem 4, and Section 12.2) to show that the recursion relations, (6.10.21) through (6.10.23), may be written as

$$S(i) = A^T(i)\{[S(i+1)]^{-1} + B(i)\}^{-1} A(i) + C(i),$$
$$R(i) = A^T(i)\,[I + S(i+1)\,B(i)]^{-1}\,R(i+1),$$
$$Q(i) = Q(i+1) - R^T(i+1)\,B(i)\,[I + S(i+1)\,B(i)]^{-1}\,R(i+1),$$

where

$$A(i) = f_x^i - f_u^i(H_{uu}^i)^{-1}H_{ux}^i, \qquad B(i) = f_u^i(H_{uu}^i)^{-1}(f_u^i)^T,$$
$$C(i) = H_{xx}^i - H_{xu}^i(H_{uu}^i)^{-1}H_{ux}^i.$$

(See Problem, Section 2.2, for the case with no terminal constraints.)

6.11 Sufficient conditions for a local minimum for multistage systems

In the previous section, the existence of a neighboring minimal path depended on three conditions:

$$Z_{uu}^i = H_{uu}^i + (f_u^i)^T S(i+1) f_u^i > 0, \qquad (6.11.1)$$

$$Q(i) < 0, \qquad (6.11.2)$$

$$S(i+1)f_x^i - R(i+1)Q^{-1}(i)R^T(i) \text{ finite}; \qquad i = 0, \ldots, N-1. \qquad (6.11.3)$$

These are the analogous conditions to the convexity, normality, and conjugate-point conditions discussed in Section 6.3. Note that the convexity condition is not simply $H_{uu} > 0$, as it was in the continuous-time case. In other words, it is not necessary that H even be a local minimum for minimality. The discrete minimum (maximum) principle is not true without additional qualifications. This is because, in discrete systems, the finite step size prevents us from making variations with arbitrary amplitude, while we must do so to establish the minimality of H. Equation (6.11.3) is, in a sense, redundant since, for finite Q, the equations for S, R cannot become infinite in a finite number of steps.

Numerical solution of optimal programming and control problems

7

7.1 Introduction

Unless the system equations, the performance index, and the constraints are quite simple, we must employ numerical methods to solve optimal programming and control problems. However, the amount of numerical computation required for even a relatively simple problem is forbidding if it must be done by hand. This is why the calculus of variations found very little use in engineering and applied science until quite recently. The development of economical, high-speed computers in the mid-1950's has dramatically changed this situation. It is now possible to solve complicated optimal programming and control problems in a reasonable length of time and at a reasonable cost.

High-speed computers solve *initial-value problems* for sets of differential equations very readily. However, as we have seen, optimal programming and control problems are at least *two-point boundary-value problems* and, in some cases, multipoint boundary-value problems (e.g., when there are interior point constraints or state variable inequality constraints). Finding solutions to these nonlinear two-point boundary-value problems is, in many cases, *not* a trivial extension of finding solutions to initial-value (one-point boundary-value) problems.

The nonlinear two-point boundary-value problem encountered in a large class of optimal programming problems was summarized in Section 2.8. The problem is to find

(a) the n state variables, $x(t)$,
(b) the n influence functions, $\lambda(t)$,
(c) the m control variables, $u(t)$,

to satisfy, simultaneously,

(i) the n system differential equations (involving x, u),
(ii) the n influence (adjoint, Euler-Lagrange) differential equations (involving λ, x, u),
(iii) the m optimality conditions (involving λ, x, u),
(iv) the initial and final boundary conditions (involving x, λ).

All numerical methods for the solution of such problems necessarily involve either flooding or iterative procedures.

Flooding (or dynamic programming), as applied to two-point boundary-value problems, can be described as a process of generating many solutions satisfying the specified boundary conditions at one end, using the unspecified boundary conditions as parameters. If the correct range of parameters is chosen, some of the solutions will pass through (or near) the desired boundary conditions at the other end.

At the present time, all the proposed iterative procedures use "successive linearization." A nominal solution is chosen that satisfies none, one, two, or three of the four conditions (i) through (iv) above; then this nominal solution is modified by successive linearization so that the remaining conditions are also satisfied. Only three of the fifteen possible approaches have been used extensively so far, namely, the three shown in Table 7.1.1.

Table 7.1.1 Iterative Procedures

	Nominal Solution Satisfies:			
	(i) System equations	(ii) Influence equations	(iii) Optimality conditions	(iv) Boundary conditions
Neighboring extremal methods	Yes	Yes	Yes	No
Gradient methods	Yes	Yes	No	No
Quasilinearization methods	No	No	Yes	Yes or No

When using neighboring extremal methods and quasilinearization methods, we must solve a succession of *linear* two-point boundary-value problems. Such problems can be solved by (a) finding the transition matrix between unspecified boundary conditions at one end and specified boundary conditions at the other end, or by (b) "sweeping" the boundary conditions from one end point to the other end point, which involves solving a matrix Riccati equation (see Sections 6.2 and 6.6).

For all three classes of iterative procedures, it is possible to handle terminal constraints either by (a) gradient projection (linear penalty functions) or by (b) nonlinear (usually quadratic) penalty functions.

7.2 Extremal field methods; dynamic programming

One method of solving optimal programming problems is systematically to choose values for the unspecified initial (or terminal) conditions and compute the corresponding optimal solutions from the initial (or terminal) point. Computation continues until the region of the state space in and around the opposite point is so well covered with optimal solutions that it is possible to interpolate the desired optimal solution. Clearly, this is one way of solving the Hamilton-Jacobi-Bellman (HJB) equation in a certain domain of the state space (by the method of "characteristics")† and it would be useful for generating the optimal nonlinear feedback law for terminal control if all the optimal paths were computed *backward* from the terminal hypersurface. Relatively simple examples of nonlinear optimal feedback control schemes computed in this way were given in Sections 4.1 and 4.3.

Another possibility is to solve the HJB partial differential equation directly, starting at the terminal hypersurface. This is the procedure called dynamic programming, discussed in Chapter 4. For problems with more than two or three state variables, this procedure is not feasible even on the larger present-day computers. Even storing the answer (the whole extremal field) for a problem with three or more state variables usually requires an impractically large amount of space.

7.3 Neighboring extremal algorithms

Introduction

These methods are characterized by iterative algorithms for improving estimates of the unspecified initial (or terminal) conditions so as to satisfy the specified terminal (or initial) conditions.

The main difficulty with these methods is *getting started*; i.e., finding a first estimate of the unspecified conditions at one end that produces a solution reasonably close to the specified conditions at the other end. The reason for this peculiar difficulty is that extremal solutions are often *very sensitive* to small changes in the unspecified boundary conditions. This extraordinary sensitivity is a direct result of the nature of the Euler-Lagrange equations, which, as we have seen

†R. Courant and D. Hilbert, *Methods of Mathematical Physics*, Vol. II, New York: Interscience, 1962, Chapter 2.

in Chapter 2, are influence function equations. They are, in fact, adjoint differential equations to the linear perturbation system equations, where the linearization is made about an extremal path. If the fundamental solutions to the linear perturbation system equations decrease in magnitude with increasing time, the fundamental solutions of the adjoint (Euler-Lagrange) equations increase in magnitude with increasing time. Thus, the solutions $x(t)$ and $\lambda(t)$ of the differential equations tend to become widely different in orders of magnitude as the integration proceeds in either direction. Since the number of significant digits that can be carried in a computer is limited regardless of whether fixed-point or floating-point arithmetic operations are used, the different growth of $x(t)$ and $\lambda(t)$ contributes greatly to the loss of accuracy.† One manifestation of this problem is that transition matrix solutions become ill-conditioned as the elements take on widely different ranges of value.‡ Since inversion of the transition matrix is necessarily part of the numerical method, the resultant accuracy is very poor. Another aspect of the same problem is the fact that small errors in guessing the influence functions at the initial time may produce enormous errors in the influence functions at the terminal time. This will be especially noticeable in highly dissipative systems, such as systems with friction or drag. Since the system equations and the Euler-Lagrange equations are coupled together, it is not unusual for the numerical integration, with poorly guessed initial conditions, to produce "wild" trajectories in the state space. These trajectories may be so wild that values of $x(t)$ and/or $\lambda(t)$ exceed the numerical range of the computer!

In view of this starting difficulty, the direct integration method is usually practical only for finding neighboring extremal solutions after one extremal solution is obtained by some other method, such as a gradient method.

Some state variables specified at a fixed terminal time.

To set forth the basic ideas of this method, we consider first the relatively simple class of problems treated in Section 2.4: Find $u(t)$ to minimize

$$J = \phi[x(t_f)] + \int_{t_0}^{t_f} L[x(t),u(t),t]\, dt\,, \tag{7.3.1}$$

where

$$\dot{x} = f(x,u,t)\,, \tag{7.3.2}$$

$$x(t_o) \quad \text{is specified,} \tag{7.3.3}$$

†See Kalman (1965).

‡Large values all appear to be equal and small values become zero.

$$\psi[x(t_f)] = \begin{bmatrix} x_1(t_f) - x_1^f \\ \cdot \\ \cdot \\ \cdot \\ x_q(t_f) - x_q^f \end{bmatrix} = 0\,, \qquad x_1^f, \ldots, x_q^f \ \text{specified}, \tag{7.3.4}$$

$$t_o, t_f \ \text{specified}. \tag{7.3.5}$$

First-order necessary conditions for an extremal solution are

$$\dot{\lambda} = -\left(\frac{\partial f}{\partial x}\right)^T \lambda - \left(\frac{\partial L}{\partial x}\right)^T, \tag{7.3.6}$$

$$\lambda_j(t_f) = \left(\frac{\partial \phi}{\partial x_j}\right)_{t=t_f}, \qquad j = q+1, \ldots, n, \tag{7.3.7}$$

$$0 = \left(\frac{\partial f}{\partial u}\right)^T \lambda + \left(\frac{\partial L}{\partial u}\right)^T. \tag{7.3.8}$$

The differential equations (7.3.2) and (7.3.6) are to be solved with the n initial conditions (7.3.3) and the n final conditions (7.3.4) plus (7.3.7), using (7.3.8) to determine $u(t)$. There are n unspecified initial conditions, $\lambda(t_o)$, and n unspecified terminal conditions, $[\lambda_1(t_f), \ldots, \lambda_q(t_f), x_{q+1}(t_f), \ldots, x_n(t_f)]$.

A Transition-Matrix Algorithm. This class of problems may be treated as follows:

STEP (a). Guess the n unspecified initial conditions, $\lambda(t_o)$. (Alternatively, guess the n unspecified final conditions with obvious modifications to the succeeding steps.)

STEP (b). Integrate (7.3.2) and (7.3.6) forward from t_o to t_f, using (7.3.8) to determine $u(t)$.

STEP (c). Record $x_1(t_f), \ldots, x_q(t_f), \lambda_{q+1}(t_f), \ldots, \lambda_n(t_f)$.

STEP (d). Determine the $(n \times n)$ transition matrix $[\partial\mu(t_f)/\partial\lambda(t_o)]$, where

$$\delta\mu(t_f) = \begin{bmatrix} \delta x_1(t_f) \\ \cdot \\ \cdot \\ \cdot \\ \delta x_q(t_f) \\ \delta\lambda_{q+1}(t_f) \\ \cdot \\ \cdot \\ \cdot \\ \delta\lambda_n(t_f) \end{bmatrix} = \frac{\partial\mu(t_f)}{\partial\lambda(t_o)}\,\delta\lambda(t_o)\,.$$

(See two methods for doing this below.)

STEP (e). Choose $\delta\mu(t_f)$ so as to bring the next solution *closer* to desired values of $\mu(t_f)$. For example, one might choose $\delta\mu(t_f) = -\epsilon[\mu(t_f) - \mu^f], 0 < \epsilon \leq 1$.

STEP (f). With chosen values of $\delta\mu(t_f)$ from Step (e), invert the transition matrix of Step (d) to find $\delta\lambda(t_o)$:

$$\delta\lambda(t_o) = \left[\frac{\partial\mu(t_f)}{\partial\lambda(t_o)}\right]^{-1} \delta\mu(t_f).$$

STEP (g). Using $\lambda(t_o)_{\text{new}} = \lambda(t_o)_{\text{old}} + \delta\lambda(t_o)$, repeat Steps (a) through (f) until $\mu(t_f)$ has the specified values to the desired accuracy.

If the changes $\delta\mu(t_f)$ in Step (e) are too large, the iterative procedure may not converge. One way to check "size" is to compare $\mu(t_f)_{\text{new}} - \mu(t_f)_{\text{old}}$ with desired $\delta\mu(t_f)$; if they differ by more than, say, 10 to 20%, Step (e) should be repeated with smaller values of $\delta\mu(t_f)$.

The transition matrix in Step (d) above may be found in two different ways: (1) direct numerical differentiation, and (2) determination of unit solutions for the linear perturbation equations.

Direct numerical differentiation requires n additional integrations of the nonlinear system (7.3.2) and (7.3.6) using (7.3.8). On each of these integrations, one of components $\lambda_i(t_o)$ is changed by a small amount from the original guess in Step (a). The n quantities $\delta\mu(t_f)$ are recorded for each integration and divided by $\delta\lambda_i(t_o)$. In this way the transition matrix $[\partial\mu(t_f)/\partial\lambda(t_o)]$ is found. The difficulty with this straightforward approach is the following: If $\delta\lambda_i(t_o)$ is chosen too small, truncation error in integrating the nonlinear differential equations makes the determination of $\delta\mu(t_f)$ very inaccurate; if $\delta\lambda_i(t_o)$ is chosen too large, the linearity assumption will not be valid.

Determination of unit solutions requires n integrations of the $2n$th-order linear perturbation equations (6.1.21) and (6.1.22). On each of these integrations, one of the components of $\delta\lambda(t_o)$ is placed equal to unity, with all the other components zero and $\delta x(t_o) = 0$. This method is more accurate than direct numerical differentiation, but it requires additional computer programming. Furthermore, it may still produce an ill-conditioned transition matrix if unit solutions differ widely in numerical magnitude; the inversion required in Step (f) is then very inaccurate.†

A Backward-Sweep Algorithm. An effective way around the difficulty of an ill-conditioned transition matrix is offered by the following adaptation of the sweep method discussed in Section 6.2:

†See Problem 5, Section 5.2 for a way around the inversion of ill-conditioned matrices.

STEP (a). Estimate the q parameters $\nu^T = [\lambda_1(t_f), \ldots, \lambda_q(t_f)]$ and the $n - q$ unspecified terminal state variables $[x_{q+1}(t_f), \ldots, x_n(t_f)]$.

STEP (b). Integrate (7.3.2) and (7.3.6) *backward* from t_f to t_0 using (7.3.8) to determine $u(t)$, with boundary conditions (7.3.4), (7.3.7), and the estimated quantities in Step (a).

STEP (c). Simultaneously with Step (b), integrate (6.2.11) through (6.2.13) backward with boundary conditions (6.2.3) through (6.2.5), which in this case are

$$S(t_f) = \phi_{xx}\big|_{t=t_f}, \quad Q(t_f) = 0, \quad \text{and}$$

$$R_{ij}(t_f) = \begin{cases} 1, & i = j, \\ 0, & i \neq j; \end{cases} \qquad i = 1, \ldots, n, \qquad j = 1, \ldots, q.$$

STEP (d). Record $x(t_0)$, $\lambda(t_0)$, and $(S - R^T Q^{-1} R)_{t=t_0}$. Choose $\delta x(t_0)$ to come closer to the specified values of $x(t_0)$. Then, from (6.2.15), we have $\delta\lambda(t_0) = (S - R^T Q^{-1} R)_{t=t_0} \delta x(t_0)$.

STEP (e). Integrate the perturbation equations (6.1.21) and (6.1.22) *forward* with boundary conditions $\delta x(t_0)$ and $\delta\lambda(t_0)$ from Step (d). Record $d\nu^T = [\delta\lambda_1(t_f), \ldots, \delta\lambda_q(t_f)]$ and $[\delta x_{q+1}(t_f), \ldots, \delta x_n(t_f)]$.

STEP (f). Using $\nu_{\text{new}} = \nu_{\text{old}} + d\nu$ and $(x_i(t_f))_{\text{new}} = [x_i(t_f)]_{\text{old}} + \delta x_i(t_f)$, $i = q + 1, \ldots, n$, repeat Steps (a) through (f) until $x(t_0)$ has the specified values to the desired accuracy.

Functions of the state variables specified at an unspecified terminal time.

We now consider the more general problem of finding $u(t)$ to minimize

$$J = \phi[x(t_f), t_f] + \int_{t_0}^{t_f} L[x(t), u(t), t]\, dt \qquad \text{(performance index)}, \quad (7.3.9)$$

where

$$\dot{x} = f(x, u, t) \qquad (n \text{ equations}). \qquad (7.3.10)$$

$$x(t_0) = x^0; \qquad t_0 \text{ and } x^0 \text{ specified} \qquad (n \text{ initial conditions}), \quad (7.3.11)$$

$$\psi[x(t_f), t_f] = 0 \qquad (q \text{ terminal conditions}). \qquad (7.3.12)$$

The terminal time, t_f, is determined *implicitly* by the terminal conditions (7.3.12).

First-order necessary conditions for an extremal solution are

$$\dot{\lambda} = -\left(\frac{\partial H}{\partial x}\right)^T, \qquad (7.3.13)$$

$$0 = \frac{\partial H}{\partial u}; \qquad (7.3.14)$$

$$\lambda^T(t_f) = \left(\frac{\partial \Phi}{\partial x}\right)_{t=t_f}, \tag{7.3.15}$$

$$\Omega[x,u,\nu,t]_{t=t_f} \equiv \left(\frac{d\Phi}{dt} + L\right)_{t=t_f} = 0, \tag{7.3.16}$$

where

$$\Phi(x,\nu,t) = \phi(x,t) + \nu^T\psi(x,t), \qquad \frac{d\Phi}{dt} = \frac{\partial \Phi}{\partial t} + \frac{\partial \Phi}{\partial x} f.$$

The $2n$ differential equations (7.3.10) and (7.3.13) must be solved, and the $q+1$ parameters ν and t_f must be determined to satisfy the n initial conditions (7.3.11) and the $q+n+1$ terminal conditions (7.3.12), (7.3.15), and (7.3.16), using (7.3.14) to determine $u(t)$.

A Transition-Matrix Algorithm. The following procedure may be used.

STEP (a). Guess the n terminal conditions $x(t_f)$, the q parameters ν, and the terminal time t_f.

STEP (b). Determine $\psi[x(t_f),t_f]$, and $\lambda(t_f)$, $\Omega[x(t_f),u(t_f),\nu,t_f]$ from (7.3.15) and (7.3.16); $u(t_f)$ must be determined from (7.3.14) evaluated at $t = t_f$, using $\lambda(t_f)$ and $x(t_f)$.

STEP (c). Integrate (7.3.10) and (7.3.13) *backward* from t_f to t_o, using (7.3.14) to determine $u(t)$, and using the terminal conditions $x(t_f)$, $\lambda(t_f)$ from Steps (a) and (b).

STEP (d). Record $x(t_o)$.

STEP (e). Find the $[(n+q+1) \times (n+q+1)]$ transition matrix

$$\frac{\partial[x(t_o),\psi,\Omega]}{\partial[x(t_f),\nu,t_f]},$$

where

$$\begin{bmatrix} \delta x(t_o) \\ d\psi \\ d\Omega \end{bmatrix} = \frac{\partial[x(t_o),\psi,\Omega]}{\partial[x(t_f),\nu,t_f]} \begin{bmatrix} \delta x(t_f) \\ d\nu \\ dt_f \end{bmatrix}.$$

STEP (f). Choose values of $\delta x(t_o)$, $d\psi$, and $d\Omega$ so as to bring the next solution closer to the desired values of $x(t_o)$, $\psi = 0$, and $\Omega = 0$. For example, one might choose

$$\begin{bmatrix} \delta x(t_o) \\ d\psi \\ d\Omega \end{bmatrix} = -\epsilon \begin{bmatrix} x(t_o) - x^o \\ \psi[x(t_f),t_f] \\ \Omega[x(t_f),t_f] \end{bmatrix}, \qquad 0 < \epsilon \leq 1.$$

STEP (g). With chosen values of $\delta x(t_o)$, $d\psi$, and $d\Omega$ from Step (f), invert the transition matrix of Step (e) to find $\delta x(t_f)$, $d\nu$, and dt_f.

STEP (h). Using

$$\begin{bmatrix} x(t_f) \\ \nu \\ t_f \end{bmatrix}_{\text{new}} = \begin{bmatrix} x(t_f) \\ \nu \\ t_f \end{bmatrix}_{\text{old}} + \begin{bmatrix} dx(t_f) \\ d\nu \\ dt_f \end{bmatrix},$$

repeat Steps (a) through (h) until $x(t_0) = x^o$, $\psi[x(t_f),t_f] = 0$, and $\Omega[x(t_f),u(t_f),\nu,t_f] = 0$ to the desired accuracy. Note that $dx(t_f) = \delta x(t_f) + \dot{x}(t_f)\, dt_f$.

If the changes in Step (f) are too large, the iterative procedure will not converge. One way to avoid this is to compare the actual changes in $x(t_0)$, ψ, and Ω with desired changes; if they differ by more than, say, 10 to 20%, Step (h) should be repeated with smaller values of $\delta x(t_0)$, $d\psi$, and $d\Omega$.

The transition matrix in Step (e) may be found in two different ways: (1) direct numerical differentiation and (2) determination of unit solutions for the linear perturbation equations.

Direct numerical differentiation requires $n + q + 1$ additional backward integrations of the nonlinear system (7.3.10) and (7.3.13) using (7.3.24). On each of these integrations, one of the components of $x(t_f)$, ν, and t_f is changed by a small amount from the original guess in Step (a). The $n + q + 1$ quantities $\delta x(t_0)$, $d\psi$, and $d\Omega$ are recorded for each choice and divided by the change from the original guess. In this way the transition matrix in Step (e) can be found. The difficulty with this straightforward approach is the same as that mentioned in the previous section.

Determination of unit solutions requires $n + q + 1$ backward integrations of the $2n$th-order linear perturbation equations (6.1.21) and (6.1.22). On each of these integrations, one of the components of $[\delta x(t_f),d\nu,dt_f]$ is placed equal to unity with all the other components zero. The determination of $\delta\lambda(t_f)$, $d\psi$, and $d\Omega$ is made by considering the first-order perturbations of the terminal conditions (7.3.15), (7.3.12), and (7.3.16) which were given in (6.5.9) through (6.5.11). Equations (6.5.10) and (6.5.11) form part of the transition matrix in Step (e), whereas $\{\partial x(t_0)/\partial[x(t_f),\nu,t_f]\}$ must be found by integrating the perturbation equations backwards $n + q + 1$ times with $\delta\lambda(t_f)$ obtained from (6.5.9) for unit values of $\delta x(t_f)$, $d\nu$, and dt_f. While this method is more accurate than direct numerical differentiation, it obviously requires more computer programming and is prone to the same difficulties as described in the previous section.

Note that a necessary condition for a minimum is

$$\left(\frac{d\Omega}{dt}\right)_{t=t_f} \geq 0. \tag{7.3.17}$$

If the inequality holds in Equation (7.3.17), we may solve (6.5.11) for dt_f in terms of $\delta x(t_f)$ and $d\nu$ and substitute the result into (6.5.9) and (6.5.10); in this case, only $n + q$ unit solutions need be found to determine the transition matrix.

A Backward-Sweep Algorithm. One of the possible difficulties with the transition-matrix approach is that sufficient numerical accuracy may not be attainable even with the technique of finding unit solutions for the linear perturbation equations. This is particularly true for dissipative systems, for reasons mentioned earlier in this section. Usually, an effective way to avoid this difficulty is to use the following adaptation of the ***backward-sweep approach***† presented in Section 6.6:

STEPS (a), (b), and (c). Same as in the Transition-Matrix Algorithm.

STEP (d). Simultaneously with Step (c), integrate (6.6.8) through (6.6.13) with boundary conditions given there.

STEP (e). Record x, λ, S, R, Q, m, n, α at $t = t_0$. Choose $\delta x(t_0)$, $d\psi$, and $d\Omega$ as in Step (f) of the Transition Matrix Algorithm. Then, from (6.6.14), (6.6.15), and (6.6.18), determine $d\nu$, dt_f, and $\delta\lambda(t_0)$. Record $d\nu$ and dt_f.

STEP (f). Integrate the perturbation equations (6.1.21) and (6.1.22) *forward* with boundary conditions $\delta x(t_0)$ and $\delta\lambda(t_0)$. Record $dx(t_f) = \delta x(t_f) + \dot{x}(t_f)\, dt_f$.

STEP (g). Using

$$\begin{bmatrix} x(t_f) \\ \nu \\ t_f \end{bmatrix}_{\text{new}} = \begin{bmatrix} x(t_f) \\ \nu \\ t_f \end{bmatrix}_{\text{old}} + \begin{bmatrix} dx(t_f) \\ d\nu \\ dt_f \end{bmatrix},$$

repeat Steps (a) through (g) until $x(t_0) = x^0$, $\psi[x(t_f),t_f] = 0$, and $\Omega[x(t_f),u(t_f),\nu,t_f] = 0$ to the desired accuracy.

7.4 First-order gradient algorithms

Introduction

Gradient methods were developed to surmount the "initial guess" difficulty associated with direct integration methods (see Introduction of Section 7.3). They are characterized by iterative algorithms for improving estimates of the control histories, $u(t)$, so as to come closer to satisfying the optimality conditions and the boundary conditions.

†The reason for this is that there is less chance of difference in growth in the elements of $S(t)$ than in $X(t)$ and $\Lambda(t)$, used in the transition matrix algorithm.

First-order gradient methods usually show great improvements in the first few iterations but have poor convergence characteristics as the optimal solution is approached. Second-order gradient methods, discussed in the next section, have excellent convergence characteristics as the optimal solution is approached but may have starting difficulties associated with picking a "convex" nominal solution.

Problems with Some State Variables Specified at a Fixed Terminal Time

Consider the same class of problem treated under this heading in Section 7.3. A first-order gradient algorithm for solving this class of problems is presented below:

STEP (a). Estimate a set of control variable histories, $u(t)$.

STEP (b). Integrate the system equations $\dot{x} = f(x,u,t)$ forward with the specified initial conditions $x(t_o)$ and the control variable histories from Step (a). Record $x(t)$, $u(t)$, and $\psi[x(t_f)]$.

STEP (c). Determine an n-vector of influence functions $p(t)$, and an $(n \times q)$ matrix of influence functions, $R(t)$, by backward integration of the influence equations, using the $x(t_f)$ obtained in Step (b) to determine the boundary conditions.

We have

$$\dot{p} = -\left(\frac{\partial f}{\partial x}\right)^T p - \left(\frac{\partial L}{\partial x}\right)^T; \qquad p_i(t_f) = \begin{cases} 0 & i = 1, \ldots, q, \\ \left(\dfrac{\partial \phi}{\partial x_i}\right)_{t=t_f} & i = q+1, \ldots, n, \end{cases}$$

$$\dot{R} = -\left(\frac{\partial f}{\partial x}\right)^T R; \qquad R_{ij}(t_f) = \begin{cases} 1, & i = j, \\ 0, & i \neq j, \end{cases} \qquad \begin{array}{l} i = 1, \ldots, n, \\ j = 1, \ldots, q. \end{array}$$

STEP (d). Simultaneously with Step (c), compute the following integrals:

$$I_{\psi\psi} = \int_{t_o}^{t_f} R^T \frac{\partial f}{\partial u} W^{-1} \left(\frac{\partial f}{\partial u}\right)^T R\, dt \qquad [(q \times q)\text{-matrix}],$$

$$I_{J\psi} = I_{\psi J}^T = \int_{t_o}^{t_f} \left(p^T \frac{\partial f}{\partial u} + \frac{\partial L}{\partial u}\right) W^{-1} \left(\frac{\partial f}{\partial u}\right)^T R\, dt \qquad (q\text{-row vector}),$$

$$I_{JJ} = \int_{t_o}^{t_f} \left(p^T \frac{\partial f}{\partial u} + \frac{\partial L}{\partial u}\right) W^{-1} \left[\left(\frac{\partial f}{\partial u}\right)^T p + \left(\frac{\partial L}{\partial u}\right)^T\right] dt \qquad (\text{scalar}),$$

where W is an $(m \times m)$ positive-definite matrix (see comments below).

STEP (e). Choose values of $\delta\psi$ to cause the next nominal solution to be closer to the desired values $\psi[x(t_f)] = 0$. For example, one might choose $\delta\psi = -\epsilon\psi[x(t_f)]$, $0 < \epsilon \leq 1$. Then determine ν from $\nu = -[I_{\psi\psi}]^{-1}(\delta\psi + I_{\psi J})$.

STEP (f). Repeat Steps (a) through (f), using an improved estimate of $u(t)$, where

$$\delta u(t) = -[W(t)]^{-1}\left[\frac{\partial L}{\partial u} + [p(t) + R(t)\nu]^T\frac{\partial f}{\partial u}\right]^T .$$

Stop when $\psi[x(t_f)] = 0$ and $I_{JJ} - I_{J\psi}I_{\psi\psi}^{-1}I_{\psi J} = 0$ to the desired degree of accuracy.

Comments on the algorithm. This algorithm is based on the influence functions introduced in Section 2.4. The quantities $p(t)$ and $R(t)$ predict how changes in the control histories, $\delta u(t)$, will change the performance index and the q terminal conditions $\psi^T \triangleq [x_1(t_f), \ldots, x_q(t_f)]$:

$$\delta J = \int_{t_o}^{t_f}\left(p^T\frac{\partial f}{\partial u} + \frac{\partial L}{\partial u}\right)\delta u(t)\, dt , \tag{7.4.1}$$

$$\delta\psi \triangleq \begin{bmatrix} \delta x_1 \\ \cdot \\ \cdot \\ \cdot \\ \delta x_q \end{bmatrix}_{t=t_f} = \int_{t_o}^{t_f} R^T\frac{\partial f}{\partial u}\delta u(t)\, dt . \tag{7.4.2}$$

Backward integration of the influence equations in Step (c) is *uncoupled* from the system equations [except for determination of $\partial f/\partial x$ and $\partial L/\partial x$], which makes the integration quite stable numerically, for reasons mentioned in Section 7.2.

Since (7.4.1) and (7.4.2) are *linearized* relations, there is no minimum for δJ subject to constraints on the size of $\delta\psi$. A mathematically simple device for creating a minimum is to add a quadratic integral penalty function in $\delta u(t)$ to (7.4.1):

$$\delta J_1 = \delta J + \frac{1}{2}\int_{t_o}^{t_f}(\delta u)^T W\, \delta u\, dt , \tag{7.4.3}$$

where $W(t)$ is an arbitrary $(m \times m)$ positive-definite weighting matrix.

The minimization of δJ_1 subject to constraints (7.4.2), where $\delta\psi$ is specified, is a linear-quadratic optimization problem of the type discussed in Section 5.3. It is readily solved by adjoining (7.4.2) to (7.4.3) with constant Lagrange multipliers ν:

$$\delta\bar{J} = \delta J_1 + \nu^T \left[\int_{t_0}^{t_f} R^T \frac{\partial f}{\partial u} \delta u(t)\, dt - \delta\psi \right]. \tag{7.4.4}$$

If we neglect the change in coefficients,† the first variation of (7.4.4) is given by

$$\delta(\delta\bar{J}) = \int_{t_0}^{t_f} \left[\frac{\partial L}{\partial u} + (p + R\nu)^T \frac{\partial f}{\partial u} + (\delta u)^T W \right] \delta(\delta u)\, dt\,,$$

from which it is clear that a minimum in $\delta\bar{J}$ occurs if

$$\delta u = -W^{-1} \left[\frac{\partial L}{\partial u} + (p + R\nu)^T \frac{\partial f}{\partial u} \right]^T. \tag{7.4.5}$$

Substituting (7.4.5) into (7.4.2), we find that

$$\delta\psi = -I_{\psi J} - I_{\psi\psi}\nu\,, \tag{7.4.6}$$

where $I_{\psi J}$ and $I_{\psi\psi}$ are as defined in Step (d), above. If $I_{\psi\psi}$ is nonsingular, we can solve (7.4.6) for the required values of ν :

$$\nu = -[I_{\psi\psi}]^{-1}(\delta\psi + I_{\psi J})\,. \tag{7.4.7}$$

The predicted change in δJ can be found by substituting (7.4.5) and (7.4.7) into (7.4.1):

$$\delta J = -(I_{JJ} - I_{J\psi} I_{\psi\psi}^{-1} I_{\psi J}) + I_{J\psi} I_{\psi\psi}^{-1} \delta\psi\,, \tag{7.4.8}$$

where I_{JJ} is as defined in Step (d) above.

As the optimum solution is approached and $\delta\psi = 0$, it is clear from (7.4.8), (7.4.7), and (7.4.5) that

$$I_{JJ} - I_{J\psi} I_{\psi\psi}^{-1} I_{\psi J} \to 0\,, \tag{7.4.9}$$

$$\nu \to -I_{\psi\psi}^{-1} I_{\psi J}\,, \tag{7.4.10}$$

$$\frac{\partial L}{\partial u} + (p + R\nu)^T \frac{\partial f}{\partial u} \to 0 \quad \text{for} \quad t_0 \le t \le t_f\,. \tag{7.4.11}$$

Note that (7.4.11) may be interpreted as $(\partial H/\partial u)$, where

$$H \triangleq L(x,u,t) + \lambda^T f(x,u,t)\,, \tag{7.4.12}$$

$$\lambda(t) = p(t) + R(t)\nu\,. \tag{7.4.13}$$

Equation (7.4.13) is useful for estimating $\lambda(t)$ to find solutions by the methods of Sections 7.3, 7.5, and 7.6.

The choice of the weighting matrix in Step (d) should be made to limit the size of the first step in the algorithm. This may be done by comparing actual $\delta\psi$ and δJ with the predicted values from (7.4.6)

†The change in coefficients is taken into account in Sections 7.3 and 7.5.

and (7.4.8); if there is too large a discrepancy, W should be increased; if there is too small a discrepancy, it is possible to take larger steps, and W should be reduced. After a satisfactory first step, W may be maintained as is through subsequent iterations. In view of (6.1.16), a very satisfactory value of W is $\epsilon(\partial^2 H/\partial u^2)$, where $0 < \epsilon \leq 1$, *provided* that $(\partial^2 H/\partial u^2)$ is positive definite.

Problems with functions of the state variables specified at an unspecified terminal time.

Consider the same class of problems treated under this heading in Section 7.3. The steps in a first-order gradient algorithm are outlined below:

STEP (a). Estimate a set of control variable histories $u(t)$ and a terminal time t_f.

STEP (b). Integrate the system equations forward with the specified initial conditions using $u(t)$ and t_f from Step (a):

$$\dot{x} = f(x,u,t)\,, \qquad x(t_0) \qquad \text{specified.}$$

Record $x(t)$, $u(t)$, $\psi[x(t_f),t_f]$, $[(d\phi/dt) + L]_{t=t_f}$, and $(d\psi/dt)_{t=t_f}$.

STEP (c). Determine an n-vector of influence functions $p(t)$, and an $(n \times q)$-matrix of influence functions, $R(t)$, by backward integration of the differential equations

$$\dot{p} = -\left(\frac{\partial f}{\partial x}\right)^T p - \left(\frac{\partial L}{\partial x}\right)^T, \qquad p(t_f) = \left(\frac{\partial \phi}{\partial x}\right)_{t=t_f},$$

$$\dot{R} = -\left(\frac{\partial f}{\partial x}\right)^T R\,, \qquad R(t_f) = \left(\frac{\partial \psi}{\partial x}\right)_{t=t_f}.$$

STEP (d). Identical to Step (d) in previous algorithm.

STEP (e). Identical to Step (e) in previous algorithm, except that $\delta\psi$ is replaced by $d\psi$ and

$$\nu = -\left[I_{\psi\psi} + \frac{1}{b}\frac{d\psi}{dt}\left(\frac{d\psi}{dt}\right)^T\right]^{-1}\left[d\psi + I_{\psi J} + \frac{1}{b}\left(\frac{d\phi}{dt} + L\right)\left(\frac{d\psi}{dt}\right)\right],$$

where b is a scalar weighting factor (see comment below).

STEP (f). Repeat Steps (a) through (e) using improved estimates of $u(t)$ and t_f obtained by adding $\delta u(t)$ and dt_f to the previous estimates of $u(t)$ and t_f, where

$$\delta u(t) = -\,[W(t)]^{-1}\left[\frac{\partial L}{\partial u} + (p + R\nu)^T \frac{\partial f}{\partial u}\right]^T,$$

$$dt_f = -\frac{1}{b}\left(\frac{d\phi}{dt} + \nu^T \frac{d\psi}{dt} + L\right)_{t=t_f} .$$

Stop when $\psi[x(t_f),t_f] = 0$, $[(d\phi/dt) + \nu^T(d\psi/dt) + L)_{t=t_f} = 0$ and $I_{JJ} - I_{J\psi}I_{\psi\psi}^{-1}I_{\psi J} = 0$ to the desired degree of accuracy.

Comments on the algorithm. This algorithm is based on the influence functions $p(t)$ and $R(t)$ introduced in Section 2.7, where

$$dJ = \left(\frac{d\phi}{dt} + L\right)_{t=t_f} dt_f + \int_{t_0}^{t_f}\left(\frac{\partial L}{\partial u} + p^T\frac{\partial f}{\partial u}\right)\delta u(t)\,dt , \qquad (7.4.14)$$

$$d\psi = \left(\frac{d\psi}{dt}\right)_{t=t_f} dt_f + \int_{t_0}^{t_f}\left(R^T\frac{\partial f}{\partial u}\right)\delta u(t)\,dt . \qquad (7.4.15)$$

Quadratic penalty functions in $\delta u(t)$ and dt_f are added to (7.4.14), and (7.4.15) is adjoined to (7.4.14) with multipliers ν:

$$d\bar{J} = dJ + \frac{1}{2}b(dt_f)^2 + \int_{t_0}^{t_f}\frac{1}{2}(\delta u)^T W\,\delta u\,dt$$
$$+ \nu^T\left[\left(\frac{d\psi}{dt}\right)_{t=t_f} dt_f + \int_{t_0}^{t_f}\left(R^T\frac{\partial f}{\partial u}\right)\delta u\,dt\right], \qquad (7.4.16)$$

where b is an arbitrary positive constant and $W(t)$ is an arbitrary $(m \times m)$ positive-definite matrix.

Neglecting the change in the coefficients,† the first variation of (7.4.16) is

$$d(d\bar{J}) = \left[\frac{d\phi}{dt} + L + \nu^T\frac{d\psi}{dt} + b\,dt_f\right]_{t=t_f} d(dt_f)$$
$$+ \int_{t_0}^{t_f}\left[\frac{\partial L}{\partial u} + (p + R\nu)^T\frac{\partial f}{\partial u} + (\delta u)^T W\right]\delta(\delta u)\,dt , \qquad (7.4.17)$$

from which it is clear that a minimum in $d\bar{J}$ occurs if

$$\delta u(t) = -W^{-1}\left[\frac{\partial L}{\partial u} + (p + R\nu)^T\frac{\partial f}{\partial u}\right]^T , \qquad (7.4.18)$$

$$dt_f = -\frac{1}{b}\left[\frac{d\phi}{dt} + \nu^T\frac{d\psi}{dt} + L\right]_{t=t_f} . \qquad (7.4.19)$$

Substituting (7.4.18) and (7.4.19) into (7.4.15), we find that

$$d\psi = -\frac{1}{b}\left[\left(\frac{d\psi}{dt}\right)\left(\frac{d\phi}{dt} + \nu^T\frac{d\psi}{dt} + L\right)\right]_{t=t_f} - (I_{\psi J} + I_{\psi\psi}\nu) , \qquad (7.4.20)$$

†The change in coefficients is taken into account in Sections 7.3 and 7.5.

where $I_{\psi J}$ and $I_{\psi\psi}$ are as defined in Step (d). If $I_{\psi\psi} + (1/b)(d\psi/dt)(d\psi/dt)^T$ is nonsingular, we can solve (7.4.20) for the required values of ν:

$$\nu = -\left[I_{\psi\psi} + \frac{1}{b}\frac{d\psi}{dt}\left(\frac{d\psi}{dt}\right)^T\right]^{-1}\left[d\psi + I_{\psi J} + \frac{1}{b}\left(\frac{d\phi}{dt} + L\right)\left(\frac{d\psi}{dt}\right)\right]. \quad (7.4.21)$$

The predicted change in dJ can be found by substituting (7.4.18) and (7.4.19) into (7.4.14):

$$dJ = -\frac{1}{b}\left[\left(\frac{d\phi}{dt} + L\right)\left(\frac{d\phi}{dt} + \nu^T\frac{d\psi}{dt} + L\right)\right]_{t=t_f} - (I_{JJ} - I_{J\psi}\nu), \quad (7.4.22)$$

where I_{JJ} is as defined in Step (d).

As the optimum solution is approached, it follows from (7.4.18) and (7.4.19) that

$$\frac{\partial L}{\partial u} + (p + R\nu)^T\frac{\partial f}{\partial u} \to 0 \qquad \text{for} \qquad t_o \le t \le t_f, \quad (7.4.23)$$

$$\left(\frac{d\phi}{dt} + \nu^T\frac{d\psi}{dt} + L\right)_{t=t_f} \to 0. \quad (7.4.24)$$

If, in addition, $d\psi = 0$, Equation (7.4.25) may be used in (7.4.20) and then (7.4.22) to show that

$$\nu \to -I_{\psi\psi}^{-1}I_{\psi J}, \quad (7.4.25)$$

$$I_{JJ} - I_{J\psi}I_{\psi\psi}^{-1}I_{\psi J} \to 0. \quad (7.4.26)$$

Note that (7.4.23) and (7.4.24) may be interpreted as $(\partial H/\partial u)$ and Ω, where

$$H \triangleq L + \lambda^T f, \quad (7.4.27)$$

$$\lambda(t) = p(t) + R(t)\nu. \quad (7.4.28)$$

Equation (7.4.28) is useful for estimating $\lambda(t)$ to find solutions by the methods of Sections 7.3, 7.5, and 7.6.

The choice of the weighting constant b and the weighting matrix $W(t)$ in Step (d) should be made to limit the size of the first step in the algorithm. This may be done by comparing actual $d\psi$ and dJ with the predicted values from (7.4.20) and (7.4.22); if there is too large a discrepancy, b and W should be increased; if there is too small a discrepancy, it is possible to take larger steps, and b and W should be reduced.

Min-H Algorithms. A number of "min-H" first-order gradient algorithms have been proposed. One of the most recent is due to Gottlieb

(1967)† where reference to previous work is also given. See also H. Halkin, *Method of Convex Ascent*, pp. 211–239 of Balakrishnan and Neustadt (1964). In this algorithm the control is determined by minimizing H, as defined in Equations (7.4.27) and (7.4.28), with respect to the control u, holding x and λ fixed. This permits large changes in u and appears to accelerate the convergence of the algorithm near the optimum.

Problem. Consider

$$J = \phi(x(T),T) + \int_{t_0}^{T} L(x,u,t)\,dt\,,$$

$$\dot{x} = f(x,u,t)\,,\; x(t_0) = x_0 \quad \text{given,} \quad T \quad \text{fixed,}$$

and a given control $u^{(1)}(t)$ which is claimed to minimize J (at least relatively).

(a) Write out (do not derive) a detailed step-by-step *flow chart* (including all necessary computational steps that must be followed by you or a computer) which will prove or disprove this claim. If the procedure in certain cases is not known in the literature, then say so.

(b) Let

$$J = \frac{1}{3}x^3(T) + \frac{1}{2}\int_0^T u^2(t)\,dt\,, \qquad \dot{x} = u\,, \qquad x(0) = 2 \qquad (x, u \text{ scalars})\,,$$

and $u^{(1)}(t) = -2\,,\; 0 \le t \le T = 1$. Is $u^{(1)}(t)$ optimal? If not, find a $u^{(2)}(t)$ which is better by carrying out either one step of the first-order gradient method or any other method treated in this chapter.

(c) Find a $u^{o}(t)$ for (b) which is at least a weak minimizing control [i.e., it must pass nearly all tests in (a)].

7.5 Second-order gradient algorithms

Introduction

As discussed in the introduction of Section 7.4, first-order gradient algorithms show great improvements in the first few iterations but have poor convergence characteristics as the optimal solution is approached. Second-order gradient algorithms require the nominal solution to be "convex" [$(\partial^2 H/\partial u^2) > 0$ over the whole time period for a minimization problem; $(\partial^2 H/\partial u^2) < 0$ for a maximization

†R. G. Gottlieb, "Rapid Convergence to Optimum Solutions using a Min-H Strategy," *AIAA Jour.*, Vol. 5, No. 2, pp. 322–329, February 1967.

problem]. It is sometimes difficult to pick a convex nominal solution; in such cases, one may be able to use first-order gradient methods to improve the nominal solution to the point where it is convex. Note that the first-order gradient algorithm described in the previous section provides its own convexity by choice of a positive-definite W matrix and a positive constant b (negative W and b for a maximization problem).

Problems with functions of the state variables specified at an unspecified terminal time, including minimum-time problems.

We again consider the class of problems described under this heading in Section 7.3. As in that section, we shall describe two approaches to the solution: the transition-matrix approach and the backward-sweep approach.

A Transition-Matrix Algorithm.† This approach proceeds as follows:

STEP (a). Guess a set of control variable histories, $u(t)$, and a terminal time, t_f.

STEP (b). Integrate the system equations $\dot{x} = f(x,u,t)$ forward with the specified initial conditions $x(t_0)$ and the guessed control variable histories to the guessed final time t_f. Record $x(t)$, $u(t)$, and $\psi[x(t_f),t_f]$.

STEP (c). Guess a set of multipliers ν, with the constraint that they make

$$\left(\frac{\partial \Omega}{\partial u}\right)_{t=t_f} = 0\,, \qquad \text{where} \qquad \Omega = \frac{d\phi}{dt} + \nu^T \frac{d\psi}{dt} + L\,.$$

Record $\Omega(t_f)$.

STEP (d). Determine an n-vector of influence functions, $\lambda(t)$, by backward integration of

$$\dot{\lambda}^T = -\frac{\partial L}{\partial x} - \lambda^T \frac{\partial f}{\partial x}\,, \qquad \lambda^T(t_f) = \left(\frac{\partial \phi}{\partial x} + \nu^T \frac{\partial \psi}{\partial x}\right)_{t=t_f}.$$

Record $H_u(t) = L_u + \lambda^T f_u$.

STEP (e). Simultaneously with Step (d), determine $n + q + 1$ homogeneous solutions and one particular solution to the perturbation equations.

$$\begin{bmatrix} \delta\dot{x} \\ \delta\dot{\lambda} \end{bmatrix} = \begin{bmatrix} A\,, & -B \\ -C\,, & -A^T \end{bmatrix} \begin{bmatrix} \delta x \\ \delta\lambda \end{bmatrix} + \begin{bmatrix} v(t) \\ w(t) \end{bmatrix}, \qquad \begin{matrix} (7.5.1) \\ (7.5.2) \end{matrix}$$

†See H. J. Kelley, R. Kopp, and G. Moyer, "A Trajectory Optimization Technique Based on the Second Variation," in *Progress in Astronautics*, Vol. 14. New York: Academic Press, 1964.

where A, B, C are as given in (6.1.23) through (6.1.25), and

$$v(t) = -f_u H_{uu}^{-1} \, \delta H_u^T , \tag{7.5.3}$$

$$w(t) = H_{xu} H_{uu}^{-1} \, \delta H_u^T , \tag{7.5.4}$$

and $\delta H_u(t)$ is chosen to bring $H_u(t)$ closer to zero; e.g., $\delta H_u = -\epsilon H_u$, $0 < \epsilon \leq 1$.

$n + q + 1$ unit solutions must be found for these equations with $\delta H_u(t) = 0$. The particular solution must be found for $\delta H_u(t) \neq 0$, with $\delta x(t_f) = 0$, $d\nu = 0$, $dt_f = 0$:

$$\begin{bmatrix} \delta x(t) \\ \\ \delta \lambda(t) \end{bmatrix} = \frac{\partial[x(t),\lambda(t)]}{\partial[x(t_f),\nu,t_f]} \begin{bmatrix} \delta x(t_f) \\ d\nu \\ dt_f \end{bmatrix} + \begin{bmatrix} \delta x^{(p)}(t) \\ \\ \delta \lambda^{(p)}(t) \end{bmatrix} . \tag{7.5.5}$$

STEP (f). Choose values of $\delta x(t_o)$, $d\psi$, and $d\Omega$ so as to bring the next solution closer to the desired values of $x(t_o)$, ψ, and $\Omega = 0$. For example, one might choose

$$\begin{bmatrix} \delta x(t_o) \\ d\psi \\ d\Omega \end{bmatrix} = -\epsilon \begin{bmatrix} x(t_o) - x^o \\ \psi[x(t_f),t_f] \\ \Omega[x(t_f),t_f] \end{bmatrix} , \qquad 0 < \epsilon \leq 1 . \tag{7.5.6}$$

STEP (g). Using the chosen values of $\delta x(t_o)$, $d\psi$, and $d\Omega$ with the expression for $\delta x(t_o)$ from (7.5.5), and the expressions for $d\psi$ and $d\Omega$ from the terminal boundary conditions, determine $\delta x(t_f)$, $d\nu$, and dt_f from

$$\begin{bmatrix} \delta x(t_o) \\ \\ d\psi \\ \\ d\Omega \end{bmatrix} = \begin{bmatrix} \dfrac{\partial x(t_o)}{\partial x(t_f)} , & \dfrac{\partial x(t_o)}{\partial \nu} , & \dfrac{\partial x(t_o)}{\partial t_f} \\ \dfrac{\partial \psi}{\partial x(t_f)} , & 0 , & \dfrac{d\psi}{dt_f} \\ \dfrac{\partial \Omega}{\partial x(t_f)} , & \left(\dfrac{d\psi}{dt_f}\right)^T , & \dfrac{d\Omega}{dt_f} \end{bmatrix} \begin{bmatrix} \delta x(t_f) \\ \\ d\nu \\ \\ dt_f \end{bmatrix} + \begin{bmatrix} \delta x^{(p)}(t_o) \\ \\ 0 \\ \\ 0 \end{bmatrix} . \begin{matrix} (7.5.7) \\ \\ (7.5.8) \\ \\ (7.5.9) \end{matrix}$$

STEP (h). The desired change in $u(t)$ is then given by

$$\delta u(t) = -\left(\frac{\partial^2 H}{\partial u^2}\right)^{-1} [-\delta H_u^T(t) + H_{ux} \, \delta x + H_{u\lambda} \, \delta\lambda] , \tag{7.5.10}$$

where $\delta x(t)$, $\delta\lambda(t)$ are obtained from Step (e), using $\delta x(t_f)$, $d\nu$, dt_f from Step (g). The desired changes in t_f and ν are given by dt_f and $d\nu$ from Step (g).

STEP (i). The procedure is repeated over and over, letting ϵ gradually become one and $(\partial H/\partial u) = 0$, $\psi[x(t_f),t_f] = 0$, $\Omega[x(t_f),t_f] = 0$, and $x(t_o) = x^o$ to the desired accuracy.

Comments on the algorithm. This algorithm may be interpreted as the result of solving the following linear-quadratic problem:

Minimize

$$\delta J = -\int_{t_0}^{t_f} \delta H_u \, \delta u \, dt + \frac{1}{2}\int_{t_0}^{t_f} [\delta x^T \, \delta u^T] \begin{bmatrix} H_{xx} & H_{xu} \\ H_{ux} & H_{uu} \end{bmatrix} \begin{bmatrix} \delta x \\ \delta u \end{bmatrix} dt$$

$$+ \frac{1}{2}(\delta x^T \Phi_{xx} \delta x)_{t_f} + (\dot{\Phi} + L)_{t_f} dt_f ,$$

subject to

$$\delta \dot{x} = f_x \, \delta x + f_u \, \delta u , \qquad \delta x(t_0) \text{ specified},$$

$$d\psi = \psi_x \, dx + \psi_t \, dt_f , \qquad d\psi \text{ specified},$$

$$d\Omega = \Omega_x \, dx + \Omega_t \, dt_f , \qquad d\Omega \text{ specified}.$$

Here, δH_u is the specified change in H_u. If we let $-\delta H_u = H_u$, then δJ is the expansion of J to second order along a nominal trajectory.
A Backward-Sweep Algorithm. The procedure is the same as in the transition matrix approach except for Steps (e) and (g). Instead of finding $n + q + 1$ homogeneous solutions and one particular solution to the second-order influence equations, we "sweep" the terminal boundary conditions for the second-order influence equations backward to the initial time, t_0, taking into account the fact that the differential equations are not homogeneous; that is, $\delta H_u(t) \neq 0$. This is done by introducing a nonhomogeneous version of (6.6.1) through (6.6.3):

$$\begin{bmatrix} \delta\lambda(t) \\ d\psi \\ d\Omega \end{bmatrix} = \begin{bmatrix} S(t) & , R(t) & , m(t) \\ R^T(t) & , Q(t) & , n(t) \\ m^T(t) & , n^T(t) & , \alpha(t) \end{bmatrix} \begin{bmatrix} \delta x(t) \\ d\nu \\ dt_f \end{bmatrix} + \begin{bmatrix} h(t) \\ g(t) \\ \beta(t) \end{bmatrix} . \qquad \begin{matrix} (7.5.11) \\ (7.5.12) \\ (7.5.13) \end{matrix}$$

These latter relations are now differentiated with respect to time, noting that $d\psi$, $d\Omega$, $d\nu$, and dt_f are constants:

$$\begin{bmatrix} \delta\dot{\lambda} \\ 0 \\ 0 \end{bmatrix} = \begin{bmatrix} \dot{S} & \dot{R} & \dot{m} \\ \dot{R}^T & \dot{Q} & \dot{n} \\ \dot{m}^T & \dot{n}^T & \dot{\alpha} \end{bmatrix} \begin{bmatrix} \delta x(t) \\ d\nu \\ dt_f \end{bmatrix} + \begin{bmatrix} S \\ R^T \\ m \end{bmatrix} [\delta \dot{x}] + \begin{bmatrix} \dot{h} \\ \dot{g} \\ \dot{\beta} \end{bmatrix} . \qquad \begin{matrix} (7.5.14) \\ (7.5.15) \\ (7.5.16) \end{matrix}$$

Now the perturbation equations (7.5.1) and (7.5.2) are used to eliminate $\delta\dot{\lambda}$ and $\delta\dot{x}$ from (7.5.14) through (7.5.16), and (7.5.11) is used to eliminate $\delta\lambda$. The result is

$$\begin{bmatrix} 0 \\ 0 \\ \vdots \\ 0 \end{bmatrix} = \begin{bmatrix} \dot{S} + SA + A^T S - SBS + C & , \dot{R} + (A^T - SB)R & , \dot{m} + (A^T - SB)m \\ \cdot & \cdots \; , \dot{Q} - R^T BR & , \dot{n} - R^T Bm \\ \cdot & & \\ \cdot & \cdots & , \dot{\alpha} - m^T Bm \end{bmatrix} .$$

$$\begin{bmatrix} \delta x(t) \\ d\nu \\ dt_f \end{bmatrix} + \begin{bmatrix} \dot{h} + (A^T - SB)h + Sv + w \\ \dot{g} - R^T(Bh - v) \\ \dot{\beta} - m^T(Bh - v) \end{bmatrix} \cdot \quad \begin{matrix} (7.5.17) \\ (7.5.18) \\ (7.5.19) \end{matrix}$$

If Equations (7.5.17) through (7.5.19) are to be identities, true for all $\delta x(t)$, $d\nu$, and dt_f, and if Equations (7.5.11) through (7.5.13) are to be true at $t = t_f$, we must have (6.6.8) through (6.6.13) plus the additional relations

$$\dot{h} = -(A^T - SB)h - Sv - w\,, \qquad h(t_f) = 0\,, \tag{7.5.20}$$

$$\dot{g} = R^T(Bh - v)\,, \qquad g(t_f) = 0\,, \tag{7.5.21}$$

$$\dot{\beta} = m^T(Bh - v)\,, \qquad \beta(t_f) = 0\,. \tag{7.5.22}$$

The set of equations for S, Q, R, m, n, α, h, g, and β must be integrated backward from $t = t_f$ to $t = t_0$. Then Equations (7.5.12) and (7.5.13) may be used at $t = t_0$ to determine $d\nu$ and dt_f in terms of $\delta x(t_0)$, $d\psi$, and $d\Omega$:

$$d\nu = \overline{Q}^{-1}\left[d\psi - g - \overline{R}^T\,\delta x - \frac{n}{\alpha}(d\Omega - \beta)\right]_{t=t_0}, \tag{7.5.23}$$

$$dt_f = \frac{-1}{\alpha}\left[(m^T + n^T\overline{Q}^{-1}\overline{R}^T)\,\delta x + n^T\overline{Q}^{-1}(d\psi - g) - \left(1 + \frac{n^T\overline{Q}^{-1}n}{\alpha}\right)(d\Omega - \beta)\right]_{t=t_0}, \tag{7.5.24}$$

where $\overline{Q}$ and $\overline{R}$ are as defined in (6.6.16) and (6.6.17).

Having these values of $d\nu$ and dt_f, we can substitute them into (7.5.11) to obtain $\delta\lambda(t_0)$ in terms of $\delta x(t_0)$ and integrate the differential equations (7.5.1) and (7.5.2) forward once to obtain $\delta u(t)$ and $\delta x(t_f)$, since

$$\delta u(t) = -\left(\frac{\partial^2 H}{\partial u^2}\right)^{-1}\left[-\delta H_u^T(t) + \frac{\partial^2 H}{\partial u\,\partial x}\,\delta x + \frac{\partial^2 H}{\partial u\,\partial\lambda}\,\delta\lambda\right]. \tag{7.5.25}$$

This brings us back to Step (h) in the transition-matrix algorithm.

One advantage of the sweep procedure over the transition-matrix procedure is greater numerical accuracy. The unit solutions of the second-order influence equations may differ by orders of magnitude, producing a transition matrix that is ill-conditioned. This results in inaccurate determination of $\delta x(t_f)$, $d\nu$, and dt_f in (7.5.7) through (7.5.9). Usually no such difficulty occurs for the sweep procedure. Another advantage is the somewhat greater ease in generating feedback gains for neighboring optimal trajectories, and the implicit checks of the convexity, normality, and conjugate-point conditions (see Section 6.3).

Differential dynamic programming. Another version of the backward-sweep algorithm has been proposed by Jacobson (1968);† he calls it "differential dynamic programming" (DDP). In DDP, the Hamiltonian is first minimized with respect to u, holding x and λ fixed, yielding an improved control u°. Then variations in x and λ and the corresponding further variations in u away from u° are considered; a second-order expansion in δx and δu is minimized with respect to δu, giving rise to a linear feedback $\delta u = -C(t)\,\delta x$. The new control on the next iteration is then $u^\circ - C(t)\,\delta x$. This is a refinement of the "min-H" algorithms discussed in Section 7.4. It permits large changes in u and overcomes the difficulty of nonconvex nominal solutions that may occur with the backward sweep algorithm.

Jacobson's algorithm also includes a "step-size adjustment method" that appears quite attractive; if an iteration yields no improvement in performance (or results in too large a change in x from the nominal), it is repeated, using "improved" control only in a time period $t_1 \le t \le t_f$, where $t_1 > t_0$. The iteration is repeated and the time t_1 is increased until some improvement in J is achieved (or the changes in x are sufficiently small); t_1 is then decreased in subsequent iterations until $t_1 = t_0$.

Problem 1. For the fixed-terminal-time case, show that the first-order gradient procedure is a special case of the second-order gradient procedure *if* one does the following things:

(a) Approximate

$$H_{xx} = 0\,, \qquad H_{ux} = 0\,, \qquad H_{uu} = W\,, \qquad \phi_{xx}\big|_{t=t_f} = \psi_{xx}\big|_{t=t_f} = 0\,.$$

(b) Set nominal values of $\nu = 0$.
(c) Identify $\lambda(t)$ as $p(t)$, and $-\delta H_u(t)$ as the nominal value of

$$H_u(t) \equiv p^T \frac{\partial f}{\partial u} + \frac{\partial L}{\partial u}\,.$$

In particular, show that

$$S(t) = 0\,, \qquad R(t) \text{ for both methods is identical,}$$

$$w(t) = 0\,, \qquad h(t) = 0\,,$$

$$Q(t) = I_{\psi\psi}(t)\,, \qquad g(t) = -I_{\psi J}(t)\,.$$

Problem 2. If an admissible nominal solution very close to the optimal solution is available, show that the sweep procedure of Section 7.5

†D. H. Jacobson, "New Second Order and First Order Algorithms for Determining Optimal Control: A Differential Dynamic Programming Approach," *J. Optimization Theory and Appl.*, December 1968 (to appear).

can be used to obtain the optimal solution in one iteration, using $\nu = 0$ as the prior estimate. Note that this implies that

$$\delta H_u(t) = -\left(\lambda^T \frac{\partial f}{\partial u} + \frac{\partial L}{\partial u}\right), \qquad \lambda^T(t_f) = \left(\frac{\partial \phi}{\partial x}\right)_{t=t_f}, \qquad S(t_f) = \left(\frac{\partial^2 \phi}{\partial x^2}\right)_{t=t_f},$$

and

$$\nu_{\text{new}} = -[Q(t_o)]^{-1} g(t_o), \qquad \lambda(t)_{\text{new}} = \lambda(t) + R(t)\nu_{\text{new}} + h(t) + S(t)\delta x(t),$$

$$\begin{aligned} \delta u(t) &= -H_{uu}^{-1}[f_u^T\lambda + L_u^T + H_{ux}\,\delta x + f_u^T(S\,\delta x + R\nu_{\text{new}} + h)] \\ &= -H_{uu}^{-1}[(H_{ux}\,\delta x + f_u^T\lambda_{\text{new}} + L_u^T)]\,. \end{aligned}$$

If the nominal is optimal, then we have

$$(H_u)_{\text{new}} = f_u^T\lambda_{\text{new}} + L_u^T = 0\,, \qquad t_o \le t \le t_f,$$

which, with $\delta x(t_o) = 0$, implies that $\delta u(t) = 0$, $t_o \le t \le t_f$.

7.6 A quasilinearization algorithm

Introduction

As outlined in Section 7.1, one variation of quasilinearization involves choosing nominal functions for $x(t)$ and $\lambda(t)$ that satisfy as many of the boundary conditions as possible. The nominal control vector, $u(t)$, is then determined by use of the optimality conditions. The system equations and the influence equations are linearized about the nominal and a succession of nonhomogeneous, linear two-point boundary-value problems are solved to modify the solution until it satisfies the system and influence equations to the desired accuracy.

These methods are attractive for several reasons. First, it is often easier to guess nominal-state-variable histories than nominal-control-variable histories. Second, these methods converge rapidly near the optimum solution (just as the second-order gradient methods do).

Problems with some state variables specified at a fixed terminal time

We consider the problems described under this heading in Section 7.3 to describe the method. For a nonsingular problem, $u(t)$ is determined from $x(t)$ and $\lambda(t)$ by (7.3.8):

$$\frac{\partial H}{\partial u} = 0 \Rightarrow u = u(x,\lambda,t)\,; \tag{7.6.1}$$

thus, the system equations and the influence equations may be written as

$$\dot{x} = f(x,\lambda,t) \qquad (n \text{ equations}), \tag{7.6.2}$$

$$\dot{\lambda} = g(x,\lambda,t) \qquad (n \text{ equations}), \tag{7.6.3}$$

and the boundary conditions (7.3.3), (7.3.4), and (7.3.7) may be written as

$$x(t_0) \text{ specified} \qquad (n \text{ equations}), \tag{7.6.4}$$

$$h[x(t_f),\lambda(t_f)] = 0 \qquad (n \text{ equations}). \tag{7.6.5}$$

Equations (7.6.2) through (7.6.5) constitute a nonlinear two-point boundary-value problem for $x(t)$, $\lambda(t)$.

Let $x^{(i)}(t)$ and $\lambda^{(i)}(t)$ be the values of $x(t)$ and $\lambda(t)$ on the ith iteration, and let us assume that they do *not* satisfy any of the relations (7.6.2) through (7.6.5). Thus, we seek values of $x^{(i+1)}(t)$ and $\lambda^{(i+1)}(t)$ that *come closer* to satisfying (7.6.2) through (7.6.5); i.e., we would like to have

$$\begin{bmatrix} \dot{x}^{(i+1)} - f[x^{(i+1)},\lambda^{(i+1)},t] \\ \dot{\lambda}^{(i+1)} - g[x^{(i+1)},\lambda^{(i+1)},t] \\ x^{(i+1)}(t_0) - x(t_0) \\ h[x^{(i+1)}(t_f),\lambda^{(i+1)}(t_f)] \end{bmatrix} = (1-\epsilon) \begin{bmatrix} \dot{x}^{(i)} - f[x^{(i)},\lambda^{(i)}] \\ \dot{\lambda}^{(i)} - g[x^{(i)},\lambda^{(i)}] \\ x^{(i)}(t_0) - x(t_0) \\ h[x^{(i)}(t_f),\lambda^{(i)}(t_f)] \end{bmatrix}, \qquad \begin{matrix} (7.6.6) \\ (7.6.7) \\ (7.6.8) \\ (7.6.9) \end{matrix}$$

where $0 < \epsilon \leq 1$.

In order to accomplish (7.6.6) through (7.6.9), we linearize the left-hand sides about $x^{(i)}(t)$ and $\lambda^{(i)}(t)$; i.e., we let

$$x^{(i+1)}(t) = x^{(i)}(t) + \delta x(t), \tag{7.6.10}$$

$$\lambda^{(i+1)}(t) = \lambda^{(i)}(t) + \delta\lambda(t). \tag{7.6.11}$$

Substituting (7.6.10) and (7.6.11) into (7.6.6) through (7.6.9) and retaining only first-order terms in the Taylor expansion of the left-hand sides, we have

$$\begin{bmatrix} \dfrac{d}{dt}(\delta x) - \dfrac{\partial f}{\partial x}\delta x - \dfrac{\partial f}{\partial \lambda}\delta\lambda \\ \dfrac{d}{dt}(\delta \lambda) - \dfrac{\partial g}{\partial x}\delta x - \dfrac{\partial g}{\partial \lambda}\delta\lambda \\ \delta x(t_0) \\ \left(\dfrac{\partial h}{\partial x}\delta x + \dfrac{\partial h}{\partial \lambda}\delta\lambda\right)_{t=t_f} \end{bmatrix} = -\epsilon \begin{bmatrix} \dot{x}^{(i)} - f[x^{(i)},\lambda^{(i)}] \\ \dot{\lambda}^{(i)} - g[x^{(i)},\lambda^{(i)}] \\ x^{(i)}(t_0) - x(t_0) \\ h[x^{(i)}(t_f),\lambda^{(i)}(t_f)] \end{bmatrix} \qquad \begin{matrix} (7.6.12) \\ (7.6.13) \\ (7.6.14) \\ (7.6.15) \end{matrix}$$

Equations (7.6.12) through (7.6.15) constitute a nonhomogeneous *linear* two-point boundary-value problem for $\delta x(t)$, $\delta\lambda(t)$, which may be solved by transition-matrix or sweep methods as in Section 7.5.

As successive iterations are made, ϵ should gradually be increased to one. If the method converges, it will do so quadratically similar to other second-order methods.

Problems with functions of the state variables specified at an unspecified terminal time, including minimum-time problems.

For these problems, not only must one guess $x(t)$ and $\lambda(t)$, but also ν and t_f. Improvements $d\nu$ and dt_f must be found in a manner similar to that of Section 7.5.

7.7 A second-order gradient algorithm for multistage systems†

As mentioned in Section 2.6, proper multistage formulation of continuous problems contributes significantly to the speed of convergence of iterative computations. In this section, a backward-sweep algorithm analogous to the one presented in Section 7.5 for continuous systems is presented. A transition-matrix algorithm could also be used in some cases, but is not presented here. Only the case with a *fixed number of stages* is considered, as in Sections 2.6 and 6.10. The difference between this section and Section 6.10 is that the initial nominal does *not* have to satisfy the optimality conditions $\partial H^i/\partial u^{(i)} = 0$.

A Backward-Sweep Algorithm. Consider the class of problems described in Section 2.6. The backward-sweep algorithm consists of the following steps:

STEP (a). Estimate a sequence of control vectors $u(i)$ and solve the system equations

$$x(i+1) = f^i[x(i),u(i)]\,, \qquad x(0) \ \text{specified}, \qquad i = 0,\ldots,N-1\,,$$

sequentially in ascending order of the index. Record $x(i)$, $u(i)$, and $\psi[x(N)]$.

STEP (b). Estimate a set of multipliers ν, and solve the *first-order influence equations*

$$\lambda^T(i) = L_x^i + \lambda^T(i+1)f_x^i, \qquad \lambda^T(N) = [\phi_x + \nu^T\psi_x]_{x=x(N)}, \qquad i = N-1,$$
$$i = N-1,\ldots,0,$$

sequentially in descending order of the index, evaluating the coefficients on the nominal path of Step (a). Record $H_u^i = L_u^i + \lambda^T(i+1)f_u^i$. Simultaneously, solve the following *second-order influence equations* sequentially in descending order of the index:

$$S(i) = Z_{xx}(i) - Z_{xu}(i)\,Z_{uu}^{-1}(i)\,Z_{ux}(i)\,; \qquad S(N) = [\phi_{xx} + \nu^T\psi_{xx}]_{x=x(N)}\,;$$
$$R(i) = f_x^i R(i+1) - Z_{xu}(i)\,Z_{uu}^{-1}(i)f_u^i R(i+1)\,; \qquad R(N) = [\psi_x^T]_{x=x(N)}\,;$$

†This section is based on the Ph.D. Thesis of S. R. McReynolds, Harvard University, Cambridge, Massachusetts, 1966.

$$Q(i) = Q(i+1) - R^T(i+1)(f_u^i)^T Z_{uu}^{-1}(i) f_u^i R(i+1)\,; \qquad Q(N) = 0\,;$$

$$h(i) = f_x^i h(i+1) - Z_{xu}(i) Z_{uu}^{-1}(i)\,[f_u^i h(i+1) - dH_u^i]\,; \qquad h(N) = 0\,;$$

$$g(i) = g(i+1) - R^T(i+1)(f_u^i)^T Z_{uu}^{-1}(i)\,[f_u^i h(i+1) - dH_u^i]\,; \qquad g(N) = 0\,;$$

where Z_{xx}, Z_{xu}, Z_{uu} are as defined in (6.10.19), (6.10.15), and (6.10.14), and $dH_u^i = -\epsilon H_u^i$, $0 < \epsilon \le 1$. Record $Z_{uu}^{-1}(i)\, Z_{ux}(i)$, $Z_{uu}^{-1}(i)\, f_u^i\, R(i+1)$, $Z_{uu}^{-1}(i)\,[f_u^i h(i+1) - dH_u^i]$, $Q(0)$, and $g(0)$.

STEP (c). Choose values of $d\psi$ to cause the next nominal solution to be closer to the desired values of $\psi[x(N)] = 0$. For example, we might choose

$$d\psi = -\epsilon\psi[x(N)]\,, \qquad 0 < \epsilon \le 1\,.$$

Then determine and record $d\nu$ from

$$d\nu = -\,[Q(0)]^{-1}\,[d\psi - g(0)]\,.$$

STEP (d). Repeat Steps (a) through (c), using improved values for $u(i)$ and ν obtained by adding $du(i)$ and $d\nu$ to the values of $u(i)$ and ν on the previous nominal, where

$$du(i) = -Z_{uu}^{-1}(i)\,[Z_{ux}(i)\,dx(i) + f_u^i R(i+1)\,d\nu + f_u^i\, h(i+1) - dH_u^i]\,,$$

and $dx(i) = x(i)_{\text{new}} - x(i)_{\text{old}}$. *Stop* when $\psi[x(N)]$ and H_u^i are zero to the desired accuracy.

Comments on the algorithm. The only part of the algorithm that requires further explanation beyond that given in Section 6.10 is the handling of the nonhomogeneous terms involving dH_u^i. This requires a *nonhomogeneous* version of (6.10.7) and (6.10.8):

$$\begin{bmatrix} d\lambda(i) \\ d\psi \end{bmatrix} = \begin{bmatrix} S(i) & , R(i) \\ R^T(i) & , Q(i) \end{bmatrix} \begin{bmatrix} dx(i) \\ d\nu \end{bmatrix} + \begin{bmatrix} h(i) \\ g(i) \end{bmatrix}$$

Proceeding as in Section 6.10, it is straightforward to show that the sequences $h(i)$ and $g(i)$ are determined by the recursive relations given in Step (b) above. Note that $h(i+1)$ occurs in $du(i)$ of Step (d) and $g(0)$ occurs in $d\nu$ of Step (c).

7.8 A conjugate-gradient algorithm

Let u be the vector of parameters or functions that is to be chosen to optimize a criterion $J(u)$. If $u^{(i)}$ denotes the value of u at the ith iteration, the algorithms discussed in earlier sections can be placed in one of the following two categories:

First-order methods

$$\Delta u^{(i)} \triangleq u^{(i+1)} - u^{(i)} = -\epsilon(J_u)_{u=u^{(i)}}$$

where ϵ is chosen according to some constraints on the size of Δu step.

Second-order methods

$$\Delta u^{(i)} = u^{(i+1)} - u^{(i)} = -([J_{uu}]^{-1}J_u^T)_{u=u^{(i)}}$$

Some of the advantages and disadvantages of the two categories are described in Table 7.8.1.

Table 7.8.1 Types of Algorithms

	Advantages	Disadvantages
First-order methods	Useful for starting solution. Simplicity in computing J_u	Slow convergence near optimum
Second-order methods	Fast convergence near optimum	Must compute J_{uu}^{-1}, which is quite involved, particularly if u is a function (that is, ∞ dimensions) Initial J_{uu}^{-1} may not exist or bear any relationship to value near optimum; divergence of algorithm may result

The conjugate-gradient method attempts to combine the advantages of both methods while eliminating their disadvantages. Initially, the algorithm behaves like a first-order method but, as iteration proceeds, it behaves more like a second-order method. At no time, however, is it necessary to compute J_{uu}^{-1}. There are several versions of this algorithm, but they are all built around two key ideas. *First*, a sequence of directions, $p_1, p_2, \ldots, p_n$ is generated which has the *orthogonality* or *conjugate* property with respect to J_{uu}; that is, we have

$$[p_i^T J_{uu} p_j] = 0, \qquad i \neq j. \tag{7.8.1}$$

Second, a sequence of one-dimensional searches is made along each of the conjugate directions to find the optimum in that direction. The optimum point along one direction serves as the starting point for the search in the next direction:

$$u^{(i+1)} = u^{(i)} - \alpha_i p_i, \tag{7.8.2}$$

where

$$\alpha_i = \arg\ \min J(u^{(i)} - \alpha_i p_i)\,. \tag{7.8.3}$$

This is shown pictorially in Figure 7.8.1.

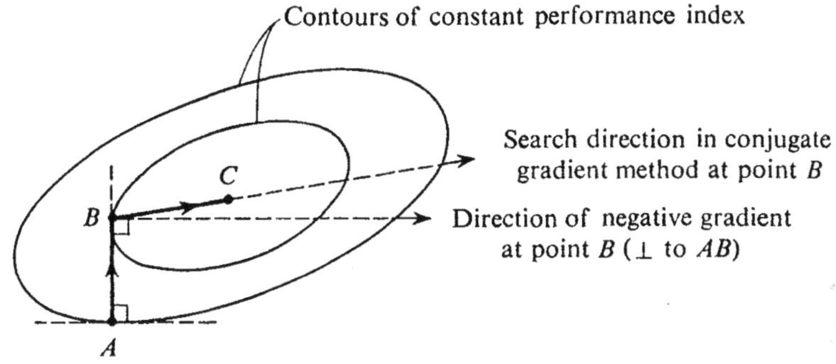

Figure 7.8.1. Conjugate gradient directions in two-dimensional minimization for a quadratic surface.

If the criterion is quadratic, so that J_{uu} in (7.8.1) is constant and positive definite, the directions $p_1, \ldots, p_n$ form a set of orthogonal, linearly-independent basis vectors in the optimization space (we assume u is an n-vector here). We have

$$u^{(N+1)} = u^{(1)} - \sum_{i=1}^{N} \alpha_i p_i\,. \tag{7.8.4}$$

Because of the orthogonality of the p_i's, it can be shown that the separate determination of the α's is equivalent to simultaneous determination. In other words, $u^{(N+1)}$ will yield the exact minimum if J is quadratic. For detailed justification of these statements, see Beckman, 1960.

The flow diagram of one version of the conjugate-gradient method is shown in Figure 7.8.2.

When applying the algorithm to nonlinear criteria other than quadratic, finite convergence is no longer assured. However, if the criterion near the optimum is reasonably approximated by a quadratic function, rapid convergence can still be expected.

For extension of this method to control problems see Lasdon, Warren, and Mitter, *Trans. IEEE-GAC*, Vol. 12, 1967, pp. 132–138.

Problem. For the algorithm stated above, with $J = \frac{1}{2}(u - \bar{u})^T A(u - \bar{u})$, verify Equation (7.8.1), and show that

$$\alpha_i = J_u(i) p_i / p_i^T A p_i\,.$$

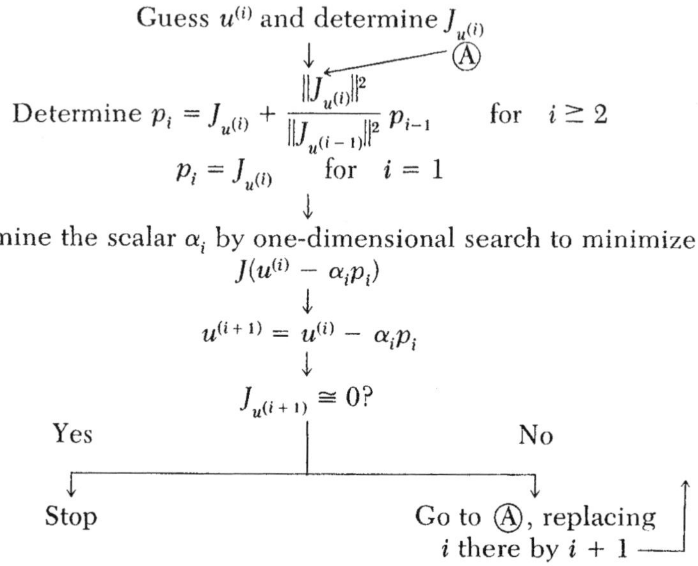

Guess $u^{(i)}$ and determine $J_{u^{(i)}}$
Ⓐ
Determine $p_i = J_{u^{(i)}} + \dfrac{\|J_{u^{(i)}}\|^2}{\|J_{u^{(i-1)}}\|^2} p_{i-1}$ for $i \geq 2$
$p_i = J_{u^{(i)}}$ for $i = 1$
Determine the scalar α_i by one-dimensional search to minimize $J(u^{(i)} - \alpha_i p_i)$
$u^{(i+1)} = u^{(i)} - \alpha_i p_i$
$J_{u^{(i+1)}} \cong 0?$
Yes
No
Stop
Go to Ⓐ, replacing i there by $i + 1$

Figure 7.8.2. Flow diagram of a conjugate-gradient algorithm.

7.9 Problems with inequality constraints on the control variables

The algorithms discussed so far in this chapter apply to problems in which there are no inequality constraints on the control and/or state variables. Thus they apply only to problems in which the performance index and/or the system equations are nonlinear. Linear optimal programming problems must have control and/or state variable constraints to be well posed; for such problems the solution is *always* on a constraint boundary, so the numerical problem is one of determining when to switch from one constraint boundary to another and which one to switch to (if there are more than two constraint boundaries). For nonlinear problems with control and/or state variable constraints, part of the solution may be on the constraint boundaries (constrained arcs) and part may be inside the constraint boundaries (unconstrained arcs).

Integral penalty functions. The simplest, but not necessarily the most effective, approach to solving such problems is to use integral penalty functions. If the inequality constraint

$$C(x,u,t) \leq 0 \tag{7.9.1}$$

is specified for $t_0 \leq t \leq t_f$, the performance index, J, may be augmented as follows:

$$\bar{J} = J + \mu \int_{t_0}^{t_f} [C(x,u,t)]^2 \, 1(C) \, dt \,, \tag{7.9.2}$$

where

$$1(C) = \begin{cases} 0 & C < 0 \,, \\ 1 & C > 0 \,. \end{cases} \tag{7.9.3}$$

By suitable choice of the constant μ (positive if J is to be minimized, negative if J is to be maximized), the constraint (7.9.1) may be satisfied *approximately*. In general, the larger $|\mu|$ is taken, the smaller will be the value of the integral in (7.9.2). However, if $|\mu|$ is taken too large, the iterative algorithms discussed in Sections 7.3 through 7.8 will tend to concentrate more on satisfying the constraint than on minimizing (or maximizing) the original performance index, J. As a result, convergence to a satisfactory solution is very slow. (See Section 1.10 for the analogous situation in parameter optimization problems.)

Joining together constrained and unconstrained arcs. A more effective approach to solving such problems is to join together constrained and unconstrained arcs, making use of the necessary conditions described in Sections 3.10 and 3.11. Unlike the integral penalty function approach, this approach is capable of finding the exact solution and uses less computer time. However, one must guess beforehand the sequence of constrained and unconstrained arcs, and the computer programming is more complicated.

To describe the approach, we consider a fairly general example. Find $u(t)$ to minimize

$$J = \phi[x(t_f)] + \int_{t_0}^{t_f} L(x,u,t) \, dt \tag{7.9.4}$$

subject to

$$\dot{x} = f(x,u,t) \,, \qquad x(t_0) \text{ specified} \,, \qquad t_f \text{ specified} \,, \tag{7.9.5}$$

and

$$C(x,u,t) \leq 0 \,, \tag{7.9.6}$$

where both C and u are scalar functions. Suppose that we have reason to believe that the optimal trajectory is made up of three arcs, as follows:

(a) Unconstrained arc ($C < 0$) for $t_0 \leq t \leq t_1$,
(b) Constrained arc ($C = 0$) for $t_1 \leq t \leq t_2$,
(c) Unconstrained arc ($C < 0$) for $t_2 \leq t \leq t_f$.

However, we do not know, *a priori*, the values of $x(t_1), t_1, x(t_2)$, and t_2. We assume that a nominal trajectory (nonoptimal) can be found that satisfies the constraints (7.9.5) and (7.9.6). We then attempt to find a small change from the nominal control program, $\delta u(t)$, to decrease J and still satisfy the constraints. A first-order gradient procedure like the one described in Section 7.3 may be used, with the modifications (see Section 3.10) that on $C = 0$ we have

$$C(x,u,t) = 0 \Rightarrow u(x,t), \tag{7.9.7}$$

$$\dot{\lambda}^T = -L_x - \lambda^T f_x - \mu C_x, \tag{7.9.8}$$

where

$$\mu = -C_u^{-1}(L_u + \lambda^T f_u) \geq 0. \tag{7.9.9}$$

The time t_1 is determined as the time at which $C = 0$, whereas t_2 is determined as the time at which $\mu = 0$ and is turning negative.† Note that $\mu > 0$ on $C = 0$ is a necessary condition. The control may be discontinuous at t_1 and/or t_2 if H has two or more minima with respect to u; however, H must be continuous at t_1 and t_2. The algorithm becomes a split-interval algorithm, where $\delta u = -\epsilon H_u$ on the unconstrained arcs and Equations (7.9.8) and (7.9.9) are used to compute $\lambda(t)$ across the constrained arc, noting that $\lambda(t)$ is continuous at t_1 and t_2. For further discussion, see Bryson, Denham, and Dreyfus (1963), and Denham and Bryson (1964).

7.10 Problems with inequality constraints on the state variables

The discussion and the integral penalty function method presented in Section 7.9 apply to this case also. However, the joining together of constrained and unconstrained arcs using a first-order gradient algorithm is more complicated because, in general, the influence functions, $\lambda(t)$, are discontinuous at the entry and exit points of any constrained arc. These discontinuities occur because unconstrained arcs must be tangent to constrained arcs at their juncture points, giving interior point constraints (see Section 3.11).

If the constraint is

$$S(x,t) \leq 0, \tag{7.10.1}$$

then on constrained arcs ($S = 0$), Equations (7.9.7) through (7.9.9) apply if we replace $C(x,u,t)$ by $S^{(q)}(x,u,t)$. As in Section 7.9, the algorithm may be regarded as a split-interval algorithm, with the

†This is the usual case. However, a more complicated behavior may result at the exit point t_2 with $\mu(t_2) \neq 0$ (see e.g., Section 8.6).

added complication that it deals with a multipoint boundary-value problem. For further discussion, see Bryson, Denham, and Dreyfus (1963), and Denham and Bryson (1964).

Some problems with state variable inequality constraints may be split into two completely uncoupled two-point boundary-value problems. This is a great simplification; it is treated in Speyer, Mehra, and Bryson (1968) (see Section 3.12).

7.11 Mathematical programming approach

If we accept the fact that most optimal control problems must be solved numerically using a digital computer, one can always regard the determination of $u(t)$, $t_0 \leq t \leq t_f$ as equivalent to the determination of $u(0), u(1), \ldots, u(N-1)$, the discrete equivalent of $u(t)$. Thus, consider the system

$$x(i+1) = f^i[x(i), u(i)], \tag{7.11.1}$$

the constraints

$$C^i[x(i), u(i)] \leq 0, \qquad i = 0, 1, \ldots, N-1, \tag{7.11.2}$$

and the criterion function

$$J = [x(N)] + \sum_{i=0}^{N-1} L^i[x(i), u(i)]. \tag{7.11.3}$$

Now (according to Section 1.7) let

$$y^T = [x(1), \ldots, x(N); u(0), \ldots, u(N-1)], \qquad L(y) = J.$$

We may let

$$f(y) = \begin{bmatrix} x(1) - f^0[x(0), u(0)] \\ \cdot \\ \cdot \\ \cdot \\ x(N) - f^{N-1}[x(N-1), u(N-1)] \\ C^0[x(0), u(0)] \\ \cdot \\ \cdot \\ \cdot \\ C^{N-1}[x(N-1), u(N-1)] \end{bmatrix}.$$

The problem has now been reduced to that of Section 1.7; i.e., determine y such that $L(y)$ is minimized subject to $f(y) \leq 0$. If we apply the necessary conditions (1.7.12) and (1.7.13) to the above equation, the usual necessary condition for discrete-time optimal control problems results. Clearly, more complicated discrete optimization problems can be similarly converted and their corresponding necessary conditions derived from the basic equations (1.7.12) and (1.7.13). Once they have been determined, numerical solutions consist of successively finding a $u(i)$ sequence which satisfies the necessary conditions. The two steps outlined in Section 1.9, namely, finding a feasible solution and finding a feasible direction of improvement, may be easy or difficult, depending on the problem and the method of discretization or parametrization.

Example. *Minimum terminal error control*

Let the system be

$$x(i+1) = \Phi(i)x(i) + d(i)u(i), \qquad i = 0, \ldots, N-1, \tag{7.11.4}$$

the constraint

$$|u(i)| < 1 \Leftrightarrow \begin{cases} u(i) - 1 \leq 0, \\ -1 - u(i) \leq 0; \end{cases} \qquad i = 0, \ldots, N-1. \tag{7.11.5}$$

the criterion

$$J = \tfrac{1}{2}\|x(N)\|^2. \tag{7.11.6}$$

Since any $u(i)$ sequence satisfying (7.11.5) is a feasible sequence, the first part of the numerical solution is trivial. Now define

$$H^\circ(i) = \lambda^T(i+1)[\Phi(i)x(i) + d(i)u(i)] \tag{7.11.7}$$

and

$$\lambda^T(i) = \lambda^T(i+1)\Phi^T(i); \qquad \lambda^T(n) = x^T(n). \tag{7.11.8}$$

It then follows, by the usual reasoning, that

$$\frac{\partial J}{\partial u(i)} = \frac{\partial H^\circ(i)}{\partial u(i)} = \lambda^T(i+1)\, d(i) \tag{7.11.9}$$

if we ignore the constraint (7.11.5). Hence, a feasible improvement would be given by the vector

$$v(i) = \begin{cases} 1 - u(i), & \dfrac{\partial H^\circ}{\partial u(i)} \leq 0, \\ -1 - u(i), & \dfrac{\partial H^\circ}{\partial u(i)} > 0, \end{cases} \tag{7.11.10}$$

which simply indicates the direction of the vertex of the hypercube $|u(i)| \leq 1$ in the solution space that lies in the same quadrant as the gradient $[\partial H^\circ(i)/\partial u(i)]$. This is always a feasible direction, since a feasible solution must always lie in or on the hypercube. How far to proceed along the feasible direction to get the maximum improvement can either be determined by a one-dimensional search or analytically expressed as

$$\delta u(i) = \alpha v(i)\,, \tag{7.11.11}$$

where

$$\alpha = \text{sat}\left[\frac{x(N)^T K v(i)}{v(i)^T K^T K v(i)}\right]$$

and

$$K = [\Phi(N-1), \ldots, \Phi(1)d(0) \cdots \Phi(N-1)d(N-2) \cdots d(N-1)]\,.$$

Note that in this case, because the components of constraint, Equation (7.11.5), are uncoupled (i.e., choice of $u(i)$ does not affect $u(j)$ for $i \neq j$), the feasible improvement direction is easy to find. When the constraint involves state variables, this is no longer true. The problem becomes considerably harder.

Singular solutions of optimization and control problems

8

8.1 Introduction

In some optimization problems, extremal arcs ($H_u = 0$) occur on which the matrix H_{uu} is singular. Such arcs are called *singular arcs*; they satisfy the necessary condition on convexity [Equation (6.3.12)] but not the strengthened condition [Equation (6.3.1)]; that is, H_{uu} is only *semidefinite*. Additional tests are needed to determine if a singular arc is optimizing or not.

We shall discuss only the case in which the Hamiltonian is *linear* in one or more of the control variables (but *nonlinear* in one or more of the state variables), since this appears to be the most common case in which singular arcs are encountered in applications. For such systems, the coefficient of the linear control term in H vanishes identically on a singular arc; thus, the control is *not* determined in terms of the state and adjoint variables, x and λ, by the necessary condition $H_u = 0$ (or minimizing H) along the singular arc. Instead, the control is determined by the requirement that the coefficient of these linear terms remain zero on the singular arc; i.e., the time derivatives of H_u must be zero.

Recently, an additional necessary condition, analogous to the convexity condition [Equation (6.3.12)], has been derived for singular arcs.† For a minimum problem with a single control variable (u is a scalar) it is easily stated:

$$(-1)^k \frac{\partial}{\partial u}\left[\left(\frac{d}{dt}\right)^{2k} H_u\right] \geq 0\,, \qquad k = 0,1,2,\ldots . \tag{8.1.1}$$

Conditions analogous to the "no-conjugate-point condition" [Equation (6.3.3)] have not as yet been developed for singular arcs; hence,

†Tait (1965); Kelley, Kopp, and Moyer (1966); and Robbins (1965).

no sufficient conditions for optimality of singular arcs are available. Equation (8.1.1) is derived in Section 8.4 for the case $k = 1$.

8.2 Singular solutions of optimization problems for linear dynamic systems with quadratic criteria

Consider the terminal control problem of Section 5.2 with $B(t) = 0$. The performance index is then quadratic in the state, x, but independent of the control, u:

$$J = \frac{1}{2} x^T(t_f) S_f x(t_f) + \frac{1}{2} \int_{t_0}^{t_f} x^T(t) A(t) x(t)\, dt\,. \tag{8.2.1}$$

We assume that S_f and $A(t)$ are positive semidefinite. The system equations are linear:

$$\dot{x} = F(t)x + G(t)u\,, \qquad x(t_0)\,, t_0\,, t_f \text{ given.} \tag{8.2.2}$$

Therefore, the Hamiltonian is linear in u:

$$H = \tfrac{1}{2} x^T A x + \lambda^T (Fx + Gu)\,, \tag{8.2.3}$$

$$\Rightarrow \frac{\partial H}{\partial u} = \lambda^T G\,, \qquad \text{where} \qquad \dot{\lambda} = -F^T \lambda - Ax\,, \qquad \lambda(t_f) = S_f x(t_f)\,. \tag{8.2.4}$$

If u is bounded, the minimum of H with respect to u *may* occur on the boundary (as it *must* in problems linear in the control and the state; see Section 3.9). A necessary condition, then, is that

$$\lambda^T G\, \delta u \geq 0 \tag{8.2.5}$$

for all admissible variations δu.

However, it may be possible to find intervals where a function $u(t)$ *inside* the bounded region will yield a $\lambda(t)$ such that

$$\frac{\partial H}{\partial u} = \lambda^T G = 0\,, \tag{8.2.6}$$

that is, a stationary solution. Such arcs are called *singular arcs*, and they may or may not be minimizing.

If *u is unbounded*, the system can be moved instantaneously to certain other states by the use of *impulses* in the control. Such impulses do *not* affect the performance index. Thus, if it is possible to move the system to $x = 0$ by an impulse, this obviously minimizes J, since then $J = 0$! If this is not possible, an impulse may be used to move the system to a minimizing singular arc, then travel along the singular arc until a state is reached from which an impulse will take the system to $x = 0$ (or, less restrictively, to a state $x(t_f)$, where

$x^T(t_f)S_f x(t_f) = 0$ if S_f is only semidefinite). In any case, note that (8.2.6) does *not* supply information directl, for the determination of such controls.

Example 1. *Autonomous linear system of second order with one control variable, quadratic performance index in state variables only.*† A special case of the above, with two state variables and one control variable, is

$$J = \frac{1}{2}\int_0^{t_f} x_1^2 \, dt \,, \tag{8.2.7}$$

where

$$\dot{x}_1 = x_2 + u \,, \qquad \dot{x}_2 = -u \,; \qquad (x_1 \,, x_2 \,, u \quad \text{are scalar functions}); \tag{8.2.8}$$

$$x_1(0) \,, \quad x_2(0) \,, \quad t_f \qquad \text{given,} \qquad x_1(t_f) = x_2(t_f) = 0 \,.$$

This problem is linear in u but nonlinear in x_1 through the performance index. The Hamiltonian is

$$H = \lambda_1(x_2 + u) + \lambda_2(-u) + \tfrac{1}{2} x_1^2 \,, \tag{8.2.9}$$

where

$$\dot{\lambda}_1 = -\frac{\partial H}{\partial x_1} = -x_1 \,, \qquad \dot{\lambda}_2 = -\frac{\partial H}{\partial x_2} = -\lambda_1 \,. \tag{8.2.10}$$

Singular arcs must be such that

$$\frac{\partial H}{\partial u} = \lambda_1 - \lambda_2 = 0 \tag{8.2.11}$$

for a finite interval of time. During this interval, then, we have

$$\frac{d}{dt}\left(\frac{\partial H}{\partial u}\right) = \dot{\lambda}_1 - \dot{\lambda}_2 = 0 \,, \tag{8.2.12}$$

$$\Rightarrow -x_1 + \lambda_1 = 0 \,.$$

Since H is not an explicit function of t, it must be constant for an optimal solution:

$$H = \tfrac{1}{2} x_1^2 + \lambda_1 x_2 + (\lambda_1 - \lambda_2)u = \text{const.} \tag{8.2.13}$$

On a singular arc then, using $(\partial H/\partial u) = (d/dt)/(\partial H/\partial u) = 0$, we have

$$H = \tfrac{1}{2} x_1^2 + x_1 x_2 = \text{const,} \tag{8.2.14}$$

†This is similar to an example treated by C. D. Johnson and J. E. Gibson, "Singular Solutions in Problems of Optimal Control," *IEEE Trans. Automatic Control*, January 1963.

which represents a one-parameter family of singular arcs (hyperbolas) in the (x_1x_2)-space (see Figure 8.2.1).

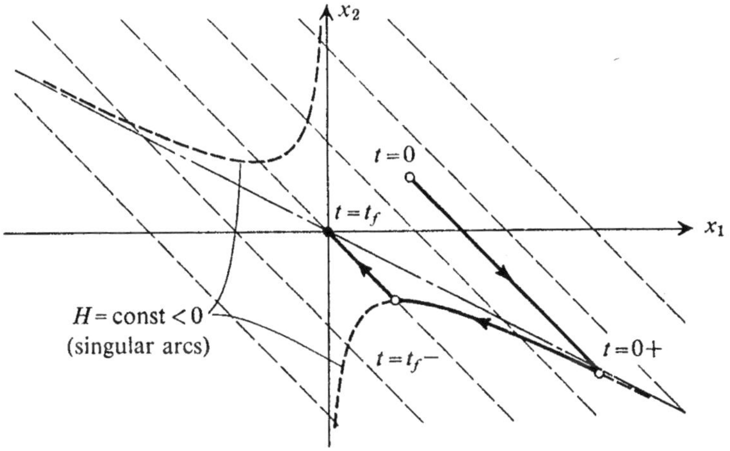

Figure 8.2.1. Optimal path involving a singular arc for Example 1.

If we also use

$$\frac{d^2}{dt^2}\left(\frac{\partial H}{\partial u}\right) = -\dot{x}_1 + \dot{\lambda}_1 = 0\,, \tag{8.2.15}$$

it follows that

$$-x_2 - u - x_1 = 0\,,$$

or

$$u = -(x_1 + x_2)\,, \tag{8.2.16}$$

which is a linear feedback-control law, applicable on singular arcs. Note that

$$-\frac{\partial}{\partial u}\frac{d^2}{dt^2}\left(\frac{\partial H}{\partial u}\right) = 1 > 0\,, \tag{8.2.17}$$

so that the generalized convexity condition mentioned in Section 8.1 is satisfied.

For *unbounded u*, it is clear that the state may be shifted along lines of constant $x_1 + x_2$ *instantaneously* by the use of Dirac function impulses in u; positive impulses move the state down and right (see Figure 8.2.1), while negative impulses move the state up and left. Such motions have *no effect* on the performance index directly, since $u(t)$ does not enter into the criterion function.

On a singular arc, using (8.2.16), we have

$$\dot{x}_1 = x_2 - (x_1 + x_2)\,, \qquad \dot{x}_1 + x_1 = 0\,,$$

$$\dot{x}_2 = x_1 + x_2\,, \qquad \dot{x}_2 - x_2 = x_1\,,$$

$$\Rightarrow x_1 = (\text{const})\, e^{-t}\,; \tag{8.2.18}$$

that is, the magnitude of x_1 decreases exponentially with time; the arrows in Figure 8.2.1 indicate the direction of motion along the two branches of the singular arc. Thus, a typical extremal solution (see Figure 8.2.1) involves an initial impulse to move the state to the singular arc (at $t = 0+$), then motion along the singular arc until the line $x_1 + x_2 = 0$ is reached, then another impulse to move the state to the origin. The fact that the state must arrive at $x_1 + x_2 = 0$ at $t = t_f$ determines the value of the constant H, which picks out the particular singular arc in the one-parameter family of possible arcs.† It is straightforward to show that

$$H = -2c^2 \frac{e^{-2t_f}}{(1 - e^{-2t_f})^2} \leq 0\,, \tag{8.2.19}$$

where

$$c = x_1(0) + x_2(0)\,.‡$$

It is also straightforward to show that

$$x_1(0+) = \frac{2c}{1 - e^{-2t_f}}\,, \qquad x_2(0+) = \frac{-c}{\tanh t_f}\,, \tag{8.2.20}$$

$$x_1(t) = x_1(0+)e^{-t}\,, \qquad x_2(t) = x_1(0+)\sinh t + x_2(0+)e^{t}\,. \tag{8.2.21}$$

If *u is bounded* then, instead of using Dirac function impulses, we must use the maximum or minimum values of u to get onto and off the singular arc; this was the problem treated by Johnson and Gibson for $t_f \to \infty$. The possible optimal solution thus involves a *combination of bang-bang control and linear feedback control* (singular arcs).

Example 2. *General stationary linear system, quadratic performance index in state variables only*

$$J = \frac{1}{2}(x^T S_f x)_{t=t_f} + \frac{1}{2}\int_0^{t_f} x^T A x \, dt\,, \tag{8.2.22}$$

---

†Of course, to prove optimality one has to compare this with the possibility of using impulse-different singular arc-impulse combinations that may arrive at the origin before t_f and remain there. The x value will be larger for the latter case but integrated over a shorter time interval. See Problem in this section.

‡In Johnson and Gibson's problem we have $t_f \to \infty$, so that $H \to 0$. The singular arcs are then degenerate hyperbolas, namely, the two lines $x_1 + 2x_2 = 0$ and $x_1 = 0$.

$$\dot{x} = Fx + Gu\,; \qquad x(0)\,,\quad t_f \quad \text{given,}$$

$$A, F, G, \textit{constant}\ \text{matrices}, \qquad x \text{ an } n\text{-vector}, \qquad u \text{ an } m\text{-vector}. \tag{8.2.23}$$

$$H = \tfrac{1}{2}x^TAx + \lambda^T(Fx + Gu) = \text{const for optimal solution} \quad (1 \text{ equation}), \tag{8.2.24}$$

where

$$\dot{\lambda}^T = -\lambda^TF - x^TA\,, \qquad \lambda(t_f) = S_fx(t_f)\,, \tag{8.2.25}$$

On singular arcs, we must have:

$$\frac{\partial H}{\partial u} = \lambda^TG = 0 \qquad (m \text{ equations}), \tag{8.2.26}$$

$$\frac{d}{dt}\left(\frac{\partial H}{\partial u}\right) = \dot{\lambda}^TG = -(\lambda^TF + x^TA)G = 0 \qquad (m \text{ equations}), \tag{8.2.27}$$

$$\frac{d^2}{dt^2}\left(\frac{\partial H}{\partial u}\right) = -\dot{\lambda}^TFG - \dot{x}^TAG = (\lambda^TF + x^TA)FG - (x^TF^T + u^TG^T)AG = 0\,,$$

$$\Rightarrow u = -(G^TAG)^{-1}G^T[(AF - F^TA)x - F^TF^T\lambda]\,, \tag{8.2.28}$$

$$\Rightarrow G^TAG \qquad \text{must be nonsingular.}$$

Now (8.2.24), (8.2.26), and (8.2.27) represent $2m + 1$ equations in λ and x which define the locus of possible singular arcs in the $2n$ dimensional (x,λ) space. Equation (8.2.28) is the linear feedback law that obtains on a singular arc. The generalized convexity condition

$$-\frac{\partial}{\partial u}\left[\frac{d^2}{dt^2}\left(\frac{\partial H}{\partial u}\right)^T\right] = G^TAG \geq 0 \tag{8.2.29}$$

is satisfied if A is positive semidefinite. For a more detailed treatment of this problem see Johnson and Wonham (1964).

Problem. The system in Example 1 is completely controllable and hence can be moved to the origin in an arbitrarily short time, again by using arbitrarily large control; for example,

$$u(t) = \begin{cases} u(0)\,, & 0 \leq t < \Delta\,, \\ u(\Delta)\,, & \Delta \leq t < 2\Delta\,, \\ 0\,, & 2\Delta \leq t < t_f\,, \end{cases}$$

will take the system to the origin in time 2Δ, where $\Delta \to 0$. Show that such control is not as good as the singular control, despite the fact that the criterion does not include u.

8.3 Singular solutions of optimization problems for nonlinear dynamic systems

In the previous section we restricted our attention to cases in which the Hamiltonian was linear in the control variables and quadratic in the state variables. In this section we shall relax the restriction on the type of nonlinearity of the Hamiltonian in the state variables, but we shall continue to assume that the Hamiltonian is linear in the control variables.† We shall treat the problem of minimizing

$$\phi[x(t_f)] \tag{8.3.1}$$

subject to the constraints

$$\dot{x} = f(x) + g(x)u\,, \qquad t_o \leq t \leq t_f\,, \qquad x(t_o) \text{ given}, \tag{8.3.2}$$

$$\psi[x(t_f)] = 0\,, \tag{8.3.3}$$

where x is an n-vector, u is a *scalar*,‡ and ψ is a q-vector.

The Hamiltonian is linear in u and is assumed to be nonlinear in x:

$$H = \lambda^T[f(x) + g(x)u]\,. \tag{8.3.4}$$

Necessary conditions for a stationary solution include

$$H_u = \lambda^T g = 0\,; \tag{8.3.5}$$

$$\dot{\lambda}^T = -[\lambda^T(f_x + g_x u)]\,, \qquad \lambda^T(t_f) = (\phi_x + \nu^T\psi_x)_{t=t_f}\,. \tag{8.3.6}$$

Now (8.3.5) does *not* directly determine $u(x,\lambda)$, since u does not appear in the equation. Nonetheless, it may be possible to find $u(t)$ over a finite time period so that (8.3.5) is satisfied; if this is possible, it follows that

$$\frac{d}{dt}(H_u) = \lambda^T\dot{g} + \dot{\lambda}^T g = (\lambda^T g_x)\dot{x} + \dot{\lambda}^T g = 0\,. \tag{8.3.7}$$

Substituting (8.3.2) and (8.3.6) into (8.3.7) yields

$$\frac{d}{dt}(H_u) = (\lambda^T g_x)(f + gu) - [\lambda^T(f_x + g_x u)]g = \lambda^T q(x) = 0\,, \qquad \text{where}$$

$$q(x) = g_x f - f_x g\,. \tag{8.3.8}$$

Note that the terms in u cancelled out of (8.3.8), so that (8.3.8) does *not* determine $u(x,\lambda)$ any more than (8.3.5) did. Hence, we try again to find some relations that determine u by differentiating (8.3.8) with respect to time:

---

†This is not the most general case but includes most of the important cases that have appeared in applications to date.

‡This is not a serious restriction since the developments in this and the following section apply individually to components of u in the vector case.

$$\frac{d^2}{dt^2}(H_u) = \lambda^T\dot{q} + \dot{\lambda}^T q = \lambda^T q_x(f + gu) - \lambda^T(f_x + g_x u)q$$

$$= \lambda^T(q_x f - f_x q) + \lambda^T(q_x g - g_x q)u = 0\,. \tag{8.3.9}$$

Now, provided that we have $\lambda^T(q_x g - g_x q) \neq 0$ [see (8.1.1)], Equation (8.3.9) *does* determine u:

$$u = -\frac{\lambda^T(q_x f - f_x q)}{\lambda^T(q_x g - g_x q)}\,. \tag{8.3.10}$$

The control law (8.3.10) maintains the stationarity condition (8.3.5) *if* (8.3.5) and (8.3.8) were satisfied at the beginning (or the end) of the singular arc. This is reminiscent of the state variable inequality constraint of Section 3.11. Thus, singular arcs (with a scalar control variable) are not possible at all points of the $2n$-dimensional (x,λ)-space; they are restricted, via (8.3.5) and (8.3.7), to a hypersurface of dimension $2n - 2$, called a *singular surface*. For stationary systems with unspecified terminal time, the singular surface is of dimension $2n - 3$ since the Hamiltonian is zero over the whole time period:

$$H = \lambda^T(f + gu) = 0\,,$$

or, using (8.3.5),

$$H = \lambda^T f = 0\,. \tag{8.3.11}$$

For stationary systems with unspecified terminal time and $n = 3$, Equations (8.3.5), (8.3.8), and (8.3.11) are linear and homogeneous in λ_1, λ_2, λ_3. Consistency among these equations requires the vanishing of the determinant of the coefficients of λ_1, λ_2, and λ_3. This yields a relation that defines a singular surface *in the state space* (see sounding rocket example below).

Example. *Thrust program for maximum altitude of a sounding rocket.*† A simple version of the problem can be stated as follows: Given a single-stage rocket vehicle with a fixed amount of propellant, how should the thrust be programmed to maximize the final altitude? The equations of motion are

$$\dot{v} = (1/m)\,[F(t) - D(v,h)] - g\,, \tag{8.3.12}$$

†This famous problem was proposed by R. H. Goddard in 1919 and formulated clearly by G. Hamel "Über eine mit dem Problem der Rakete Zusammenhängende Aufgabe der Variationsrechnung," *ZAMM*, Vol. 7, No. 6, 1927, pp. 451–452. H. S. Tsien and R. C. Evans gave an important special solution: "Optimum Thrust Programming for a Sounding Rocket," *Amer. Rocket Soc. J.*, Vol. 21, No. 5, 1951, pp. 99–107; a complete report was given by B. Garfinkel, "A Solution of the Goddard Problem," *SIAM J. Control*, Vol. 1, No. 3, pp. 349–368.

$$\dot{h} = v\,, \tag{8.3.13}$$

$$\dot{m} = -(1/c)\,F(t)\,, \tag{8.3.14}$$

where

v = vertical velocity,
h = altitude,
m = mass of vehicle,
F = thrust, the control variable,
D = drag, a given function of v and h,
g = gravity force per unit mass (taken as constant here for simplicity),
c = specific impulse (impulse per unit mass of fuel burned).

The problem is to find $F(t)$ to maximize $h(t_f)$ with

$$v(0) = 0\,, \qquad h(0) = 0\,, \qquad m(0) \ \text{given}, \qquad m(t_f) \ \text{given}\,, \tag{8.3.15}$$

$$0 \le F(t) \le F_{\max}\,. \tag{8.3.16}$$

Since we are maximizing h, and not specifying $v(t_f)$, we have

$$\lambda_h(t_f) = 1\,, \qquad \lambda_v(t_f) = 0\,. \tag{8.3.17}$$

The Hamiltonian is

$$H = \lambda_v\left(\frac{F-D}{m} - g\right) + \lambda_h v - \lambda_m \frac{F}{c}\,, \tag{8.3.18}$$

which is *linear* in the control variable $F(t)$. The problem is stationary, so H is constant for an optimal path; since the final time is not specified, the constant must be zero:

$$H(t) = 0\,. \tag{8.3.19}$$

The equations for the influence functions are

$$\dot{\lambda}_v = \lambda_v \frac{1}{m}\frac{\partial D}{\partial v} - \lambda_h\,, \tag{8.3.20}$$

$$\dot{\lambda}_h = \lambda_v \frac{1}{m}\frac{\partial D}{\partial h}\,, \tag{8.3.21}$$

$$\dot{\lambda}_m = \lambda_v \frac{(F-D)}{m^2}\,. \tag{8.3.22}$$

The maximum of H with respect to F is given by maximizing

$$\left(\frac{\lambda_v}{m} - \frac{\lambda_m}{c}\right)F\,,$$

which has three possible solutions:

$$F = F_{\max} \qquad \text{if} \qquad \frac{\lambda_v}{m} - \frac{\lambda_m}{c} > 0\,, \tag{8.3.23}$$

$$0 < F < F_{\max} \qquad \text{if} \qquad \frac{\lambda_v}{m} - \frac{\lambda_m}{c} = 0 \qquad \text{(singular arc)}, \tag{8.3.24}$$

$$F = 0 \qquad \text{if} \qquad \frac{\lambda_v}{m} - \frac{\lambda_m}{c} < 0\,. \tag{8.3.25}$$

On a singular arc, we have

$$I = c\lambda_v - m\lambda_m = 0\,. \tag{8.3.26}$$

If I is zero, clearly we have $\dot{I} = 0$ also on a singular arc. Differentiating (8.3.26) and substituting (8.3.14), (8.3.20), (8.3.22), and (8.3.26), we have

$$\left(\frac{\partial D}{\partial v} + \frac{D}{c}\right)\lambda_v - m\lambda_h = 0\,. \tag{8.3.27}$$

Similarly we have $\ddot{I} = 0$ on a singular arc. Differentiating (8.3.27) and substituting from (8.3.12) through (8.3.14), (8.3.20) through (8.3.22), (8.3.26), and (8.3.27), we obtain

$$F = D + mg + \frac{m}{D + 2c(\partial D/\partial v) + c^2(\partial^2 D/\partial v^2)}\left[-g\left(D + c\frac{\partial D}{\partial v}\right)\right.$$
$$\left. + c(c - v)\frac{\partial D}{\partial h} - vc^2\frac{\partial^2 D}{\partial v \partial h}\right], \tag{8.3.28}$$

where λ_v cancelled out of both sides of (8.3.28). Thus (8.3.28) is a *nonlinear feedback law* for the thrust on singular arcs.

The locus of possible singular arcs is defined in the state space by requiring consistency among Equations (8.3.19), (8.3.26), and (8.3.27):

$$\begin{vmatrix} \dfrac{F - D}{m} - g\,, v & , -\dfrac{F}{c} \\ c \quad ,0 & , -m \\ \dfrac{\partial D}{\partial v} + \dfrac{D}{c} \quad , -m\,, & 0 \end{vmatrix} = 0\,,$$

or

$$D + mg - \frac{v}{c}D - v\frac{\partial D}{\partial v} = 0\,, \tag{8.3.29}$$

which is a *surface* in the state space, i.e., the (v,h,m)-space.

If $D(v,h)$ is monotonic in both v and h (and it usually is), the solution usually contains only three arcs:

(a) $F = F_{max}$ until (8.3.29) is satisfied;
(b) Singular arc using the feedback law (8.3.28) until $m = m(t_f)$;
(c) $F = 0$ until $v = 0$.

The solution could be described as "bang-singular-bang," which is the same as the solution of Example 1 in Section 8.2. Note that $m = m(t_f)$ could be reached on the maximum thrust arc (a) *before* (8.3.29) was satisfied, so that no singular arc would appear in the solution. This is exactly what happens if $D = 0$ (no atmosphere) for (8.3.29) then reduces to $m = 0$.

The case considered by *Tsien and Evans* was

$$D = \frac{1}{2}\rho_o v^2 C_D S e^{-\beta h} \Rightarrow \frac{\partial D}{\partial v} = \frac{2D}{v}, \tag{8.3.30}$$

$$\frac{\partial^2 D}{\partial v^2} = \frac{2D}{v^2}, \qquad \frac{\partial D}{\partial h} = -\beta D, \qquad \frac{\partial^2 D}{\partial h \partial v} = -2\beta\frac{D}{v}.$$

The singular surface is then

$$mg = \left(1 + \frac{v}{c}\right)D. \tag{8.3.31}$$

The nonlinear feedback law on the singular arc is

$$F = D + mg + \frac{mg}{1 + 4(c/v) + 2(c^2/v^2)}\left[\frac{\beta c^2}{g}\left(1 + \frac{v}{c}\right) - 1 - 2\frac{c}{v}\right]. \tag{8.3.32}$$

Tsien and Evans also had no upper limit on F; that is, $F_{max} \to \infty$. As a result, an *impulse* (instantaneous burning of part of the propellant) is indicated at $t = 0$ of sufficient magnitude to bring the vehicle to the critical surface (8.3.31). Note that $me^{v/c}$ is constant during an impulse. Equation (8.3.32) is then used until the remainder of the propellant is consumed, followed by a coasting arc to maximum altitude.

Problem 1. Another approach to the problem of singular arcs is to eliminate the control variable and change state variables in such a way that one of the original state variables becomes a control variable. In the sounding rocket example of this section, the equations of motion are

$$\dot{v} = (1/m)(F(t) - D(v,h)) - g, \qquad \dot{h} = v, \qquad \dot{m} = -(1/c)F(t).$$

(a) Eliminate the control variable $F(t)$ and choose h and $me^{v/c}$ as new state variables. Show that this leaves $v(t)$ as a control variable.
(b) State the first-order necessary conditions for maximum altitude, using these new variables.

(c) State the convexity (Legendre-Clebsch) condition, using these new variables.

(d) Discuss how you think this approach could be used to determine the entire trajectory if $0 \leq F(t) \leq F_{max}$.

Problem 2. Consider the problem

$$J = \phi[x(t_f)], \qquad \dot{x} = Fx + Gu; \qquad |u(t)| \leq 1, \qquad t_f \text{ specified.}$$

Show that, if (F,G) is controllable and $\phi_{x(t_f)} \neq 0$, then singular arcs cannot exist. [HINT: See Appendix B2.]

Problem 3. *Thrust, angle-of-attack, and bank-angle programs for minimum fuel turn of a jet (or rocket) airplane at constant altitude* (see Problem 8, Section 1.3, for quasisteady version and nomenclature). The equations of motion are

$$m\dot{V} = -C_{D_0}qS - \eta C_{L_\alpha}\bar{\alpha}^2 qS + T(t), \qquad mg = C_{L_\alpha}\bar{\alpha}qS\cos\sigma,$$

$$mV\dot{\beta} = C_{L_\alpha}\bar{\alpha}qS\sin\sigma, \qquad \dot{m} = -\frac{T(t)}{c},$$

where

$$q = \tfrac{1}{2}\rho V^2 S.$$

We wish to determine $T(t)$, $\bar{\alpha}(t)$, $\sigma(t)$ to maximize $m(t_f)$ for given values of $\beta(t_f) - \beta(t_0)$ and $m(t_0)$. *Show that*, on optimizing singular arcs, σ and V are determined by:

$$\tan^2\sigma = 1 + \left(\frac{V}{V_0}\right)^4, \qquad \left(\frac{V}{V_0}\right)^4\left(3 + \epsilon\frac{V}{V_0}\right)\Big/\left(1 + \epsilon\frac{V}{V_0}\right) = \sec^2\sigma,$$

where

$$V_0 = \sqrt{\frac{2g\ell}{\delta}}, \qquad \ell = \frac{2m\eta}{C_{L_\alpha}\rho S}, \qquad \delta = 2\sqrt{\frac{\eta C_{D_0}}{C_{L_\alpha}}} = \left(\frac{D}{L}\right)_{\min}$$

and $$\epsilon = \frac{V_0}{c}.$$

Note that α and T are then determined from the equations of motion.

8.4 A generalized convexity condition for singular arcs†

A necessary condition for minimality, discussed in Section 6.3, is the convexity condition

†This section is based on the Ph.D. Thesis of K. Tait, Harvard University, 1965.

$$H_{uu} \geq 0 . \tag{8.4.1}$$

For singular arcs, $H_{uu} = 0$, so condition (8.4.1) is satisfied but yields little information. A more useful necessary condition, mentioned in the introduction to this chapter, can be derived, which looks very much like Equation (8.4.1), namely,

$$(-1)^k \frac{\partial}{\partial u}\left[\left(\frac{d}{dt}\right)^{2k}\left(\frac{\partial H}{\partial u}\right)\right] \geq 0, \qquad k = 0,1,2,\ldots . \tag{8.4.2}$$

We shall now proceed to derive (8.4.2) for the case $k = 1$. To do this we refer back to Section 6.1, where the second variation of the performance index was given:

$$\delta^2 J = \frac{1}{2}[\delta x^T \Phi_{xx}\, \delta x]_{t_f} + \frac{1}{2}\int_{t_o}^{t_f} [\delta x^T, \delta u]\begin{bmatrix} H_{xx} & H_{xu} \\ H_{ux} & H_{uu}\end{bmatrix}\begin{bmatrix}\delta x \\ \delta u\end{bmatrix} dt , \tag{8.4.3}$$

with constraints

$$\delta \dot{x} = H_{\lambda x}\, \delta x + H_{\lambda u}\, \delta u , \qquad \delta x(t_o) = 0 ; \tag{8.4.4}$$

$$\delta \dot{\lambda} = -H_{x\lambda}\, \delta\lambda - H_{xx}\, \delta x - H_{xu}\, \delta u , \qquad \delta\lambda(t_f) = [\Phi_{xx}\, \delta x]_{t_f} . \tag{8.4.5}$$

Consider the identically zero integral obtained by integrating (8.4.5) as follows:

$$\frac{1}{2}\int_{t_o}^{t_f} (\delta\dot{\lambda}^T + \delta\lambda^T f_x + \delta u H_{ux} + \delta x^T H_{xx})\, \delta x\, dt \equiv 0 .$$

Integrating the first term by parts gives

$$0 = \frac{1}{2}[\delta\lambda^T \delta x]_{t_o}^{t_f} + \frac{1}{2}\int_{t_o}^{t_f} [-\delta\lambda^T \delta\dot{x} + (\delta\lambda^T f_x + \delta u H_{ux} + \delta x^T H_{xx})\, \delta x]\, dt .$$

Substituting for $\delta\dot{x}$ from (8.4.4) and using $\delta x(t_o) = 0$, $\delta\lambda^T(t_f) = [\delta x^T \Phi_{xx}]_{t_f}$ give

$$0 = \frac{1}{2}[\delta x^T \Phi_{xx}\, \delta x]_{t_f} + \frac{1}{2}\int_{t_o}^{t_f} [-\delta\lambda^T H_{\lambda u}\, \delta u + \delta u H_{ux}\, \delta x + \delta x^T H_{xx}\, \delta x]\, dt . \tag{8.4.6}$$

Subtracting Equation (8.4.6) from Equation (8.4.3), we have

$$\delta^2 J = \frac{1}{2}\int_{t_o}^{t_f} (\delta x^T H_{xu} + \delta\lambda^T H_{\lambda u} + \delta u H_{uu})\, \delta u\, dt . \tag{8.4.7}$$

By straightforward differentiation and substitution from (8.4.4) and (8.4.5), it can be shown that

$$\frac{d}{dt}(\delta x^T H_{xu} + \delta\lambda^T H_{\lambda u} + \delta u H_{uu}) = \delta x^T (\dot{H}_u)_x + \delta\lambda^T (\dot{H}_u)_\lambda + \delta u (\dot{H}_u)_u , \tag{8.4.8}$$

$$\frac{d^2}{dt^2}(\delta x^T H_{xu} + \delta\lambda^T H_{\lambda u} + \delta u H_{uu}) = \delta x^T(\ddot{H}_u)_x + \delta\lambda^T(\ddot{H}_u)_\lambda + \delta u(\ddot{H}_u)_u\,. \quad (8.4.9)$$

Now we may integrate (8.4.7) by parts, using (8.4.8), to obtain

$$\delta^2 J = -\frac{1}{2}\int_{t_0}^{t_f}[\delta x^T(\dot{H}_u)_x + \delta\lambda^T(\dot{H}_u)_\lambda + \delta u(\dot{H}_u)_u]\,\delta u_1(t)\,dt + \frac{1}{2}[(\delta x^T H_{xu} + \delta\lambda^T H_{\lambda u} + \delta u H_{uu})\,\delta u_1]_{t_0}^{t_f}, \quad (8.4.10)$$

where

$$\delta u_1(t) = \int_{t_0}^{t}\delta u(t)\,dt\,. \quad (8.4.11)$$

Similarly, we may integrate (8.4.10) by parts, using (8.4.8) and (8.4.9), to obtain

$$\delta^2 J = \frac{1}{2}\int_{t_0}^{t_f}[\delta x^T(\ddot{H}_u)_x + \delta\lambda^T(\ddot{H}_u)_\lambda + \delta u(\ddot{H}_u)_u]\,\delta u_2(t)\,dt + \frac{1}{2}[(\delta x^T H_{xu} + \delta\lambda^T H_{\lambda u} + \delta u H_{uu})\,\delta u_1]_{t_0}^{t_f} - \frac{1}{2}[(\delta x^T(\dot{H}_u)_x + \delta\lambda^T(\dot{H}_u)_\lambda + \delta u(\dot{H}_u)_u)\,\delta u_2]_{t_0}^{t_f}, \quad (8.4.12)$$

where

$$\delta u_2(t) = \int_{t_0}^{t}\delta u_1(t)\,dt\,. \quad (8.4.13)$$

Note that the integrand in Equation (8.4.12) is very similar to the integrand in (8.4.7). The term analogous to $\delta u^T H_{uu}\,\delta u$ is $\delta u^T(\ddot{H}_u)_u\,\delta u_2$. In the nonsingular case, the convexity condition $H_{uu} \geq 0$ is necessary for a minimum, since the term can always be made dominant by choosing δu to be a positive impulse followed immediately by a negative impulse. Let us try this same special variation here for the singular case $H_u = H_{uu} = 0$. It is clear that, again, all terms can be neglected in (8.4.10), since all δx and $\delta\lambda$ are effectively zero as a result of the double impulse and $(\dot{H}_u)_u = 0$ from (8.3.7). On the other hand, we have, for δu_1, δu_2, and δu, the situation in Figure 8.4.1. If we consider $(\ddot{H}_u)_u$ constant during the period of variation of δu in Figure 8.4.1, it is clear that

$$\int_{t_0}^{t_f}(\delta u\,\delta u_2)\,dt < 0\,. \quad (8.4.14)$$

Thus, in order to assure that $\delta^2 J \geq 0$ for this special variation, it is necessary that

$$(\ddot{H}_u)_u \leq 0\,. \quad (8.4.15)$$

It may happen that $\ddot{H}_u$ is *independent* of u, or, more generally, that

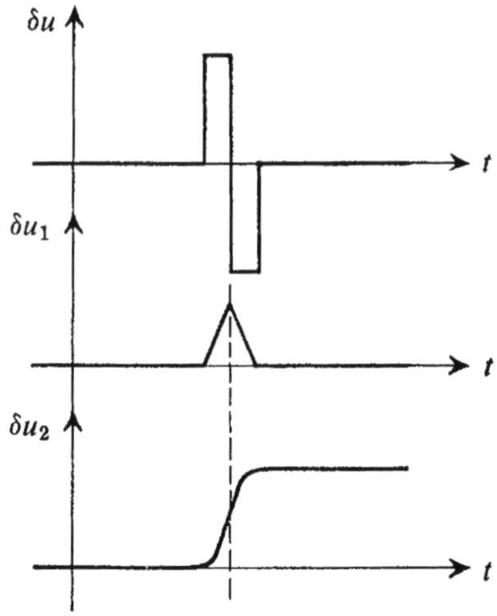

Figure 8.4.1. Special variation in $u(t)$ used in deriving generalized convexity condition ($k = 1$).

$$\left(\frac{d}{dt}\right)^i H_u = 0\,, \qquad i = 0,1,\ldots,m-1 \tag{8.4.16}$$

and

$$\left(\frac{d}{dt}\right)^m (H_u) = a(x,\lambda) + b(x,\lambda)u\,. \tag{8.4.17}$$

The procedure to be followed for the solution of such problems is conceptually the same as in the case above, where $m = 2$. The following facts can be shown [(see Tait (1965); Kelley, Kopp, and Moyer (1966); Robbins (1966)]:

(i) The variable m is always even. We shall call the problem a singular problem of order m if it fulfills (8.4.16) and (8.4.17).
(ii) The singular surface in the $(x - \lambda)$-space in this case is of dimension $2n - m$.
(iii) The generalization of (8.4.15) is

$$(-1)^{\frac{m}{2}} \frac{\partial}{\partial u}\left(\frac{d^m}{dt^m} H_u\right) \geq 0\,. \tag{8.4.18}$$

Note that the singularity property is invariant with respect to a one-to-one transformation of the control variable. Suppose that we consider $u = h(v)$; then, in the transformed system, we have

$$(\partial/\partial v)\, H(x,\lambda,h(v)) = H_u h_v = 0\,,$$

$$(\partial^2/\partial v^2)\, H(x,\lambda,h(v)) = H_{uu} h_v^2 + H_u h_{vv} = 0$$

and, furthermore,

$$\frac{d}{dt} H_v = \dot{H}_u h_v + H_u \frac{d}{dt} h_v = 0 ,$$

$$\frac{d^2}{dt^2} H_v = \ddot{H}_u h_v + \dot{H}_u \dot{h}_v + \dot{H}_u \dot{h}_v + H_u \ddot{h}_v = \ddot{H}_u h_v ,$$

and, hence,

$$h_v^2 \frac{\partial}{\partial u} (\ddot{H}_u) = \frac{\partial}{\partial v} (\ddot{H}_v)$$

$$\Rightarrow \operatorname{sgn}\left(\frac{\partial}{\partial v} \frac{d^2}{dt^2} H_v\right) = \operatorname{sgn}\left(\frac{\partial}{\partial u} \frac{d^2}{dt^2} H_u\right) .$$

Problem. Let $\dot{u} = h(u)v$, where v is now the control variable, and

$$J = \phi(x(t_f),t_f) + \int_0^{t_f} L(v)\, dt .$$

Show that this does not change the character of the singular problem.

8.5 Conditions at a junction

When one considers the joining of a nonsingular arc to a singular arc, or vice versa, an additional necessary condition of similar appearance to (8.4.18) can be derived. We proceed as follows:

Suppose that an optimal $x(t)$ joins a singular surface at $t = t_2$; then, from previous discussions (Section 8.3), we have

$$H_u(t_2) = 0 , \qquad \dot{H}_u(t_2) = 0 ,$$

and

$$\ddot{H}_u(t_2) = \lambda^T(q_x f - f_x q) + \lambda^T(q_x g - g_x q)u .$$

For $t < t_2$ and $t_2 - t$ very small, we may expand H_u in Taylor series as

$$H_u(t) = H_u(t_2) - \dot{H}_u(t_2)(t_2 - t) + \tfrac{1}{2}\ddot{H}_u(t_2)(t_2 - t)^2 - \cdots ,$$

$$= \tfrac{1}{2}[\lambda^T(q_x f - f_x q) + \lambda^T(q_x g - g_x q)u](t_2 - t)^2 - \cdots .$$

Since, at t, the trajectory is by definition nonsingular, $u(t)$ must be at a constraint boundary, and $H_u(t) \neq 0$. If $u(t)$ is at the upper constraint boundary ($\Rightarrow H_u(t) < 0$) , we have

$$[\lambda^T(q_x f - f_x q) + \lambda^T(q_x g - g_x q)u_{\text{upper}}] < 0 . \qquad (8.5.1)$$

Similarly if u is at u_{lower} , we have

$$[\lambda^T(q_x f - f_x q) + \lambda^T(q_x g - g_x q)u_{\text{lower}}] > 0 . \qquad (8.5.2)$$

Taking the difference of the above two expressions, we arrive at the necessary condition that, for transition to be possible from both constraint boundaries,

$$\lambda^T(q_x g - g_x q) = \frac{\partial}{\partial u}\frac{d^2}{dt^2}\left(\frac{\partial H}{\partial u}\right)\bigg|_{\text{junction time}} \leq 0\,. \tag{8.5.3}$$

If the order of the singularity m is higher than two, this argument is straightforwardly extended, using (8.4.17), to yield

$$\frac{\partial}{\partial u}\left(\frac{d^m}{dt^m}\frac{\partial H}{\partial u}\right)\bigg|_{\text{junction time}} \leq 0\,, \qquad m = 4,6,\ldots\,. \tag{8.5.4}$$

The junction phenomenon in singular problems is quite complex. For another possible form of joining nonsingular and singular arcs for which the solution "chatters" at infinite frequency onto a singular arc, see Kelley, Kopp, and Moyer (1966), p. 84, and Johnson (1965).

Problem. Show that:

(i) If $m/2$ is odd, a discontinuity in control from an optimal nonsingular arc to a singular arc at $t = t_2$ is permitted.

(ii) If $m/2$ is even, discontinuity in control at a junction point is not permitted. [HINT: Compare (8.5.4) with (8.4.18).]

8.6 A resource allocation problem involving inequality constraints and singular arcs

Having discussed inequality constraints in Chapter 3 and singular arcs in this chapter, it seems appropriate at this point to present an example that fully utilizes the theories involved.

Problem formulation. In a simplified model of resource planning for a national economy, there are two state variables of interest:

r = total capital/total number of workers (represented by buildings, machines, land, etc.),

w = number of employed workers/total number of workers (the employment rate).

The output (gross national product) per employed worker, f, is a function of capital per employed worker, r/w:

$$f(r/w) = \text{output/employed worker},$$

with the properties $f(\alpha) > 0$, $(df/d\alpha) > 0$, $(d^2f/d\alpha^2) < 0$, for $\alpha \geq 0$. The output per capita, wf, is divided three ways by the choice of two control variables, s and e, where

s = fraction of wf invested in *capital* improvement,

e = fraction of wf devoted to *training* of workers (education),

$1 - s - e =$ fraction of wf left for *consumption.*

The relationships among r, w, s, and e are:

$$\dot{r} = swf(r/w) - (n + \delta)r\,, \qquad r(t_0) = r_0\,, \tag{8.6.1}$$

$$\dot{w} = \frac{e}{d}\,wf(r/w) - (n + \mu)w\,, \qquad w(t_0) = w_0\,, \tag{8.6.2}$$

where

$n =$ rate of growth of labor force,
$\delta =$ rate of depreciation of capital,
$\mu =$ rate of mortality or retirement of workers,
$d =$ training cost/worker.

The inequality constraints on the state and control variables, for all t, are:

$$1 \geq s + e\,, \tag{8.6.3}$$

$$s \geq 0\,, \qquad e \geq 0\,, \tag{8.6.4}$$

$$w \leq 1\,, \tag{8.6.5}$$

$$w \geq 0\,. \tag{8.6.6}$$

It turns out that $w > 0$ for the problem we consider below, so (8.6.6) will be omitted henceforth.

A reasonable criterion of performance can be taken as

$$J = \int_{t_0}^{t_f} (1 - s - e)wf(r/w)\exp(-\gamma t)\,dt\,, \tag{8.6.7}$$

where

$$\gamma = \text{interest rate}\,;$$

that is, we wish to maximize the overall consumption over the planning period from t_0 to t_f. The term $\exp(-\gamma t)$ indicates that consumption *now* is valued more highly than consumption *later.*

This is a problem with control and state variable inequality constraints in which the control variables enter into the system equation and criterion linearly (implying possible singular arcs).

SOLUTION AND ANALYSIS. Following the practice established in previous chapters, we define the Hamiltonian.†

†We could have defined the Hamiltonian by adjoining the control variable inequality constraints (8.6.3) and (8.6.4), in which case, we would require $H_u = 0$ even when we are off the state variable constraint (8.6.5). However, it was felt that direct maximization of the Hamiltonian in (8.6.8) offers more insight in this case.

$$H(r,w,\lambda_r,\lambda_w,s,e,\eta,t) = (1 - s - e)wf\exp(-\gamma t) + \lambda_r[swf - (n+\delta)r] + (\lambda_w + \eta)\left[\frac{e}{d}wf - (n+\mu)w\right], \quad (8.6.8)$$

where

$$\eta = \begin{cases} \leq 0, & w = 1, \\ 0, & w < 1, \end{cases} \quad (8.6.9)$$

and where $w - 1 \leq 0$ plays the role of $S(x,t) \leq 0$ in Chapter 3, which is a first-order state variable inequality constraint in this case. We shall consider the case $w < 1$ first, then the case $w = 1$.

Case where $w < 1$. Since the Hamiltonian is linear in s and e, its maximization can be studied through examination of the gradient:

$$H_e = wf\left(\frac{\lambda_w}{d} - \exp(-\gamma t)\right), \quad (8.6.10)$$

$$H_s = wf(\lambda_r - \exp(-\gamma t)). \quad (8.6.11)$$

Geometrically, there are seven subcases concerning the direction of grad H in the solution space. This is illustrated in Figure 8.6.1.

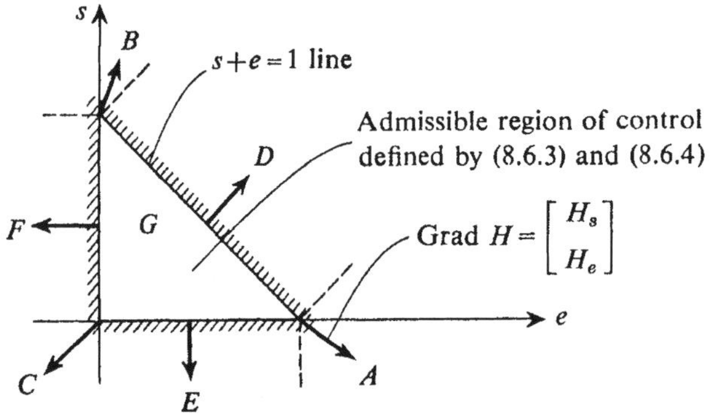

Figure 8.6.1. Control space illustration of constraints and possible directions of Grad H when $w < 1$.

(i) CASE A: $H_e > H_s \Rightarrow (\lambda_w/d) > \lambda_r$. It is clear that the Hamiltonian is maximized by our choosing $e = 1$ and $s = 0$. The resultant system and Euler equations are

$$\dot{r} = -(n+\delta)r, \qquad \dot{w} = \left[\frac{f}{d} - (n+\mu)\right]w,$$

$$\dot{\lambda}_r = -H_r = -\frac{f'(r/w)}{d}\lambda_w + (n+\delta)\lambda_r,$$

$$\dot{\lambda}_w = -H_w = \left[-\frac{(f - f'(r/w))}{d} + (n + \mu)\right]\lambda_w ,$$

where prime denotes differentiation with respect to the argument.

(ii) CASE B. $H_s > H_e \Rightarrow \lambda_r > (\lambda_w/d) \Rightarrow s = 1$ and $e = 0$, and the system and Euler equations can be similarly determined.

(iii) CASE C. $H_s < 0$, $H_e < 0 \Rightarrow \lambda_r < \exp(-\gamma t)$, $(\lambda_w/d) < \exp(-\gamma t)$ $\Rightarrow s = 0$, and $e = 0$ with the corresponding equations of motion.

(iv) CASE D. $H_s = H_e > 0 \Rightarrow (\lambda_w/d) = \lambda_r \Rightarrow s + e = 1$. This is an interesting singular case. The maximization of H yielded one equation for the two control variables s and e. Following the recipe established in earlier sections of this chapter, we differentiate the defining singular equation $\lambda_r - (\lambda_w/d) = 0$ with respect to time twice. This results in

$$(f - (r/w)f') - df' + (\delta - \mu)\,d = 0 , \tag{8.6.12}$$

$$f(sw - e(r/d)) - (\delta - \mu)\,r = 0 . \tag{8.6.13}$$

Solving (8.6.13) and $s + e = 1$ together, we get

$$s = \frac{wf - (\delta - \mu)r}{f(w + (r/d))}, \qquad e = \frac{f(r/d) + (\delta - \mu)r}{f(w + (r/d))}, \tag{8.6.14}$$

from which the resultant motion can be determined. It is easily verified for this case that the singular arc is a straight line going through the origin of the state space.

(v) CASE E. $H_s < 0$, $H_e = 0 \Rightarrow s = 0$, $\lambda_w = d\exp(-\gamma t)$. This case is singular in the control variable e. Differentiating $\lambda_w - d \exp(-\gamma t) = 0$ twice, we have

$$\left[f - \left(\frac{r}{w}\right)f'\right] = (n + \mu + \gamma)d , \tag{8.6.15}$$

$$e = \frac{d(\mu - \delta)}{f} . \tag{8.6.16}$$

(vi) CASE F. $H_s = 0$ and $H_e < 0 \Rightarrow e = 0$, $\lambda_r = \exp(-\gamma t)$

For this singular case, we obtain

$$f' = (n + \delta + \gamma) , \tag{8.6.17}$$

$$s = \frac{(\delta - \mu)r}{fw} . \tag{8.6.18}$$

Note that, depending on the relative magnitude of μ and δ, one of the two cases above cannot be sustained due to the constraints (8.6.4). In either case, the resultant motions are straight lines through the origin of the state space. Henceforth, we shall assume $\mu > \delta$.

(vii) CASE G. $H_s = 0$, $H_e = 0 \Rightarrow \lambda_r = \exp(-\gamma t)$ and $\lambda_w = d\exp(-\gamma t)$. Differentiation of these two relationships leads to (8.6.15) and (8.6.17), which, in general, can be satisfied only for different values of r/w. Hence, we conclude that this doubly singular case can not be sustained.

Case where $w = 1$. If the trajectory remains on $w = 1$, it is clear that we must have $\dot{w} = 0$, which implies

$$e = \frac{(n + \mu)\, d}{f}. \tag{8.6.19}$$

Furthermore, we have

$$H_e = \frac{\lambda_w + \eta}{d} - \exp(-\gamma t) = 0, \tag{8.6.20}$$

which not only determines η but also implies

$$\frac{\lambda_w}{d} - \exp(-\gamma t) \geq 0 \qquad (\text{since } \eta \leq 0), \tag{8.6.21}$$

meaning that further improvement would be possible if the constraint $w \leq 1$ were removed. The value for s is determined according to the sign of H_s. There are three subcases, which are shown in Figure 8.6.2.

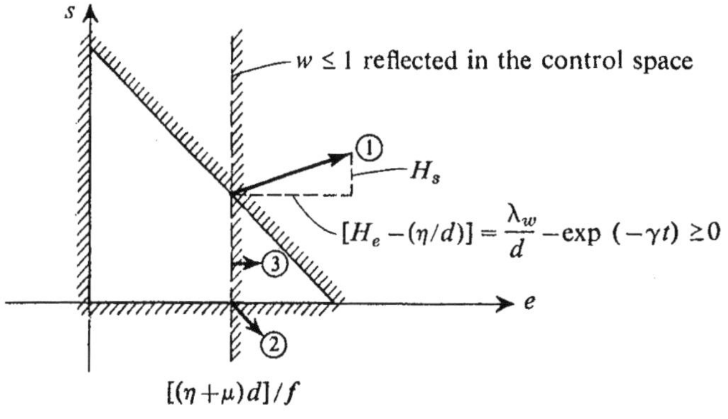

Figure 8.6.2. Control space illustration of constraints and possible directions of Grad H when $w = 1$.

(viii) CASE (1). $H_s > 0 \Rightarrow s = 1 - e = [f - (n + \mu)\, d/f]$;

(ix) CASE (2). $H_s < 0 \Rightarrow s = 0$;

(x) CASE (3). $H_s = 0$. This is another singular case resulting in

$$s = [(n + \mu)\, r]/f\,, \tag{8.6.22}$$

which, together with (8.6.19), leads to $\dot{r} = 0$, $\dot{w} = 0$, an equilibrium point! This completes the catalog of possible cases.

Corner conditions, terminal conditions, and transition sequences. The remaining problem is to determine the sequence of cases (i.e., to piece together a path in the state space) that satisfies the given initial conditions

$$r(0) = r_o\,, \qquad w(0) = w_o\,, \tag{8.6.23}$$

intermediate corner conditions, and the terminal conditions. The corner conditions occur at the time of entry, t_1 , onto the constraint boundary $w = 1$. They are [see (3.13.4) and (3.13.5)]

$$w(t_1) = 1 \qquad \text{and} \qquad H(t_1^-) = H(t_1^+)\,, \tag{8.6.24}$$

$$\lambda_w(t_1^-) = \lambda_w(t_1^+) + \pi\,, \tag{8.6.25}$$

$$\lambda_r(t_1^-) = \lambda_r(t_1^+)\,. \tag{8.6.26}$$

Two sets of terminal conditions are considered.

Terminal state and time fixed. Here we have

$$r(t_f) = r_T\,, \qquad w(t_f) = w_T\,, \tag{8.6.27}$$

which implies that

$$\begin{aligned} \lambda_r(t_f) &= \nu_r = \text{constant to be determined,} \\ \lambda_w(t_f) &= \nu_w = \text{constant to be determined.} \end{aligned} \tag{8.6.28}$$

Terminal time infinite and states free. In this case we have

$$\lambda_r(\infty) = 0\,, \qquad \lambda_w(\infty) = 0\,. \tag{8.6.29}$$

It is readily verified that this case can be realized only by the equilibrium case, Case (3), since the λ_r and λ_w equations are homogeneous and asymptotically stable. This is a standard result for models of optimal saving.

Finally, the possible transitions from case to case must be considered. The fact that both H_e and H_s are continuous functions of time (except at the entry corner t_1) helps to eliminate certain transition sequences. Figure 8.6.3 shows the various possible transitions. One type of transition, that of getting off the constraint $w = 1$, deserves further discussion. Referring to Figure 8.6.2 once again, we

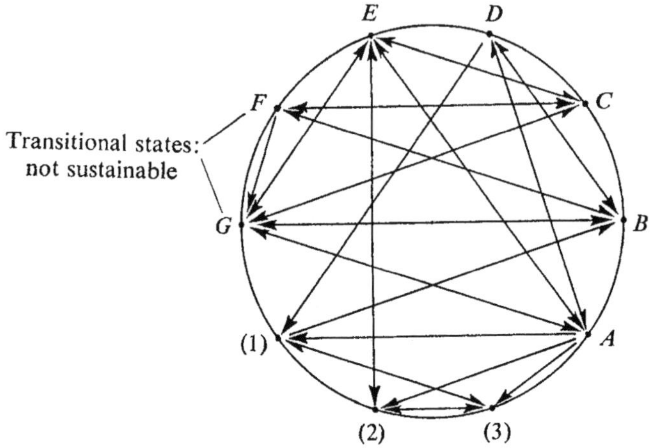

Figure 8.6.3. Possible transitions between cases.

note that the transition from Case (1) to Case B occurs whenever we have

$$\lambda_r - \exp(-\gamma t) > \left|\frac{n}{d}\right|,$$

implying that a *discontinuity* in η is possible at an exit corner. On the other hand, transition from Cases (2) and (3) to Cases E and G can take place only when

$$\eta(t) = d\exp(-\gamma t) - \lambda_w(t)$$

is turning positive.

For the infinite-time, free-terminal-state problem, possible sequences that are economically meaningful and that satisfy all the conditions are

$$\begin{array}{lcl} B \to D \to 1 \to 3 & \text{or} & A \to 1 \to 3 \\ A \nearrow & \text{or} & A \to 2 \to 3 \\ & \text{or} & A \to 3 \to . \end{array}$$

The trajectory in state space is shown in Figure 8.6.4.

Optimality of the extremals and numerical results. There are three possible approaches to the establishment of the optimality of the extremals discussed above. One approach would use arguments based on the concavity of the system and the criterion function and show that the stationary extremals are maximizing. A second approach would involve checking the Jacobi (conjugate-point) condition and the Weierstrass condition along the extremals. A third approach would be to attempt a *numerical* solution of the problem, using dy-

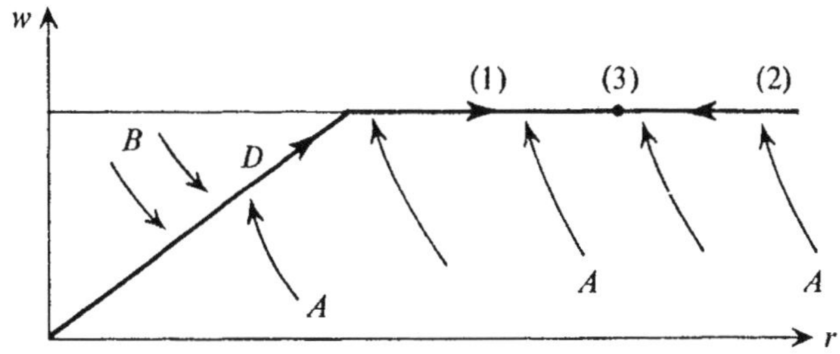

Figure 8.6.4. State space trajectories of the economic planning example.

namic programming. This is the particular approach chosen for this problem in view of the analytical complications and the two-dimensional nature of the problem.

A straightforward dynamic-programming solution was carried out for

$$f(r/w) = (r/w)^\alpha ; \qquad \alpha = 0.3 ; \qquad n = .03 , \qquad \mu = .15 , \qquad \delta = .05 ,$$
$$\gamma = .05 , \qquad d = 2 .$$

The range and quantization of variables involved is

$$\begin{array}{lll} .5 \le w(t) \le 1 & (10 \text{ divisions}), & .8 \le w(t_f) \le 1 , \\ 0 \le r(t) \le 4 & (40 \text{ divisions}), & r(t_f) = 4 , \\ 5 \le t \le 15 = t_f & (30 \text{ divisions}). & \end{array}$$

First-order difference approximations are used for the system differential equations.

A typical trajectory is shown in Figure 8.6.5. It is clear from inspection that it agrees well with the qualitative trajectory obtained in Figure 8.6.4.

Problem 1. Assuming $f > 0$, $f' > 0$, and $f'' < 0$, show that the generalized Legendre-Clebsch condition $(\partial/\partial u)(d^2/dt^2)H_u \ge 0$ is satisfied for cases D, E, and F. [HINT: Case D can be reduced to one variable by eliminating e through the use of the $s + e = 1$ equation.]

Problem 2. Let $\dot{x} = u\sqrt{x} - nx$ and

$$J = \max_u \int_0^T (1 - u)\sqrt{x}\, dt , \qquad |u| \le 1 , \qquad x, u \quad \text{scalar.}$$

(i) Show that this problem has a singular solution
(ii) Now consider $\dot{u} = v$, where v is the control variable and

$$J = \max_v \int_0^T [(1 - u)\sqrt{x} - \tfrac{1}{2} v^2]\, dt .$$

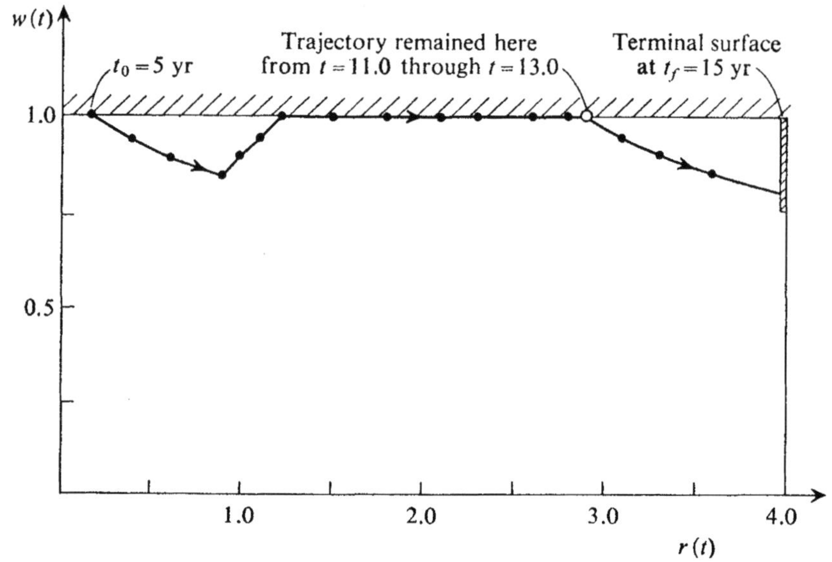

Figure 8.6.5. Numerical example of an optimal trajectory for the economic planning problem.

Show that this problem still possesses the singular character even though $H_{vv} < 0$.

One interpretation of the above problem is

x = capital goods/population, $\sqrt{x}$ = output/capital goods,
u = fraction of output devoted to reinvestment in capital goods,
$(1 - u)\sqrt{x}$ = consumption/population, v = rate of change of u,
n = depreciation constant for capital goods/population.

Differential games

9

9.1 Discrete games

Elementary game theory is concerned with discrete optimization problems involving two players with conflicting interests. In a typical matrix game there are two players, u and v, and a selection of strategies, $u_i, i = 1,2,\ldots,m$; and $v_j, j = 1,2,\ldots,n$, for each player. For each pair of strategies there is a corresponding payoff, $J = L_{ij}$. Player u attempts to minimize the payoff, while v attempts to maximize it. This is a "perfect information game" in the sense that each player has all the information above and that each player knows the other player's choice of strategies. Now, if v (the maximizer) plays first, he should obviously pick the column with the largest minimum, since he knows u will subsequently pick the row with the minimum. Similarly, if u (the minimizer) plays first, he should pick the row with the smallest maximum, since he knows v will subsequently pick the column with the maximum. An example of a 2×2 game is shown in Figure 9.1.1.

	v_1	v_2 ← v is maximizing	
u_1	$L_{11} = 2$	$L_{12} = 7$	← row with smaller maximum
u_2	$L_{21} = 5$	$L_{22} = 9$	

↗ u is minimizing ↖ column with larger minimum

Figure 9.1.1. A simple discrete game.

The optimal choices for the game of Figure 9.1.1 are u_1 and v_2, with the payoff 7, regardless of who plays first; i.e., we have

$$\max_{v_j} \min_{u_i} L_{ij} = 7 = \min_{u_i} \max_{v_j} L_{ij}$$

(v plays first) (u plays first)

or

$$L(u_1,v_j) \leq L(u_1,v_2) \leq L(u_i,v_2)\,.$$

The choice u_1, v_2 is called the *minimax solution* of the game.

However, the choice is not always so simple; suppose that the value of L_{11} is changed from 2 to 11, as in Figure 9.1.2.

	v_1	v_2	← v is maximizing
u_1	$L_{11} = 11$	$L_{12} = 7$	
u_2	$L_{21} = 5$	$L_{22} = 9$	
↑ u is minimizing			

Figure 9.1.2. A discrete game where order of play makes a difference.

Then we have

$$\underset{v}{\max}\,\underset{u}{\min}\,L_{ij} = 7 \leq \underset{u}{\min}\,\underset{v}{\max}\,L_{ij} = 9\,.$$
$$(v \text{ plays first}) \qquad (u \text{ plays first})$$

If v (the maximizer) plays first, he should pick v_2, since this is the column with the larger minimum, namely, 7. If u (the minimizer) plays first, he should pick u_2, since this is the row with the smaller maximum, namely, 9. Thus, it makes a difference who plays first. This dilemma may be resolved by having each side make a random selection of strategies on each play according to some *fixed* probability. Thus, if v plays a fixed choice while u uses a random choice, the expected payoff for various probability mixes of u_1 and u_2 is as shown in Figure 9.1.3(a) [and similarly for v in Figure 9.1.3(b)].

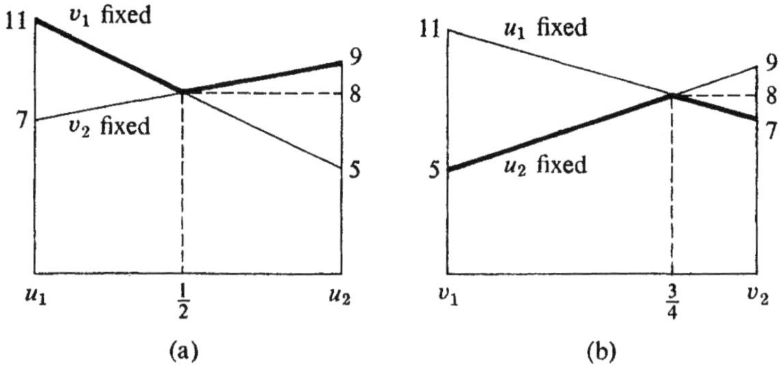

Figure 9.1.3. Illustrations of the minimax solution of the discrete game of Figure 9.1.2.

In Figure 9.1.3(a), if u plays any probability mix other than half of the time u_1, half of the time u_2, v can obtain a higher average payoff by playing the fixed strategy indicated by the solid line. Similarly, we see that v must play the probability mix one-quarter of the time v_1, three-quarters of the time v_2 to realize maximal expected payoff. It is no accident that

$$E[\min_p \max_q L_{ij}] = 8 = E[\max_p \min_q L_{ij}],$$

where the expectation and optimization are taken over the two probability mixes p and q. This is the essence of the celebrated minimax principle of Von Neumann and Morganstern, which says that, by randomization, the difference between minimax and maximin can be equalized on an expected value basis.

Problem 1. Find the minimax solution for the following payoff matrix, Figure 9.1.4, where u is minimizing and v is maximizing.

	v_1	v_2	v_3
u_1	3	1	6
u_2	2	4	9
u_3	12	7	8
u_4	9	3	10

Figure 9.1.4. Game matrix for Problem 1.

Problem 2. (a) Consider the typical path-cost problem of dynamic programming (see Chapter 4, Section 4.2).

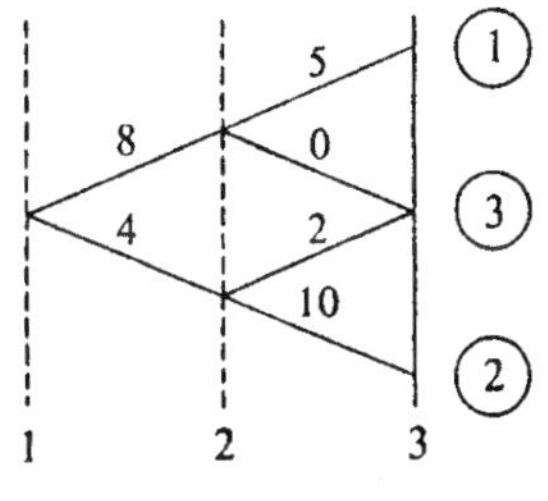

Figure 9.1.5. Path-cost grid for Problem 2.

Let player u (minimizing) control decisions at level 2 and player v (maximizing) control level 1. Compute the minimax cost of the game that corresponds to the case of u announcing his strategies first. Also compute the maximin cost and show that it is less than the minimax cost.

(b) Reduce this two-stage game to a matrix game, as described in this section.

Problem 3. (a) Consider the path problem shown in Figure 9.1.6.

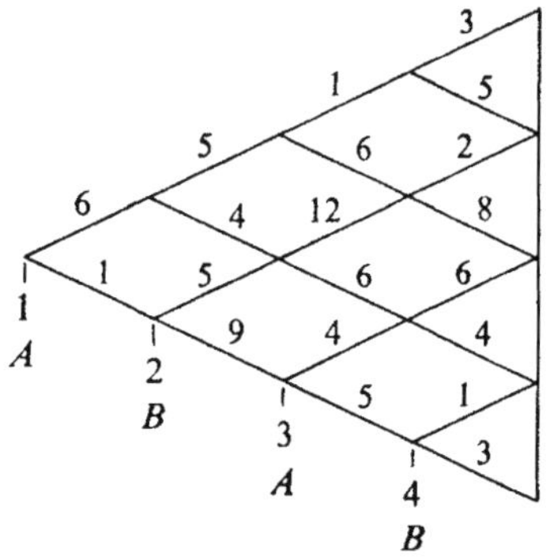

Figure 9.1.6. Path-cost grid for Problem 3.

There are two players, A and B, taking turns alternately in choosing the path at each step. Player A controls steps 1 and 3 and wishes to maximize the total cost of the path, while player B controls steps 2 and 4 and wishes to minimize. Determine the optimal cost and strategy at all junction points, in particular the starting point.

(b) Solve the same problem with this added complication. At each decision point the player has a choice:

(i) Choose the path deterministically but pay an additional cost of 2 units (i.e., for player A, 2 units are to be *subtracted* from his optimal cost; for player B, 2 units are to be *added* to his optimal cost).

(ii) Flip an unbiased coin and let it decide the choice of the path. Determine the optimal expected cost and the strategy (deterministic or random) at each point.

9.2 Continuous games

If the choices of u and v are continuous instead of discrete, there must be a continuous payoff function, $L(u,v)$, instead of a payoff matrix L_{ij}. We look for a pair of choices, u^o, v^o, such that

$$L(u^o,v) \leq L(u^o,v^o) \leq L(u,v^o)\,; \qquad \text{for all} \quad u,v\,. \tag{9.2.1}$$

Following Section 1.1, we claim that necessary conditions for u^o and v^o are

$$\frac{\partial L}{\partial u} = 0, \qquad \frac{\partial L}{\partial v} = 0; \tag{9.2.2}$$

$$\frac{\partial^2 L}{\partial u^2} \geq 0, \qquad \frac{\partial^2 L}{\partial v^2} \leq 0; \tag{9.2.3}$$

and sufficient conditions are (9.2.2) and (9.2.3) with the equalities changed to inequalities. Any u^o, v^o satisfying the sufficient conditions is called a *game-theoretic saddle point.* It should be pointed out that (9.2.2) and (9.2.3) are *not* equivalent to the usual conditions for a *calculus saddle point*, which are

$$\frac{\partial L}{\partial u} = 0, \qquad \frac{\partial L}{\partial v} = 0, \tag{9.2.2'}$$

$$\frac{\partial^2 L}{\partial u^2}\frac{\partial^2 L}{\partial v^2} - \left(\frac{\partial^2 L}{\partial u\, \partial v}\right)^2 \leq 0, \tag{9.2.3'}$$

as the following examples show.

Example 1. Consider the case

$$J = L(u,v) = \tfrac{1}{2}(u^2 - v^2); \qquad -1 \leq u \leq 1; \qquad -1 \leq v \leq 1,$$

$$\frac{\partial L}{\partial u} = 0, \qquad \frac{\partial L}{\partial v} = 0 \Rightarrow u^o = 0, \qquad v^o = 0,$$

$$\frac{\partial^2 L}{\partial u^2} = 1 > 0, \qquad \frac{\partial^2 L}{\partial v^2} = -1 < 0, \qquad \frac{\partial^2 L}{\partial u^2}\frac{\partial^2 L}{\partial v^2} - \left(\frac{\partial^2 L}{\partial u\, \partial v}\right)^2 = -1 < 0.$$

This is both a game-theoretic and a calculus saddle point.

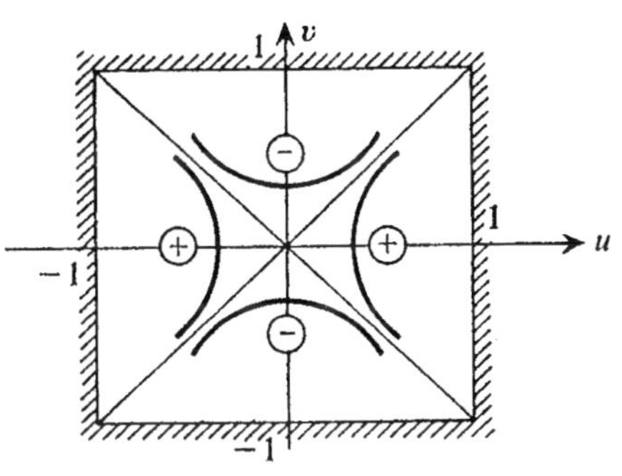

Figure 9.2.1. Saddle point geometry for Example 1.

Example 2. Consider the case

$$L = u^2 - 3uv + 2v^2\,; \qquad -1 \leq u \leq 1\,; \qquad -1 \leq v \leq 1\,,$$

$$\frac{\partial L}{\partial u} = 0\,, \qquad \frac{\partial L}{\partial v} = 0 \Rightarrow u^o = 0\,, \qquad v^o = 0\,,$$

$$\frac{\partial^2 L}{\partial v^2}\frac{\partial^2 L}{\partial u^2} - \left(\frac{\partial^2 L}{\partial u\, \partial v}\right)^2 = -1 < 0 \Rightarrow \text{calculus saddle point (as shown in Figure 9.2.2),}$$

but $$\frac{\partial^2 L}{\partial v^2} = 4 \nless 0 \Rightarrow \textit{not}\text{ a game-theoretic saddle point.}$$

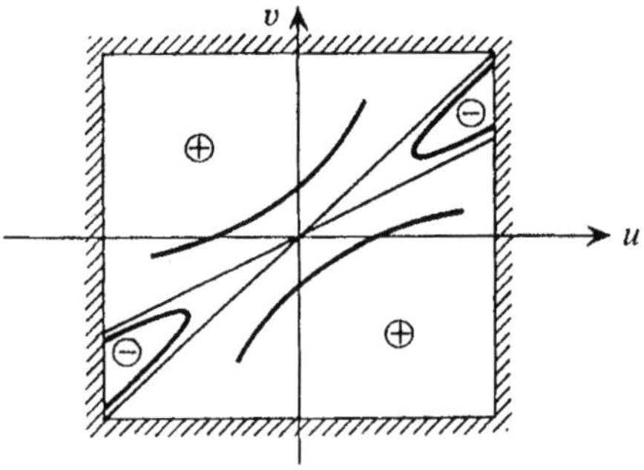

Figure 9.2.2. Saddle point geometry for Example 2.

In fact, we can verify

$$\max_v [\min_u L(u,v)] = \max [-\tfrac{1}{4}v^2 \quad \text{with} \quad u = \tfrac{3}{2}v] = 0 \quad \text{with} \quad v = 0\,;$$

$$\min_u [\max_v L(u,v)] = \min_u [u^2 + 3|u| + 2 \quad \text{with } v = -\operatorname{sgn} u] = 2 \quad \text{with} \quad u = 0\,;$$

that is, $\max_v \min_u L < \min_u \max_v L$.

On the other hand, if we write $L = u^2 - 3uv + 2v^2 = \frac{1}{4}(2u - 3v)^2 - \frac{1}{4}v^2$, then $u^o = 0$, $v^o = 0$ is a game-theoretic saddle point in the $(2u - 3v) - v$ coordinate system. The difference between the two types of saddle points arises because of the cross-product terms in the payoff function $L(u,v)$. If $\partial^2 L/\partial u \partial v = 0$, the two types of saddle points are equivalent. Such problems are called *separable* problems, and $\min_u \max_v L$ always equals $\max_v \min_u L$ in separable problems. In later sections, this separability condition is always assumed to hold when minimaximizing functions of two sets of variables. Consequently, we do not consider the possibility of mixed or random strategies to equalize the difference between minimax and maximin.

Problem. Verify that $L = uv$ has both a game-theoretic and a calculus saddle point.

9.3 Differential games†

The natural extension of the material in Sections 9.1 and 9.2 to the dynamic case yields what is known as differential games.‡ In the present context we visualize a general setup as follows: Given the dynamic system

$$\dot{x} = f(x,u,v,t)\,, \qquad x(t_0) = x_0\,, \tag{9.3.1}$$

the terminal constraints

$$\psi(x(t_f),t_f) = 0\,, \tag{9.3.2}$$

and the performance criterion

$$J = \phi(x(t_f),t_f) + \int_{t_0}^{t_f} L(x,u,v,t)\,dt\,. \tag{9.3.3}$$

find u^o and v^o such that

$$J(u^o,v) \leq J(u^o,v^o) \leq J(u,v^o)\,. \tag{9.3.4}$$

If we examine the arguments of Chapter 2, we find that the derivation of first-order necessary conditions depends only on the stationarity and not on the maximization or minimization of J. It is natural to expect that necessary conditions for the minimax problem posed above can be obtained in exactly the same way to yield

$$H = \lambda^T f + L\,, \tag{9.3.5}$$

$$\dot{\lambda}^T = -H_x\,, \qquad \lambda^T(t_f) = \Phi_{x(t_f)}\,, \tag{9.3.6}$$

$$H_u = 0\,, \qquad H_v = 0\,, \tag{9.3.7}$$

or

$$H^o = \max_v \min_u H\,. \tag{9.3.7$'$}$$

While a straightforward application of (9.3.5) through (9.3.7) often yields useful results (see Section 9.4), several precautions should be observed. First of all, Equations (9.3.7) or (9.3.7′) involve the minimaximization of a function of u and v. Based on the discussion of Section 9.2, we know that a game-theoretic saddle point generally

†The theory of differential games was created by Isaacs concurrently with but independent of the development of control theory. See Isaacs (1965) and Ho (1965).

‡Another natural extension would be sequential or multistage games. These are not discussed here in this brief coverage.

does not exist unless we explicitly assume that H is separable. Fortunately, for most applications, we have $f \equiv f_1(x,u,t) + f_2(x,v,t)$ and $L \equiv L_1(x,u,t) + L_2(x,v,t)$, which we shall assume henceforth. However, it should be noted that separability of H generally does not imply separability of J, which is the item of our real interest. The validity of the latter is difficult to check and most of the time simply not true. For example, let $J = x^2(3) + \Sigma_{i=0}^{2} [u(i)^2 - v(i)^2]$ and the dynamics be $x(i+1) = x(i) + u(i) + v(i)$. Direct substitution will show that J is not separable ($u(i)v(j)$ terms), while H is separable for this problem. This means, in general, that solutions obtained from solving the two-point boundary-value problem of Equations (9.3.5) through (9.3.7) may *not* satisfy the saddle-point condition of (9.3.4). This, however, does not completely destroy the usefulness of the calculus of variations approach since, in many control situations, one is willing to take the position that one side will play first. For example, in "worst case designs," we assume that nature is perverse enough to determine the worst $v(t)$; but we do not assume that nature is perverse enough to actually change $v(t)$ as the game evolves. In other words, we are implicitly assuming in our worst case design that nature plays first by announcing $v^o(t)$ ahead of time. Thus, the answers maximin or minimax, while not equal, may still be useful.

The second point is related to the interpretation of u^o and v^o in (9.3.4) as open-loop or closed-loop strategies. While it is clear that, in the deterministic one-sided control problem, there is no difference between open- and closed-loop control, the situation is different in the game case.

A simple example illustrates this point. Consider a two-dimensional pursuit problem involving simple kinematics with the square of the terminal miss distance at a fixed final time, t_f, as the criterion (see Figure 9.3.1).

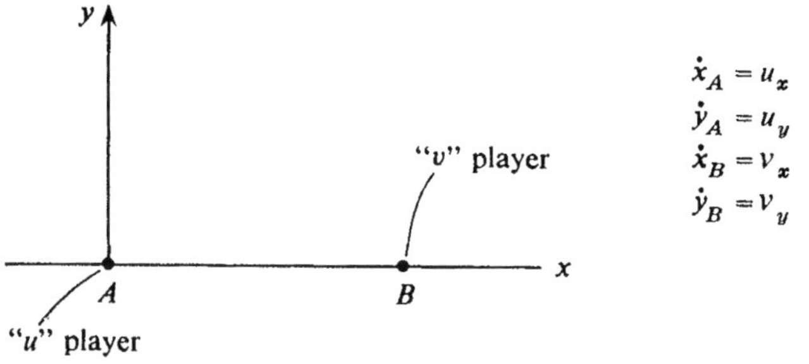

Figure 9.3.1. A simple pursuit-evasion game in two dimensions.

The constraints are $\|u\|^2 \leq 1$ and $\|v\|^2 \leq \frac{1}{4}$. By inspection, we have

$$u^o(t) \triangleq \begin{bmatrix} u_x \\ u_y \end{bmatrix} = \begin{bmatrix} 1 \\ 0 \end{bmatrix}, \tag{9.3.8}$$

$$v^o(t) \triangleq \begin{bmatrix} v_x \\ v_y \end{bmatrix} = \begin{bmatrix} \frac{1}{2} \\ 0 \end{bmatrix}, \tag{9.3.9}$$

and $J(u^o,v^o) = [t_f - (x_B(0) + \frac{1}{2}t_f)]^2$.† In feedback form, we have

$$u^o(t) = \begin{bmatrix} \dfrac{x_B - x_A}{\sqrt{(x_B - x_A)^2 + (y_B - y_A)^2}} \\ \dfrac{y_B - y_A}{\sqrt{(x_B - x_A)^2 + (y_B - y_A)^2}} \end{bmatrix} = k_u(x,t) \tag{9.3.8'}$$

and, similarly,

$$v^o(t) = \tfrac{1}{2} u^o(0) = k_v(x,t)\,. \tag{9.3.9'}$$

It is obvious that $v^o(t) = [{}^{1/2}_{0}]$ is optimal whether or not $u^o(t)$ is generated by (9.3.8) or (9.3.8′) or any other process, provided that $y_B(0) - y_A(0) = 0$ and $x_B(0) - x_A(0) > 0$. On the other hand, it is not at all clear that $u^o(t) = [{}^1_0]$ will be optimal if $v^o(t)$ is generated by the feedback process of (9.3.9′) with the same initial conditions. In fact, by simple construction, one can show that $u^o(t) = [{}^1_0]$ will guarantee a value for $J \leq [t_f - (x_B(0) + \frac{1}{2}t_f)]^2$ only if the control interval $t_f \leq x_B(0) - x_A(0)$. Even so, the value of the criterion using $u^o(t) = [{}^1_0]$ will be larger than that which could be obtained if (9.3.8′) were used when v does not play optimally.

Mathematically, this phenomenon is explained by the fact that the second inequality in (9.3.4) can be viewed in two ways:

$$\min_{u(t)} J[u;v^o(t)] = J[u^o(t;x_0,t_0);v^o(t)]\,, \tag{9.3.10}$$

$$\min_{u(t)} J[u,k_v(x,t)] = J[u^o(t;x_0,t_0);k_v(x,t)]\,. \tag{9.3.10'}$$

Equations (9.3.10) and (9.3.10′) constitute *two different* one-sided control problems from the viewpoint of u. Equation (9.3.10′) is a much more stringent type of optimality. It means that u^o must be optimal against an opponent whose control $v(t)$ is produced in a feedback fashion; that is, v can immediately take advantage of any nonoptimal play made by u. The optimal control $u^o(t)$, if any, obtained from (9.3.10) and (9.3.10′) are generally different (see Section 9.4 for more explicit results).

Thus, a general procedure for solving differential games consists essentially of two steps:

†We assume that capture does not occur in t_f.

(i) Solve for u^o and v^o via the two-point boundary-value problem (tpbvp) [Equations (9.3.1), (9.3.2), (9.3.5) through (9.3.7)] or via dynamic programming (see Problem, this section).
(ii) Verify the inequalities in (9.3.4) separately by solving two one-sided problems, using u^o and v^o in open- or closed-loop form.

It should be emphasized that the verification in (ii) is necessary to establish the saddle-point property of the solution. Existence of a solution to (i) does not generally imply that a saddle point is reached, as should be clear from the above discussions. The verification step (ii) yields various second-order necessary conditions, which are expected. We have

$$H^o_{uu} \geq 0 , \qquad H^o_{vv} \leq 0 , \tag{9.3.11}$$

or

$$H(x,\lambda,t) = \min_{u \epsilon U} \max_{v \epsilon V} H(x,\lambda,u,v,t) , \tag{9.3.11'}$$

no conjugate point for the problem

$$J(u^o,v^o) = \min_u J(u,v^o) ,$$

$$\text{where} \qquad v^o = \begin{cases} v^o(t;x_0,t_0) , \\ k_v(x,t) , \end{cases} \tag{9.3.12}$$

and

no conjugate point for the problem

$$J(u^o,v^o) = \max_v J(u^o,v) ,$$

$$\text{where} \qquad u^o = \begin{cases} u^o(t;x_0,t_0) , \\ k_u(x,t) . \end{cases} \tag{9.3.13}$$

Finally, to establish the saddle point, we must show that u^o, v^o in (9.3.12) and (9.3.13) are the same.

Example. *Minimax terminal miss with bounded acceleration.* In a pursuit-evasion game, the pursuer's control is his acceleration, $a_p(t)$, normal to the initial line of sight (ILOS) to the evader. The evader's control is his acceleration, $a_e(t)$, also normal to the ILOS. The relative velocity along the ILOS is such that the normal time of closest approach is t_f. If $v(t)$ is the relative velocity normal to the ILOS, and $y(t)$ is the relative displacement normal to the ILOS, the equations of motion are†

†This is a differential game version of Example 2, Section 5.2, with bounded acceleration instead of a quadratic penalty on acceleration.

$$\dot{v} = a_p - a_e\,, \qquad v(t_o) = v_o\,; \tag{9.3.14}$$

$$\dot{y} = v\,, \qquad y(t_o) = 0\,. \tag{9.3.15}$$

The pursuer wishes to minimize the terminal miss, $|y(t_f)|$, whereas the evader wishes to maximize it, so the performance index may be taken as

$$J = \tfrac{1}{2}[y(t_f)]^2\,. \tag{9.3.16}$$

The accelerations of the pursuer and the evader are limited:

$$\left.\begin{aligned} |a_p| &\leq a_{pm}\,, \\ |a_e| &\leq a_{em}\,, \end{aligned}\right\} \qquad \text{where} \qquad a_{pm} > a_{em}\,. \qquad \begin{aligned}(9.3.17)\\(9.3.18)\end{aligned}$$

The solution proceeds by first forming the Hamiltonian

$$H = \lambda_v(a_p - a_e) + \lambda_y v\,. \tag{9.3.19}$$

The adjoint equations are then

$$\dot{\lambda}_v = -\lambda_y\,, \qquad \lambda_v(t_f) = 0\,, \tag{9.3.20}$$

$$\dot{\lambda}_y = 0\,, \qquad \lambda_y(t_f) = y(t_f)\,, \tag{9.3.21}$$

and the optimality conditions are

$$a_p(t) = -a_{pm}\,\mathrm{sgn}\,\lambda_v\,, \tag{9.3.22}$$

$$a_e(t) = -a_{em}\,\mathrm{sgn}\,\lambda_v\,. \tag{9.3.23}$$

The adjoint equations are easily integrated to yield

$$\lambda_v(t) = (t_f - t)y(t_f)\,, \tag{9.3.24}$$

$$\lambda_y(t) = y(t_f) = \text{const.} \tag{9.3.25}$$

It is therefore clear that

$$\mathrm{sgn}\,\lambda_v(t) = \mathrm{sgn}\,y(t_f) = \text{const.} \tag{9.3.26}$$

Substituting (9.3.26) into (9.3.22) and (9.3.23) and those equations into (9.3.14) and (9.3.15) yields a simple set of differential equations whose solution may be written as

$$y(t_f) = v_o(t_f - t_o) - \tfrac{1}{2}(a_{pm} - a_{em})\,(t_f - t_o)^2\,\mathrm{sgn}\,y(t_f)\,, \tag{9.3.27}$$

which is a relation for determining $y(t_f)$. Thus, we have

$$y(t_f) = \begin{cases} \tfrac{1}{2}(t_f - t_o)^2\left[\dfrac{2v_o}{t_f - t_o} - (a_{pm} - a_{em})\right] & \text{if } \dfrac{2v_o}{(t_f - t_o)(a_{pm} - a_{em})} > 1 \\[2ex] -\tfrac{1}{2}(t_f - t_o)^2\left[\dfrac{-2v_o}{t_f - t_o} - (a_{pm} - a_{em})\right] & \text{if } \dfrac{2v_o}{(t_f - t_o)(a_{pm} - a_{em})} < -1\,. \end{cases} \tag{9.3.28}$$

However, *no solution* of (9.3.27) exists for

$$-1 < \frac{2v_o}{(t_f - t_o)(a_{pm} - a_{em})} < 1 . \tag{9.3.29}$$

In fact, for this range of initial conditions, it is always possible for the pursuer to bring the terminal miss to zero, i.e., to make $y(t_f) = 0$. He can do this, for example, by choosing $a_p(t)$ in such a way that

$$a_p(t) = a_e(t) + \frac{2v(t)}{t_f - t} . \tag{9.3.30}$$

Problem 1. Verify the saddle point property of the solution (9.3.22), (9.3.23), and (9.3.28) in the example above.

Problem 2. Show the validity of the equation

$$-J_t^o = \min_u \max_v H(x, J_x^o, v, u, t) ,$$

which is the analog of Equation (4.2.13) in Chapter 4, Section 4.2.

9.4 Linear-quadratic pursuit-evasion games†

Consider two dynamic systems,

$$\dot{x}_p = F_p x_p + G_p u , \qquad x_p(t_o) \text{ given} , \tag{9.4.1}$$

$$\dot{x}_e = F_e x_e + G_e v , \qquad x_e(t_o) \text{ given} , \tag{9.4.2}$$

where the subscripts p and e stand for the pursuer and the evader, respectively, and where the matrices F_p, F_e, G_p, and G_e have the usual definition associated with general linear systems. The pursuer uses the control $u(t)$ to attempt capture of the evader; the evader uses control $v(t)$ to attempt to avoid capture.

A particularly simple game for such a linear system has as objective the minimization of the final miss by the pursuer and the maximization of the final miss by the evader, where the miss is defined as a weighted quadratic form:

$$\|x_p(t_f) - x_e(t_f)\|^2_{A^TA} . \tag{9.4.3}$$

To make the game meaningful, however, we must also put some limit on the control variables. A simple limitation is an integral quadratic constraint:

†S. Baron, Ph.D. Thesis, Harvard University, 1965.

$$\int_{t_o}^{t_f} \|u\|^2_{\bar{R}_p}\, dt \le E_p\,, \tag{9.4.4}$$

$$\int_{t_o}^{t_f} \|v\|^2_{\bar{R}_e}\, dt \le E_e\,, \tag{9.4.5}$$

where $\bar{R}_p > 0\,, \bar{R}_e > 0$ and E_p, E_e are positive constants. For simplicity, we have also assumed a fixed final time, t_f.

Clearly, both pursuer and evader will use all the control they can in the case of finite minimax terminal miss, so the constraints (9.4.4) and (9.4.5) will be *equalities*. Adjoining these constraints to the performance index (9.4.3) yields

$$J = \frac{1}{2}\|x_p(t_f) - x_e(t_f)\|^2_{A^TA} + \frac{1}{2}\int_{t_o}^{t_f} [\|u\|^2_{R_p} - \|v\|^2_{R_e}]\, dt\,, \tag{9.4.6}$$

where $R_p = c_p\bar{R}_p\,, R_e = c_e\bar{R}_e$, and c_p, c_e are positive constants to be determined so as to satisfy (9.4.4) and (9.4.5) as equalities. Note that the second constraint, (9.4.5), was subtracted from (9.4.3), since the evader is attempting to maximize (9.4.3).

Introducing the definition

$$\hat{x}_p(t) \triangleq \Phi_p(t_f,t)x_p(t)\,, \tag{9.4.7}$$

$$\hat{x}_e(t) \triangleq \Phi_e(t_f,t)x_e(t)\,, \tag{9.4.8}$$

$$z(t) \triangleq A(\hat{x}_p(t) - \hat{x}_e(t))\,, \tag{9.4.9}$$

where $\Phi_p(t_f,t)$ and $\Phi_e(t_f,t)$ are the fundamental matrices for F_p and F_e, respectively, we find that the problem may be expressed more compactly as

$$J = \min_u \max_v \left\{\frac{1}{2}\|z(t_f)\|^2 + \frac{1}{2}\int_{t_o}^{t_f}[\|u\|^2_{R_p} - \|v\|^2_{R_e}]\, dt\right\}, \tag{9.4.10}$$

where

$$\dot{z} = \mathscr{P}(t)u - \mathscr{E}(t)v\,, \tag{9.4.11}$$

and

$$\mathscr{P}(t) = A\Phi_p(t_f,t)G_p(t)\,, \qquad \mathscr{E}(t) = A\Phi_e(t_f,t)G_e(t)\,,$$

$$z(t_o) = z_o \triangleq A[\Phi_p(t_f,t_o)x_p(t_o) - \Phi_e(t_f,t_o)x_e(t_o)]\,. \tag{9.4.12}$$

Necessary conditions for a stationary solution are

$$\dot{\lambda} = 0\,; \qquad \lambda(t_f) = z(t_f)\,, \tag{9.4.13}$$

$$H_u = 0 \Rightarrow u = -R_p^{-1}\mathscr{P}^T\lambda = -R_p^{-1}\mathscr{P}^T z(t_f) \tag{9.4.14}$$

$$H_v = 0 \Rightarrow v = -R_e^{-1}\mathscr{E}^T\lambda = -R_e^{-1}\mathscr{E}^T z(t_f)\,, \tag{9.4.15}$$

where

$$H = \tfrac{1}{2}(u^T R_p u - v^T R_e v) + \lambda^T[\mathscr{P}(t)u - \mathscr{E}(t)v] .$$

The two-point boundary-value problem of Equations (9.4.11) and (9.4.13) through (9.4.15) is linear and very simple. An instructive way to solve it is to use the backward-sweep method discussed in Sections 6.2 and 6.3. We define a matrix $S(t)$ by

$$\lambda(t) = S(t)z(t) . \tag{9.4.16}$$

Since, in this problem, $\dot{\lambda} = 0$ from (9.4.13), it follows from (9.4.11) that

$$\dot{S}z + S[\mathscr{P}(t)u - \mathscr{E}(t)v] = 0 . \tag{9.4.17}$$

Using (9.4.14) and (9.4.15) with (9.4.16), we may express u and v in terms of z:

$$u = -R_p^{-1}\mathscr{P}^T Sz , \qquad v = -R_e^{-1}\mathscr{E}^T Sz . \tag{9.4.18}$$

Substituting (9.4.18) into (9.4.17), it is apparent that $S(t)$ must satisfy

$$\dot{S} = S[\mathscr{P}R_p^{-1}\mathscr{P}^T - \mathscr{E}R_e^{-1}\mathscr{E}^T]S \tag{9.4.19}$$

or

$$\frac{d}{dt}(S^{-1}) = -\mathscr{P}R_p^{-1}\mathscr{P}^T + \mathscr{E}R_e^{-1}\mathscr{E}^T ,$$

and, from (9.4.13), the boundary condition for S is

$$S(t_f) = I . \tag{9.4.20}$$

Integrating (9.4.19) with (9.4.20) gives

$$S^{-1}(t) = I + M_p(t_f,t) - M_e(t_f,t) , \tag{9.4.21}$$

where

$$M_p(t_f,t) = \int_t^{t_f} \mathscr{P}(t)R_p^{-1}(t)\mathscr{P}^T(t)\, dt , \tag{9.4.22}$$

$$M_e(t_f,t) = \int_t^{t_f} \mathscr{E}(t)R_e^{-1}(t)\mathscr{E}^T(t)\, dt . \tag{9.4.23}$$

Now, Equations (9.4.18) are *feedback strategies* for u and v in terms of the current state. The matrices in (9.4.22) and (9.4.23) are the reduced controllability matrices of the components of the pursuit and evasion systems that are picked out by A (see Section 5.3). For verification of the saddle-point condition [step (ii), Section. 9.3], we consider two auxiliary problems:

$$\left.\begin{array}{ll}\max\limits_{v}\left\{\frac{1}{2}\|z(t_f)\|^2+\frac{1}{2}\int_{t_0}^{t_f}(\|u\|^2_{R_p}-\|v\|^2_{R_e})\,dt\right\} & \\ \text{subject to} \quad \dot{z}=\mathscr{P}(t)u-\mathscr{E}(t)v & \\ \text{and} \quad u=-R_p^{-1}\mathscr{P}^TSz\,, & \end{array}\right\} \qquad \text{(P1)}$$

where S is as given by (9.4.19).

$$\left.\begin{array}{ll}\min\limits_{u}\left\{\frac{1}{2}\|z(t_f)\|^2+\frac{1}{2}\int_{t_0}^{t_f}(\|u\|^2_{R_p}-\|v\|^2_{R_e})\,dt\right\} & \\ \text{subject to} \quad \dot{z}=\mathscr{P}(t)u-\mathscr{E}(t)v & \\ \text{and} \quad v=-R_e^{-1}\mathscr{E}^TSz\,, & \end{array}\right\} \qquad \text{(P2)}$$

where S is as given by (9.4.19).

When $u=-R_p^{-1}\mathscr{P}^TSz$ and $v=-R_e^{-1}\mathscr{E}^TSz$ are substituted into the criteria of (P1) and (P2), respectively, both problems reduce to the standard one-sided linear-quadratic problems of Chapter 5. For problem (P1) we find that

$$v=-R_e^{-1}\mathscr{E}^TS^{(1)}z\,,$$

where

$$\dot{S}^{(1)}=S^{(1)}\mathscr{P}R_p^{-1}\mathscr{P}^TS+S\mathscr{P}R_p^{-1}\mathscr{P}^TS^{(1)}-S\mathscr{P}R_p^{-1}\mathscr{P}^TS-S^{(1)}\mathscr{E}R_e^{-1}\mathscr{E}^TS^{(1)},$$
$$S^{(1)}(t_f)=I\,;$$

and for (P2) we find that

$$u=-R_p^{-1}\mathscr{P}S^{(2)}z\,,$$

where

$$\dot{S}^{(2)}=-S\mathscr{E}R_e^{-1}\mathscr{E}^TS^{(2)}-S^{(2)}\mathscr{E}R_e^{-1}\mathscr{E}^TS+S^{(2)}\mathscr{P}R_p^{-1}\mathscr{P}S^{(2)}+S\mathscr{E}R_e^{-1}\mathscr{E}^TS\,,$$
$$S^{(2)}(t_f)=I\,.$$

Note that $S^{(1)}\equiv S^{(2)}=S$ [Equation (9.4.19)] since $S^{(1)}(t_f)=S^{(2)}(t_f)=S(t_f)$. Hence, we have verified that the feedback strategies (9.4.18) do, indeed, constitute a saddle point for the minimax problem.

Now we may also attempt to verify the optimality of u^o and v^o considered as open-loop strategies. Problems (P3) and (P4) then result.

$$\left.\begin{array}{ll}\max\limits_{v}\left\{\frac{1}{2}\|z(t_f)\|^2+\frac{1}{2}\int_{t_0}^{t_f}(\|u\|^2_{R_p}-\|v\|^2_{R_e})\,dt\right\} & \\ \text{subject to} \quad \dot{z}=\mathscr{P}(t)u+\mathscr{E}(t)v & \\ \text{and} \quad u=-R_p^{-1}\mathscr{P}(t)S(t_o)z(t_o)\,, \quad \text{a time function}\,. & \end{array}\right\} \qquad \text{(P3)}$$

$$\left.\begin{aligned}&\min_u \left\{\frac{1}{2}\|z(t_f)\|^2 + \frac{1}{2}\int_{t_0}^{t_f} (\|u\|^2_{R_p} - \|v\|^2_{R_e})\,dt\right\}\\ &\text{subject to} \qquad \dot z = \mathscr{P}(t)u + \mathscr{E}(t)v\\ &\text{and} \qquad v = -R_e^{-1}\mathscr{E}(t)S(t_0)z(t_0)\,, \qquad \text{a time function.}\end{aligned}\right\} \quad \text{(P4)}$$

Problems (P3) and (P4) are linear-quadratic one-sided optimization problems with the added complications that known input functions $u^o(t)$ or $v^o(t)$ are present. The solutions are again straightforward. We have, for (P3),

$$v(t) = -R_e^{-1}\mathscr{E}(t)[S^{(3)}(t)z(t) + a(t)]\,,$$

where

$$\dot a = S^{(3)}\mathscr{P}R_p^{-1}\mathscr{P}^TS(t_0)z(t_0) - S^{(3)}\mathscr{E}R_e^{-1}\mathscr{E}^Ta\,, \qquad a(t_f) = 0\,,$$

$$\dot S^{(3)} = -S^{(3)}\mathscr{E}R_e^{-1}\mathscr{E}^TS^{(3)}\,, \qquad S^{(3)}(t_f) = I\,,$$

and, for (P4),

$$u(t) = -R_p^{-1}\mathscr{P}(t)[S^{(4)}(t)z(t) + b(t)]$$

and

$$\dot b = -S^{(4)}\mathscr{E}R_e^{-1}\mathscr{E}^TS(t_0)z(t_0) + S^{(4)}\mathscr{P}R_p^{-1}\mathscr{P}^Tb\,, \qquad b(t_f) = 0\,,$$

$$\dot S^{(4)} = S^{(4)}\mathscr{P}R_p^{-1}\mathscr{P}^TS^{(4)}\,, \qquad S^{(4)}(t_f) = I\,.$$

Note that $S^{(3)} \neq S^{(1)}$ and $S^{(4)} \neq S^{(2)}$. These are, explicitly, cases of Equations (9.3.10) and (9.3.10′), discussed in Section 9.3. Based on our knowledge of the Riccati equation (see Section 6.3), the following statements are self-evident:

(i) If $S^{(1)} = S^{(2)} = S$ is finite for $t_0 \leq t \leq t_f$, the feedback strategies (9.4.14) and (9.4.15) constitute a saddle point for J with $J^o = \frac{1}{2}\|z(t_0)\|^2_{S(t_0)}$.
(ii) The open-loop strategy $u^o(t) = -R_p^{-1}\mathscr{P}(t)S(t_0)z_0$ is optimal only if $S^{(3)}$ remains finite. $S^{(3)}$ will always blow up if $t_0 - t_f$ is too long and $\mathscr{E}R_e^{-1}\mathscr{E}^T > 0$.
(iii) The open-loop strategy $v^o(t) = -R_e^{-1}(t)S(t_0)z_0$ is always optimal since $S^{(4)} < \infty$ always.

Statements (ii) and (iii) are assertions for the general class of linear-quadratic problems analogous to those made for the simple example in Section 9.3. Statement (i) can be given a further interpretation. From (9.4.21) we see that S finite implies that

$$[I + M_p(t_f,t) - M_e(t_f,t)] > 0\,, \qquad t_0 \leq t \leq t_f\,. \tag{9.4.24}$$

Since M_p and M_e are controllability matrices for the pursuer and the evader, respectively, Equation (9.4.20) says that saddle point is ensured if $M_p > M_e$ or the pursuer is more controllable than the evader.

As the amount of control energy in (9.4.4) and (9.4.5) becomes very large, it is interesting to see how the terminal miss behaves. Clearly, the adjoining constants c_p and c_e tend to zero [see (9.4.6)] as E_p and E_e tend to infinity. This, in turn, means that $M_p \to \infty$, $M_e \to \infty$; if $M_p - M_e \to \infty$, then $S(t) \to 0$, $t \to t_f$ [see (9.4.21)]. From the fact that $J^o = \frac{1}{2}\|z(t_0)\|^2_{S(t_0)}$, (9.4.10) and c_e, $c_p \to 0$ imply that the terminal miss tends to zero. Thus, the pursuer must be more controllable than the evader.

Example. *Guidance law for target interception.* A special case† of the class of problems treated above can be formulated as follows: The equations of motion in space for an interceptor and a target are

$$\dot{\mathbf{v}}_p = \mathbf{f}_p + \mathbf{a}_p, \qquad \dot{\mathbf{r}}_p = \mathbf{v}_p, \qquad \dot{\mathbf{v}}_e = \mathbf{f}_e + \mathbf{a}_e, \qquad \dot{\mathbf{r}}_e = \mathbf{v}_e, \tag{9.4.25}$$

where

v = velocity of a body in three dimensions,
r = position vector of a body in three dimensions,
f = gravitational force per unit mass exerted on the body,
a = control acceleration of a body.

We assume that the position difference between the pursuer and the evader is small enough that $\mathbf{f}_p \cong \mathbf{f}_e$. Hence, if we are only interested in the difference $\mathbf{r}_p(t) - \mathbf{r}_e(t)$, the effect of the external forces can be ignored. Now consider the criterion

$$J = \frac{b}{2}\,[\mathbf{r}_p(t_f) - \mathbf{r}_e(t_f)] \cdot [\mathbf{r}_p(t_f) - \mathbf{r}_e(t_f)] + \frac{1}{2}\int_0^{t_f} [c_p^{-1}(\mathbf{a}_p \cdot \mathbf{a}_p) - c_e^{-1}(\mathbf{a}_e \cdot \mathbf{a}_e)]\,dt, \tag{9.4.26}$$

where c_p and c_e are constants related to the energy of the pursuer and the evader, respectively. Applying the results of this section, it can be directly verified that Equations (9.4.14) and (9.4.15) become, in this case,

$$\mathbf{a}_p = \frac{-c_p(t_f - t)\,[\mathbf{r}_p(t) - \mathbf{r}_e(t) + (\mathbf{v}_p(t) - \mathbf{v}_e(t))\,(t_f - t)]}{(1/b) + (c_p - c_e)\,[(t_f - t)^3/3]}, \tag{9.4.27}$$

$$\mathbf{a}_e = \frac{c_e}{c_p}\,\mathbf{a}_p. \tag{9.4.28}$$

We note immediately that:

†This is the differential game analog of Example 2, Chapter 5, Section 5.2. Coordinate-free vector notation in three-space is used here.

(i) If $c_p > c_e$, the feedback-control gain is always of one sign.
(ii) If $c_p < c_e$, the feedback gain will change sign at

$$\frac{1}{b} + (c_p - c_e)\,[(t_f - t)^3/3] = 0 \tag{9.4.29}$$

for t_f sufficiently large.

But Equation (9.4.29) is simply the conjugate-point condition (9.4.25) of this section specialized for this problem. Hence, for case (ii), Equation (9.4.27) is no longer optimal for large t_f. This fact is, of course, obvious to start with, particularly in the case in which $b = \infty$. In that case, interception is not possible when $c_p < c_e$ (see $M_p < M_e$). Assuming (i) and letting $b = \infty$, the control strategy for the pursuer simplifies to

$$\mathrm{a}_p = \frac{-3}{[1 - (c_e/c_p)]\,(t_f - t)^2}\,[\mathrm{r}_p(t) - \mathrm{r}_e(t) + (\mathrm{v}_p(t) - \mathrm{v}_e(t))\,(t_f - t)]\,. \tag{9.4.30}$$

Let the pursuer and the target be on a nominal collision course with range R and closing velocity $V = R/(t_f - t)$. Let $x_p - x_e$ represent the lateral deviation from the collision course as shown in Figure 9.4.1.

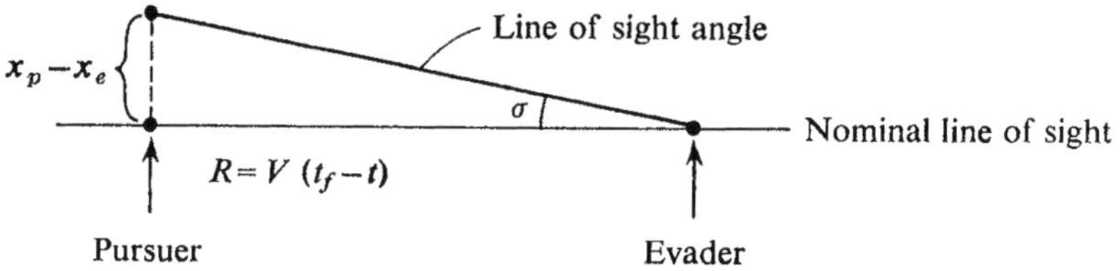

Figure 9.4.1. Proportional navigation guidance.

Then the lateral control acceleration to be applied by the pursuer according to (9.4.30) is

$$\mathrm{a}\ (\text{lateral}) = \frac{3}{[1 - (c_e/c_p)]}\,V\dot{\sigma}\,, \tag{9.4.31}$$

which is simply proportional navigation with the effective navigation constant $K_e = 3/[1 - (c_e/c_p)]$. From experience, it has been found that the best value for K_e ranges from 3 to 5 (Ramo and Puckett, 1959).† In view of Equation (9.4.28), we see that the value of 3 corresponds to the case in which the target is not maneuverable ($c_e = 0$), while the value of 5 corresponds to $(c_e/c_p) = \frac{2}{5}$.

†S. Ramo and A. Pucket, *Guided Missile Engineering.* New York: McGraw-Hill, 1959, pp. 176-180.

Problem 1. Verify the results of (P1) and (P2).

Problem 2. Verify the results of (P3) and (P4).

Problem 3. Consider the more general linear-quadratic minimax problem,

$$J = \frac{1}{2}(x^T S_f x)_{t=t_f} + \frac{1}{2}\int_{t_o}^{t_f} [x^T u^T v^T] \begin{bmatrix} A & 0 & 0 \\ 0 & B & 0 \\ 0 & 0 & -C \end{bmatrix} \begin{bmatrix} x \\ u \\ v \end{bmatrix} dt ,$$

$$\dot{x} = Fx + Gu + Dv , \quad R_f x(t_f) = 0 .$$

Assuming that $B > 0$, $C > 0$, derive the abnormality and conjugate-point conditions for the problem.

ANSWER.

$$S - RQ^{-1}R^T \text{ finite;}$$

$$Q < 0 \quad \text{or} \quad Q > 0 ; \quad t_o \leq t < t_f ,$$

where

$$\dot{S} = -SF - F^T S - A + S(GB^{-1}G^T - DC^{-1}D^T)S , \quad S(t_f) = S_f ;$$

$$\dot{R}^T = -R^T F + R^T(GB^{-1}G^T - DC^{-1}D^T)S , \quad R(t_f) = R_f ;$$

$$\dot{Q} = +R^T(GB^{-1}G^T - DC^{-1}D^T)R , \quad Q(t_f) = 0 .$$

9.5 A minimax-time intercept problem with bounded controls

The problem considered here is a game analog of the "bang-bang" control problem in Section 3.9. There are two systems with bounded controls:

$$\dot{x}_p = F_p x_p + g_p u , \quad |u| \leq 1 , \tag{9.5.1}$$

$$\dot{x}_e = F_e x_e + g_e v , \quad |v| \leq 1 . \tag{9.5.2}$$

The p system (pursuer) wishes to intercept the e system (evader) in minimum time, whereas the e system attempts to maximize the time of intercept. Let the condition for intercept be

$$Ax_p(t_f) = Ax_e(t_f) , \tag{9.5.3}$$

where A = row vector. This *scalar* intercept condition defines the final time, t_f, implicitly. Using the nomenclature of Section 9.4, we define the scalar quantity z:

$$z \triangleq A[\Phi_p(t_f,t)x_p(t) - \Phi_e(t_f,t)x_e(t)] . \tag{9.5.4}$$

Then (9.5.1) and (9.5.2) reduce to the scalar equation

$$\dot{z} = p(t)u - e(t)v\,, \tag{9.5.5}$$

where

$$p(t) = A\Phi_p(t_f,t)g_p\,, \qquad e(t) = A\Phi_e(t_f,t)g_e\,. \tag{9.5.6}$$

Introducing the Hamiltonian function

$$H(z,u,v,\lambda,t) = \lambda[p(t)u - e(t)v] + 1\,, \tag{9.5.7}$$

the necessary conditions† for stationarity according to Section 9.3 become

$$\dot{\lambda} = -H_z = 0\,, \tag{9.5.8}$$

$$\lambda(t_f) = \nu\,, \tag{9.5.9}$$

$$H(z,u^o,v^o,\lambda,t) = \min_{|u|\le 1}\max_{|v|\le 1} H(z,u,v,\lambda,t)\,, \tag{9.5.10}$$

which implies that

$$u^o(t) = -\operatorname{sgn}(\nu p(t))\,, \tag{9.5.11}$$

$$v^o(t) = -\operatorname{sgn}(\nu e(t))\,. \tag{9.5.12}$$

Substituting (9.5.11) and (9.5.12) into (9.5.5), with the requirement that $z(t_f) = 0$, we get

$$z(t_0;t_f) = \operatorname{sgn}\nu \int_{t_0}^{t_f} Q(t_f,t)\,dt\,, \qquad \text{where} \qquad Q(t_f,t) = |p(t)| - |e(t)|\,. \tag{9.5.13}$$

The smallest value of t_f that satisfies Equation (9.5.13) then qualifies as a possible minimax time to intercept. Since every term of the scalar equation (9.5.13) is either known or can be precomputed, the determination of t_f and the sign of ν can be made very quickly. Consequently, the optimal control action $u^o(t)$ and $v^o(t)$ for the p and e systems can be found in terms of $z(t_0)$. Equations (9.5.11) and (9.5.12) are, effectively, feedback strategies. If a multidimensional intercept is required, we have more than one Equation (9.5.13) to solve for t_f and the signs of the vector ν, but the concept is essentially the same. In fact, most of the results in Section 3.9 for the one-sided problem can be extended to the present problem.

Problem 1. Consider the minimax time-to-rendezvous problem with an uncooperative opponent; that is,

†There is actually another condition, $H(t_f) = -\Phi_t = 0$, which, however, we shall not need, since only the sign of $\lambda(t_f)$ is important.

$$\ddot{x} = u\,, \qquad |u| \le a\,,$$
$$\ddot{y} = v\,, \qquad |v| \le b\,; \qquad a > b\,.$$

Derive the feedback minimax strategies for both u and v.

Problem 2. (See S. Baron, Ph.D. Thesis, Harvard University, 1966, Chapter 5, Section 5.) Consider the pursuit-evasion problem defined by

$$\ddot{x} = u - v \quad (x \text{ denotes the predicted miss}), \qquad |u| \le 1\,, \quad |v| \le 1\,,$$

$$\int_0^{t_f} u^2\,dt \le c_p\,, \qquad \int_0^{t_f} v^2\,dt \le c_e\,, \qquad J = \tfrac{1}{2}\|x(t_f)\|^2\,.$$

It is clear that $\max \int_0^{t_f} u^2\,dt = \max \int_0^{t_f} v^2\,dt = (t_f - t_0) \triangleq B$. Thus we shall assume c_p and $c_e \le B$.

(i) In the case in which capture is impossible, show that

$$\int_0^{t_f} u^2\,dt = c_p \qquad \text{and} \qquad \int_0^{t_f} v^2\,dt = c_e$$

under optimal play.

(ii) Show that, if $c_e < B/3$, then $v(t) < 1$ for all t under optimal play.

(iii) For $x(t_0) \ne 0$, $B/3 \le c_p \le B$, $c_e \le B/3$, determine the optimal u and v strategies as well as the surface separating the region of capture and noncapture in $(x(t_0), c_p, c_e)$-space.

We adopt the convention in the case of capture that the pursuer will only use the minimum $\int_{t_0}^{t_f} u^2\,dt$ to effect capture. (This renders the optimal u strategy unique.)

ANSWER TO (iii). We have that

$$\left(1 - \frac{c_p^2}{B}\right)^2 = \frac{4}{3}\left[1 - 2\left[\frac{z}{B^2} + \sqrt{\frac{c_e^2}{3B}}\right]\right]$$

is the capture surface and

$$u = \text{linear control when} \quad \frac{z}{B^2} + \sqrt{\frac{c_e^2}{3B}} \le \frac{1}{3}\,,$$

$$u = \text{sat control} = -\,\text{sat}\left\{3\left[1 - 2\left[\frac{z}{B^2} + \sqrt{\frac{c_e^2}{3B}}\right]\right]^{-1/2}\right\}$$

when $\dfrac{z}{B^2} + \sqrt{\dfrac{c_e^2}{3B}} > \dfrac{1}{3}$.

Problem 3. (Suggested by E. Gilbert, University of Michigan.) Let the equations of motion of the pursuer and evader be

$$\dot{x}_{p_1} = \frac{8}{\pi} u, \qquad x_{p_1}(t_o) = -11;$$

$$\dot{x}_{e1} = x_{e2}, \qquad x_{e1}(t_o) = 1;$$

$$\dot{x}_{e2} = -x_{e1} + v, \qquad x_{e2}(t_o) = 0;$$

where $|u| \leq 1$, $|v| \leq 1$. Let the condition for intercept be

$$x_{p_1}(t) = x_{e_1}(t), \qquad x_{e_2}(t) = 0.$$

Show that the minimax capture time does not exist but can be made as close to 2π as possible.

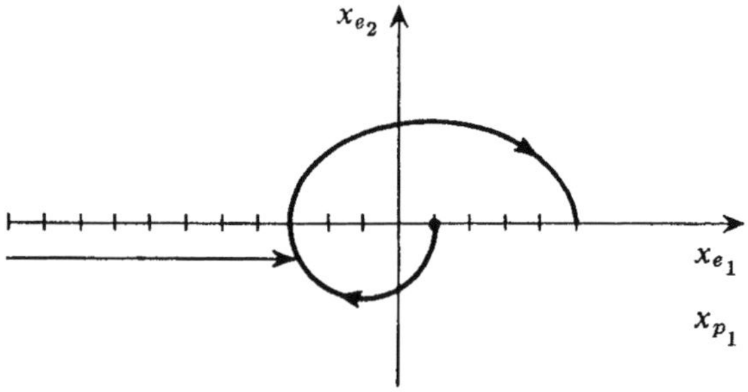

Figure 9.5.1. State space trajectories of Problem 3.

Problem 4. (See Isaacs, pp. 28, 273.) In a planar minimax intercept-time problem, the pursuer and the evader have constant velocity magnitudes, V_p and V_e, respectively ($V_p > V_e$). The evader has direct control of the direction of his velocity, whereas the pursuer has control over his lateral acceleration, which is bounded. The pursuer thus has a minimum radius of turn, R.

In a coordinate frame centered in the pursuer (see Figure 9.5.2) with y-axis always kept parallel to the pursuer's velocity vector, the relative position of the evader is (x,y), where

$$\dot{x} = V_e \sin\psi - V_p \frac{y}{R} u, \qquad \dot{y} = V_e \cos\psi - V_p + V_p \frac{x}{R} u; \qquad -1 \leq u \leq 1.$$

Here ψ is the evader's control (unbounded) and u is the pursuer's control (bounded, $|u| \leq 1$).

The intercept time, t_f, is determined by

$$(x^2 + y^2)_{t=t_f} = \ell^2,$$

and the initial conditions $x(0)$, $y(0)$ are specified. *Show that* the minimax strategies (under some conditions) are

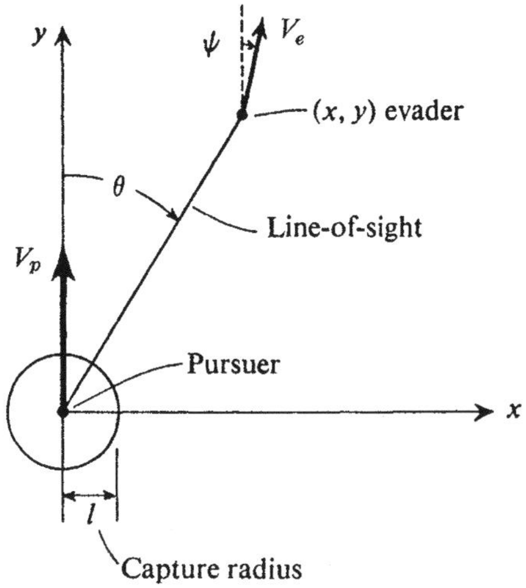

Figure 9.5.2. Nomenclature for Problem 4.

$$u = -\operatorname{sgn}(\theta - \psi), \qquad \text{where} \qquad \theta = \tan^{-1}\frac{x}{y}; \qquad \dot{\psi} = \frac{V_p}{R}\operatorname{sgn}(\theta - \psi),$$

and that these strategies give rise to a one-parameter (ψ_f) family of extremal paths given parametrically (parameter ψ) by

$$x = \left[\ell - \frac{V_e}{V_p}R(\psi - \psi_f)\right]\sin\psi + R[1 - \cos(\psi - \psi_f)],$$

$$y = \left[\ell - \frac{V_e}{V_p}R(\psi - \psi_f)\right]\cos\psi + R\sin(\psi - \psi_f),$$

$$t_f - t = \frac{R}{V_p}(\psi - \psi_f).$$

[HINT: Note that $\lambda_x^2 + \lambda_y^2 = \text{const.}$]

9.6 A discussion of differential games

The previous sections demonstrated that the usual variational techniques can be applied to solve two-sided optimization problems. Some general remarks on such problems are appropriate at this point.

(i) In an optimization problem with an unpredictable forcing function or parameter, it is natural to attempt to model the unknown as a random process, and then to find the control to minimize the expected value of the performance index, as will be done in

Chapter 14. However, it may happen that so little is known about the statistics of the forcing function or the parameter that an adequate stochastic formulation is not available. In this case, a conservative control design can be made assuming the *worst possible* forcing function; i.e., we assume the forcing function to be manipulated by an intelligent adversary who is trying to maximize whatever we are trying to minimize. This is another way of replacing a stochastic problem by a deterministic one. (The usual way is to work with the mean and covariance of the state variables, which are deterministically related as in Chapters 11 through 14. For example, the problems discussed in previous sections can be formulated as stochastic problems in which the action of the evader is considered to be a random process.)

(ii) There is another aspect of differential games that makes them operationally different from the usual optimal control problems. For example, consider the one-sided optimal control problem of bringing the first component of a linear dynamic system to zero in minimal time. Such a problem is simple to solve and not very interesting. However, its differential game analogue may be quite interesting, and yet, as has been demonstrated, it is no more difficult to solve. In this sense, most differential game problems are easier, since they take place in lower-dimensional space (one, two, or three). Mainly for this reason, Isaacs (1965) was able to solve many interesting problems and illustrate their feedback strategies graphically.

(iii) On the other hand, the introduction of an opposing control variable also results in complications. First of all, in game problems we must consider feedback strategies or, what is equivalent, continuous real-time solution of open-loop strategies. The problem of open-loop differential games is not too useful from a practical viewpoint (see, however, the discussion in Section 9.3). Hence, one either solves a problem completely or not at all. It also may not be sound to consider linearized feedback strategies around some open-loop nominal as we have done for the usual control problem. This is because the opponent may choose a strategy (open- or closed-loop, not necessarily optimal) such that it can induce a control through our linearized feedback strategy that will cause us to deviate significantly from the nominal. In such a case, we will no longer be able to claim any local minimax property of the controller.

(iv) The presence of the opposing control also brings about a richness of the behavior of the systems under minimax control. We have seen in Section 9.4 how conjugate points arise naturally. In fact, questions such as existence (see Problem 3, Section 9.5), uniqueness (two distinct strategies leading to same cost), singularity (H_u identically zero) of solutions are the rule rather than the exception in differential games. A large part of the effort in the solution of a game is the identification and determination of the regions and surfaces governing these situations [see Isaacs (1965) and Problem 2, Section 9.5].

(v) Finally, one may consider the very natural and interesting extensions of problems of this chapter to cases where there are random effects and/or where the interests of the players are not directly opposite; i.e., stochastic differential games and nonzero sum differential games. Relatively little is known about such problems at present. We shall not pursue them here [see, however, Ho (1966), Behn and Ho (1968), and Starr and Ho (1969)].†

†Y. C. Ho, "Optimal Terminal Maneuver & Evasion Strategy," *SIAM J. Control*, Vol. 4, No. 3, 1966; R. Behn and Y. C. Ho, "On a Class of Stochastic Differential Games," *Trans. Automatic Control IEEE*, June 1968; A. Starr and Y. C. Ho, "Nonzero Sum Differential Games," *J. Optimization Theory and Applications*, 1969 (to appear).

Some concepts of probability

10

10.1 Discrete-valued random scalars

Almost everyone has some intuitive notions about a *random scalar* quantity and the *probabilities* associated with it. A common example is the *outcome* of the throw of a die; the probability of a 1, 2, 3, 4, 5, or 6 occurring on the throw is exactly the same for each possibility, called *elementary* or *disjoint* events. For a "loaded" die, some outcomes are more likely than others. If the die is thrown N times and N_j denotes the number of times of the occurrence of the jth value, it is intuitively reasonable to define the *probability* of the event j as

$$p(j) = \lim_{N\to\infty} \frac{N_j}{N} . \tag{10.1.1}$$

The *compound event*, "either j or k occurs on a given throw of the die," is often of interest and has associated with it the probability that

$$p(j \quad \text{or} \quad k) = \lim_{N\to\infty} \left(\frac{N_j + N_k}{N}\right) . \tag{10.1.2}$$

It is clear, from the above definitions, that a probability function $p(j)$ satisfies

$$0 \le p(j) \le 1 , \qquad \sum_j p(j) = 1 , \quad p(j \quad \text{or} \quad k) = p(j) + p(k) . \tag{10.1.3}$$

A discrete-valued random scalar, x, is defined as a discrete-valued function $x(j)$ with the probability of occurrence of the jth value of x given by $p(j)$. $p(j)$ is called the probability mass function of the random variable $x(j)$. It is customary to drop the explicit indication of $x(j)$ as a function of j. We write x and $p(x)$ whenever this would not result in an ambiguity. Thus, to characterize a discrete random variable x, the probability mass function $p(x)$ must be specified. The random variable x may be characterized approximately by specifying a finite number of moments of $p(j)$. The first two moments are:

(i) The *mean of* x (first moment of $p(x)$) ,

$$\bar{x} \triangleq \sum_j x(j)p(j) . \tag{10.1.4}$$

(ii) The *variance of* x (second central moment of $p(x)$) ,

$$\sigma^2 = \sum_j (x(j) - \bar{x})^2 p(j) , \tag{10.1.5}$$

where σ is also known as the ***standard deviation*** of x.

If $p(x)$ is considered as a distribution of discrete masses on a straight line, then $\bar{x}$ and σ represent the location of the center of mass and the radius of gyration about the center of mass, respectively. Obviously, this is the origin of the terminology "mass function" for $p(x)$.

The "expected value" of a function of x is defined as

$$E[f(x)] \triangleq \sum_j f(x(j))p(j) . \tag{10.1.6}$$

Clearly, from (10.1.4) and (10.1.5), *the mean is the expected value of* x and the *variance is the expected value of* $(x - \bar{x})^2$. Note that the expected value operator is a *linear operator.*

10.2 Discrete-valued random vectors

A random vector is a vector whose components are scalar random variables as described in Section 10.1. Let us consider a random vector x with components x_i , $i = 1,2, \ldots , n$. If each component of the vector can take on a discrete set of values $x_i(j_i)$, where $j_i =$ $1,2, \ldots , m_i$, clearly there are $m_1 m_2 \cdots m_n$ possible vectors. For a complete characterization of the random vector, the *joint probability mass function* $p(j_1,j_2, \ldots ,j_n)$ must be specified. Here, $p(j_1,j_2, \ldots ,j_n)$ is the probability that x_1 has its j_1th value, *and* x_2 has its j_2th value, . . . , *and* x_n has its j_nth value. The function $p(j_1,j_2, \ldots ,j_n)$ is often written as $p(x_1,x_2, \ldots ,x_n)$ when this will not result in any ambiguity. If one is interested in only one of the components of the random vector, say, x_1 , the *marginal probability mass function* is given by

$$p(j_1) = \sum_{j_2=1}^{m_2} \cdots \sum_{j_n=1}^{m_n} p(j_1,j_2, \ldots ,j_n) . \tag{10.2.1}$$

Here, $p(j_1)$ is the probability of a *compound* event, that x_1 takes on its j_1th value while $x_2, \ldots , x_n$ take on any possible values. Thus, a random variable may represent the outcome of an elementary event or a compound event, depending on the underlying space of events. More generally, we have

$$p(j_1, \ldots, j_i) = \sum_{j_{i+1}=1}^{m_{i+1}} \cdots \sum_{j_n=1}^{m_n} p(j_1, \ldots, j_n). \tag{10.2.2}$$

The marginal probability functions are sufficient to characterize individual components or groups of components of x. However, to characterize x completely, we must specify $p(x_1, \ldots, x_m)$.

As in the scalar case, it is possible to characterize x approximately by specifying moments of $p(x)$:

(i) *The mean of x*

$$\bar{x} \triangleq E(x) \triangleq \sum_{j_1=1}^{m_1} \cdots \sum_{j_n=1}^{m_n} \begin{bmatrix} x_1(j_1) \\ \cdot \\ \cdot \\ \cdot \\ x_n(j_n) \end{bmatrix} p(j_1, j_2, \ldots, j_n). \tag{10.2.3}$$

(ii) *The covariance of x*. Instead of the single variance associated with a random scalar, a random vector has n variances and $n(n-1)/2$ quantities called covariances associated with it. The *covariance matrix*, P, is defined as

$$P \triangleq E(x-\bar{x})(x-\bar{x})^T = E \begin{bmatrix} (x_1-\bar{x}_1)^2, \ldots, (x_1-\bar{x}_1)(x_n-\bar{x}_n) \\ \cdot \qquad\qquad \cdot \\ \cdot \qquad\qquad \cdot \\ \cdot \qquad\qquad \cdot \\ (x_n-\bar{x}_n)(x_1-\bar{x}_1), \ldots, (x_n-\bar{x}_n)^2 \end{bmatrix}$$

$$= \sum_{j_1=1}^{m_1} \cdots \sum_{j_n=1}^{m_n} \begin{bmatrix} (x_1-\bar{x}_1)^2, \ldots, (x_1-\bar{x}_1)(x_n-\bar{x}_n) \\ \vdots \\ (x_n-\bar{x}_n)(x_1-\bar{x}_1), \ldots, (x_n-\bar{x}_n)^2 \end{bmatrix} p(j_1, j_2, \ldots, j_n). \tag{10.2.4}$$

where it is understood that the expectation operator E applies to each element of the matrix. The terms on the diagonal of the P matrix are the variances of the components, and the off-diagonal elements are the covariances. Note that the matrix is symmetric, so that there are only $n(n+1)/2$ distinct quantities.

Example. Consider a vector with two components $x = \begin{bmatrix} x_1 \\ x_2 \end{bmatrix}$. The first component can take on only two values, $x_1(1) = 3$, $x_1(2) = 4$. The second component has three possible values, $x_2(1) = 0$, $x_2(2) = -1$, $x_2(3) = 2$. Thus, there are six possible elementary events or vectors. Suppose that the probabilities of these vectors are

$$p(1,1) = .1, \quad p(1,2) = .2, \quad p(1,3) = .3;$$

$$p(2,1) = .2, \quad p(2,2) = .1, \quad p(2,3) = .1.$$

The expected value of the vector is then

$$E\begin{bmatrix} x_1 \\ x_2 \end{bmatrix} = .1\begin{bmatrix} 3 \\ 0 \end{bmatrix} + .2\begin{bmatrix} 3 \\ -1 \end{bmatrix} + .3\begin{bmatrix} 3 \\ 2 \end{bmatrix} + .2\begin{bmatrix} 4 \\ 0 \end{bmatrix} + .1\begin{bmatrix} 4 \\ -1 \end{bmatrix} + .1\begin{bmatrix} 4 \\ 2 \end{bmatrix}$$

or

$$\begin{bmatrix} \bar{x}_1 \\ \bar{x}_2 \end{bmatrix} = \begin{bmatrix} 3.4 \\ .5 \end{bmatrix}.$$

The variances and covariances are given by

$$E\begin{bmatrix} (x_1 - \bar{x}_1)^2 , (x_1 - \bar{x}_1)(x_2 - \bar{x}_2) \\ (x_2 - \bar{x}_2)(x_1 - \bar{x}_1), (x_2 - \bar{x}_2)^2 \end{bmatrix}$$

$$= .1\begin{bmatrix} -.4 \\ -.5 \end{bmatrix}[-.4,-.5] + .2\begin{bmatrix} -.4 \\ -1.5 \end{bmatrix}[-.4,-1.5] + .3\begin{bmatrix} -.4 \\ 1.5 \end{bmatrix}[-.4,1.5]$$

$$+ .2\begin{bmatrix} .6 \\ -.5 \end{bmatrix}[.6,-.5] + .1\begin{bmatrix} .6 \\ -1.5 \end{bmatrix}[.6,-1.5] + .1\begin{bmatrix} .6 \\ 1.5 \end{bmatrix}[.6,1.5].$$

or

$$P = \begin{bmatrix} .240 , - .10 \\ -.10 \ , \ 1.650 \end{bmatrix}.$$

It may be verified directly that the P-matrix in this case is positive definite.

Problem. Show that any covariance matrix is at least positive semi-definite. [HINT: See Appendix A3.]

10.3 Correlation, independence, and conditional probabilities

We have seen in the previous section that, if one deals with more than one random variable at a time, it is *not* sufficient to know just the marginal probability functions of the individual random variables. When we are characterizing random variables approximately, this fact manifests itself in the off-diagonal covariance terms in the P-matrix. If we have $P_{ij} \neq 0$ for $i \neq j$, we say that the random variables x_i and x_j are *correlated.* In the more general case, we say that random variables $x_1, \ldots, x_n$ are *dependent* if knowledge of $p(x_1), p(x_2), \ldots, p(x_n)$ does not determine $p(x_1, \ldots, x_n)$ completely. If, on the other hand,

$$p(x_1, \ldots, x_n) = p(x_1)p(x_2) \cdots p(x_n) \quad \text{for all possible values of } x_1, \ldots, x_n, \tag{10.3.1}$$

the random variables are said to be *independent.* Note that pairwise independence of $x_1, \ldots, x_n$ does not imply independence.† On the

†See, for example, Feller, Vol. I, p. 126.

other hand, pairwise independence is sufficient to ensure no correlation, i.e., a diagonal P-matrix. Note, also, that lack of correlation does not imply independence.

Let two random vectors x and y be dependent. Given the fact that x has taken on a particular value, we should be able to predict the value of y better than if this information were not available. In this connection, it is useful to introduce the concept of a *conditional probability mass function*,

$$p(y/x) = \frac{p(y,x)}{p(x)} \qquad \text{for} \qquad p(x) \neq 0\,, \tag{10.3.2}$$

where $p(y/x)$ is the probability of y, conditioned on a given value of x. The conditional mean and conditional covariance are defined in a manner similar to Equations (10.2.3) and (10.2.4). The joint probability mass function is simply replaced by the conditional probability mass function. Note, however, that the conditional mean and covariance are not constants but random variables, since they are functions of the conditioning random variable x. Also, since $p(y,x) \triangleq p(x/y)p(y)$, we have

$$p(y/x) = \frac{p(x/y)p(y)}{p(x)}\,. \tag{10.3.3}$$

This is known as the *Bayes Formula.* In (10.3.3) we can regard $p(y)$ as the prior probability of y without knowledge of x, while $p(y/x)$ is the posterior probability of y given the fact that x has a certain value. If x and y are independent, (10.3.3) reduces to $p(y/x) = p(y)$, which implies that knowledge of x does not contribute to the prediction of y. This, of course, is the intuitive basis for independence.

10.4 Continuous-valued random variables

For the purposes of this book, the concepts described in Sections 10.1, 10.2, and 10.3 can be extended in a straightforward manner to continuous-valued random vectors.† The function $p(x_1,x_2,\ldots,x_n)$ becomes a *probability density function,* where $p(x_1,\ldots,x_n)$ $dx_1,\ldots,dx_n$ is the probability that the value of the random vector x will lie in the differential volume $dx_1,\ldots,dx_n$ with center at $(x_1,\ldots,x_n)$. Thus, we have

$$\int_{-\infty}^{\infty} \cdots \int_{-\infty}^{\infty} p(x_1,\ldots,x_n)\,dx_1 \cdots dx_n = 1\,. \tag{10.4.1}$$

†There are various technical conditions and additional assumptions that must be imposed to make the extension to the continuous case rigorous. For a readable account see Parzen (1960) and (1962).

The expected value of the vector is defined as

$$E(x) = \int_{-\infty}^{\infty} \cdots \int_{-\infty}^{\infty} \begin{bmatrix} x_1 \\ \cdot \\ \cdot \\ \cdot \\ x_n \end{bmatrix} p(x_1, \ldots, x_n)\, dx_1 \cdots dx_n = \begin{bmatrix} \bar{x}_1 \\ \cdot \\ \cdot \\ \cdot \\ \bar{x}_n \end{bmatrix} = \bar{x}\,. \quad (10.4.2)$$

The variances and covariances of the vector are given by

$$P = E[(x - \bar{x})(x - \bar{x})^T] = \int_{-\infty}^{\infty} \cdots \int_{-\infty}^{\infty} \begin{bmatrix} (\bar{x}_1 - \bar{x}_1)^2, \ldots (x_1 - \bar{x}_1)(x_n - \bar{x}_n) \\ \cdot \\ \cdot \\ \cdot \\ (x_1 - \bar{x}_1)(x_n - \bar{x}_n), \ldots (x_n - \bar{x}_n)^2 \end{bmatrix} p(x_1, \ldots x_n)\, dx_1 \cdots dx_n\,. \quad (10.4.3)$$

In short, to find expected values for discrete-valued (continuous-valued) random variables, probability mass (density) functions are used with summations (integrals) over all possible values of the random variables.

A mass function can be represented by a density function through the use of Dirac delta functions:

$$p_{\text{density}}(x) = [p_{\text{mass}}(x(j))]\, \delta[x - x(j)]\,. \quad (10.4.4)$$

Transformations of density functions. If x is a random vector with density function $p(x)$ and $y = f(x)$ is a one-to-one transformation that is everywhere differentiable, it can be shown that the density function of y is given by

$$p(y) = p(x)\, |J|^{-1} \qquad \text{with} \qquad x = f^{-1}(y)\,, \quad (10.4.5)$$

where $|J|$ is the determinant of the Jacobian matrix J whose elements are

$$J_{ik} = \partial f_i / \partial x_k\,. \quad (10.4.6)$$

This is readily shown by considering corresponding differential volumes in the x and y spaces.

The distribution function. Another useful probability function is the *distribution function, $P(x)$.*†

$P(x_1, x_2, \ldots, x_n) =$ probability that the sample value of the first component is less than or equal to $x_1, \ldots$ and the sample value of the nth is less than or equal to x_n.

†There should be no confusion between this symbol and the covariance matrix P, since these two quantities usually appear in different contexts.

Obvious properties of a distribution function are

$P(-\infty) = 0$,

$P(+\infty) = 1$,

$P(x)$ is nondecreasing for increasing values of each component,

$P(x)$ may be discontinuous.

When $P(x)$ is continuous and possesses a derivative, we have

$$\frac{\partial^n P(x)}{\partial x_1 \partial x_2 \cdots \partial x_n} = p(x_1, x_2, \ldots, x_n) \tag{10.4.4}$$

and

$$\int_{-\infty}^{x_1} \cdots \int_{-\infty}^{x_n} p(\xi_1, \ldots, \xi_n)\, d\xi_1 \cdots d\xi_n = P(x_1, \ldots, x_n). \tag{10.4.5}$$

We will not have occasion to use distribution functions very often in this book.

Problem 1. If x is a two-dimensional vector with density function $p(x_1,x_2)$ and $y = f(x_1,x_2)$ is a unique transformation from (x_1,x_2)-space to the scalar y-space, show that the density function of y is given by

$$p(y) = \int_\Gamma p(x_1,x_2) \frac{|[(\partial f/\partial x_2)\, dx_1] - [(\partial f/\partial x_1)\, dx_2]|}{\sqrt{(\partial f/\partial x_1)^2 + (\partial f/\partial x_2)^2}},$$

where $\int_\Gamma$ is a *line integral* along the y = const contour in the (x_1x_2)-space.

Problem 2. Let x be a scalar random variable and let y be a scalar quantity such that

$$y = ax^2 \qquad (\text{note } \textit{double}\text{-valued mapping } y \to x).$$

(a) If $p(x) = \dfrac{1}{\sigma\sqrt{2\pi}} \exp\left[-\dfrac{x^2}{2\sigma^2}\right]$, show that the probability density of y is given by

$$p(y) = \begin{cases} \dfrac{1}{\sigma\sqrt{2\pi a y}} \exp\left[-\dfrac{y}{2a\sigma^2}\right], & y \geq 0, \\ 0, & y < 0. \end{cases}$$

(b) Show, also, that

$$E(y) \triangleq \bar{y} = a\sigma^2, \qquad E[y - \bar{y}]^2 = 2a^2\sigma^4,$$

$$P(y) = \begin{cases} \operatorname{erf}\left(\frac{1}{\sigma}\sqrt{\frac{y}{2a}}\right), & y \geq 0, \\ 0, \quad y < 0; & \text{distribution function.} \end{cases}$$

Problem 3. Let x be a scalar random variable and let y be a scalar such that

$$y = ax + bx^2 .$$

If

$$p(x) = \frac{1}{\sigma\sqrt{2\pi}} \exp\left[-\frac{x^2}{2\sigma^2}\right],$$

show the probability density of y is given by

$$p(y) = \begin{cases} \frac{1}{\sigma\sqrt{2\pi(a^2+4by)}}\left\{\exp\left[-\frac{1}{2\sigma^2}\left(\frac{y}{b}+\frac{a^2}{2b^2}+\frac{a}{b}\sqrt{\frac{y}{b}+\frac{a^2}{4b^2}}\right)\right]\right. \\ \left. + \exp\left[-\frac{1}{2\sigma^2}\left(\frac{y}{b}+\frac{a^2}{2b^2}-\frac{a}{b}\sqrt{\frac{y}{b}+\frac{a^2}{4b^2}}\right)\right]\right\}, & y \geq -\frac{a^2}{4b}; \\ 0, \quad y < -\frac{a^2}{4b}. \end{cases}$$

10.5 Common probability mass functions

Uniform mass function. The simplest mass function for a random scalar is the uniform distribution, where all possible values are equiprobable. If there are N possible values of x, namely, $x(1), \ldots, x(N)$, then

$$p(j) = \frac{1}{N}, \qquad j = 1, \ldots, N. \tag{10.5.1}$$

Obviously, we have

$$\sum_{j=1}^{N} p(j) = 1, \tag{10.5.2}$$

$$\bar{x} = E(x) = \frac{1}{N}\sum_{j=1}^{N} x(j), \tag{10.5.3}$$

$$E(x-\bar{x})^2 = \frac{1}{N}\sum_{j=1}^{N} (x(j)-\bar{x})^2 . \tag{10.5.4}$$

Binomial mass function. Let p be the probability that a certain event will occur on every trial. The probability of this event *not* occurring is $1 - p$. In n successive trials, the probability of the event not occurring on any of the trials is

$$p_0 = (1 - p)^n . \tag{10.5.5}$$

The probability of the event happening *exactly once* in n trials is

$$p_1 = np(1 - p)^{n-1} , \tag{10.5.6}$$

since it could happen on each of the n trials, and the probability of its happening on any specific trial and not on all the others is $p(1 - p)^{n-1}$. Similarly, the probability of the event happening exactly k times in n trials is

$$p_k = \frac{n!}{k!(n-k)!} p^k (1 - p)^{n-k} \tag{10.5.7}$$

since $n!/k!(n - k)!$ is the number of ways that n distinct items can be taken k at a time. Direct computation yields the mean and variance of the random variable, k (the number of times that a certain event will occur in n trials):

$$E(k) = \sum_{k=0}^{n} kp_k = np , \tag{10.5.8}$$

$$E(k - np)^2 = \sum_{k=0}^{n} (k - np)^2 p_k = np(1 - p) . \tag{10.5.9}$$

A quick way of summarizing the above results is to define a *generating function* of the random variable k by

$$G(t) = (pt + q)^n = \sum_{k=0}^{n} \frac{n!}{k!(n-k)!} (pt)^k q^{n-k} , \qquad q = 1 - p . \tag{10.5.10}$$

The coefficient of t^k is p_k. The generating function $G(t)$ of a random variable has the properties $G(1) = 1$, $(d/dt)G(t)|_{t=1} = E(k)$, $G''(1) + G'(1) - (G'(1))^2 = \text{var}(k)$, which frequently provides a quick way to compute the mean and variance.

Example 1. The probability of rolling a 1 on each throw of a die is $\frac{1}{6}$. The probability of rolling exactly k 1's in n successive throws is

$$\frac{n!}{k!(n-k)!} \left(\frac{1}{6}\right)^k \left(\frac{5}{6}\right)^{n-k} .$$

Suppose that $n = 6$; then we have

$$p_0 = .335 ,$$
$$p_1 = .401 ,$$
$$p_2 = .201 ,$$
$$p_3 = .053 ,$$

$p_4 = .008$,
$p_5 = .0006$,
$p_6 = .00002$,
and $\quad E(k) = 1$, $\quad E(k-1)^2 = \frac{5}{6}$.

Poisson mass function. Consider the binomial mass function with n very large and p very small. In fact, let $p = \mu/n$, where μ is the average number of times that the event occurred in n trials. The generating function becomes

$$\left(\frac{\mu t}{n} + 1 - \frac{\mu}{n}\right)^n = \left[1 + \frac{\mu}{n}(t-1)\right]^n. \tag{10.5.11}$$

As $n \to \infty$, this expression tends to $e^{\mu(t-1)}$, which may be written

$$e^{-\mu} \sum_{k=0}^{\infty} \frac{(\mu t)^k}{k!}. \tag{10.5.12}$$

Thus, the probability of an event happening exactly k times in a great many trials, when it happens μ times on the average, is

$$p_k = \frac{\mu^k}{k!} e^{-\mu}. \tag{10.5.13}$$

Instead of considering events occurring during a great many trials, we may consider events occurring during a given length of time. Thus, p_k, above, is the probability of an event happening exactly k times in a given time interval when it happens μ times on the average in this interval.

Example 2. The average number of cars passing a certain point in a unit time is μ; $(\mu t)^k e^{-\mu t}/k!$ is the probability that exactly k cars will pass that point in the time t.

We may also think of p_k, above, as the probability that exactly k objects will occupy a certain space when, on the average, μ objects occupy that space.

Example 3. In sowing grass seed, we spread μ seeds per unit area on the average; $(\mu A)^k e^{-\mu A}/k!$ is the probability that exactly k seeds will fall on an area A.

Problem 1. Show that the mean of the Poisson mass function is μ and that the variance is also μ.

Problem 2. If we could actually count the number of molecules in a given volume of air at one atmosphere pressure and a temperature of 68°F, how small could we take a sample cubic volume before the standard deviation in density would exceed one-thousandth of the mean density? Assume that the number of particles in a given volume V is a random variable with a Poisson distribution; that is,

$$p_k = \frac{(\mu V)^k}{k!} e^{-\mu V}$$

is the probability of finding exactly k particles in the volume V, where there are μ particles per unit volume on the average. For air at 68°F and one atmosphere pressure, $\mu = 2.7 \times 10^{19}$ particles per cubic centimeter.

ANSWER. The sample would be a cube with side equal to 3.3×10^{-5} cm.

10.6 Common probability density functions

Uniform density function. The simplest density function for a random scalar is the uniform distribution:

$$p(x) = \begin{cases} \frac{1}{c}, & b - \frac{c}{2} \le x \le b + \frac{c}{2}, \\ 0, & x > b + \frac{c}{2}, x < b - \frac{c}{2}. \end{cases} \tag{10.6.1}$$

Obviously, we have

$$\int_{-\infty}^{\infty} p(x)\, dx = 1, \tag{10.6.2}$$

$$E(x) = b, \tag{10.6.3}$$

$$E(x - b)^2 = c^2/12. \tag{10.6.4}$$

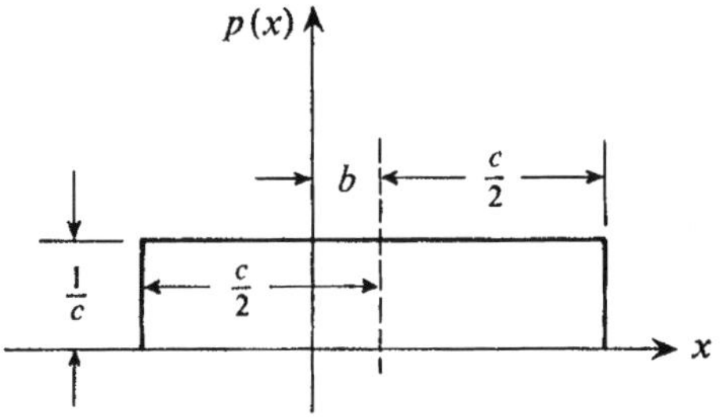

Figure 10.6.1. The uniform density function.

Gaussian density function for a random scalar. Perhaps the most common distribution for a random scalar is the gaussian distribution:

$$p(x) = \frac{1}{(2\pi)^{1/2}\sigma} \exp\left[-\frac{(x-\bar{x})^2}{2\sigma^2}\right]. \qquad (10.6.5)$$

It is easily shown that

$$\int_{-\infty}^{\infty} p(x)\, dx = 1\,, \qquad (10.6.6)$$

$$E(x) = \bar{x}\,, \qquad (10.6.7)$$

$$E(x-\bar{x})^2 = \sigma^2\,. \qquad (10.6.8)$$

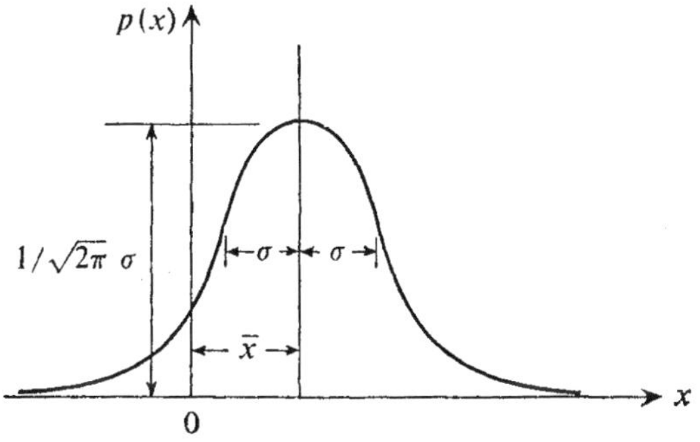

Figure 10.6.2. The gaussian density function.

The justification for representing many complicated phenomena by gaussian density functions lies in the *central limit theorem*,† which states that, if x is the sum of N independent random quantities having nonidentical density functions, then x tends to have a gaussian density function as $N \to \infty$ (see Problems 1 and 2). The probability that x lies between $\bar{x} - \xi$ and $\bar{x} + \xi$ is given by

$$\int_{\bar{x}-\xi}^{\bar{x}+\xi} p(x)\, dx = \frac{1}{(2\pi)^{1/2}\sigma} \int_{-\xi}^{\xi} e^{-(t^2/2\sigma^2)}\, dt$$

$$= \frac{2}{\sqrt{\pi}} \int_0^{\xi/\sqrt{2}\sigma} e^{-n^2}\, dn \triangleq \operatorname{erf}(\xi/\sqrt{2}\sigma)\,. \qquad (10.6.9)$$

Tables of this "normal probability integral" or "error function" are found in many places. Of particular interest are the values for $\xi = \sigma$, 2σ, and 3σ, given below:

†See, for example, Cramer (1946), p. 317.

ξ	Value
σ	.683
2σ	.955
3σ	.997

The "three sigma" (3σ) value is often used in practical problems as virtually the upper bound on the variation from the mean, since the probability that x lies between -3σ and $+3\sigma$ is .997. Analogous to the concept of a generating function for the mass function, a *characteristic function* for the density function of a random variable is defined by

$$M_x(jv) \triangleq E(e^{jvx}) = \int_{-\infty}^{\infty} e^{jvx} p(x)\, dx\,, \qquad j = \sqrt{-1}\,, \tag{10.6.10}$$

which is just the Fourier transform of the density function. It can be easily verified that

$$E(x^n) = (-j)^n \left. \frac{d^n M_x(jv)}{dv^n} \right|_{v=0} \tag{10.6.11}$$

Problem 1. Using the results of Problem 1, Section 10.4, consider the case in which x_1 and x_2 are independent random scalars, each uniformly distributed on the interval $(-\frac{1}{2}, \frac{1}{2})$. With $y = x_1 + x_2$, show that

$$p(y) = \begin{cases} 1 - |y|\,, & |y| < 1\,, \\ 0\,, & |y| > 1\,. \end{cases}$$

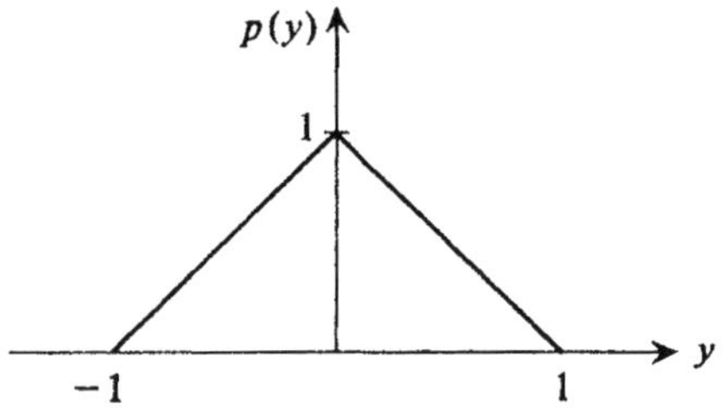

Figure 10.6.3. Density function of the sum of two uniformly distributed random variables.

Problem 2. Using the results of Problem 1 (and Problem 1 of Section 10.4 again), consider the case in which x_1, x_2, and x_3 are independent random scalars, each uniformly distributed on the interval $(-\frac{1}{2}, \frac{1}{2})$. With $y = x_1 + x_2 + x_3$, show that

$$p(y) = \begin{cases} \frac{3}{4} - y^2, & 0 \le |y| \le \frac{1}{2}, \\ \frac{1}{2}\left(\frac{3}{2} - |y|\right)^2, & \frac{1}{2} \le |y| \le \frac{3}{2}, \\ 0, & |y| \ge \frac{3}{2}. \end{cases}$$

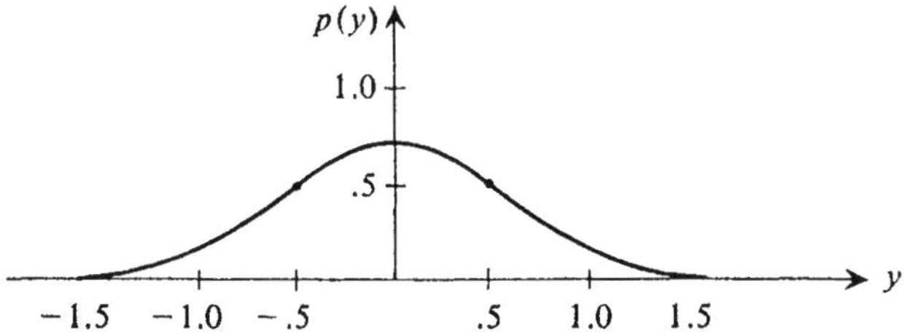

Function 10.6.4. Density function of the sum of three uniformly distributed random variables.

Note that $p(y)$ is tending toward a gaussian distribution, as indicated by the central limit theorem.

Problem 3. Show that the characteristic function of a gaussian random variable is

$$M_x(jv) = \exp\left[jv\bar{x} - \frac{v^2\sigma^2}{2}\right].$$

10.7 Gaussian density function for a random vector

If x is a random n-vector, where the components can take on a continuous set of values, the most common probability density encountered in practice, and certainly the most important for this book, is the gaussian or normal distribution:†

$$p(x) = \frac{1}{(2\pi)^{n/2}|P|^{1/2}} \exp\left[-\frac{1}{2}(x - \bar{x})^T P^{-1}(x - \bar{x})\right]. \qquad (10.7.1)$$

It can be shown that‡

$$\int_{-\infty}^{\infty} \cdots \int_{-\infty}^{\infty} p(x)\, dx_1 \cdots dx_n = 1, \qquad (10.7.2)$$

$$E(x) = \bar{x} = \text{mean value of vector}, \qquad (10.7.3)$$

$$E[x - \bar{x})(x - \bar{x})^T] = P = \text{covariance matrix of vector}, \qquad (10.7.4)$$

†A convenient shorthand notation for this is "x is $N(\bar{x},P)$."
‡Raiffa and Schlaifer (1961), pp. 246–251.

where $|P|$ is the determinant of P, P^{-1} is the matrix inverse of P. Note that $p(x)$ is completely characterized by giving only $\bar{x}$ and P.

If P is a diagonal matrix, then $x - \bar{x}$ has components that are statistically independent, since $p(x)$ may then be factored into a product of n scalar normal distributions. In other words, if the components of a gaussian random vector are uncorrelated, they are statistically independent. By virtue of its definition, P is a nonnegative definite matrix; i.e., it has positive (or zero) eigenvalues. Hence, by an orthogonal transformation, S,

$$y = S(x - \bar{x})\,, \tag{10.7.5}$$

it is always possible to diagonalize P. Another way of saying this is that the hypersurfaces of constant *likelihood* (constant values of probability density) in the x-space are hyperellipsoids, and, by a rotation of axes, it is possible to use the principal axes of these hyperellipsoids as coordinate axes.

We are often interested in the probability that x lies inside the hyperellipsoid:

$$(x - \bar{x})^T P^{-1} (x - \bar{x}) = l^2\,, \tag{10.7.6}$$

where l is a constant. By transforming to principal axes, this expression becomes

$$\frac{y_1^2}{\sigma_1^2} + \frac{y_2^2}{\sigma_2^2} + \cdots + \frac{y_n^2}{\sigma_n^2} = l^2\,. \tag{10.7.7}$$

By another transformation, $z_i = (y_i/\sigma_i)$, this expression becomes the equation for a hypersphere in n dimensions:

$$z_1^2 + z_2^2 + \cdots + z_n^2 = r^2. \tag{10.7.8}$$

The probability of finding z inside this hypersphere is

$$\int\int_V \cdots \int \frac{1}{(2\pi)^{n/2}} \exp\left\{-\frac{1}{2}[z_1^2 + \cdots + z_n^2]\right\} dz_1, \ldots, dz_n\,, \tag{10.7.9}$$

where the integration is carried out over the volume V of the hypersphere, r, where

$$r^2 = z_1^2 + z_2^2 + \cdots + z_n^2\,. \tag{10.7.10}$$

In the z space $|P| = 1$, since all the variances are unity and all covariances are zero. Thus the probability of finding x inside the hyperellipsoid $(x - \bar{x})^T P^{-1} (x - \bar{x}) = l^2$ is

$$\left[\frac{1}{(2\pi)^{n/2}}\right] \int_0^l \exp\left(-\frac{1}{2}r^2\right) f(r)\, dr\,, \tag{10.7.11}$$

where $f(r)\,dr$ is the spherically symmetric volume element in an n-dimensional space. For $n = 1,2,3$, this probability is given by

$$
\begin{aligned}
n = 1: &\quad \sqrt{2/\pi}\int_0^l \exp(-\tfrac{1}{2}r^2)\,dr = \operatorname{erf}(l/\sqrt{2}), \\
n = 2: &\quad \int_0^l \exp(-\tfrac{1}{2}r^2)r\,dr = 1 - \exp(-\tfrac{1}{2}l^2), \qquad (10.7.12)\\
n = 3: &\quad \sqrt{2/\pi}\int_0^l \exp(-\tfrac{1}{2}r^2)r^2\,dr = \operatorname{erf}(l/\sqrt{2}) - \sqrt{2/\pi}\,l\exp(-\tfrac{1}{2}l^2).
\end{aligned}
$$

Of particular interest are the values for $l = 1,2,3$:

n/l	1	2	3
1	.683	.955	.997
2	.394	.865	.989
3	.200	.739	.971

These are often called the one-, two-, or three-sigma probabilities.

Example. Consider a normally distributed two-dimensional vector with $\bar{x} = 0$ and

$$P = \begin{bmatrix} P_{11} & P_{12} \\ P_{12} & P_{22} \end{bmatrix} = \begin{bmatrix} 4, & 1 \\ 1, & 1 \end{bmatrix}.$$

The eigenvalues of this covariance matrix are given by

$$\begin{vmatrix} 4 - \sigma^2, & 1 \\ 1, & 1 - \sigma^2 \end{vmatrix} = 0$$

or

$$\sigma^4 - 5\sigma^2 + 3 = 0, \qquad \Rightarrow \sigma_1^2 = 4.30, \qquad \sigma_2^2 = .70,$$

and the eigenvectors are proportional to

$$\begin{bmatrix} 1 \\ .30 \end{bmatrix}, \quad \begin{bmatrix} 1 \\ -3.30 \end{bmatrix}.$$

The likelihood ellipses

$$(x_1, x_2)\begin{bmatrix} 4, & 1 \\ 1, & 1 \end{bmatrix}^{-1}\begin{bmatrix} x_1 \\ x_2 \end{bmatrix} = l^2$$

are shown in Figure 10.7.1 for $l = 1,2,3$. The probability of finding x inside the $l = 1$ ellipse is .394, inside the $l = 2$ ellipse is .865, and inside the $l = 3$ ellipse is .989.

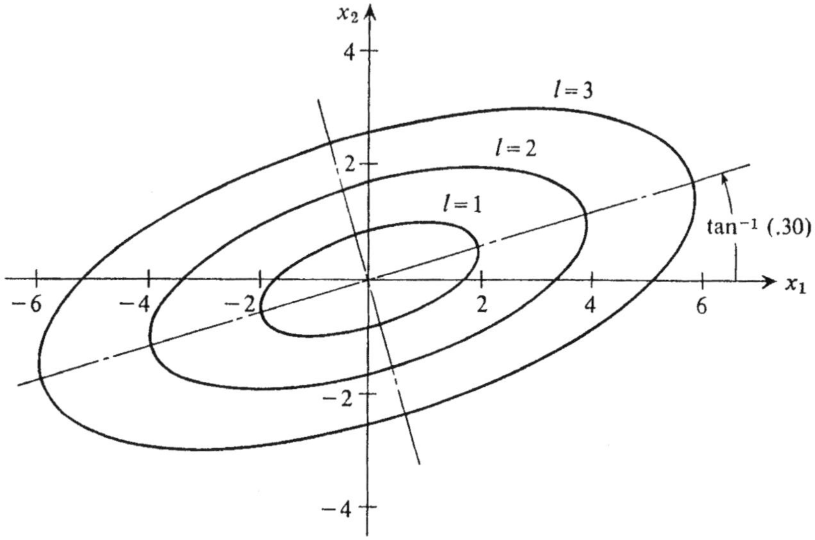

Figure 10.7.1. Likelihood ellipses for the example of two dimensional gaussian random vectors.

An important property of gaussian random vectors. The remaining part of this book depends heavily on one important property of gaussian random vectors; that is, *a linear combination of gaussian random vectors is also a gaussian random vector.* Stated analytically, if x is a gaussian random vector with mean $\bar{x}$ and covariance P_x, and $y = Ax + b$ where A is a constant matrix and b is a constant vector, then y is a gaussian random vector with mean $\bar{y}$ and covariance P_y, where

$$\bar{y} = A\bar{x} + b\,, \tag{10.7.13}$$

$$P_y = AP_xA^T\,. \tag{10.7.14}$$

The relations (10.7.13) and (10.7.14) follow very simply from the definition of expected values:

$$\begin{aligned}\bar{y} = E(y) &= \int_{-\infty}^{\infty} \cdots \int (Ax + b)p(x)\,dx_1 \cdots dx_n \\ &= A\int_{-\infty}^{\infty} \cdots \int xp(x)\,dx_1, \ldots, dx_n + b\int_{-\infty}^{\infty} \cdots \int p(x)\,dx_1 \cdots dx_n \\ &= A\bar{x} + b\,,\end{aligned}$$

and

$$P_y = E[(y - \bar{y})(y - \bar{y})^T] = \int_{-\infty}^{\infty} \cdots \int A(x - \bar{x})(x - \bar{x})^T A^T p(x)\, dx_1 \cdots dx_n$$
$$= AP_x A^T .$$

To show that y is a gaussian random vector is also quite simple if A is a nonsingular matrix.† The probability that y lies in a certain region of the y-space, R_y, is equal to the probability that x lies in the corresponding region R_x of the x-space, that is, we have

$$\int_{R_y} \cdots \int p(y)\, dy_1 \cdots dy_n = \int_{R_x} \cdots \int p(x)\, dx_1 \cdots dx_n . \qquad (10.7.15)$$

Changing variables of integration in the right-hand integral, using

$$dx_1 \cdots dx_n = |AA^T|^{-1/2}\, dy_1 \cdots dy_n , \qquad (10.7.16)$$

gives the result

$$p(y) = |AA^T|^{-1/2}\, p(x) = \frac{|(AA^T)|^{-1/2}}{(2\pi)^{n/2}|P_x|^{1/2}} \times$$
$$\exp\left\{-\frac{1}{2}(y - \bar{y})^T (A^{-1})^T P_x^{-1} A^{-1}(y - \bar{y})\right\} \qquad \text{with} \qquad x = A^{-1}(y - b)$$

or

$$p(y) = \frac{1}{(2\pi)^{n/2}|P_y|^{1/2}} \exp\left\{-\frac{1}{2}(y - \bar{y})^T P_y^{-1}(y - \bar{y})\right\}, \qquad (10.7.17)$$

which was to be shown.

Problem 1. Define the joint characteristic function of a random vector analogous to (10.6.10) via the use of multidimensional Fourier transform, and show that, for gaussian x,

$$M_x(jv) = \exp(jv^T\bar{x} - \tfrac{1}{2}v^T P v) .$$

(See Davenport and Root, *Introduction to Random Signals and Noise*, McGraw-Hill, 1958, p. 153.)

Problem 2. Prove (10.7.13) and (10.7.14) for arbitrary A by using the result of Problem 1 (see Cramer (1946), p. 312).

Problem 3. If b is a gaussian random vector independent of x, with mean $\bar{b}$ and covariance P_b, show that Equations (10.7.13) and (10.7.14) are modified to

---

†For a proof of the case in which A is singular, try Problem 2 or see Cramer (1946).

$$E(y) = A\bar{x} + \bar{b}\,, \qquad \text{cov}(y) = P_y = AP_xA^T + P_b\,.$$

[HINT: Consider the vector $\left[\begin{smallmatrix} x \\ b \end{smallmatrix}\right]$.]

Problem 4. Let v be a three-dimensional gaussian random vector with rectangular components v_1, v_2, and v_3 with zero mean, and suppose that the components are uncorrelated and have equal variances, σ^2. Show that the probability density function for the *magnitude* of v is

$$p(v) = \left(\frac{2}{\pi}\right)^{1/2} \frac{v^2}{\sigma^3} \exp\left(-\frac{v^2}{2\sigma^2}\right),$$

where $v = (v_1^2 + v_2^2 + v_3^2)^{1/2}$ is, by definition, nonnegative. [HINT: Change to spherical coordinates. In the kinetic theory of gases, $p(v)$ is called the *Maxwell-Boltzmann density function*, where v is the velocity magnitude of molecules and $\sigma^2 = (kT/m)$, with T = temperature, k = Boltzmann's const, m = mass of a molecule. In statistics, $p(v)$ is called the "*chi-squared*" *density function* for $n = 3$; obviously, it can be extended to any number of dimensions.

Introduction to random processes

11

11.1 Random sequences and the markov property

A *random sequence* is a collection of continuous-valued random variables (scalar or vector) indexed by a discrete-valued parameter such as

$$x(0),x(1),x(2),\ldots,x(N).$$

For convenience, the values of the index parameter are usually taken as integers and, in applications, often symbolize units of time or distance.†

Examples of vector random sequences are quite common: (a) The velocity of the wind (magnitude and direction) at successive time instants at some fixed location; (b) the position and velocity of the mass center of a vehicle at evenly-spaced time intervals; (c) the shear, bending moment, slope, and deflection at a set of points along a beam at a given instant of time; (d) amplitude and phase of a radio wave at different instants.

To describe a random sequence completely, the joint probability density function

$$p[x(N),x(N-1),\ldots,x(0)]$$

of all elements in the sequence must be specified. In general, this involves an enormous amount of information. Fortunately, most random sequences encountered in applications have a special property that makes them simpler to specify than the most general random sequence; this is the markov property.

Markov sequences. A random sequence $x(k)$, $k = 0,1,\ldots,N$ is said to be *markovian* if

†Unfortunately, the nomenclature above is identical to that for the N values of a discrete-valued random vector. However, it is usually clear which interpretation is meant from the context in which the symbols are used.

$$p[x(k+1)/x(k),x(k-1),\ldots,x(0)] = p[x(k+1)/x(k)] \quad (11.1.1)$$

for all k; that is, the probability density function of $x(k+1)$ depends only on knowledge of $x(k)$ and not on $x(k-l)$, $l = 1,2,\ldots$. The knowledge of $x(k)$ that is required can be either deterministic [exact value of $x(k)$ known] or probabilistic ($p[x(k)]$ known). In words, *the markov property implies that a knowledge of the present separates the past and the future.*

The joint probability density function of a markov random sequence can be described completely by specifying the initial density function $p[x(0)]$ and the *transition density functions* $p[x(k+1)/x(k)]$. This is easily seen as follows:

$$\begin{aligned} p[x(N),x(N-1),\ldots,x(0)] &= p[x(N)/x(N-1),\ldots,x(0)] \\ &\qquad p[x(N-1),\ldots,x(0)] \quad (11.1.2) \\ &= p[x(N)/x(N-1),\ldots,x(0)] \\ &\qquad p[x(N-1)/x(N-2),\ldots,x(0)]\cdots p[x(1)/x(0)]p[x(0)]. \end{aligned}$$

From (11.1.1), however, this last expression may be simplified to

$$\begin{aligned} p[x(N),\ldots,x(0)] &= p[x(N)/x(N-1)]p[x(N-1)/x(N-2)]\cdots \\ &\qquad p[x(1)/x(0)]p[x(0)]; \quad (11.1.3) \end{aligned}$$

that is, the joint probability density function is equal to the product of the transition density functions and the initial density function.

Purely random sequences. If $p[x(k+1)/x(k)] = p[x(k+1)]$ for all possible values of k, the sequence is said to be a *purely random sequence.*

The successive outcomes of spinning a balanced wheel, and letting friction gradually bring it to a stop, form a purely random sequence. The outcome of a spin is the distance measured clockwise along the perimeter of the wheel from a reference line on the wheel to a fixed line not on the wheel (see Figure 11.1.1). The probability density

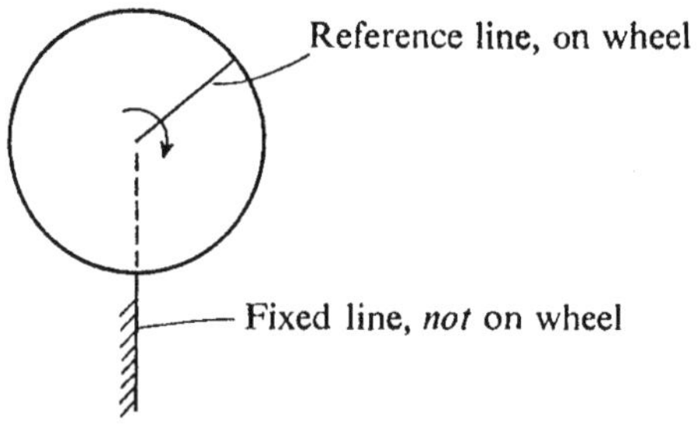

Figure 11.1.1. Spinning wheel as an example of a purely random sequence.

function of the outcome is uniform and does *not* depend on previous spins. If we always use the same wheel, the successive outcomes form a *stationary purely random sequence.* Suppose, however, that we have many wheels of different diameters and we change wheels after each spin; the successive outcomes now form a *nonstationary purely random sequence.*† The probability density function of the outcome is still uniform on each spin, but it is a different constant on each spin.

The *cumulative score* for the wheel-spinning game is *not* a purely random sequence, since it depends on the previous score; the cumulative score is, in fact, a markov sequence. A slight generalization of the cumulative score is the scalar markov sequence $x_1(k)$, where

$$x_1(k+1) = c(k)x_1(k) + w(k)\,, \tag{11.1.4}$$

$w(k)$ is a scalar purely random sequence and $c(k)$ is a known sequence ($c = 1$ in the cumulative score).

The generality of markov sequences. The markov property may seem restrictive as presented above. For example, if Equation (11.1.4) is replaced by a scalar second-order difference equation

$$x_1(k+1) = c_1(k)x_1(k) + c_2(k)x_1(k-1) + w(k)\,, \tag{11.1.5}$$

where $w(k)$ is a purely random markov sequence and $c_1(k)$ and $c_2(k)$ are known sequences, clearly the sequence $x_1(k)$ is *not markovian.* However, we can make it into a component of a *vector* markov sequence using the state vector $\left[\begin{smallmatrix} x_1(k) \\ x_2(k) \end{smallmatrix}\right]$, where $x_2(k+1) = x_1(k)$, since

$$\begin{bmatrix} x_1(k+1) \\ x_2(k+1) \end{bmatrix} = \begin{bmatrix} c_1(k)\,, & c_2(k) \\ 1\,, & 0 \end{bmatrix} \begin{bmatrix} x_1(k) \\ x_2(k) \end{bmatrix} + \begin{bmatrix} 1 \\ 0 \end{bmatrix} w(k)\,;$$

that is, the two-component vector $\left[\begin{smallmatrix} x_1(k+1) \\ x_2(k+1) \end{smallmatrix}\right]$ depends only on a knowledge of $\left[\begin{smallmatrix} x_1(k) \\ x_2(k) \end{smallmatrix}\right]$, the previous element in the sequence, and the purely random sequence $w(k)$.

Generalizing the above procedure, we see that, given any vector random sequence that depends *on the finite past,* in the sense of (11.1.5) we can always convert it to an equivalent markov random sequence by properly enlarging the state vector. Thus, markov random sequences with finite dimensional state vectors include, in this sense, all random sequences between the two extreme cases, namely, the purely random sequence and the random sequence that depends on the infinite past. Consequently, markov random sequences can be used to represent a great variety of physical phenomena. For this

†Alternatively, we could use the same wheel but change the unit of distance measurement along the perimeter after each spin.

reason, we shall henceforth restrict our study of random sequences to markov random sequences.

Markov chains. If the quantity x in a markov sequence is a *discrete-valued* random vector with a finite number of possible values, the markov sequence is usually called a *markov chain*. Most of the discussion above applies to markov chains, except that a *probability transition matrix* is required instead of a *transition density function* and the other probability density functions are replaced by probability mass functions.

A markov chain may be used as an approximation to a scalar continuous-valued markov sequence. The probability transition matrix (see Problem 3 of this section) is a discrete approximation to $p[x(t+1)/x(t)]$, the transition density function of the two variables $x(t+1)$ and $x(t)$. The probability vector at each instant is a discrete approximation to the density function $p[x(t)]$.

An example of a purely random chain is the successive spinning of a wheel whose circumference is marked off into a finite number, N, of equal lengths and each length is given an integer number from 1 to N. The outcome of the spin is then one of these numbers. The probability mass function is uniform and equal to $1/N$; it does not depend on previous spins.

Problem 1. Consider the markov sequence generated by keeping the cumulative score of a game like the wheel-spinning game, where

$$s(k+1) = s(k) + w(k)\,; \qquad s(0) = 0\,,$$

and $s(k)$ is the score after the kth spin and $w(k)$ is a purely random sequence with a uniform density function equal to $1/N$ in the range $(0,N)$. Show that the mean and variance of the score are given by

$$\bar{s}(k) = k\frac{N}{2}, \qquad E[s(k) - \bar{s}(k)]^2 = k\frac{N^2}{12}.$$

Show, also, that the density function for $s(k)$ tends toward a gaussian distribution as k increases (see Problems 1 and 2 of Section 10.6).

Problem 2. Consider the markov chain generated by keeping the cumulative score of a game like the wheel-spinning game as in Problem 1, except that $w(k)$ is a purely random chain with possible values 1,2, ..., N, with a uniform mass function equal to $1/N$. Show that the mean and variance are given by

$$\bar{s}(k) = k\frac{N+1}{2}, \qquad E[s(k) - \bar{s}(k)]^2 = k\frac{N^2-1}{12}.$$

Problem 3. Suppose that we characterize the weather in a certain rain forest by only three states: State 1 is cloudy, state 2 is rainy, and state 3 is sunny. The daily weather then forms a markov chain. The probability-transition diagram, as observed over a long period, is shown in Figure 11.1.2.

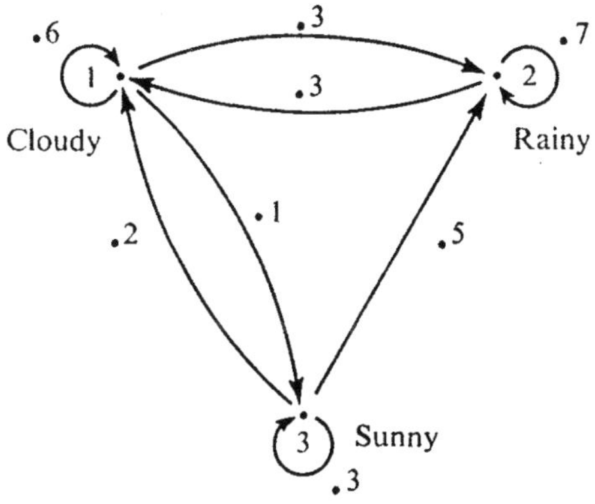

Figure 11.1.2. Probability-transition diagram of day-to-day weather changes in a rain forest (Problem 3).

By this we imply the probability transition matrix,

$$\begin{bmatrix} p_1(k+1) \\ p_2(k+1) \\ p_3(k+1) \end{bmatrix} = \begin{bmatrix} .6 & .3 & .2 \\ .3 & .7 & .5 \\ .1 & 0 & .3 \end{bmatrix} \begin{bmatrix} p_1(k) \\ p_2(k) \\ p_3(k) \end{bmatrix}.$$

Show that a stationary probability mass function exists (as $k \to \infty$) for this system, with $p_1 = .42$, $p_2 = .52$, $p_3 = .06$.

Given that a certain day is sunny ($p_3 = 1$, $p_1 = p_2 = 0$), predict the weather for the next few days.

Problem 4. A fair coin is tossed many times successively. Find the probability of tossing heads on the sixth toss *and* the seventh toss *without* having tossed two successive heads previously. [HINT. This is a markov chain with three states: State 1 is tails, state 2 is heads, state 3 is the second successive head without two successive heads having been tossed previously.]

The probability-transition diagram is shown in Figure 11.1.3, and the initial mass function is $p_1(1) = p_2(1) = \frac{1}{2}$, $p_3(1) = 0$. The problem asks for $p_3(7)$. Note that no stationary mass function exists for this system. Why?

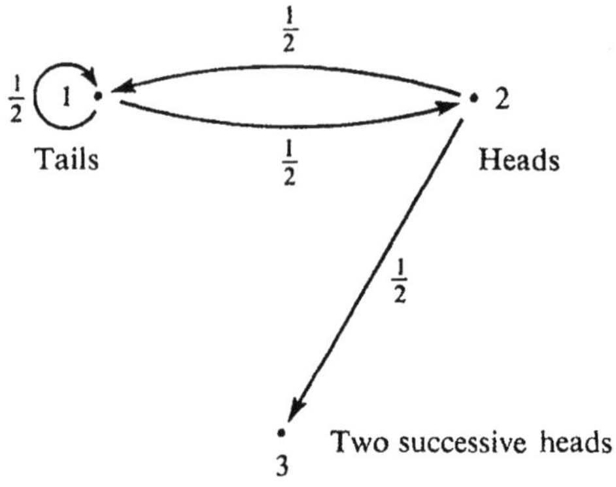

Figure 11.1.3. Probability-transition diagram of coin tossing problem.

11.2 Gauss-markov random sequences

A *gauss-markov random sequence* is a markov random sequence with the added restriction that $p[x(k)]$ and $p[x(k+1)/x(k)]$ are gaussian probability density functions for all k.

The density function $p[x(k)]$ of a gauss-markov random sequence is, therefore, completely described by giving two deterministic sequences, the mean-value vector $\bar{x}(k) = E[x(k)]$ and the covariance matrix $X(k) = E\{[x(k) - \bar{x}(k)]\,[x(k) - \bar{x}(k)]^T\}$.

Many natural and man-made dynamic phenomena may be approximated quite accurately by gauss-markov random sequences. In addition, it is often expedient to approximate a nongaussian markov random sequence by a gauss-markov random sequence because of the limited statistical knowledge available concerning the actual sequence.

Since linear transformations of a gaussian vector preserve its gaussian character (see Section 10.7), *a gauss-markov random sequence can always be represented by the state vector of a multistage linear dynamic system forced by a gaussian purely random sequence in which the initial state vector is gaussian:*

$$x(k+1) = \Phi(k)x(k) + \Gamma(k)w(k)\,, \qquad (11.2.1)$$

where

$$x \text{ is an } n\text{-vector}, \; w \text{ is an } m\text{-vector}\,,$$

$$E[w(k)] = \overline{w}(k)\,, \qquad (11.2.2)$$

$$E\{[w(k) - \overline{w}(k)][w(\ell) - \overline{w}(\ell)]^T\} = \begin{cases} \mathrm{X}(k)\,, & k = \ell\,, \\ 0\,, & k \neq \ell\,, \end{cases} \qquad (11.2.3)$$

$$E[x(0)] = \bar{x}(0), \tag{11.2.4}$$

$$E\{[x(0) - \bar{x}(0)][x(0) - \bar{x}(0)]^T\} = X(0), \tag{11.2.5}$$

$$E\{[x(0) - \bar{x}(0)][w(k) - \overline{w}(k)]^T\} = 0. \tag{11.2.6}$$

A block diagram representation is shown in Figure 11.2.1.

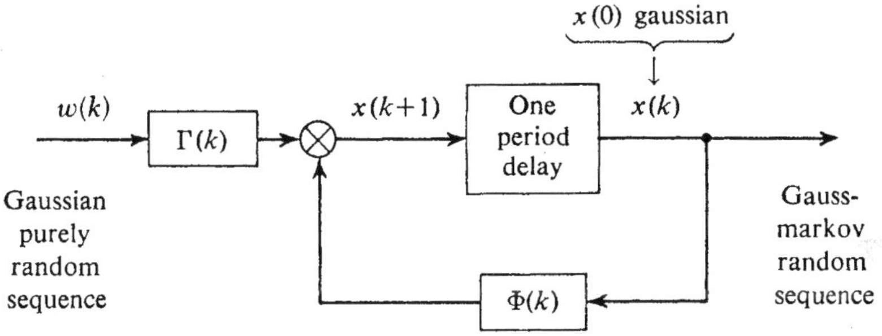

Figure 11.2.1: Representation of a gauss-markov sequence.

The initial state is gaussian with mean value $\bar{x}(0)$ and covariance $X(0)$. The transition density function $p[x(k+1)/x(k)]$ is gaussian with mean value $\Phi(k)x(k) + \Gamma(k)\overline{w}(k)$ and covariance $\Gamma(k)\chi(k)\Gamma^T(k)$.†

$$p[x(k+1)/x(k)] = \frac{1}{(2\pi)^{n/2}|\Gamma(k)\chi(k)\Gamma^T(k)|^{1/2}}$$

$$\exp\left\{-\frac{1}{2}[x(k+1) - \Phi(k)x(k) - \Gamma(k)\overline{w}(k)]^T\right.$$

$$\left.\cdot\,[\Gamma(k)\chi(k)\Gamma^T(k)]^{-1}[x(k+1) - \Phi(k)x(k) - \Gamma(k)\overline{w}(k)]\right\}.$$

The relations for determining the sequences $\bar{x}(k)$ and $X(k)$ are readily found from the data of Equations (11.2.1) through (11.2.6). Taking the expected value of (11.2.1), we have

$$\boxed{\bar{x}(k+1) = \Phi(k)\bar{x}(k) + \Gamma(k)\overline{w}(k), \qquad \bar{x}(0) \text{ and } \overline{w}(k) \text{ given.}} \tag{11.2.7}$$

Subtracting (11.2.7) from (11.2.1), we have

$$x(k+1) - \bar{x}(k+1) = \Phi(k)[x(k) - \bar{x}(k)] + \Gamma(k)[w(k) - \overline{w}(k)]. \tag{11.2.8}$$

Multiplying (11.2.8) by its transpose and then taking the expected value of both sides yields

†If $\Gamma\chi\Gamma^T$ is singular, some linear combinations of components of $x(k+1)$ are known exactly, given $x(k)$.

$$X(k+1) = \Phi(k)X(k)\Phi^T(k) + \Gamma(k)\chi(k)\Gamma^T(k), \qquad X(0) \text{ and } \chi(k) \text{ given,} \tag{11.2.9}$$

where we have used the relation

$$E[x(k) - \bar{x}(k)]\,[w(k) - \bar{w}(k)]^T = 0\,, \tag{11.2.10}$$

which results from the purely random character of $w(k)$.

Equations (11.2.7) and (11.2.9) are linear difference equations for the mean value vector and covariance matrix. Note that they are *not coupled* and, hence, the sequences $\bar{x}(k)$ and $X(k)$ may be calculated separately. They completely specify the evolution of the density function $p[x(k)]$. Pictorially, the one-dimensional case is shown in Figure 11.2.2 [see example of Equation (11.2.20)]. One of the principal results of the study of stochastic processes is the quantitative statement of Figure 11.2.2, e.g., Equations (11.2.7) and (11.2.9).

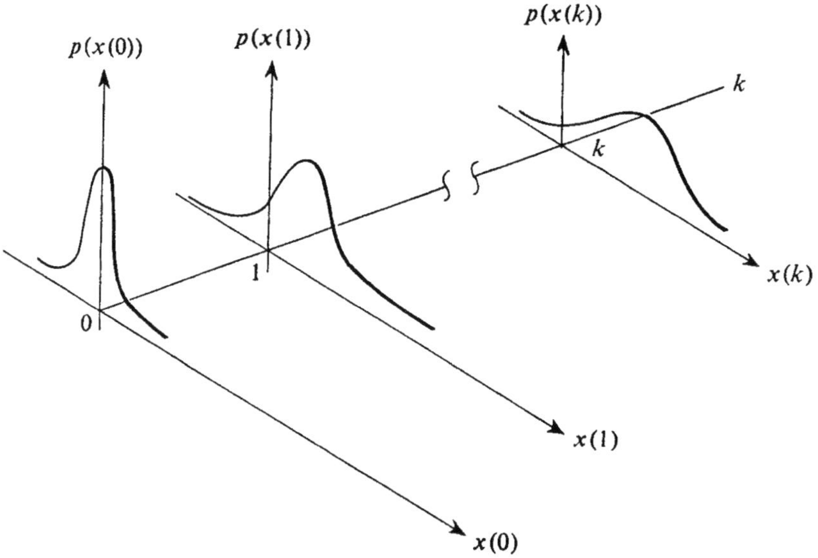

Figure 11.2.2. Evolution of the probability density of a random sequence.

Correlation matrix of a gauss-markov random sequence. The *correlation matrix* of a random sequence $x(k)$ is defined as

$$C(k,\ell) \triangleq E\{[x(k) - \bar{x}(k)]\,[x(\ell) - \bar{x}(\ell)]^T\}\,. \tag{11.2.11}†$$

†In some places, the correlation matrix is defined as $E[x(k)x^T(\ell)] = C^\circ$. Clearly, we have $C = C^\circ - \bar{x}(k)\bar{x}^T(\ell)$.

It is a deterministic double sequence of matrices, since it depends on two indexing parameters, k and ℓ. The covariance matrix sequence is, of course, a special case of (11.2.11):

$$X(k) \equiv C(k,k)\,. \tag{11.2.12}$$

Using Equation (11.2.1) and the transition matrix $\Phi(k+\ell,k) = \Phi(k+\ell-1)\ldots\Phi(k)$, the value of $x(k+\ell)$ may be expressed in terms of $x(k)$ and the sequence $w(n)\,, n = k,\ldots,k+\ell-1$:

$$x(k+\ell) = \Phi(k+\ell,k)x(k) + \sum_{n=k}^{k+\ell-1} \Phi(k+\ell,n+1)\Gamma(n)w(n)\,, \tag{11.2.13}$$

where $\Phi(m,m) = I =$ unit matrix. Subtracting mean values, postmultiplying by $[x(k) - \bar{x}(k)]^T$, and taking the expected value of (11.2.13) gives

$$C(k+\ell,k) = \Phi(k+\ell,k)X(k) + \sum_{n=k}^{k+\ell-1} \Phi(k+\ell,n+1)\Gamma(n) \\ E\{[w(n) - \bar{w}(n)]\;[x(k) - \bar{x}(k)]^T\}\,.$$

However, we have $E\{[w(n) - \bar{w}(n)]\,[x(k) - \bar{x}(k)]^T = 0$ for $n = k,\ldots,k+\ell-1$, since $w(n)$ is a purely random sequence. Therefore, the correlation matrix of a gauss-markov random sequence is given simply by

$$\boxed{C(k+\ell,k) = \Phi(k+\ell,k)X(k)}\;; \qquad \ell = 1,2,\ldots, \tag{11.2.14}$$

where $X(k)$ can be found from (11.2.9).

From the definition of $C(k+\ell,k)$ in Equation (11.2.11), it is evident that

$$\boxed{C(k,k+\ell) = X(k)\Phi^T(k+\ell,k)}\;; \qquad \ell = 1,2,\ldots. \tag{11.2.15}$$

Subtracting mean values, postmultiplying by $[w(k) - \bar{w}(k)]^T\Gamma^T(k)$, and taking the expected value of (11.2.13) gives the cross-correlation between output and input of a gauss-markov random sequence:

$$\boxed{\begin{aligned} &E\{[x(k+\ell) - \bar{x}(k+\ell)]\,[w(k) - \bar{w}(k)]^T\Gamma^T(k)\} \\ &= \begin{cases} \Phi(k+\ell,k+1)\Gamma(k)\chi(k)\Gamma^T(k)\,, & \ell = 1,2,\ldots, \\ 0\,, & \ell = 0,-1,-2,\ldots. \end{cases} \end{aligned}} \tag{11.2.16}$$

The relations (11.2.16) or (11.2.15) may be used experimentally to find components of the transition matrix for an unknown linear system. To find all components of Φ from (11.2.16), it is necessary that $(\Gamma\chi\Gamma^T)^{-1}$ exist.†

Statistically stationary sequences. If we have $\Phi(k+\ell,k)=\Phi(\ell)$, $\Gamma(\ell)=\Gamma=$ constant matrix, and $\chi(k)=\chi=$ constant matrix, it is possible that $X(k)\to X$, a constant matrix, as $k\to\infty$ (see Problem 2, Appendix B4). The correlation matrix and the cross-correlation above also become stationary in this case; that is, $C(k+\ell,k)\to C(\ell)$. From (11.2.14), (11.2.15), and (11.2.16), we have, for $\ell=0,1,2,\ldots$,

$$C(\ell)=\Phi(\ell)X, \tag{11.2.17}$$

$$C(-\ell)=X\Phi^T(\ell); \tag{11.2.18}$$

$$E\{[x(k+\ell)-\bar{x}(k+\ell)][w(k)-\bar{w}(k)]^T\Gamma^T(k)\}\to\begin{cases}\Phi(\ell-1)\Gamma\chi\Gamma^T, & \ell=1,2,\ldots,\\ 0, & \ell=0,-1,-2,\ldots.\end{cases} \tag{11.2.19}$$

A first-order gauss-markov sequence. Consider the special case of Equation (11.2.1), where $x(k)$ is a scalar, $\Phi=\alpha=$ const. and $\Gamma=1$:

$$x(k+1)=\alpha x(k)+w(k); \qquad E[w(k)]=0,$$

$$E[w(k)w(\ell)]=\begin{cases}q, & k=\ell,\\ 0, & k\neq\ell,\end{cases} \tag{11.2.20}$$

$$E[x(0)]=0, \qquad E[x(0)]^2=X_0 \qquad \text{(all quantities scalars).} \tag{11.2.21}$$

The transition matrix is a scalar, $\Phi(k+\ell,k)=\alpha^\ell$:

$$\Rightarrow X(k+1)=\alpha^2X(k)+q; \qquad X(0)=X_0. \tag{11.2.22}$$

The solution of this difference equation is

$$X(k)=\alpha^{2k}X_0+q\frac{1-\alpha^{2k}}{1-\alpha^2}. \tag{11.2.23}$$

From (11.2.14), (11.2.15), and (11.2.16), then, we have

$$C(k+\ell,k)=\begin{cases}\alpha^\ell\left[\alpha^{2k}X_0+q\dfrac{1-\alpha^{2k}}{1-\alpha^2}\right] & \ell\geq 0,\\[2ex] \alpha^{-\ell}\left[\alpha^{2(k+\ell)}X_0+q\dfrac{1-\alpha^{2(k+\ell)}}{1-\alpha^2}\right] & \ell<0,\end{cases} \tag{11.2.24}$$

$$E[x(k+\ell)w(k)]=\begin{cases}\alpha^{\ell-1}q, & \ell=1,\ldots,\\ 0, & \ell=0,-1,\ldots.\end{cases} \tag{11.2.25}$$

As $k\to\infty$, a statistically stationary sequence results if we have $0\leq\alpha<1$:

$$X(k)\to\frac{q}{1-\alpha^2}, \tag{11.2.26}$$

†This is one approach to the "identification" problem. More refined methods exist but are not treated in this book.

$$C(k+\ell,k) \to C(\ell) = \frac{q}{1-\alpha^2}\alpha^{|\ell|}; \qquad \text{as} \quad k \to \infty, \qquad 0 \le \alpha < 1. \tag{11.2.27}$$

Note that, if X_0 had been equal to $q/(1-\alpha^2)$, then $X(k)$ would remain constant; i.e., it would start in the statistically stationary state.

Problem 1. Show that the mean value vector and the covariance matrix may be expressed in terms of the transition matrix $\Phi(k,\ell)$ as follows:

$$\bar{x}(k) = \Phi(k,0)\bar{x}(0) + \sum_{n=0}^{k-1} \Phi(k,n+1)\Gamma(n)\overline{w}(n),$$

$$X(k) = \Phi(k,0)X(0)\Phi^T(k,0) + \sum_{n=0}^{k-1} \Phi(k,n+1)\Gamma(n)\chi(n)\Gamma^T(n)\Phi^T(k,n+1),$$

where

$$\Phi(k,\ell) = \Phi(k,k-1)\Phi(k-1,k-2)\cdots\Phi(\ell+1,\ell)$$

and

$$\Phi(k,k) = I = \text{unit matrix}.$$

Problem 2. Consider the stationary correlation of the first-order gauss-markov sequence, Equation (11.2.27), for the case in which there are N steps in the sequence. The correlation matrix, Equation (11.2.27), is then an $(N \times N)$-matrix:

$$C = \frac{q}{1-\alpha^2}\begin{bmatrix} 1 & ,\alpha, & \alpha^2, & \cdots & \alpha^{N-1} \\ \alpha & ,1, & & \cdots & \cdot \\ & & & & \cdot \\ \alpha^2 & ,\alpha, & & \cdots & \cdot \\ \cdot & & & & \\ \cdot & \cdots & & 1 & ,\alpha \\ \cdot & & & & \\ \alpha^{N-1} & \cdots & & \alpha & ,1 \end{bmatrix}.$$

Show that the *inverse* of the correlation matrix is the *tridiagonal* matrix:

$$C^{-1} = \frac{1}{q}\begin{bmatrix} 1, & -\alpha, & 0, & \cdots & 0 \\ & & & & \cdot \\ -\alpha, & 1+\alpha^2, & -\alpha, & \cdots & \cdot \\ & & & & \cdot \\ 0, & -\alpha, & 1+\alpha^2, & \cdots & 0 \\ \cdot & & & & \\ \cdot & \cdots & -\alpha & ,1+\alpha^2, & -\alpha \\ \cdot & & & & \\ 0 & \cdots & & -\alpha, & 1 \end{bmatrix}.$$

Problem 3. The volume of water in a reservoir is recorded each year on the same date. Let $x(k)$ be the volume in the kth year, and let $u(k-1)$ and $v(k-1)$ be the inflow and the outflow, respectively, during the previous year. Records indicate that $u(k)$ may be well-approximated by a statistically stationary first-order gauss-markov sequence, with mean value $\bar{u}$ and correlation

$$E[(u(k)-\bar{u})(u(\ell)-\bar{u})] = \frac{q}{1-\alpha^2}\alpha^{|k-\ell|}, \qquad 0<\alpha<1,$$

whereas the outflow is regulated to be constant equal to the mean inflow; that is, $v(k)=\bar{u}$ for all k. Show that the variance of the volume increases without limit as k increases. [HINT. Form a vector gauss-markov process with state vector $x(k)$, $u(k)-\bar{u}$ and show that

$$X_{xx}(k+1) = X_{xx}(k) + \frac{q}{(1-\alpha)^2} \qquad \text{as} \qquad k \to \infty,$$

where X_{xx} is the variance of the volume.]

11.3 Random processes and the markov property

A *random process* is a collection of continuous-valued random variables indexed by a continuous-valued parameter such as

$$x(t), \qquad t_o \le t \le t_f.$$

In applications, the continuous-valued parameter (or independent variable), t, is often a time or distance measure.

All continuous dynamic processes that occur in natural and man-made systems are random processes, since the state of such systems can never be known exactly. Examples of vector random processes are: (a) the velocity of the wind (magnitude and direction) at a fixed location as a function of time; (b) the position and velocity of a vehicle as a function of time; (c) the shear, bending moment, slope, and deflection as a function of distance along a beam at a given time instant; (d) amplitude and phase of a radio wave as a function of time.

A complete description of a general random process would require knowledge of *all* possible joint density functions

$$p[x(t_1),x(t_2),\ldots,x(t_N)] \tag{11.3.1}$$

for all $t_1,t_2,\ldots,t_N$ in the interval (t_o,t_f), where N is any integer between 1 and ∞. In general, this is an impossible amount of information to supply for a given process. Fortunately, most random processes encountered in applications are markovian, and a *markov process* is completely specified by giving the joint density function

$$p[x(t), x(\tau)] \quad \text{for all} \quad t,\tau \quad \text{in the interval} \quad (t_o,t_f). \tag{11.3.2}$$

Since we have $p[x(t),x(\tau)] = p[x(t)/x(\tau)]p[x(\tau)]$, a markov process is also completely specified by giving the density functions

$$p[x(t)/x(\tau)] \quad \text{and} \quad p[x(\tau)] \quad \text{for all} \quad t,\tau \quad \text{in the interval} \quad (t_o,t_f). \tag{11.3.3}$$

For a markov process, knowledge of (11.3.2) allows us, in principle, to determine all of the joint density functions of (11.3.1) [see Section 11.1, Equations (11.1.1) through (11.1.3), replacing the $x(N)$ by $x(t_N)$, etc.].

Purely random processes. If we have $p[x(t)/x(\tau)] = p[x(t)]$ for all t,τ in (t_o,t_f), then $x(t)$ is called a *purely random (or white noise) process.* It is an imaginary process and must be considered only as a limiting case, since, in any actual process, $x(t)$ and $x(\tau)$ will be dependent when $|t - \tau|$ is sufficiently small. As we shall see, the white noise process is a very convenient fiction.† If we are dealing with a random forcing function $u(t)$, with $p[u(t)/u(\tau)] \cong p[u(t)]$ for $|t - \tau| > T$, and T is much smaller than characteristic response times of the system being forced, then $u(t)$ may be regarded as white noise relative to the system.

An example of a white noise process is the thrust, $F(t)$, of a rocket engine; the deviation of instantaneous thrust from the mean value $\overline{F}(t)$ is unpredictable from one millisecond to the next. However, the response times of the rocket vehicle are figured in seconds, or perhaps tenths of a second, so the deviation of the thrust from the mean value may be regarded as a white noise process relative to vehicle dynamics.

Another example of a white noise process is the current flowing in a wire or in an electron beam; the deviation of the instantaneous current from the nominal steady current is unpredictable from one microsecond to the next. Thus, if the current is an input to a system with characteristic response times large compared to a microsecond, the deviation of the current from the mean value may be regarded as a white noise process.

If the current in the latter example is characterized by the charge passing through a given cross section of the electron beam in successive periods of a microsecond, $\Delta q(t_j)$ where $t_{j+1} - t_j = 1\,\mu\text{sec}$, these numbers form a purely random markov sequence. Now, suppose that we keep track of the cumulative charge that has passed a given cross section of the electron beam, $q(t)$; this quantity is a scalar markov process, but it is *not* purely random. In fact, we have

†This is comparable to the convenient but fictitious "mass particle" of Newtonian mechanics, which has finite mass and zero volume.

$$\dot{q} = i(t), \qquad t_o \leq t \leq t_f, \tag{11.3.4}$$

where $i(t)$ is the current, a purely random process.

A slight generalization of the cumulative charge is the scalar markov process $x_1(t)$, where

$$\dot{x}_1 = c(t)x_1 + w(t), \tag{11.3.5}$$

where $w(t)$ is a scalar purely random process and $c(t)$ is a known function of time.

The generality of markov processes. The markov property is not as restrictive as it might seem. Suppose that Equation (11.3.5) is replaced by a second-order differential equation for the scalar $x_1(t)$:

$$\ddot{x}_1 = c_1(t)\dot{x}_1 + c_2(t)x_1 + w(t), \tag{11.3.6}$$

where $w(t)$ is a scalar purely random process and $c_1(t)$, $c_2(t)$ are known functions of time. Then $x_1(t)$ is *not* a markov process. However, we can make it into a component of a *vector* markov process using the state vector $\left[\begin{smallmatrix} x_1 \\ x_2 \end{smallmatrix}\right]$, where $x_2 = \dot{x}_1$ since

$$\frac{d}{dt}\begin{bmatrix} x_1 \\ x_2 \end{bmatrix} = \begin{bmatrix} 0 & , & 1 \\ c_2(t) & , & c_1(t) \end{bmatrix}\begin{bmatrix} x_1 \\ x_2 \end{bmatrix} + \begin{bmatrix} 0 \\ 1 \end{bmatrix} w(t). \tag{11.3.7}$$

Generalizing this procedure, we see that, given any random process involving a finite number of time derivatives, we can always convert it to an equivalent markov random process by properly enlarging the state vector. For this reason we shall henceforth restrict our study of random processes to markov processes.

11.4 Gauss-markov random processes

A *gauss-markov random process* is a markov random process with the added restriction that $p[x(\tau)]$ and $p[x(t)/x(\tau)]$ are gaussian density functions for all t,τ in the interval (t_o,t_f).

The density function $p[x(t)]$ of a gauss-markov process is, therefore, completely described by giving two deterministic functions, the mean value vector $\bar{x}(t) = E[x(t)]$ and the covariance matrix $X(t) = E\{[x(t) - \bar{x}(t)]\,[x(t) - \bar{x}(t)]^T\}$.

Many natural and man-made dynamic phenomena may be approximated quite accurately by gauss-markov random processes. In addition, it is often expedient to approximate a nongaussian markov process by a gauss-markov random process because of the limited statistical knowledge available concerning the actual process.

Since linear transformations of a gaussian vector preserve its gaussian character (see Section 10.7), a *gauss-markov random process can always be represented by the state vector of a continuous linear dynamic system forced by a gaussian purely random process where the initial state vector is gaussian:*

$$\dot{x} = F(t)x + G(t)w(t) \tag{11.4.1}$$

where x is an n-vector, w is an m-vector,

$$E[x(t_0)] = \bar{x}(t_0), \tag{11.4.2}$$

$$E\{[x(t_0) - \bar{x}(t_0)]\,[x(t_0) - \bar{x}(t_0)]^T\} = X(t_0), \tag{11.4.3}$$

and $w(t)$ is a gaussian purely random process, with

$$E[w(t)] = \overline{w}(t). \tag{11.4.4}$$

Figure 11.4.1 is a block diagram of the gauss-markov random process.

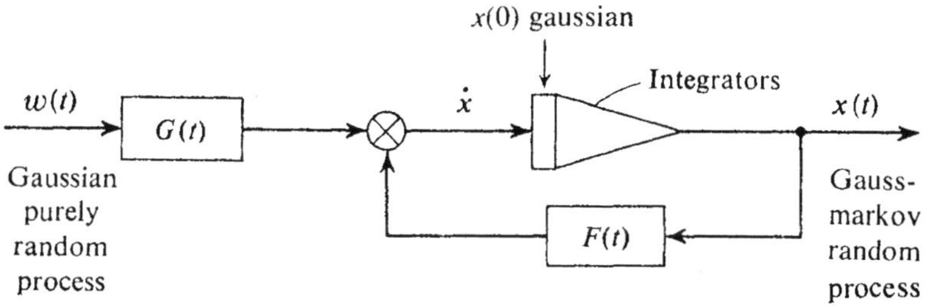

Figure 11.4.1. Representation of a gauss-markov process.

In Equations (11.4.1) through (11.4.4), we did *not* completely specify the purely random process $w(t)$. From the definition of a purely random process given in Section 11.3, we know that $w(t) - \overline{w}(t)$ is a very "jittery" random function whose value is unpredictable from one time, t, to the next, $t + \Delta t$, even as $\Delta t \to 0$. Let us see how $w(t)$ enters into the determination of the covariance matrix $X(t)$, where

$$X(t) = E[x(t) - \bar{x}(t)]\,[x(t) - \bar{x}(t)]^T. \tag{11.4.5}$$

Clearly, the mean value of $x(t)$ is determined by considering the expected value of (11.4.1):

$$\boxed{\frac{d}{dt}[\bar{x}(t)] = F(t)\bar{x}(t) + G(t)\overline{w}(t), \qquad \bar{x}(t_0) \text{ given.}} \tag{11.4.6}$$

Subtracting (11.4.6) from (11.4.1) and postmultiplying the result by $[x(t) - \bar{x}(t)]^T$ yields

$$\left\{\frac{d}{dt}[x(t)-\bar{x}(t)]\right\}[x(t)-\bar{x}(t)]^T = F(t)\,[x(t)-\bar{x}(t)]\,[x(t)-\bar{x}(t)]^T$$
$$+\,G(t)\,[w(t)-\overline{w}(t)]\,[x(t)-\bar{x}(t)]^T\,. \quad (11.4.7)$$

Adding the transpose of Equation (11.4.7) to Equation (11.4.7) gives

$$\frac{d}{dt}\{[x(t)-\bar{x}(t)]\,[x(t)-\bar{x}(t)]^T\} =$$
$$F(t)\,[x(t)-\bar{x}(t)]\,[x(t)-\bar{x}(t)]^T + [x(t)-\bar{x}(t)]\,[x(t)-\bar{x}(t)]^T F^T(t)$$
$$+\,G(t)\,[w(t)-\overline{w}(t)]\,[x(t)-\bar{x}(t)]^T + [x(t)-\bar{x}(t)]\,[w(t)-\overline{w}(t)]^T G^T(t)\,. \quad (11.4.8)$$

Taking the expected value (i.e., the ensemble average) of Equation (11.4.8) and using the definition (11.4.5) gives

$$\dot{X} = F(t)X + XF^T(t) + G(t)\,E\{[w(t)-\overline{w}(t)]\,[x(t)-\bar{x}(t)]^T\}$$
$$+\,E\{[x(t)-\bar{x}(t)]\,[w(t)-\overline{w}(t)]^T\}G^T(t)\,. \quad (11.4.9)$$

Using the transition matrix $\Phi(t,\tau)$ of the linear dynamic system (11.4.1), we know that

$$x(t)-\bar{x}(t) = \Phi(t,t_0)\,[x(t_0)-\bar{x}(t_0)] + \int_{t_0}^{t}\Phi(t,\tau)G(\tau)\,[w(\tau)-\overline{w}(\tau)]\,d\tau\,. \quad (11.4.10)$$

We assume that

$$E[x(t_0)-\bar{x}(t_0)]\,[w(t)-\overline{w}(t)]^T = 0\,; \quad (11.4.11)$$

that is, random deviations in initial conditions are uncorrelated with the random fluctuations of the forcing function. Postmultiplying (11.4.10) by $[w(t)-\overline{w}(t)]^T$, taking the expected value of the result, and using (11.4.11) yields

$$E[x(t)-\bar{x}(t)]\,[w(t)-\overline{w}(t)]^T = \int_{t_0}^{t}\Phi(t,\tau)G(\tau)$$
$$\cdot\,E\{[w(\tau)-\overline{w}(\tau)]\,[w(t)-\overline{w}(t)]^T\}\,d\tau\,. \quad (11.4.12)$$

Examining (11.4.9) and (11.4.12), it is apparent that the only knowledge of $w(t)-\overline{w}(t)$ needed to determine the covariance matrix $X(t)$ is a weighted integral of its *autocorrelation*:

$$E[w(t)-\overline{w}(t)]\,[w(\tau)-\overline{w}(\tau)]^T\,. \quad (11.4.13)$$

Now, the purely random process must be considered as a limiting case of a random process with a short correlation time (see Section 11.3). Hence, we assume a simple correlation function for (11.4.13) which will allow us to perform this limit operation:

$$E[w(t) - \bar{w}(t)]\,[w(\tau) - \bar{w}(\tau)]^T = \chi(t)\exp\left(-\frac{|t-\tau|}{T}\right), \quad (11.4.14)$$

where T is a constant, very small compared to characteristic times of the transition matrix $\Phi(t,\tau)$, and $\chi(t)$ is the covariance of $w(t)$; i.e., we have

$$\chi(t) = E[w(t) - \bar{w}(t)]\,[w(t) - \bar{w}(t)]^T. \quad (11.4.15)$$

This correlation function is shown in Figure 11.4.2 for the case in which $w(t)$ is a scalar.

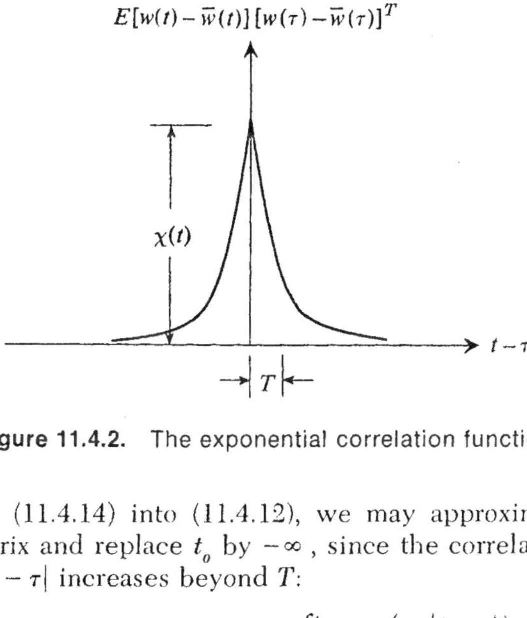

Figure 11.4.2. The exponential correlation function.

Substituting (11.4.14) into (11.4.12), we may approximate $\Phi(t,\tau) \cong I$ = unit matrix and replace t_0 by $-\infty$, since the correlation dies out rapidly as $|t-\tau|$ increases beyond T:

$$E[x(t) - \bar{x}(t)]\,[w(t) - \bar{w}(t)]^T \cong G(t)\chi(t)\int_{-\infty}^{t} \exp\left(-\frac{|t-\tau|}{T}\right)d\tau = TG(t)\chi(t). \quad (11.4.16)$$

Using (11.4.16) and its transpose in (11.4.9), we have

$$\boxed{\dot{X} = FX + XF^T + GQG^T, \qquad X(t_0) \text{ given},} \quad (11.4.17)$$

where

$$\boxed{Q(t) = 2T\chi(t).} \quad (11.4.18)$$

Thus, $Q(t)$ is a nonnegative definite matrix representing the *integral* of the correlation of the gaussian purely random process $w(t)$.

As far as (11.4.17) is concerned, we could write (11.4.14) as

$$\boxed{E[w(t) - \bar{w}(t)]\,[w(\tau) - \bar{w}(\tau)]^T = Q(t)\,\delta(t - \tau)\,,} \tag{11.4.19}$$

where $\delta(t - \tau)$ is the *Dirac delta function*, defined by

$$\delta(t) = \lim_{\epsilon \to 0} \delta_\epsilon(t)\,,$$
$$\delta_\epsilon(t) = \begin{cases} 0\,, & |t| > \epsilon\,, \\ \dfrac{1}{2\epsilon}\,, & |t| < \epsilon\,. \end{cases} \tag{11.4.20}$$

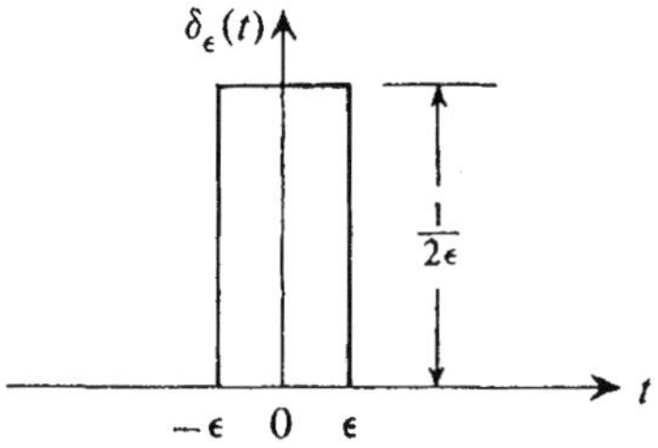

Figure 11.4.3. Definition of $\delta_\epsilon(t)$.

This definition implies that

$$\int_{t_1}^{t_2} \delta(t - \tau)\,d\tau = \begin{cases} 1\,, & t_1 < t < t_2\,, \\ 0\,, & t_1 > t \text{ or } t > t_2\,. \end{cases} \tag{11.4.21}$$

Comparing (11.4.19) with (11.4.14), the quantity $Q(t)$ may be regarded as

$$Q(t) = \lim_{T \to 0} \{2T\chi(t)\}\,, \tag{11.4.22}$$

where T is the correlation time of the random process and $\chi(t)$ is the covariance of the random process. Clearly, we have

$$\chi(t) \to \infty \qquad \text{as} \qquad T \to 0\,, \tag{11.4.23}$$

in such a way that the limit (11.4.22) is finite.

Thus, the gaussian purely random process is the limit of a gauss-markov process with very large covariance and very short correlation time.

When Q is a constant, $w(t)$ is called a *stationary* purely random process. The Fourier transform of the correlation (11.4.19) with respect to $(t - \tau)$ is simply Q; that is, the spectrum is white. For this reason, the gaussian purely random process is frequently called *white noise.*

Equations (11.4.6) and (11.4.17) are *linear differential equations* for the mean vector, $\bar{x}(t)$, and the covariance matrix, $X(t)$. They are *not* coupled and, hence, $\bar{x}(t)$ and $X(t)$ may be computed separately.

Correlation matrix of a gauss-markov random process. The correlation matrix of a random process $x(t)$ is defined as

$$\boxed{C(t,\tau) \triangleq E[x(t) - \bar{x}(t)][x(\tau) - \bar{x}(\tau)]^T .} \tag{11.4.24}$$

It is a deterministic function of the two variables t and τ. The covariance matrix, defined in (11.4.5), is a special case of (11.4.24):

$$X(t) \equiv C(t,t) . \tag{11.4.25}$$

Using (11.4.1) and (11.4.6), the value of $x(t+\tau) - \bar{x}(t+\tau)$ may be expressed in terms of $x(t) - \bar{x}(t)$ and $w(t') - \bar{w}(t')$, $t \leq t' \leq t+\tau$, through use of the transition matrix $\Phi(t+\tau,t')$:

$$\begin{aligned} x(t+\tau) - \bar{x}(t+\tau) &= \Phi(t+\tau,t)[x(t) - \bar{x}(t)] \\ &+ \int_t^{t+\tau} \Phi(t+\tau,t')G(t')[w(t') - \bar{w}(t')]\, dt' . \end{aligned} \tag{11.4.26}$$

Postmultiplying (11.4.26) by $[x(t) - \bar{x}(t)]^T$ and taking the expected value gives

$$\begin{aligned} C(t+\tau,t) = \Phi(t+\tau,t)X(t) + \int_t^{t+\tau} \Phi(t+\tau,t')G(t')E\{[w(t') - \bar{w}(t')] \cdot \\ [x(t) - \bar{x}(t)]\}\, dt' . \end{aligned} \tag{11.4.27}$$

However, the integrand in (11.4.27) is zero if $\tau > 0$ since $w(t)$ is a purely random process and is uncorrelated with $x(t)$. Therefore, we have

$$\boxed{C(t+\tau,t) = \Phi(t+\tau,t)X(t) \qquad \text{for} \qquad \tau \geq 0 .} \tag{11.4.28}$$

From the definition of the correlation matrix in (11.4.24), it is evident that

$$C(t,t+\tau) = C^T(t+\tau,t) . \tag{11.4.29}$$

Using (11.4.28) in (11.4.29), we have

$$\boxed{C(t,t+\tau) = X(t)\Phi^T(t+\tau,t) \qquad \text{for} \qquad \tau \geq 0 .} \tag{11.4.30}$$

Also, postmultiplying (11.4.26) by $[w(t) - \bar{w}(t)]^T G^T(t)$ and taking the expected value gives

$$E[x(t+\tau)-\bar{x}(t+\tau)][w(t)-\bar{w}(t)]^T G^T(t) = \begin{cases} \Phi(t+\tau,t)G(t)Q(t)G^T(t), & \tau>0, \\ 0, & \tau<0, \end{cases} \tag{11.4.31}$$

which is the cross-correlation between *input*, $G(t)w(t)$, and *output*, $x(t)$. Equations (11.4.31) or (11.4.28) may be used experimentally to find components of the transition matrix for an unknown linear system. To find all components of Φ, using (11.4.31), GQG^T must be nonsingular.

Statistically stationary processes. If we have $F(t) = F$, a constant matrix, in (11.4.1), then $\Phi(t+\tau,t)$ is just a function of τ; call it $\Phi(\tau)$, a stationary transition matrix. If, in addition, we have $G(t) = G$, a constant matrix, and $Q(t) = Q$, a constant matrix, it is possible (see Problem 2, Appendix B4) that $X(t) \to X$, a constant matrix, as $t \to \infty$. If so, we have $\dot{X} \to 0$, so that X may be determined from the linear relation

$$0 = FX + XF^T + GQG^T \Rightarrow X. \tag{11.4.32}$$

We say that such a process is *statistically stationary*; essentially, the random forcing function is balanced by the damping of the system as expressed in F.

The correlations of a statistically stationary process are also stationary; that is, $C(t+\tau,t)$ is just a function of τ, say $C(\tau)$. Therefore, we have, for $\tau \geq 0$,

$$C(\tau) = \Phi(\tau)X, \tag{11.4.33}$$

$$C(-\tau) = X\Phi^T(\tau) \tag{11.4.34}$$

and

$$E[x(t+\tau)-\bar{x}(t+\tau)][w(t)-\bar{w}(t)]^T G^T(t) = \begin{cases} \Phi(\tau)GQG^T, & \tau>0, \\ 0, & \tau<0. \end{cases} \tag{11.4.35}$$

A first-order gauss-markov process. Consider the special case of (11.4.1) in which x is a *scalar* and $F = -G = -a$, where a is a constant, $\bar{x}(t_0) = 0$, $X(t_0) = X_0$, $\bar{w}(t) = 0$, $Q(t) = q$, a constant:

$$\dot{x} = -a(x-w). \tag{11.4.36}$$

The transition matrix is stationary, $\Phi(t+\tau,t) = \Phi(\tau) = \exp(-a\tau)$, and we have

$$\dot{X} = -2aX + a^2 q, \qquad X(t_0) = X_0, \tag{11.4.37}$$

where $X(t)$ is the (scalar) variance of $x(t)$. The solution of (11.4.37) is

$$X(t) = X_0 e^{-2a(t-t_0)} + \tfrac{1}{2}qa[1-e^{-2a(t-t_0)}]. \tag{11.4.38}$$

The autocorrelation of $x(t)$ is given by

$$C(t+\tau,t)=\begin{cases} e^{-a\tau}X(t)\,, & \tau\geq 0\\ e^{a\tau}X(t+\tau)\,, & \tau<0\end{cases} \tag{11.4.39}$$

and the cross-correlation between $x(t)$ and $w(t)$ is given by

$$E[x(t+\tau)w(t)]=\begin{cases} qae^{-a\tau}\,, & \tau>0\,,\\ 0\,, & \tau<0\,.\end{cases} \tag{11.4.40}$$

As $a(t-t_0)\rightarrow\infty$, the process becomes statistically stationary, and we have

$$X(t)\rightarrow\tfrac{1}{2}qa\,, \tag{11.4.41}$$

$$C(\tau)\rightarrow\tfrac{1}{2}qae^{-a|\tau|}\,;\qquad \text{as}\quad t-t_0\rightarrow\infty\,. \tag{11.4.42}$$

The result (11.4.42) is the basis for the choice of the exponential correlation in (11.4.14). Note that the correlation time of this process is $T=1/a$ (see Figure 11.4.2).

As a becomes very large (that is, $T=1/a$ becomes very small), then $x(t)\rightarrow w(t)$; that is, $x(t)$ tends to become a gaussian purely random process with large variance, $\tfrac{1}{2}qa$, and short correlation time, $1/a$.

A second-order gauss-markov process.† Consider the stationary second-order dynamic system with a white noise forcing function:

$$\ddot{x}+2\zeta\omega\dot{x}+\omega^2x=\omega^2w(t)\,, \tag{11.4.43}$$

$$E[w(t)]=0\,,\qquad E[w(t)w(t')]=q\,\delta(t-t')\,. \tag{11.4.44}$$

In state vector nomenclature, (11.4.43) may be written as

$$\begin{bmatrix}\dot{x}_1\\ \dot{x}_2\end{bmatrix}=\begin{bmatrix}0\,, & 1\\ -\omega^2\,, & -2\zeta\omega\end{bmatrix}\begin{bmatrix}x_1\\ x_2\end{bmatrix}+\begin{bmatrix}0\\ \omega^2\end{bmatrix}w(t)\,. \tag{11.4.45}$$

The random initial conditions are assumed to be

$$E[x_1(0)]=E[x_2(0)]=0\,, \tag{11.4.46}$$

$$E\begin{bmatrix}x_1^2\,, & x_1x_2\\ x_1x_2\,, & x_2^2\end{bmatrix}_{t=0}=\begin{bmatrix}X_{11}(0)\,, & X_{12}(0)\\ X_{12}(0)\,, & X_{22}(0)\end{bmatrix}. \tag{11.4.47}$$

The transition matrix for (11.4.45) is easily obtained:

$$\Phi(t)=e^{-\zeta\omega t}\begin{bmatrix}\cos\beta t+\dfrac{\zeta\omega}{\beta}\sin\beta t\,, & \dfrac{1}{\beta}\sin\beta t\\ -\dfrac{\omega^2}{\beta}\sin\beta t\,, & \cos\beta t-\dfrac{\zeta\omega}{\beta}\sin\beta t\end{bmatrix}, \tag{11.4.48}$$

where $\beta=\omega\sqrt{1-\zeta^2}$.

†See M. C. Wang and G. E. Uhlenbeck in Wax (1954), p. 335.

From Problem 1 of this section and Equation (23), Appendix A4, we have

$$X(t) = \Phi(t)X(0)\Phi^T(t) + \int_0^t \Phi(t-\tau)gqg^T\Phi^T(t-\tau)\,d\tau\,,$$

$$\text{where} \qquad g = \begin{bmatrix} 0 \\ \omega^2 \end{bmatrix}. \tag{11.4.49}$$

Substituting (11.4.48) into (11.4.49) and integrating yields

$$\begin{aligned} X_{11}(t) = e^{-2\zeta\omega t}\Bigg[&\left(\cos\beta t + \frac{\zeta\omega}{\beta}\sin\beta t\right)^2 X_{11}(0) \\ &+ \frac{2}{\beta}\sin\beta t\left(\cos\beta t + \frac{\zeta\omega}{\beta}\sin\beta t\right)X_{12}(0) + \frac{1}{\beta^2}\sin^2\beta t\, X_{22}(0)\Bigg] \\ &+ \frac{q\omega}{4\zeta}\left[1 - \frac{e^{-2\zeta\omega t}}{\beta^2}(\omega^2 - \zeta^2\omega^2\cos 2\beta t + \zeta\omega\beta\sin 2\beta t)\right], \end{aligned} \tag{11.4.50}$$

$$\begin{aligned} X_{12}(t) = e^{-2\zeta\omega t}\Bigg[&-\frac{\omega^2}{\beta}\sin\beta t\left(\cos\beta t + \frac{\zeta\omega}{\beta}\sin\beta t\right)X_{11}(0) \\ &+ \left(\cos^2\beta t - \frac{1+\zeta^2}{1-\zeta^2}\sin^2\beta t\right)X_{12}(0) \\ &+ \frac{1}{\beta}\sin\beta t\left(\cos\beta t - \frac{\zeta\omega}{\beta}\sin\beta t\right)X_{22}(0)\Bigg] \\ &+ \frac{q\omega^2}{2(1-\zeta^2)}e^{-2\zeta\omega t}\sin^2\beta t\,, \end{aligned} \tag{11.4.51}$$

$$\begin{aligned} X_{22}(t) = e^{-2\zeta\omega t}\Bigg[&\frac{\omega^2}{1-\zeta^2}(\sin^2\beta t)X_{11}(0) \\ &- \frac{2\omega^2}{\beta}\sin\beta t\left(\cos\beta t - \frac{\zeta\omega}{\beta}\sin\beta t\right)X_{12}(0) \\ &+ \left(\cos\beta t - \frac{\zeta\omega}{\beta}\sin\beta t\right)^2 X_{22}(0)\Bigg] \\ &+ \frac{q\omega^3}{4\zeta}\left[1 - \frac{e^{-2\zeta\omega t}}{\beta^2}(\omega^2 - \zeta^2\omega^2\cos 2\beta t - \zeta\omega\beta\sin\beta t)\right]. \end{aligned} \tag{11.4.52}$$

As $\omega t \to \infty$, the covariance $X_{12}(t) \to 0$ if we have $\zeta > 0$ and the variances

$$X_{11}(t) \to (q\omega/4\zeta)\,, \qquad X_{22}(t) \to (q\omega^3/4\zeta) \qquad \text{as} \qquad \omega t \to \infty\,, \qquad \zeta > 0\,. \tag{11.4.53}$$

In Figure 11.4.4, plots of X_{11}, X_{22}, and X_{12} for $\zeta = 0.2$ are shown for

$X_{11}(0) = X_{12}(0) = X_{22}(0) = 0$. In Figure 11.4.5, two-sigma probability ellipses for different values of t are shown superimposed on the expected trajectory for the system starting in the known state $x_1(0) = 1.0$, $x_2(0) = 0$.

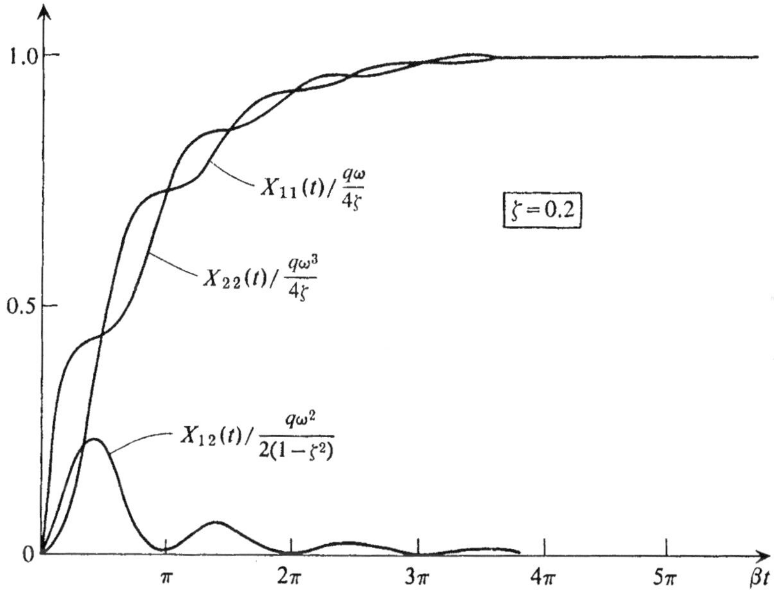

Figure 11.4.4. Covariance matrix history of an underdamped second-order gauss-markov process.

Note that in this case the two-sigma probability ellipse starts out strongly elongated in the x_2 direction. This is so since X_{22} grows faster than X_{11} at the beginning (reflecting the physical fact that velocity is more susceptible to acceleration noises than position). As time increases, the ellipse grows in size and becomes more uniform in shape and skewed. At time $t = \pi/\beta$, it becomes a circle. Thereafter, it skews till it becomes a circle again at $t = 2\pi/\beta$. This process is then repeated until, finally, the steady state values of X_{11}, X_{22}, and X_{12} are reached.

In the steady state, the trajectory of the dynamic system is expected to remain in this circle 87 per cent of the time. In Figure 11.4.5, an actual random trajectory through analog simulation is shown for such a system. Visual inspection lends credence to the analytical result.

Random forcing functions other than white noise. The result (11.4.42) shows that the output of a stationary first-order gauss-

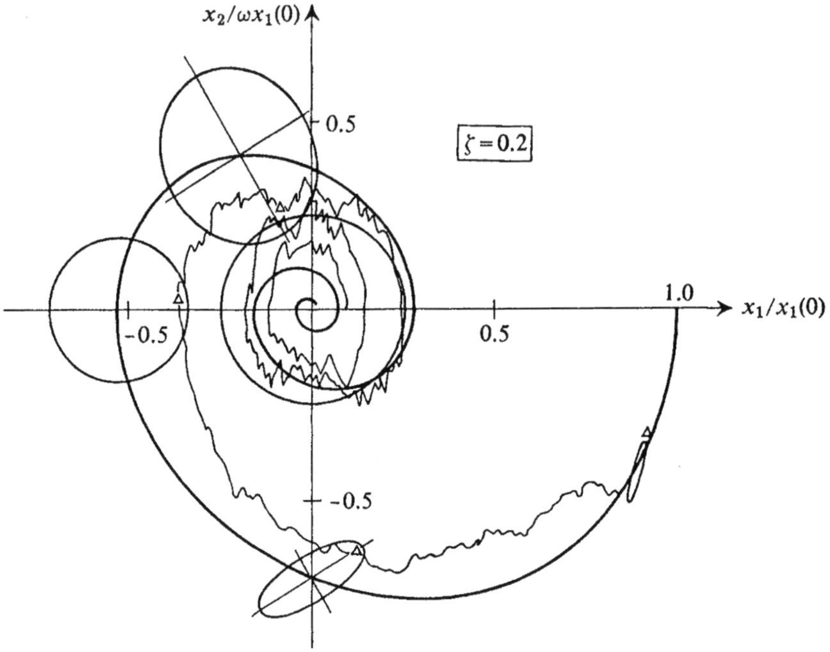

Figure 11.4.5. State space trajectory of $x(t)$ and $\bar{x}(t)$ for an underdamped second-order gauss-markov process.

markov process has an *exponential time correlation* with characteristic time equal to $1/a$. This points the way to handling problems in which the input noise is random but *not* white. By setting up shaping filters of first or higher order (stationary or nonstationary), with white noise inputs, the outputs yield *time-correlated or colored noise*, i.e., a gauss-markov process. By suitable choice of coefficients in the shaping filter, almost any noise correlation function can be approximated well enough for practical purposes. This is illustrated in Figure 11.4.6. It is clear that analysis of systems such as those in Figure 11.4.6 offer no additional difficulty, conceptually. All that is needed is to redefine the state of the entire system as $x = \left[\begin{smallmatrix} x_1 \\ x_2 \end{smallmatrix}\right]$. The enlarged state x is a gauss-markov process.

Problem 1. Show that the mean-value vector and the covariance matrix may be expressed in terms of the transition matrix $\Phi(t,\tau)$ as follows:

$$\bar{x}(t) = \Phi(t,0)\bar{x}(0) + \int_0^t \Phi(t,\tau)G(\tau)\overline{w}(\tau)\, d\tau\,,$$

$$X(t) = \Phi(t,0)X(0)\Phi^T(t,0) + \int_0^t \Phi(t,\tau)G(\tau)Q(\tau)G^T(\tau)\Phi^T(t,\tau)\, d\tau\,,$$

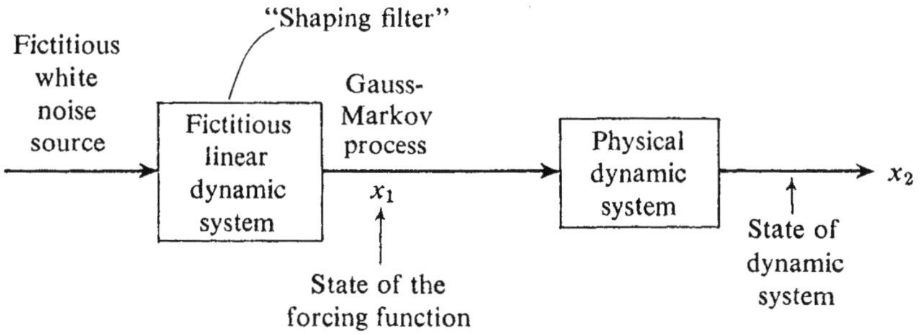

Figure 11.4.6. Time-correlated random forcing function produced by a shaping filter.

where

$$\frac{d}{d\tau}\Phi^T(t,\tau) = -F^T(\tau)\Phi^T(t,\tau)\,; \qquad \Phi(t,t) = I = \text{unit matrix.}$$

Problem 2. Show that mean value and variance of a scalar quantity. $y(t)$, which is a linear function of the vector quantity $x(t)$, say,

$$y(t) = a^T x(t)\,,$$

can be found from

$$\bar{y}(t) = \lambda^T(0)\bar{x}(0) + \int_0^t \lambda^T(\tau)G(\tau)\overline{w}(\tau)\,d\tau$$

and

$$E[y(t) - \bar{y}(t)]^2 = \lambda^T(0)X(0)\lambda(0) + \int_0^t \lambda^T(\tau)G(\tau)Q(\tau)G^T(\tau)\lambda(\tau)\,d\tau\,,$$

where

$$\frac{d\,\lambda(\tau)}{d\tau} = -F^T(\tau)\lambda(\tau)\,; \qquad \lambda(t) = a\,,$$

and x, F, G, w; and Q are as defined in this section. Note that finding $\lambda(\tau)$ requires n integrations if x is an n-vector, whereas finding $X(t)$ from (11.4.17) requires $\frac{1}{2}[n(n+1)]$ integrations, and finding $\Phi(t,\tau)$ requires n^2 integrations.

Problem 3. *A poisson process.* A scalar signal $u(t)$ is either $+u_0$ or $-u_0$ with a random time interval between changes. The average number of changes (from u_0 to $-u_0$ or vice versa) per unit time is ν. The probability of exactly n changes in a time interval t is

$$p_n(t) = \frac{(\nu|t|)^n}{n!}\,e^{-\nu|t|}\,.$$

The mean value $\bar{u}(t)$ is obviously zero. Show that the correlation is stationary and given by

$$E[u(t+\tau)u(t)] = u_0^2 e^{-2\nu|\tau|} .$$

Note that this is *not* a gaussian process, but it is markovian.

Problem 4. Often it is only possible to measure linear combinations of the elements of the state vector, corrupted by additive white noise:

$$z(t) = H(t)x(t) + v(t) ,$$

$$E[v(t)] = 0 , \qquad E[v(t)v^T(\tau)] = R(t)\,\delta(t-\tau) .$$

Show that the autocorrelation of z is

$$E[z(t+\tau)z^T(t)] = H(t+\tau)C(t+\tau,t)H^T(t) + R(t)\,\delta(\tau) .$$

Assume that $x(t)$ and $v(t)$ are independent random processes. Is $z(t)$ markovian?

Problem 5.† Consider the second-order underdamped system of Equations (11.4.43 and 11.4.44). Show that, in the statistically stationary state, the correlations are given by

$$\begin{bmatrix} C_{11}, C_{12} \\ C_{21}, C_{22} \end{bmatrix} \triangleq E\begin{bmatrix} x_1(t+\tau)x_1(t), x_1(t+\tau)x_2(t) \\ x_2(t+\tau)x_1(t), x_2(t+\tau)x_2(t) \end{bmatrix}$$

$$= \frac{\omega q e^{-\zeta\omega|\tau|}}{4\zeta} \begin{bmatrix} \cos\beta|\tau| + \dfrac{\zeta\omega}{\beta}\sin\beta|\tau| , & \dfrac{\omega^2}{\beta}\sin\beta\tau \\ -\dfrac{\omega^2}{\beta}\sin\beta\tau , & \omega^2\cos\beta|\tau| - \dfrac{\zeta\omega^3}{\beta}\sin\beta|\tau| \end{bmatrix} ,$$

$$E\begin{bmatrix} x_1(t+\tau)w(t) \\ x_2(t+\tau)w(t) \end{bmatrix} = \begin{cases} q\omega^2 e^{-\zeta\omega\tau} \begin{bmatrix} (1/\beta)\sin\beta\tau \\ \cos\beta\tau - (\zeta\omega/\beta)\sin\beta\tau \end{bmatrix} , & \tau > 0 ; \\ 0 , \quad \tau < 0 . \end{cases}$$

show also that

$$E[x_2(t+\tau) + \omega x_1(t+\tau)]\,[x_2(t) + \omega x_1(t)] = \frac{\omega^3 q e^{-\zeta\omega|\tau|}}{2\zeta}\cos\beta|\tau| .$$

The correlations C_{11}, C_{12}, and C_{22} are shown in Figure 11.4.7 for $\zeta = 0.2$.

Problem 6. Einstein, in 1905, gave a solution to the *Brownian motion* problem [see, e.g., Uhlenbeck and Ornstein in Wax (1954)]. He

†See M. C. Wang and G. E. Uhlenbeck in Wax (1954).

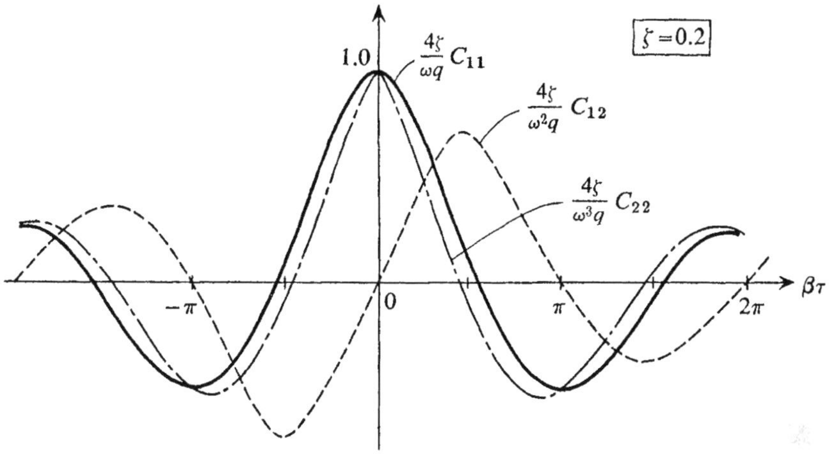

Figure 11.4.7. Statistically stationary correlation function of an underdamped second-order gauss-markov process.

assumed that the visible particles were large compared to the mean free path of the molecules of the fluid, so that the equations of motion of a visible particle would be well approximated by

$$m\dot{v} = -cv + f(t), \qquad \dot{x} = v,$$

where

m = mass of particle, $\quad v$ = velocity of particle,
c = Stokes' viscous force coefficient (a constant),
f = random force due to collision with molecules,
x = position of particle.

We can write three such equations in three mutually perpendicular directions, but it is clear that they are statistically independent from considerations of isotropy.

The mean time between collisions is very short, and $f(t)$ is well approximated by gaussian white noise; i.e., we have

$$E[f(t)] = 0, \qquad E[f(t)f(t')] = q\delta(t - t'), \qquad q = \text{const.}$$

Assuming that we have $E[x(0)] = E[v(0)] = E[x(0)]^2 = E[v(0)]^2 = E[x(0)v(0)] = 0$, show that

$$E[v(t)]^2 = (q/2cm)\,(1 - e^{-(2c/m)t}),$$

$$E[v(t)x(t)] = \frac{q}{2c^2}\,(1 - e^{-(c/m)t})^2,$$

$$E[x(t)]^2 = \frac{qt}{c^2} - \frac{qm}{2c^3}\,(3 - e^{-(ct/m)})\,(1 - e^{-ct/m}).$$

Note, as Einstein did, that for $t >> m/c$, $E[x(t)]^2 \to qt/c^2$, which is independent of the mass of the particle.

Problem 7. A large rocket is fired vertically. The altitude change of interest is small compared to the radius of the earth, so we may approximate the gravitational force per unit mass, g, as constant. The weight is large compared to the aerodynamic drag, so we will neglect the latter. Thus, the approximate equations of motion are

$$\dot{m} = -f/c\,, \qquad m\dot{v} = f - mg\,, \qquad \dot{x} = v\,,$$

where

m = mass of rocket, v = velocity of rocket,
f = thrust of rocket, c = specific impulse of rocket (a constant),
x = position of rocket.

Find the mean values and covariances of v, m, x as a function of time, given that

$$E[v(0)] = E[x(0)] = 0\,, \qquad E[m(0)] = m_0\,,$$

$$E[v(0)]^2\,, E[x(0)]^2\,, E[m(0) - m_0]^2 \quad \text{are given,}$$

$$E[v(0)x(0)] = E[v(0)\,(m(0) - m_0)] = E[x(0)\,(m(0) - m_0)] = 0\,,$$

$$E[f(t)] = f_0 = \text{const}, \qquad E[(f(t) - f_0)(f(t') - f_0)] = q\delta(t - t')\,,$$

$$q = \text{const.}$$

Problem 8. Another common type of random forcing function is a random constant that may be thought of as a forcing function with an infinite correlation time. Consider the effect of such a forcing function on an underdamped second-order system:

$$\ddot{x} + 2\zeta\omega\dot{x} + \omega^2 x = w(t)\,,$$

where

$$E[w(t)] = 0\,, \qquad E[w(t)]^2 = q = \text{const},$$

$$\zeta < 1\,, \qquad E[x(0)] = E[\dot{x}(0)] = 0\,,$$

$$E[x(0)]^2\,, \qquad E[\dot{x}(0)]^2 \qquad \text{given}, \qquad E[x(0)\dot{x}(0)] = 0\,.$$

[HINT: Think of w as another state variable, governed by $\dot{w} = 0$, with random initial conditions $E[w(0)] = 0\,, E[w(0)]^2 = q$.]

11.5 Approximation of a gauss-markov process by a gauss-markov sequence

In applications, it is usually necessary to find the mean vector $\bar{x}(t)$ and the covariance matrix $X(t)$ by *numerical integration*, using computers. If a digital computer is used, the numerical procedure essentially approximates the continuous process by a multistage process.

If the time interval (t_o, t_f) is divided into N short intervals of length $\Delta t = (t_f - t_o)/N$, we desire that

$$x(t)|_{t-t_o=k\Delta t} \cong x(k) \qquad \text{for} \qquad k = 0,1,2,\ldots,N. \tag{11.5.1}$$

A simple approximation to $\dot{x} = F(t)x + G(t)w(t)$ is

$$x(t + \Delta t) = [I + F(t)\Delta t]x(t) + G(t)w(t)\,\Delta t. \tag{11.5.2}$$

In order that this correspond to the multistage approximation

$$x(k+1) = \Phi(k)x(k) + \Gamma(k)w(k), \tag{11.5.3}$$

we must place

$$\Phi(k) = [I + F(t)\,\Delta t]_{t-t_o=k\Delta t}, \tag{11.5.4}$$

$$\Gamma(k) = G(t)\,\Delta t|_{t-t_o=k\Delta t}, \tag{11.5.5}$$

and imagine that $w(t)$ is a piece-wise constant function (staircase function):

$$w(t) = w(k) \qquad \text{for} \qquad k\,\Delta\,t \leq t - t_o < (k+1)\,\Delta t. \tag{11.5.6}$$

However, it is *not* so obvious what we should take as $\chi(k)$, the covariance of $w(k)$, in order to approximate the continuous white-noise process $w(t)$, given $Q(t)$, where

$$E[w(t) - \overline{w}(t)]\,[w(\tau) - \overline{w}(\tau)]^T = Q(t)\,\delta(t - \tau), \tag{11.5.7}$$

$$E[w(k) - \overline{w}(k)]\,[w(\ell) - \overline{w}(\ell)]^T = \chi(k)\,\delta_{k\ell}, \tag{11.5.8}$$

and

$$\delta_{k\ell} = \begin{cases} 1, & k = \ell, \\ 0, & k \neq \ell \end{cases} \quad \text{(the Kronecker delta function).}$$

The choice of $\chi(k)$ obviously depends strongly on the size of Δt and is dictated by our desire to make

$$X(t)|_{t-t_o=k\Delta t} \cong X(k), \qquad k = 0,1,2,\ldots,N. \tag{11.5.9}$$

A simple approximation to $\dot{X} = FX + XF^T + GQG^T$ is

$$X(t + \Delta t) - X(t) = F(t)X(t)\,\Delta t + X(t)F^T(t)\,\Delta t + G(t)Q(t)G^T(t)\,\Delta t. \tag{11.5.10}$$

The corresponding expression for $X(k + 1)$, from Equation (11.2.9), is

$$X(k+1) = \Phi(k)X(k)\Phi^T(k) + \Gamma(k)\chi(k)\Gamma^T(k). \tag{11.5.11}$$

Substituting (11.5.4), (11.5.5), and (11.5.9) into (11.5.11), we have

$$X(t+\Delta t) - X(t) = F(t)X(t)\,\Delta t + X(t)F^T(t)\,\Delta t + F(t)X(t)F^T(t)\,(\Delta t)^2$$
$$+ G(t)\,[\chi(k)\,\Delta t]G^T(t)\,\Delta t\,. \quad (11.5.12)$$

Now, in order that (11.5.12) correspond to (11.5.10) to first order in Δt, clearly it is necessary that

$$\boxed{Q(t)|_{t-t_0=k\Delta} = \chi(k)\,\Delta t\,.} \quad (11.5.13)$$

In other words, $\chi(k)$ is inversely proportional to the time increment Δt.

Comparing (11.5.13) with Equation (11.4.18), we notice a difference of a factor of two. This comes about because of the sharp-peaked nature of the exponential correlation used in Section 11.4 as opposed to the flat-topped nature of the piece-wise constant function used here in Equation (11.5.6).

11.6 State variables and the markov property

There is a strong similarity between the *state assumption* for deterministic processes and the *markov assumption* for stochastic processes. We show this below:

Deterministic Process	Random Process
1. At any time t there exists a finite-dimensional vector x, which, when specified, makes the future of the process independent of the past and vice versa. This is the *state* property. The set of numbers are the *state* variables.	1. At any time t there exists a probability-density function of a finite-dimensional vector x, which, when specified, makes the future probabilistic behavior of the process independent of the past and vice versa. This is the *markov* property. The density function is the *state of the random process.*
2. The dependence of the future on the present is expressed by specifying a difference (differential) equation that governs the transition (differential transition) from one instant to another; i.e., we have $x(t+1) = f(x(t), t)$ and an initial state $x(0)$.	2. The dependence of the future probabilistic behavior on the present probabilistic density is expressed by specifying a transitional mechanism (differential transitional mechanism) $p(x(t+1)) = \int p(x(t+1)/x(t))p(x(t))\,dx(t)$ and an initial density function $p(x(0))$.
3. A special but important class is when the equation is linear.	3. A special but important class is when the transitional mechanism and the initial density are gaussian.

In fact, a gauss-markov sequence (process) can be represented by a white gaussian random sequence (process) driving a linear discrete (continuous) dynamic system whose initial state is gaussian. The state of the process in the future, and at present, can be related by a differential transition mechanism rather than the general integral relationship above. From the computational viewpoint, the former representation is far more convenient. Furthermore, the state of the process is only finite dimensional, since we need only the mean and the covariances. We emphasize that Equations (11.4.6) and (11.4.17), which govern the state process, are *deterministic* and thus are amenable to deterministic treatment. This is the *key to the analysis of a stochastic process, namely, to characterize it by quantities which are themselves deterministically related and then to operate on these quantities.*

Problem 1. *Multiple choice:* A random process may be

(i) gaussian but not necessarily markov,
(ii) markov but not necessarily gaussian,
(iii) stationary but not necessarily gaussian,
(iv) gauss-markov but necessarily nonstationary.
(a) All statements are true.
(b) (i), (ii), and (iii) are true but (iv) is false.
(c) (i) and (ii) are true but (iii) and (iv) are false.
(d) All statements are false.
(e) Only (i) is true.

Problem 2. *Multiple choice:* Consider the system.

$$\dot{x} = Fx + w ,$$

where

$$E[w(t)] = 0 , \qquad E[w(t)\, w^T(\tau)] = Q\, \delta(t-\tau) ,$$

and $w(t)$ is gaussian.

Let

$$F = \begin{bmatrix} 0 & 1 \\ -\frac{3}{4} & -2 \end{bmatrix}, \qquad Q = \begin{bmatrix} 1 & 0 \\ 0 & 1 \end{bmatrix}.$$

Then the covariance of $x(t)$ as $t \to \infty$ is

(i) infinite,
(ii) zero,
(iii) finite and positive definite,
(iv) finite but semi definite.

Problem 3. *Multiple choice:* Consider the two dynamic systems shown in Figure 11.6.1(a) and (b).

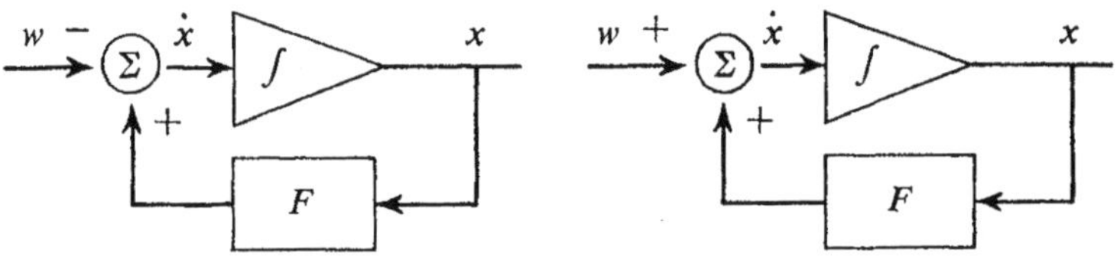

(a) (b)

Figure 11.6.1. Gauss-markov processes with opposite signs on same random forcing function.

where w is zero mean white gaussian noise. Then the covariance equation of $x(t)$ is

(i) identical for both systems,
(ii) different by a sign,
(iii) different only in off-diagonal terms of the covariance equation,
(iv) entirely different.

Problem 4. *Multiple choice:* Consider the same problem as 3 (i.e., *covariance of* $x(t)$) but with w now a correlated zero mean gaussian markov process with known correlation matrix. Initially $E(w(o)x(o)^T) = 0$. Choose your answer again from (i) to (iv).

Problem 5. Consider a scalar markov process, x_t; show that

(i) $p(x_t/x_{t+2}, x_{t+1}, x_{t-1}, x_{t-2}) = p(x_t/x_{t+1}, x_{t-1})$,
(ii) in general

$$p(x_t/x_{t+N}, \ldots, x_{t+1}, x_{t-1}, \ldots, x_{t-M}) = p(x_t/x_{t+1}, x_{t-1}).$$

Problem 6. Consider $\dot{x} = Fx + gw$ and $z = h^Tx$, where w is $N(0,1)$ and white. Determine $E(z(t+\tau)w(t)) \triangleq i(t+\tau,t)$. What is $i(t+\tau,t)$?

Problem 7. In Problem 6 let $F = \begin{bmatrix} -1 & 0 \\ 0 & -2 \end{bmatrix} g = \begin{bmatrix} 1 \\ 1 \end{bmatrix}$ determine the steady state $C(\tau)$.

Problem 8. Given $x_{t+1} = \Phi x_t + w_t$,
where w_t is white but *not* gaussian. What is the mean and covariance equation for x_t?

11.7 Processes with independent increments

From the above discussion, we are naturally led to conjecture that a more general class of stochastic processes can be represented by the output of a *nonlinear* dynamic system driven by some kind of noise. To describe such processes, it is convenient to introduce the concept of a *process with independent increments.* A process $\eta(t)$ is a stochastic process with independent increments if we have

$$p[\eta(t_2) - \eta(t_1), \eta(t_4) - \eta(t_3), \eta(t_6) - \eta(t_5), \ldots]$$
$$= p[\eta(t_2) - \eta(t_1)]p[\eta(t_4) - \eta(t_3)]p[\eta(t_6) - \eta(t_5)] \quad (11.7.1)$$

for all $t_6 > t_5 > t_4 > t_3 > t_2 > t_1$.

Two fundamental processes of independent increments are:

(a) *Wiener (or Brownian motion) process.* This is defined by the fact that $\eta(t)$ is gaussian and

$$E[\eta(t + dt) - \eta(t)] \triangleq E[d\eta(t)] = 0, \qquad E[d\eta(t)^2] = Q(t)\,dt.$$

In terms of the notation used previously in this chapter, we may represent $\eta(t)$ as

$$\dot{\eta} = w,$$

where w is white gaussian noise with zero mean and correlation $Q(t)\,\delta(t - \tau)$.

(b) *Poisson process.* This is represented by the fact that the probability that one event occurs in time dt is equal to $a\,dt$ and independent of any prior event. Once an event occurs, the amplitude of $\eta(t)$ takes a value according to a density function $p(\eta)$.

Complicated stochastic processes may then be represented by

$$dx = f(x)\,dt + g(x)\,d\eta, \qquad (11.7.2)$$

where $\eta(t)$ is a process of independent increments given by (a) or (b).† Hopefully, we can develop another differential relationship (similar to the mean and covariance equation) that is deterministic and that governs the transition of the probability density of the process. It turns out that this is, indeed, possible under suitable conditions.

For example, if we define $p(x,t/x_0,t_0)$ as the probability density of being in the state x at time t, given the system in state x_0 at t_0, we have, under suitable conditions (Wonham, 1963)††, for the Wiener process,

$$\frac{\partial p}{\partial t} = \sum_i \frac{\partial}{\partial x_i}(pf_i) + \frac{1}{2}\sum_{i,j}\frac{\partial^2}{\partial x_i \partial x_j}(pg_ig_j), \qquad p(x,t_0/x_0,t_0) = \delta(x - x_0), \qquad (11.7.3)$$

which is known as the *forward Kolmogorov* or *Fokker-Planck* equation.

In principle, partial differential equations for the probability density $p(x,t/x_0,t_0)$ may be constructed for other $\eta(t)$ processes in (11.7.2). Computationally, there is little we can do when the number of variables in x exceeds two or three.

---

†If we insisted on writing (11.7.2) as $\dot{x} = f(x,t) + g(x,t)\,(d\eta/dt)$, it would be necessary to define many different kinds of white noise corresponding to different processes of independent increments.

††W. M. Wonham, "Stochastic Problems in Optimal Control," *RIAS Technical Report* 63-14 May 1963; also IEEE Convention Records, 1963.

Optimal filtering and prediction

12

12.1 Introduction

In the previous chapters we have seen that optimal control of a dynamic system requires a knowledge of the state of that system. In practice, the individual state variables cannot be determined exactly by direct measurements; instead, we usually find that the measurements that can be made are *functions* of the state variables and that these measurements contain random errors. The system itself may also be subjected to random disturbances. In many cases, we have too few measurements at a given time to infer the state variables at that time, even if the measurements were quite precise. On occasion, we have more than enough measurements, so that the state variables are overdetermined. Thus, we are faced with the problem of making good *estimates* of the state variables from either too few or too many measurements, which are imprecise and only functions of the state variables, knowing, too, that the system itself is subjected to random disturbances.

If we believe that we understand the dynamics of the ideal system (with perfect and complete measurements and no random disturbances), and if we believe that we have some knowledge of the degree of uncertainty in the measurements and of the degree of intensity of the random disturbances to the system, then, on the basis of all the measurements up to the present time, we can determine the most likely values of the state variables. The process of determining these most likely values is called smoothing, filtering, or prediction, depending on whether we are finding past, present, or future values of the state variables, respectively. In this chapter, the filtering and prediction problems are treated. The results will be directly applicable to stochastic control problems. The smoothing problem is dealt with in Chapter 13.

12.2 Estimation of parameters using weighted least-squares

Suppose that we wish to estimate the n-component state vector x of a static system using the p-component measurement vector, z, containing random errors, v, which are independent of the state x, where

$$z = Hx + v \tag{12.2.1}$$

and

$$H = \text{a known} \quad (p \times n)\text{-matrix},$$

$$E(v) = 0\,, \tag{12.2.2}$$

$$E(vv^T) = R\,, \text{a known} \quad (p \times p) \quad \text{positive matrix.} \tag{12.2.3}$$

Let us also suppose that we had an estimate of the state, before the measurements were made, which we will call $\bar{x}$, where

$$E[(x - \bar{x})(x - \bar{x})^T] = M\,, \text{a known} \quad (n \times n) \quad \text{positive matrix.} \tag{12.2.4}$$

One very reasonable estimate of x, taking into account the measurements, z, is the *weighted-least-squares estimate*, which we shall call $\hat{x}$; for this estimate we choose $\hat{x}$ as the value of x that minimizes the quadratic form

$$J = \tfrac{1}{2}[(x - \bar{x})^T M^{-1}(x - \bar{x}) + (z - Hx)^T R^{-1}(z - Hx)]\,. \tag{12.2.5}$$

Note that the weighting matrixes, M^{-1} and R^{-1}, are the inverse matrices of the prior expected values of $(x - \bar{x})(x - \bar{x})^T$ and $(z - Hx)(z - Hx)^T$, respectively. With this choice of weighting matrices, the weighted-least-squares estimate is identical to the conditional-expected-value estimate assuming gaussian distributions of x and v; that is, we have $\hat{x} = E(x/z)$, which, in turn, is identical to the maximum-likelihood or minimum-variance estimate (see Problems 1 through 6 at the end of this section).

To determine $\hat{x}$, consider the differential of (12.2.5):†

$$dJ = dx^T[M^{-1}(x - \bar{x}) - H^T R^{-1}(z - Hx)]\,. \tag{12.2.6}$$

In order that $dJ = 0$ for arbitrary dx^T, the coefficient of dx^T in (12.2.6) must vanish:

$$\begin{aligned}(M^{-1} + H^T R^{-1} H)\hat{x} &= M^{-1}\bar{x} + H^T R^{-1} z \\ &= (M^{-1} + H^T R^{-1} H)\bar{x} + H^T R^{-1}(z - H\bar{x})\end{aligned}$$

or

$$\hat{x} = \bar{x} + PH^T R^{-1}(z - H\bar{x})\,, \tag{12.2.7}$$

†Another elementary derivation of $\hat{x}$ can be made by completing the square in Equation (12.2.5), since $\hat{x}$ can then be determined by inspection, see Problem 3.

where

$$P^{-1} = M^{-1} + H^T R^{-1} H \,. \tag{12.2.8}$$

The quantity P in Equation (12.2.8) is the covariance matrix of the error in the estimate $\hat{x}$; that is, we have

$$P = E[(\hat{x} - x)(\hat{x} - x)^T] \,. \tag{12.2.9}$$

To show this, let

$$e = \hat{x} - x = \text{error in the estimate.}$$

Then we have

$$e = \bar{x} - x + \hat{x} - \bar{x} = \bar{x} - x + K[v - H(\bar{x} - x)] \,,$$

where

$$K = PH^T R^{-1} \,,$$

or

$$e = (I - KH)(\bar{x} - x) + Kv \,. \tag{12.2.10}$$

Since $\bar{x} - x$ and v are independent, it follows from (12.2.10) that

$$E(ee^T) = (I - KH)M(I - KH)^T + KRK^T \,. \tag{12.2.11}$$

Premultiplying (12.2.8) by P and postmultiplying by M, we have

$$M = P + PH^T R^{-1} HM$$

or

$$(I - KH)M = P \,. \tag{12.2.12}$$

Substituting (12.2.12) into (12.2.11), we have

$$E(ee^T) = P - PH^T K^T + KRK^T = P - PH^T R^{-1} HP + PH^T R^{-1} HP$$

or

$$E(ee^T) = P \,. \tag{12.2.13}$$

Since M is the error covariance matrix *before* measurement, it is apparent from (12.2.8) that P, the error covariance matrix *after* measurement is never larger than M, since $H^T R^{-1} H$ is at least a positive-semidefinite matrix.† Thus, the act of measurement, *on the average*, decreases (more precisely, it never increases) the uncertainty in our knowledge of the state x.

Another noteworthy property of the estimate is the fact that

†The matrix P is said to be smaller than M if, for all nonzero vectors x, the scalar quantity $x^T P x < x^T M x$.

$$E(e\hat{x}^T) = E[(I - KH)(\bar{x} - x) + Kv][\bar{x} - KH(\bar{x} - x) + Kv]^T \\ = -(I - KH)MH^TK^T + KRK^T = -PH^TK^T + KRK^T = 0\,; \tag{12.2.14}$$

that is, the estimate and the error of the estimate are uncorrelated. In the case where x and v are gaussian, this implies that $\hat{x}$ and e are independent (see Chapter 10). We may regard (12.2.14) as a *definition* of an optimal estimate in the sense that it contains all the information in z, and no improvement in e can be obtained by knowledge of $\hat{x}$ or z.

From (12.2.5), the prior expected value of J is $(n + p)/2$. This is easily shown, since (12.2.5) can be written as

$$J = \tfrac{1}{2}tr[M^{-1}(x - \bar{x})(x - \bar{x})^T] + \tfrac{1}{2}tr[R^{-1}(z - Hx)(z - Hx)^T]\,,$$

where $tr(\quad)$ means "trace of ( )"; i.e., the sum of the elements on the principal diagonal of ( ). Then we have

$$E(J) = \tfrac{1}{2}tr\{M^{-1}E[(x - \bar{x})(x - \bar{x})^T]\} + \tfrac{1}{2}tr\{R^{-1}E[(z - Hx)(z - Hx)^T]\} \\ = \tfrac{1}{2}tr(M^{-1}M) + \tfrac{1}{2}tr(R^{-1}R)\,.$$

Now $M^{-1}M$ is an $(n \times n)$ identity matrix and $R^{-1}R$ is a $(p \times p)$ identity matrix, so we have $tr(M^{-1}M) = n$, $tr(R^{-1}R) = p$. Hence we have

$$E(J) = \tfrac{1}{2}(n + p)\,.$$

The prior knowledge of M and R is sometimes vague and uncertain. As a check, after the estimation process has been completed, the *actual* value of J in Equation (12.2.5) should be computed with $x = \hat{x}$; call it J_o. The prior expected value of J_o can be calculated and it is found to be $E(J_o) = p/2$. If the actual J_o is not close to $p/2$, the elements of M and R should be multiplied by the scale factor $J_o/p/2$, which adjusts the value of $J_o/p/2$. Note that this also multiplies P by the same scale factor [see Equation (12.2.8)], but it does *not* change $\hat{x}$ [in Equation (12.2.7) the scale factor appears in both P and R and, hence, cancels out]. For this reason it is only necessary to establish the *relative* magnitude of the elements in M and R; the scale factor $J_o/p/2$ can be applied *post facto* to obtain values of P, M, and R consistent with the scatter in the data.

Many estimation problems are *nonlinear* rather than linear; i.e., instead of (12.2.1), we have

$$z = h(x) + v\,,$$

where $h(x)$ is a known nonlinear function of x. In this case, we may apply the foregoing technique to the linearized version of $z = h(x) + v$:

$$dz \triangleq z - \bar{z} \cong \left.\frac{\partial h}{\partial x}\right|_{x=\bar{x}} (x - \bar{x}) + v \triangleq \frac{\partial h}{\partial x}dx + v\,.$$

This is illustrated in the two examples below.

In some estimation problems, the relationship between the parameters to be estimated and the available measurements is known only implicitly; i.e., we may not be able to write down explicitly the relationship $z(t) = h(x,v,t)$. On the other hand, we may still be able to determine the differential relationship $dz = (\partial h/\partial x)\, dx + v$ directly and solve the linearized estimation problem.

Also, by appropriate formulation, some dynamic estimation problems can be reduced to parameter estimation problems. Example 2 illustrates this point.

Example 1. *Position estimation from angle measurements.* We wish to estimate the location (x,y) of a point A in a plane by angle measurements z_i from several points B_i $(i = 1,2,\ldots,n)$ on a base line (see Figure 12.2.1).

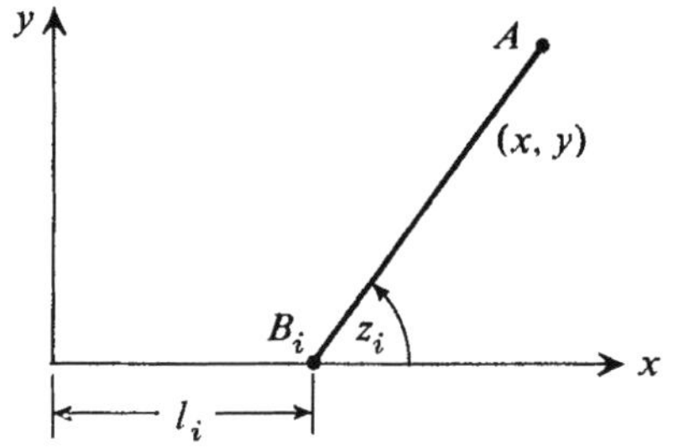

Figure 12.2.1. Position estimation using angle measurements on a base line.

The angle measurements z_i are related to the location of A and B_i by the *nonlinear* relations

$$z_i = \tan^{-1}\frac{y}{x - \ell_i} + v_i\,, \tag{12.2.15}$$

where v_i is a random error made in the angle measurement. We assume that

$$E(v_i) = 0\,, \qquad E(v_i v_j) = \begin{cases} r_i\,, & i = j\,, \\ 0\,, & i \neq j\,. \end{cases} \tag{12.2.16}$$

We now *linearize* (12.2.15) about a prior estimate of (x,y), which we will call $(\bar{x},\bar{y})$:

$$z_i - \bar{z}_i = [H_{1i} H_{2i}] \begin{bmatrix} x - \bar{x} \\ y - \bar{y} \end{bmatrix} + v_i\,,$$

where

$$\bar{z}_i = \tan^{-1}\frac{\bar{y}}{\bar{x} - \ell_i};$$

$$H_{1i} = \left(\frac{\partial z_i}{\partial x}\right)_{x=\bar{x},\, y=\bar{y}}, \qquad H_{2i} = \left(\frac{\partial z_i}{\partial y}\right)_{x=\bar{x},\, y=\bar{y}}.$$

We will let the estimate $\hat{x},\hat{y}$ be independent of prior data by considering the error covariance matrix to be infinite; i.e., we have $M \to \infty$ or $M^{-1} = 0$. Let

$$dz = \begin{bmatrix} z_1 - \bar{z}_1 \\ \cdot \\ \cdot \\ \cdot \\ z_n - \bar{z}_n \end{bmatrix}, \quad H = \begin{bmatrix} H_{1i}, H_{2i} \\ \cdot \\ \cdot \\ \cdot \\ H_{1n}, H_{2n} \end{bmatrix}, \quad dx = \begin{bmatrix} x - \bar{x} \\ \\ y - \bar{y} \end{bmatrix}, \quad v = \begin{bmatrix} v_1 \\ \cdot \\ \cdot \\ \cdot \\ v_n \end{bmatrix}.$$

Then we have

$$dz = H\,dx + v,$$

where

$$R = \begin{bmatrix} r_1 & \cdots & 0 \\ \vdots & \ddots & \vdots \\ 0 & \cdots & r_n \end{bmatrix}, \qquad P = (H^T R^{-1} H)^{-1}.$$

If $d\hat{x}$ is markedly different from zero, we should repeat the procedure, linearizing about $(\hat{x},\hat{y}) = (\bar{x} + d\hat{x}, \bar{y} + d\hat{y})$.

In principle, we should repeat the procedure until $d\hat{x} \cong 0$. The eigenvalues and eigenvectors of P determine the 39 percent likelihood ellipse around $(\hat{x},\hat{y})$.

Numerical example. Suppose that $n = 3$ and the data are

i	1	2	3	Units
ℓ_i	0	500	1,000	ft
z_i	30.1	45.0	73.6	deg
r_i	.01	.01	.04	deg².

From a rough graph (see Figure 12.2.2), we estimate $\bar{x} = 1{,}210$ ft, $\bar{y} = 700$ ft. From this, it follows that

$$dz = \begin{bmatrix} .05 \\ .40 \\ .29 \end{bmatrix} \text{deg}, \qquad H = \begin{bmatrix} -.0205, .0354 \\ -.0403, .0409 \\ -.0751, .0225 \end{bmatrix} \text{deg ft}^{-1},$$

$$P = [H^T R^{-1} H]^{-1} = \begin{bmatrix} 11.26, 10.31 \\ 10.31, 12.75 \end{bmatrix} \text{ft}^2,$$

$$K = PH^T R^{-1} = \begin{bmatrix} 13.4, -3.1, -15.3 \\ 24.0, \ 10.6, -12.2 \end{bmatrix} \text{ft(deg)}^{-1},$$

$$d\hat{x} = K\,dz = \begin{bmatrix} -5.01 \\ +1.89 \end{bmatrix}, \qquad \begin{bmatrix} \hat{x} \\ \hat{y} \end{bmatrix} = \begin{bmatrix} 1205.0 \\ 701.9 \end{bmatrix}.$$

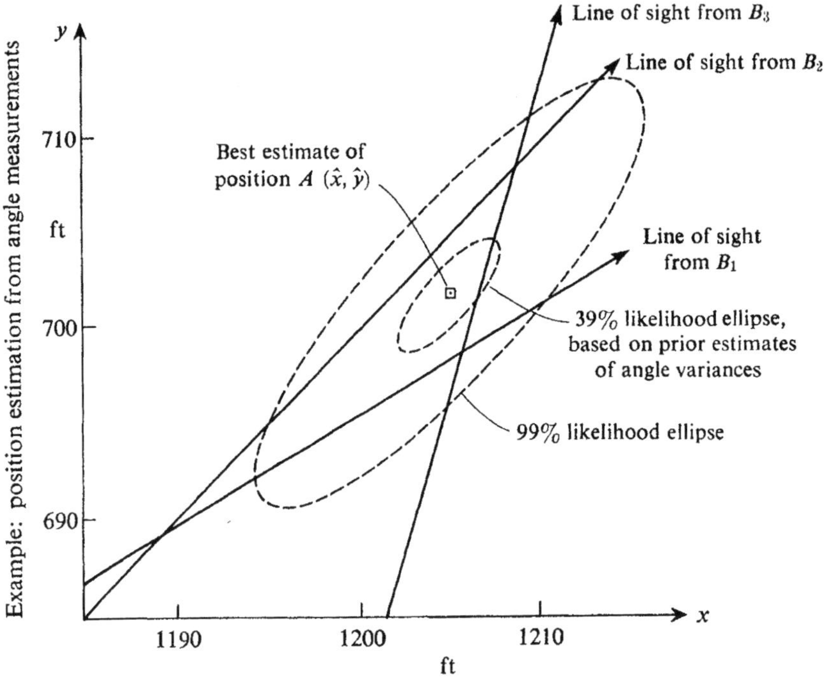

Figure 12.2.2. Numerical example of position estimation by angle measurements.

The eigenvalues and eigenvectors of P are found to be

Eigenvalue	Eigenvector direction
22.34 ft²	47.0°
1.66 ft²	−43.0°.

Using the square root of the eigenvalues (4.72 ft and 1.29 ft) as semi-axes, measured along the eigenvectors, we can sketch the 39 percent likelihood ellipse with center at $(\hat{x}, \hat{y})$ (see Figure 12.2.2). The 99 percent likelihood ellipse is three times the size of the 39 percent ellipse in linear dimension. The lines of sight from B_1, B_2, and B_3 are also shown.

Note that we have

$$(\hat{z}_1 - z_1)^2 = (30.22 - 30.1)^2 = (.12)^2 = .0144,$$
$$(\hat{z}_2 - z_2)^2 = (44.88 - 45.0)^2 = (.12)^2 = .0144,$$
$$(\hat{z}_3 - z_3)^2 = (73.73 - 73.6)^2 = (.13)^2 = .0169\,.$$

Hence, we have

$$\frac{1}{2}\sum_{i=1}^{3}\frac{(\hat{z}_i - z_i)^2}{r_i} = \frac{1.44 + 1.44 + .42}{2} = \frac{3.30}{2}\,.$$

The prior expected value of this quantity was 3.00/2; thus, based on the limited sample of three measurements, we might scale up the angle variances by the factor $3.30/3 = 1.10$. This would scale up the 39 percent likelihood ellipse by the factor $\sqrt{1.10} = 1.05$.

Example 2. *Orbit estimation from horizon sensor measurements.* An elliptic orbit in a plane is specified if the following four parameters are known (see Figure 12.2.3):

a = semi-major axis of ellipse, e = eccentricity of ellipse,

T_0 = time of perigee passage,

θ_0 = angle between perigee and a reference line.

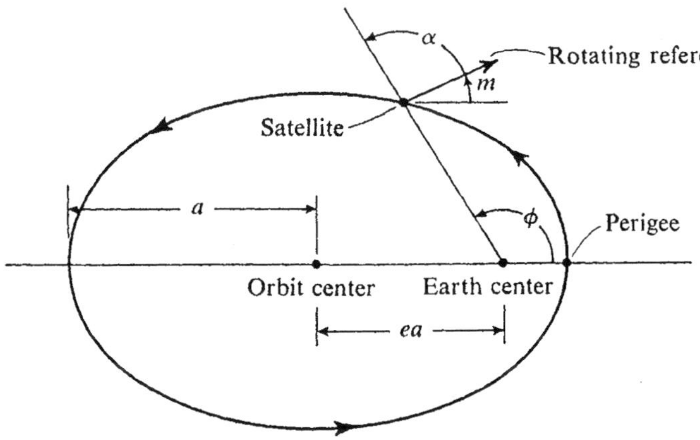

Figure 12.2.3. Nomenclature for orbit estimation using horizon sensor measurements.

A satellite is equipped with a measurement system consisting of (a) a cylinder rotating about an axis perpendicular to the orbital plane with a uniform angular velocity ω, which is set as close to $2\pi/T$ as possible, where T is the period of the orbit;

(b) a horizon sensor mounted in the cylinder, which measures the angle $\alpha(t)$ between the line of sight to the Earth's center and a reference line on the rotating cylinder. This reference line is chosen so that $\alpha(T_0) \cong 0$; that is, $\alpha \cong 0$ at perigee. In the absence of measurement error, $\alpha(t)$ is a periodic function, well defined by the parameters e, a, T_0, ω, and $\alpha(T_0)$; e.g., if the orbit is circular, then $\alpha(t) = 0$ if $\omega = 2\pi/T$ exactly. Any deviation of the orbit from a circle will cause the rotating reference line to periodically lead and lag behind the line of sight. The problem here is to improve the estimates of e, a, T_0, ω, and $\alpha(T_0)$, based on noisy measurements of $\alpha(t)$.

The relationship between α, at a time t, and the parameters a, e, T_0, ω, and $\alpha(T_0)$ is given *implicitly* as follows:†

$$\alpha = \phi - m\,; \qquad \cos\phi = \frac{\cos E - e}{1 - e\cos E}\,; \qquad M = E - e\sin E\,;$$

$$M = \frac{2\pi(t - T_0)}{T}\,; \qquad m = \omega(t - T_0) - \alpha_0\,; \qquad \alpha_0 \equiv \alpha(T_0)\,;$$

$$T = \frac{2\pi}{R\sqrt{g}}a^{3/2}\,.$$

Here the angles ϕ, M, and E are known as the true, mean, and eccentric anomalies respectively, g = acceleration of gravity at the Earth's surface, and R = radius of Earth. Note that $m \equiv M$ if $\omega = 2\pi/T$ and $\alpha_0 = 0$.

By taking *differentials* of the relationships above and eliminating $d\phi$, dM, dE, dm, and dT, the following relation may be obtained:

$$d\alpha = \frac{\partial\alpha}{\partial a}da + \frac{\partial\alpha}{\partial e}de + \frac{\partial\alpha}{\partial T_0}dT_0 + T_0\,d\omega + d\alpha_0\,,$$

where

$$\frac{\partial\alpha}{\partial a} = \frac{3\pi T_0(1 - e^2)\sin E}{aT\sin\phi(1 - e\cos E)^3}\,, \qquad \frac{\partial\alpha}{\partial e} = \frac{(1 - e^2)\sin^2 E}{\sin\phi(1 - e\cos E)^3}\,,$$

$$\frac{\partial\alpha}{\partial T_0} = \omega - \frac{2\pi}{T}\,\frac{(1 - e^2)\sin E}{\sin\phi(1 - e\cos E)^3}\,,$$

and these partial derivatives are evaluated with the best current estimates of a, e, T_0, ω, α_0 at the time of the measurement.

The measurement $z(t)$ is assumed to contain a random error, v:

†See, for example, J. M. A. Danby, *Fundamental of Celestial Mechanics*. New York: Macmillan, 1962.

$$z(t) = \alpha(t) + v\,,$$

where $E(v) = 0$, $E(v^2) = R$, and R is known. Let $\bar{\alpha}(t)$ be the *predicted* measurement, using the best current estimates of a, e, T_0, ω, α_0, at time t. Then we have

$$z(t) - \bar{\alpha}(t) \cong d\alpha(t) + v$$

or

$$z(t) - \bar{\alpha}(t) \cong \left[\frac{\partial \alpha}{\partial a}, \frac{\partial \alpha}{\partial e}, \frac{\partial \alpha}{\partial T_0}, T_0, 1\right] \begin{bmatrix} da \\ de \\ dT_0 \\ d\omega \\ d\alpha_0 \end{bmatrix} + v\,.$$

This *linear* relation can then be used to estimate da, de, dT_0, $d\omega$, and $d\alpha_0$ from the measurement $z(t) - \bar{\alpha}(t)$.

Problem 1. In Equation (12.2.1), assume that x and v are independent random vectors with gaussian density functions. Show that the joint density function $p(x,v)$ is proportional to $\exp(-J)$, where J is as defined in Equation (12.2.5). Thus, $x = \hat{x}$, $v = z - H\hat{x}$ maximize $p(x,v)$, justifying the name "maximum likelihood estimate."

Problem 2. Establish the relations

$$P = M - MH^T(HMH^T + R)^{-1}HM\,, \tag{a}$$

$$PH^TR^{-1} = MH^T(HMH^T + R)^{-1}\,. \tag{b}$$

Note that these relations involve inverting matrices of smaller dimension to determine P and K than Equation (12.2.8) if R is of smaller dimension than P; i.e., if $p < n$. Equations (a) and (12.2.8) are known as the "matrix inversion pair" (see Problem 4, Section 1.3).

Problem 3. Complete the square in Equation (12.2.5) and show that

$$\begin{aligned} J = {} & \tfrac{1}{2}[x - \bar{x} - PH^TR^{-1}(z - H\bar{x})]^TP^{-1}[x - \bar{x} - PH^TR^{-1}(z - H\bar{x})] \\ & + \tfrac{1}{2}(z - H\bar{x})^TR^{-1}(R - HPH^T)R^{-1}(z - H\bar{x})\,. \end{aligned}$$

Clearly, J is minimized by choosing $x = \hat{x}$, where

$$\hat{x} = \bar{x} + PH^TR^{-1}(z - H\bar{x})\,,$$

which is in agreement with Equation (12.2.7).

Problem 4. Given two correlated gaussian random vectors x and z, with mean values $\bar{x}, \bar{z}$ and covariance matrices P_{xx}, P_{zz}, respectively, and correlation $E[(x - \bar{x})(z - \bar{z})^T] = P_{xz}$, *show that* the conditional density function $p(x/z)$ is gaussian, with

$$E(x/z) = \bar{x} + P_{xz}P_{zz}^{-1}(z - \bar{z}) = \hat{x},$$
$$E\{[(x - \hat{x})(x - \hat{x})^T]/z\} = P_{xx} - P_{xz}P_{zz}^{-1}P_{xz}^T.$$

Problem 5. In Problem 4, let $z = Hx + v$, where H is a known matrix and v is independent of x, with mean value zero and covariance R. Let $P_{xx} = M$ and *show that*

$$P_{zz} = R + HMH^T, \qquad P_{xz} = MH^T, \qquad \bar{z} = H\bar{x}.$$

Using these relations in Problem 4, verify Equations (12.2.7) and (12.2.8) (you will also need the results of Problem 2). Note that $K = P_{xz}P_{zz}^{-1}$, a most reasonable result!

Problem 6. In Problem 4, show that the gaussian random vectors $e = E(x/z) - x$ and $z - \bar{z}$ are independent; i.e., we have $E[e(z - \bar{z})^T] = 0$.

Problem 7. Suppose that the number of theoretical relationships among measured variables z and state variables x is less than the number of measured variables; for example, we have

$$Az = Hx + Av,$$

where

$$A \text{ is a } (q \times p)\text{-matrix}, \qquad q < p, \qquad H \text{ is a } (q \times n)\text{-matrix},$$
$$E(v) = 0, \qquad E(vv^T) = R \text{ is a } (p \times p)\text{-matrix}.$$

Show that the estimation procedure of this section applies with z replaced by Az and R by ARA^T.

Problem 8. Consider the usual problem of least square fit, i.e., determining x to minimize

$$J = \tfrac{1}{2}\|z - Hx\|^2.$$

Show that the error of the fit $e = z - H\hat{x}$ is orthogonal to the fit $\hat{z} = H\hat{x}$, in the sense $e^T\hat{z} = 0$.

Problem 9. In Example 2, suppose that initial estimates of an orbit are $a = 6{,}000$ miles, $\bar{e} = 1/6$, $\bar{T}_0 = 0$, $\bar{\omega} = 2\pi/T$, $\bar{\alpha}_0 = 0$. Take $R = 3{,}960$ miles, $g = 32.2$ ft sec^{-2}, and make an improved estimate of a, e, T_0, ω, and α_0, using the single measurement

$$z(t) = 16.7 \text{ deg} \qquad \text{at} \qquad t = 1{,}357 \text{ sec},$$

where

$$E(v^2) = 10^{-2}\,(\text{deg})^2\,, \qquad E(a-\bar{a})^2 = 10^{-4}\,(\text{miles})^2\,, \qquad E(e-\bar{e})^2 = 10^{-4}\,,$$
$$E(T_0-\bar{T}_0)^2 = 10^2\,(\text{sec})^2\,, \qquad E(\omega-\bar{\omega})^2 = 10^{-10}\,(\text{sec})^{-2}\,,$$
$$E(\alpha_0-\bar{\alpha}_0)^2 = 10^{-2}\,(\text{deg})^2\,,$$

and all covariances are zero. [HINT: Use Equation (a) of Problem 2, rather than Equation (12.2.8), to find P.]

12.3 Optimal filtering for single-stage linear transitions

Consider a system that makes a discrete transition from state 0 to state 1 according to the linear relation

$$x_1 = \Phi_0 x_0 + \Gamma_0 w_0\,, \tag{12.3.1}$$

where Φ_0 = a known $(n \times n)$ transition matrix, Γ_0 = a known $(n \times r)$-matrix:

$$E(w_0) = \bar{w}_0\,, \qquad E(w_0-\bar{w}_0)(w_0-\bar{w}_0)^T = Q_0\,. \tag{12.3.2}$$

The forcing vector, w_0, is thus a random vector, with mean $\bar{w}_0$ and covariance Q_0. The state x_0 is also a random vector, with mean $\hat{x}_0$ and covariance P_0; that is, we have

$$E(x_0) = \hat{x}_0\,, \qquad E(\hat{x}_0 - x_0)(\hat{x}_0 - x_0)^T = P_0\,. \tag{12.3.3}$$

Furthermore, x_0 and w_0 are independent. From this information, it follows that x_1 is also a random vector, and from Section 11.2, Equations (11.2.7) through (11.2.9), it has a mean value $\bar{x}_1$ and a covariance M_1, given by

$$\bar{x}_1 = \Phi_0\hat{x}_0 + \Gamma_0\bar{w}_0\,, \tag{12.3.4}$$

$$M_1 = \Phi_0 P_0 \Phi_0^T + \Gamma_0 Q_0 \Gamma_0^T\,. \tag{12.3.5}$$

Since $\Gamma_0 Q_0 \Gamma_0^T$ by definition (12.3.2) is a nonnegative matrix, it is apparent from (12.3.5) that, on the average, the effect of the uncertainty in w_0 in a transition of the type (12.3.1) is to *increase the uncertainty* in our knowledge of the state x_1.† This is to be contrasted with the result (12.2.8), where it was shown that measurements, on the average, *decrease the uncertainty* in our knowledge of the state.††

Suppose that we make measurements, as in Section 12.2, *after* the transition to state 1. Then, from Equations (12.2.7) and (12.2.8), the best estimate of x_1 is given by $\hat{x}_1$, where

$$\hat{x}_1 = \bar{x}_1 + P_1 H_1^T R_1^{-1}(z_1 - H_1\bar{x}_1)\,, \tag{12.3.6}$$

†More precisely, the uncertainty is increased or left unchanged. The increase in P is, of course, also dependent on the term Φ_0.

††Again, more precisely, the uncertainty is decreased or left unchanged.

$$P_1 = (M_1^{-1} + H_1^T R^{-1} H_1)^{-1} = M_1 - M_1 H_1^T (H_1 M_1 H_1^T + R_1)^{-1} H_1 M_1 . \quad (12.3.7)$$

Here $\bar{x}_1$ and M_1 are as given in (12.3.4) and (12.3.5). Note that $\bar{x}_1$ is the estimate of x_1 *before* measurement, whereas $\hat{x}_1$ is the estimate *after* measurement. Similarly, M_1 is the error covariance matrix before measurement and P_1 is the error covariance matrix after measurement. Symbolically, we can describe this process as follows:

$$\begin{array}{lccccc} & & \bar{w}_0 & & z_1 & \\ & & \downarrow & & \downarrow & \\ \text{mean:} & \hat{x}_0 & \longrightarrow & \bar{x}_1 & \longrightarrow & \hat{x}_1 , \\ & & Q_0 & & R_1 & \\ & & \downarrow & & \downarrow & \\ \text{covariance:} & P_0 & \longrightarrow & M_1 & \longrightarrow & P_1 . \end{array}$$

12.4 Optimal filtering and prediction for linear multistage processes

Consider the linear, stochastic, multistage process described by

$$x_{i+1} = \Phi_i x_i + \Gamma_i w_i , \qquad i = 0, \ldots, N-1 , \qquad (12.4.1)$$

where

$$E(x_0) = \bar{x}_0 , \qquad (12.4.2)$$

$$E(w_i) = \bar{w}_i , \qquad (12.4.3)$$

$$E(x_0 - \bar{x}_0)(x_0 - \bar{x}_0)^T = M_0 , \qquad (12.4.4)$$

$$E(w_i - \bar{w}_i)(w_j - \bar{w}_j)^T = Q_i \delta_{ij} , \qquad (12.4.5)$$

$$E(w_i - \bar{w}_i)(x_0 - \bar{x}_0)^T = 0 . \qquad (12.4.6)$$

Measurements z_i are made while the system is in stage i, and are linearly related to the state x_i by

$$z_i = H_i x_i + v_i , \qquad i = 0, \ldots, N , \qquad (12.4.7)$$

where

$$E(v_i) = 0 , \qquad (12.4.8)$$

$$E(v_i v_j^T) = R_i \delta_{ij} \qquad (12.4.9)$$

$$E(w_i - \bar{w}_i) v_j^T = 0 ,\dagger \qquad \text{and} \qquad E(x_0 - \bar{x}_0) v_i^T = 0 . \quad (12.4.10)$$

It is reasonable to expect (see the derivation in Section 12.7 or in Chapter 13) that the weighted-least-square or maximum-likelihood estimate of the state x_k, using only the measurements $z_0, \ldots, z_k$,

†The case in which w_i and v_j are correlated is considered in Chapter 13.

is given by the sequential use of the single-stage estimation procedure of the previous section:

$$\hat{x}_i = \bar{x}_i + K_i(z_i - H_i\bar{x}_i), \qquad (i = 0, \ldots, k, \quad \text{where} \quad k \leq N). \quad (12.4.11)$$

where

$$\bar{x}_{i+1} = \Phi_i\hat{x}_i + \Gamma_i\bar{w}_i, \qquad \bar{x}_0 \ \text{given}, \qquad (12.4.12)\dagger$$

$$K_i = P_iH_i^TR_i^{-1}, \qquad (12.4.13)$$

$$P_i = (M_i^{-1} + H_i^TR_i^{-1}H_i)^{-1} = M_i - M_iH_i^T(H_iM_iH_i^T + R_i)^{-1}H_iM_i, \quad (12.4.14)$$

$$M_{i+1} = \Phi_iP_i\Phi_i^T + \Gamma_iQ_i\Gamma_i^T. \qquad (12.4.15)$$

This is the *Kalman filter* for linear multistage processes (see Kalman, 1960). Note that the filter (12.4.11) and (12.4.12) is a model of the system (12.4.1), with a correction term proportional to the difference between the actual measurement z_i and the predicted measurement $H_i\bar{x}_i$. The proportionality matrix K_i in (12.4.13) is essentially the ratio between uncertainty in the state P_i and the uncertainty in the measurements R_i; the matrix H_i is simply the state-to-measurement transformation matrix of (12.4.7).

Note that:

(a) The propagation of the covariance of the error of the estimate, Equations (12.4.14) and (12.4.15), is *independent* of the measurements z_i. Thus, the covariance matrix can be computed *beforehand* and stored if the parameters of the system and the observation equations are given.

(b) The computation of the updated estimate, Equations (12.4.11) and (12.4.12), involves only the *current* measurement and error covariance. Thus, it can easily be carried out in real time.

Prediction of the state beyond the stage where measurements are available, say state m, can be done only by repeated use of (12.4.12); that is,

$$\hat{x}_{i+1} = \bar{x}_{i+1} = \Phi_i\hat{x}_i + \Gamma_i\bar{w}_i; \qquad i = m, m+1, \ldots, \quad (12.4.16)$$

where $\hat{x}_m$ is obtained from the filter (12.4.11) through (12.4.15). In other words, the best prediction we can make uses the expected value of w_i, namely, $\bar{w}_i$, in the transition relations (12.4.1), starting, however, with the filtering estimate of $\hat{x}_m$. Another way of seeing this is to consider $R_i = \infty$ for $i = m, m+1, \ldots.$ In this case, (12.4.14) and (12.4.15), (12.4.11) and (12.4.12) reduce to

$$P_{i+1} = \Phi_iP_i\Phi_i^T + \Gamma_iQ_i\Gamma_i^T, \qquad (12.4.17)$$

†The term $\bar{x}_{i+1}$ in (12.4.12) is, of course, to be understood as $E(x_{i+1}/z_i, \ldots, z_i)$ and not as $E(x_{i+1})$.

$$\hat{x}_{i+1} = \Phi_i \hat{x}_i + \Gamma_i \overline{w}_i , \tag{12.4.18}$$

which were already derived in Section 11.2, Equations (11.2.9) and (11.2.7).

The error of the multistage estimator, $e_i \equiv \hat{x}_i - x_i$, obeys an equation that can be derived from (12.4.1) and (12.4.11). We have

$$e_{i+1} = (\Phi - K_{i+1} H\Phi) e_i + (K_{i+1} H\Gamma - \Gamma) w_i + K_{i+1} v_{i+1} ; \qquad e_o = \hat{x}_o - x_o . \tag{12.4.19a}$$

Here we have dropped the subscripts on Φ, Γ, Q, and H and assumed that $\overline{w}_i = 0$, for convenience. Since $E(\hat{x}_o e_o^T) = 0$, direct calculation shows by induction on i that

$$\begin{aligned} E(\hat{x}_{i+1} e_{i+1}^T) &= E[(\Phi\hat{x}_i - K_{i+1} H\Phi e_i + K_{i+1} H\Gamma w_i + K_{i+1} v_{i+1}) \\ &\qquad ((\Phi - K_{i+1} H\Phi) e_i + (K_{i+1} H\Gamma - \Gamma) w_i + K_{i+1} v_{i+1})^T] \\ &= -K_{i+1} H [\Phi P_i (\Phi - K_{i+1} H\Phi)^T + \Gamma Q (K_{i+1} H\Gamma - \Gamma)^T] \\ &\qquad + K_{i+1} R K_{i+1}^T \\ &= K_{i+1} H M_{i+1} (H^T K_{i+1}^T - I) + K_{i+1} R K_{i+1}^T \\ &= -K_{i+1} H P_{i+1} + K_{i+1} R K_{i+1}^T = 0 \qquad \text{for all} \quad i . \end{aligned} \tag{12.4.19b}$$

This is the analogous result to Equation (12.2.14) of Section 12.2.

Statistically stationary processes. If $\Phi_i, \Gamma_i, R_i, H_i, Q_i$ are all constant matrices, the filtering process may reach a steady state in the sense that M_i and P_i become constant matrices, M and P, as $i \rightarrow \infty$. The two matrix equations for determining M and P are

$$P^{-1} = M^{-1} + H^T R^{-1} H \quad \text{or} \quad P = M - MH^T (R + HMH^T)^{-1} HM , \tag{12.4.20a}$$

$$M = \Phi P \Phi^T + \Gamma Q \Gamma^T . \tag{12.4.20b}$$

The information going out, $\Gamma Q \Gamma^T$, is balanced by the information coming in, $H^T R^{-1} H$, and any damping that the system may have (as expressed in Φ).

A simple first-order process. A stationary first-order multistage system is forced by a white random sequence, and a single measurement is made at each stage with errors that form another (independent) white random sequence:

$$x_{i+1} = \phi x_i + w_i , \qquad E(w_i) = 0 , \qquad E(w_i w_j) = q\, \delta_{ij} ; \tag{12.4.21}$$

$$z_i = x_i + v_i , \qquad E(v_i) = 0 , \qquad E(v_i v_j) = r\, \delta_{ij} , \qquad i = 1, 2, \ldots , \tag{12.4.22}$$

where

$$E(x_o) = \hat{x}_o , \qquad E(\hat{x}_o - x_o)^2 = p_o .$$

The maximum-likelihood filter is given by

$$\hat{x}_{i+1} = \phi\hat{x}_i + \frac{p_{i+1}}{r}(z_{i+1} - \phi\hat{x}_i), \qquad \hat{x}_0 \text{ given}; \tag{12.4.23}$$

$$p_{i+1} = \left(\frac{1}{r} + \frac{1}{\phi^2 p_i + q}\right)^{-1}, \qquad p_0 \text{ given}. \tag{12.4.24}$$

In the statistical steady state, we have $p_{i+1} = p_i = p$. From (12.4.24), this yields as $i \to \infty$.

$$\frac{p}{r} \to \frac{1}{2\phi^2}\left[\sqrt{\left(\frac{q}{r} + 1 - \phi^2\right)^2 + \frac{4q}{r}\phi^2} - \left(\frac{q}{r} + 1 - \phi^2\right)\right], \tag{12.4.25}$$

$$\frac{m}{r} \to \phi^2\frac{p}{r} + \frac{q}{r}. \tag{12.4.26}$$

The law of large numbers. An interesting special case of Equation (12.4.21) is when $\phi = 1$ and $q = 0$:

$$x_{i+1} = x_i, \tag{12.4.27}$$

$$z_i = x_i + v_i, \tag{12.4.28}$$

that is, we wish to determine a certain constant, x, by repeated measurements. In this case, (12.4.23) and (12.4.24) become, simply,

$$\hat{x}_{i+1} = \hat{x}_i + \frac{p_{i+1}}{r}(z_{i+1} - \hat{x}_i), \qquad \hat{x}_0 \text{ given}, \tag{12.4.29}$$

$$\frac{1}{p_{i+1}} = \frac{1}{r} + \frac{1}{p_i}, \qquad p_0 \text{ given}. \tag{12.4.30}$$

The latter recursive relation is easily solved:

$$\frac{1}{p_i} = \frac{i}{r} + \frac{1}{p_0} \qquad \text{or} \qquad p_i = \frac{p_0}{1 + [i(p_0/r)]}. \tag{12.4.31}$$

Substituting (12.4.31) into (12.4.29), we may write $\hat{x}_i$ in terms of $\hat{x}_0$, p_0, and the measurements

$$\hat{x}_1 = \hat{x}_0 + \left[\left(\frac{p_0}{r}\right)\Big/\left(1 + \frac{p_0}{r}\right)\right](z_1 - \hat{x}_0) \equiv \left(\hat{x}_0 + \frac{p_0}{r}z_1\right)\Big/\left(1 + \frac{p_0}{r}\right),$$

$$\hat{x}_2 = \hat{x}_1 + \left[\left(\frac{p_0}{r}\right)\Big/\left(1 + 2\left(\frac{p_0}{r}\right)\right)\right](z_2 - \hat{x}_1)$$

$$\vdots \qquad \equiv \left[\hat{x}_0 + \frac{p_0}{r}(z_1 + z_2)\right]\Big/\left[1 + 2\left(\frac{p_0}{r}\right)\right], \tag{12.4.32}$$

$$\vdots$$

$$\hat{x}_i = \frac{1}{1 + i(p_0/r)}\left[\hat{x}_0 + \frac{p_0}{r}\sum_{j=1}^{i} z_j\right].$$

As $i \to \infty$, Equations (12.4.31) and (12.4.32) become, simply,

$$p_i \to 0, \tag{12.4.33}$$

$$\hat{x}_i \to \frac{1}{i} \sum_{j=1}^{i} z_j \quad \text{as} \quad i \to \infty; \tag{12.4.34}$$

that is, the best estimate tends to the *average* of all the measurements and the variance tends to zero; this is *the law of large numbers.* For any finite i, the initial estimate, $\hat{x}_0$, influences the best current estimate, according to (12.4.32).

Another interesting special case of Equation (12.4.21) is when $q \to \infty$; i.e., the variable x_i changes unpredictably from one step to the next. Not surprisingly, we find, from Equation (12.4.24), that

$$p_i \to r \quad \text{for all} \quad i \quad \text{as} \quad q \to \infty. \tag{12.4.35}$$

Substituting this into Equation (12.4.23) gives

$$\hat{x}_i = z_i; \tag{12.4.36}$$

that is, the best estimate is the latest measurement.

Problem. The covariance matrix P_i is given by either

$$P_i = M_i - M_i H_i^T (H_i M_i H_i^T + R_i)^{-1} H_i M_i$$

or

$$P_i = (I - K_i H_i) M_i (I - K_i H_i)^T + K_i R_i K_i^T.$$

Discuss, from a numerical computational viewpoint, why the latter version is to be preferred. [HINT: Consider the case when $M - P$ is small compared with P.]

12.5 Optimal filtering for continuous linear dynamic systems with continuous measurements

The results of the previous section for multistage processes may be extended to continuous dynamic processes by formally letting the time between stages tend to zero. We associate x_i with $x(t_i)$ and write $\Delta = t_{i+1} - t_i$. The difference equations (12.4.1) may be written as

$$\frac{x(t_i + \Delta) - x(t_i)}{\Delta} = \frac{\Phi(t_i + \Delta, t_i) - I}{\Delta} x(t_i) + \frac{\Gamma_i}{\Delta} w(t_i). \tag{12.5.1}$$

As $\Delta \to 0$, these equations become differential equations:

$$\dot{x} = F(t)x + G(t)w(t), \tag{12.5.2}$$

where

$$F(t) = \lim_{\Delta \to 0} \frac{\Phi(t_i + \Delta, t_i) - I}{\Delta}, \tag{12.5.3}$$

$$G(t) = \lim_{\Delta \to 0} \frac{\Gamma_i}{\Delta}. \tag{12.5.4}$$

The difference equations (12.4.11) through (12.4.15) may be written as

$$\hat{x}_i - \bar{x}_i = P_i H_i^T (R_i \Delta)^{-1} (z_i - H_i \bar{x}_i) \Delta, \tag{12.5.5}$$

$$\frac{\hat{x}_{i+1} - \hat{x}_i}{\Delta} = \frac{\Phi_i - I}{\Delta} \hat{x}_i + \frac{\Gamma_i}{\Delta} \bar{w}_i + P_{i+1} H_{i+1}^T (R_{i+1} \Delta)^{-1} (z_{i+1} - H_{i+1} \bar{x}_{i+1}), \tag{12.5.6}$$

$$M_i - P_i = M_i H_i^T (R_i \Delta + H_i M_i H_i^T \Delta)^{-1} H_i M_i \Delta, \tag{12.5.7}$$

$$\frac{P_{i+1} - P_i}{\Delta} = \Delta \frac{\Phi_i - I}{\Delta} P_i \frac{\Phi_i^T - I}{\Delta} + \frac{\Phi_i - I}{\Delta} P_i + P_i \frac{\Phi_i^T - I}{\Delta}$$

$$+ \frac{\Gamma_i}{\Delta} (Q_i \Delta) \frac{\Gamma_i^T}{\Delta} - M_{i+1} H_{i+1}^T (R_{i+1} \Delta + H_{i+1} M_{i+1} H_{i+1}^T \Delta)^{-1} H_{i+1} M_{i+1}. \tag{12.5.8}$$

If we let

$$R_i \Delta \to R(t), \tag{12.5.9}$$

$$Q_i \Delta \to Q(t) \quad \text{as} \quad \Delta \to 0, \tag{12.5.10}$$

then, from (12.5.5) and (12.5.7), we have

$$\hat{x}_i \to \bar{x}_i \to \hat{x}(t), \tag{12.5.11}$$

$$P_i \to M_i \to P(t) \quad \text{as} \quad \Delta \to 0. \tag{12.5.12}$$

Using (12.5.9) and (12.5.10) in (12.5.6) and (12.5.8), we have, as $\Delta \to 0$,

$$\dot{\hat{x}} = F\hat{x} + G\bar{w} + PH^T R^{-1}(z - H\hat{x}), \qquad \hat{x}(0) = 0; \tag{12.5.13}$$

$$\dot{P} = FP + PF^T + GQG^T - PH^T R^{-1} HP, \qquad P(0) = P_0. \tag{12.5.14}$$

By comparing Equations (12.5.9) and (12.5.10) with Equation (11.5.13), it is apparent that $w(t)$ and $v(t)$ are white-noise processes, where

$$E[w(t) - \bar{w}(t)][w(t') - \bar{w}(t')]^T = Q(t)\,\delta(t - t'), \tag{12.5.15}$$

$$E[v(t) v^T(t')] = R(t)\,\delta(t - t'). \tag{12.5.16}$$

The continuous filter, Equations (12.5.13) and (12.5.14), was described by Kalman and Bucy in 1961.

The approximation of the continuous filter by a multistage filter when using a digital computer must be done with care, since R_i and Q_i in (12.5.9) and (12.5.10) depend on the size of the time step, Δ. This is discussed more fully in Section 11.5.

Prediction of the state beyond the time where measurements are available, $t > t_1$, uses (12.5.13) and (12.5.14), with $R \to \infty$ for $t > t_1$:

$$\dot{\hat{x}} = F\hat{x} + G\bar{w}, \qquad \hat{x}(t_1) \text{ given}; \tag{12.5.17}$$

$$\dot{P} = FP + PF^T + GQG^T, \qquad P(t_i) \text{ given}. \tag{12.5.18}$$

Similar to Equations (12.4.19a) and (12.4.19b), of Section 12.4 and Equation (12.2.14) of Section 12.2, we can write $e(t) = \hat{x}(t) - x(t)$ and derive

$$\dot{e} = (F - KH)e + Kv - Gw, \tag{12.5.19a}$$

$$E[e(t)\hat{x}(t)^T] = 0 \qquad \text{for all} \quad t. \tag{12.5.19b}$$

Statistically stationary processes and the steady-state Wiener filter. If F, G, H, Q, and R are all constant matrices, the filtering process *may* reach a *steady state* in the sense that P becomes a constant matrix ($\dot{P} = 0$). In principle, this steady-state matrix may be obtained by solving the $\frac{1}{2}n(n+1)$ simultaneous quadratic equations given by setting $\dot{P} = 0$ in Equation (12.5.14):

$$0 = FP + PF^T + GQG^T - PH^TR^{-1}HP. \tag{12.5.20}$$

In this steady state, the rate at which information goes out of the system, GQG^T, is just balanced by the rate at which information comes into the system, $PH^TR^{-1}HP$, and by any damping the system may have (as expressed in F).

In practice, the solution of (12.5.20) for P is impracticable for $n > 2$; instead, Equation (12.5.14) is integrated, say, with $P(0) = 0$, until $\dot{P} \to 0$.

This steady-state filter is the one considered by N. Wiener in his famous book in 1949.† He approached the problem by Fourier analysis and developed an integral equation (the Wiener-Hopf equation) for determining the impulse response matrix for the filter. Note that Equation (12.5.13) may be rewritten as

$$\dot{\hat{x}} = (F - PH^TR^{-1}H)\hat{x} + G\bar{w} + PH^TR^{-1}z.$$

†Wiener's development does not require the markov assumption.

The impulse response matrix of the Wiener filter for estimating x from z is then given by (for $\overline{w} = 0$)

$$h(t - \tau) = e^{(F - PH^TR^{-1}H)(t-\tau)}PH^TR^{-1} .$$

A simple first-order process with a single measurement. A stationary first-order dynamic system is forced by a white noise process, and a single measurement is made continuously which contains errors that form another (independent) white noise process:

$$\dot{x} = -ax + w(t) ; \qquad E[w(t)] = 0 , \qquad E[w(t)w(t')] = q\,\delta(t - t') , \tag{12.5.21}$$

$$z = x + v(t) ; \qquad E[v(t)] = 0 , \qquad E[v(t)v(t')] = r\,\delta(t - t') , \tag{12.5.22}$$

where

$$E[x(0)] = 0 , \qquad E[x(0)]^2 = p_0 ,$$

and all quantities are *scalars*, with a, q, r, p_0 constants.

The variance equation (12.5.14), for this case, becomes

$$\dot{p} = -2ap + q - (1/r)p^2 , \qquad p(0) = p_0 . \tag{12.5.23}$$

This scalar Riccati equation is readily solved, yielding

$$p(t) = p_1 + \frac{p_1 + p_2}{[(p_0 + p_2)/(p_0 - p_1)]e^{2\beta t} - 1} , \tag{12.5.24}$$

where

$$\beta = \sqrt{a^2 + (q/r)} , \qquad p_1 = r(\beta - a) , \qquad p_2 = r(\beta + a) .$$

Note that we have $p(t) \to p_1$ as $t \to \infty$.

The Kalman-Bucy filter, from (12.5.13), becomes, simply,

$$\dot{\hat{x}} = -a\hat{x} + [p(t)/r]\,[z(t) - \hat{x}] , \qquad \hat{x}(0) = 0 , \tag{12.5.25}$$

which becomes a *stationary filter* $[p(t) \to p_1]$ as $\beta t \to \infty$.

A simple second-order process with a single measurement. Consider the same constant-coefficient second-order dynamic system with white noise forcing function that was discussed in Section 11.4:

$$\ddot{x} + 2\zeta\omega\dot{x} + \omega^2 x = \omega^2 w(t) ; \qquad E[w(t)] = 0 ,$$
$$E[w(t)w(t')] = q\,\delta(t - t') . \tag{12.5.26}$$

In state vector nomenclature, $(x = x_1 , \dot{x} = x_2)$, Equation (12.5.26) becomes

$$\begin{bmatrix} \dot{x}_1 \\ \dot{x}_2 \end{bmatrix} = \begin{bmatrix} 0 , & 1 \\ -\omega^2 , & -2\zeta\omega \end{bmatrix} \begin{bmatrix} x_1 \\ x_2 \end{bmatrix} + \begin{bmatrix} 0 \\ \omega^2 \end{bmatrix} w(t) . \tag{12.5.27}$$

The random initial conditions are assumed to be

$$E[x_1(0)] = E[x_2(0)] = 0\,, \tag{12.5.28}$$

$$E\begin{bmatrix} x_1^2 & , x_1x_2 \\ x_1x_2 & , x_2^2 \end{bmatrix}_{t=0} = \begin{bmatrix} P_{11}(0)\,, P_{12}(0) \\ P_{12}(0)\,, P_{22}(0) \end{bmatrix} \triangleq P(0)\,. \tag{12.5.29}$$

A continuous measurement of the velocity variable, x_2, is made which contains errors that form another (independent) white noise process:

$$z(t) = x_2(t) + v(t)\,; \qquad E[v(t)] = 0\,, \qquad E[v(t)v(t')] = r\,\delta(t - t')\,. \tag{12.5.30}$$

From Equation (12.5.14), the two variance and one covariance equations are

$$\dot{P}_{11} = 2P_{12} - \frac{1}{r}P_{12}^2\,; \qquad P_{11}(0) \text{ given;} \tag{12.5.31}$$

$$\dot{P}_{12} = P_{22} - \omega^2 P_{11} - 2\zeta\omega P_{12} - \frac{1}{r}P_{12}P_{22}\,, \qquad P_{12}(0) \text{ given;} \tag{12.5.32}$$

$$\dot{P}_{22} = -2\omega^2 P_{12} - 4\zeta\omega P_{22} - \frac{1}{r}P_{22}^2 + \omega^4 q\,, \qquad P_{22}(0) \text{ given;} \tag{12.5.33}$$

where we have used

$$\begin{aligned} G &= \begin{bmatrix} 0 \\ \omega^2 \end{bmatrix}, \qquad H = [0,1]\,, \qquad Q = q\,, \\ R &= r\,, \qquad F = \begin{bmatrix} 0 & , & 1 \\ -\omega^2 & , & -2\zeta\omega \end{bmatrix}. \end{aligned} \tag{12.5.34}$$

It is complicated to solve the coupled Riccati equations (12.5.31) through (12.5.33) in closed form. However, the steady-state solution is quite simple; with $\dot{P}_{11} = \dot{P}_{12} = \dot{P}_{22} = 0$, we find that

$$P_{12}(t) \to 0\,, \tag{12.5.35}$$

$$P_{11}(t) \to \frac{2\zeta r}{\omega}\left[\sqrt{1 + \frac{\omega^2}{4\zeta^2}\frac{q}{r}} - 1\right], \,, \tag{12.5.36}$$

$$P_{22}(t) \to 2\zeta\omega r\left[\sqrt{1 + \frac{\omega^2}{4\zeta^2}\frac{q}{r}} - 1\right], \qquad \text{as} \quad \omega t \to \infty\,. \tag{12.5.37}$$

For $r < \infty$, P_{11} and P_{22} are *less* than the case without measurements ($r = \infty$) in Section 11.4.

The optimal filter, from (12.5.13), becomes

$$\dot{\hat{x}}_1 = \hat{x}_2 + [P_{12}(t)/r]\,[z(t) - \hat{x}_2]\,, \qquad \hat{x}_1(0) = 0\,; \tag{12.5.38}$$

$$\dot{\hat{x}}_2 = -\omega^2\hat{x}_1 - 2\zeta\omega\hat{x}_2 + [P_{22}(t)/r]\,[z(t) - \hat{x}_2]\,, \qquad \hat{x}_2(0) = 0\,. \tag{12.5.39}$$

In Figure 12.5.1, a numerical solution† of (12.5.31) through (12.5.33) is plotted for the case $\zeta = 0.2$, $\omega^2 q/r = 0.5$, $P_{11}(0) = P_{12}(0) = P_{22}(0) = 0$. Note that the steady state solution of (12.5.35) through (12.5.37) is reached, for all intents and purposes, when $\sqrt{1-\zeta^2}\,\omega t \cong 2\pi$. In Figure 12.5.2, the only change from Figure 12.5.1 is a *more* accurate measurement so that $\omega^2 q/r = 2.0$; note that the asymptotic values of P_{11} and P_{22} are lower than for Figure 12.5.1.

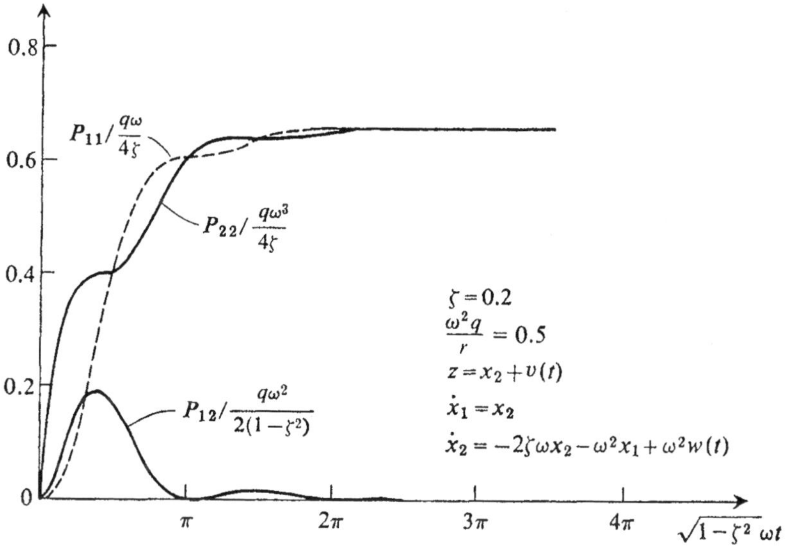

Figure 12.5.1. History of conditional covariance matrix for the second-order example with $\omega^2 q/r = 0.5$.

Observability. If in Equation (12.5.2), we have $w(t) = 0$, i.e., no process noise, then Equation (12.5.14) simplifies to

$$\dot{P} = FP + PF^T - PH^TR^{-1}HP, \qquad P(0) = P_0, \qquad (12.5.40a)$$

or

$$\dot{P}^{-1} = -F^TP^{-1} - P^{-1}F + H^TR^{-1}H, \qquad P^{-1}(0) = P_0^{-1}. \quad (12.5.40b)$$

If P_0^{-1} is singular of rank $n - r$, this implies that there is no prior information at all on r linear combinations of the state variables (i.e., r eigenvalues of P_0 are infinite). The solution of (12.5.40b), on the other hand, can be written in general as (see Appendix A4, Equation (22)).

†Calculated by use of the ASP (Automatic Synthesis Program); see Reference 4, Chapter 5.

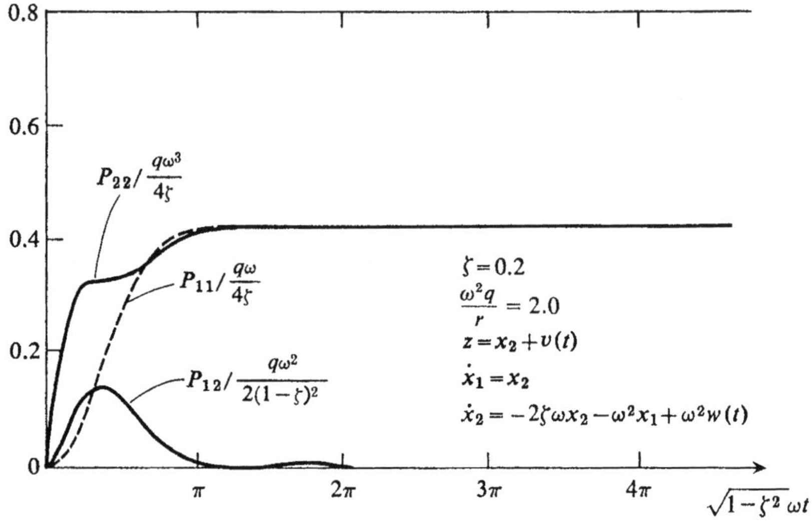

Figure 12.5.2. History of conditional covariance matrix for the second-order example with $\omega^2 q/r = 2.0$.

$$P^{-1}(t) = \Phi^T(t_0,t)P_0^{-1}\,\Phi(t_0,t) + \int_{t_0}^{t} \Phi^T(\tau,t_0)H^TR^{-1}H\Phi(\tau,t_0)\,d\tau\,, \quad (12.5.41)$$

where $\Phi(t,t_0)$ is the transition matrix corresponding to F. If the integral in (12.5.41) is positive definite for some $t > t_0$, then $P^{-1}(t) > 0$ and, hence, $\infty > P(t) > 0$. This means it is possible to acquire information on states that are initially completely uncertain via the measurements $z(t)$. Note that this ability is a property of F and H only (R is assumed to be positive definite and affects the scaling only). We say the system is *observable* whenever the integral in (12.5.41) is positive definite. This definition is in complete agreement with the deterministic definition given in Appendix B. In the deterministic case, the state of an unobservable system cannot be *determined* by operation on the measurements $z(t)$. In the stochastic case, the variance of the estimation error of the state of an unobservable system cannot be *decreased* by operation on $z(t)$.

Duality. The development of observability (above and Appendix B.3) has a strong resemblance to that of controllability (Problem 2, Section 5.3, and Appendix B.2). Both are governed by the simplified version of a Riccati equation, the P equation (12.5.14) for the former, the S equation (5.3.34), for the latter. In fact, if we consider the control problem with quadratic criterion

$$J = \frac{1}{2}x(t_f)^TS(t_f)x(t_f) + \frac{1}{2}\int_{t_0}^{t_f}[\|Hx\|_A^2 + \|u\|_B^2]\,dt\,, \quad (12.5.42)$$

and linear system equations

$$\dot{x} = Fx + Gu\,, \tag{12.5.43}$$

then the governing Riccati equation (5.3.34) for S is identical mathematically with that of P [Equation (12.5.14)] if we identify input with output and weighting matrices with covariance matrices, transpose all matrices, and reverse time. (See table below.)

Control problem	Estimation problem
G^T	H
H^T	G
B	R
A	Q
$S(t_f)$	$P(t_0)$
F	F^T
$t_f - t$	$t - t_0$.

This suggests that the optimal filter must have an interpretation as a solution of a linear-quadratic optimization problem. The details of this are discussed in Chapter 13.

The mathematical duality of these two problems can be exploited to advantage. Meaningful statements for one problem can be reinterpreted to yield meaningful results in the other problem. The first four problems below illustrate the usefulness of this duality.

Problem 1. State the conditions that guarantee the existence of a steady state P for Equation (12.5.14). [HINT: See Problem 1, Section 5.4.)

Problem 2. Why is there no analog of the conjugate point for the filtering problem of an observable system? [HINT: Can the control problem have a conjugate point with $L(x,u,t) > 0$?]

Problem 3. Show that, if there is no process noise and F is stable, $P(t)$ for an observable system approaches zero as $t \to \infty$.

Problem 4. Consider the usual filtering problem with the added complication that $E[w(t)v(\tau)^T] = T(t)\,\delta(t-\tau)$. Show that "duality" *suggests* the estimator $x(t)$ as

$$\dot{\hat{x}} = F\hat{x} + (PH^T + GT)R^{-1}(z - H\hat{x})\,, \qquad \hat{x}(t_0) = \hat{x}_0\,; \tag{12.5.44}$$

$$\dot{P} = FP + PF^T - (PH^T + GT)R^{-1}(T^TG^T + HP) + GQG^T\,, \qquad P(t_0) = P_0\,. \tag{12.5.45}$$

Verify (12.5.44) and (12.5.45) by considering an equivalent estimation problem where w^o and v^o are uncorrelated. [HINT: The analog of GT in the control problem of Chapter 5 is the cross-product matrix N in the criterion

$$J = \frac{1}{2}\|x(t_f)\|^2_{S_f} + \frac{1}{2}\int_{t_0}^{t_f} [x,u]\begin{bmatrix} A & N^T \\ N & B \end{bmatrix}\begin{bmatrix} x \\ u \end{bmatrix} dt .$$

The equivalent problem is

$$\dot{x} = (F - GTR^{-1}H)x + G[w - E(w/v)] + GTR^{-1}z , \qquad z = Hx + v .$$

Problem 5. The simplest nontrivial linear filtering problem involves the first-order stationary system

$$\dot{x} = w , \quad E[w(t)] = 0 , \quad E[x(0)] = 0 ,$$
$$E[w(t)w(t+\tau)] = q\,\delta(\tau) , \quad E[x(0)]^2 = p_0 ,$$

with the single continuous measurement $z(t)$, where

$$z = x + v , \quad E[v(t)] = 0 , \quad E[v(t)v(t+\tau)] = r\,\delta(\tau) .$$

Note that q, p_0, and r are all scalar constants.

Show that the optimum filter for estimating $x(t)$ is

$$\dot{\hat{x}} = [p(t)/r]\,(z - \hat{x}) ,$$

where

$$p(t) = \sqrt{rq}\,\frac{1 + be^{-2\alpha t}}{1 - be^{-2\alpha t}} ; \qquad \alpha = \sqrt{q/r} , \qquad b = \frac{p_0 - \sqrt{rq}}{p_0 + \sqrt{rq}} .$$

Note that, for $t >> 1/2\alpha$, the filter is stationary; that is, $p \rightarrow \sqrt{rq} =$ const.

Problem 6. Another fairly simple filtering problem involves the second-order system

$$\dot{x}_2 = w(t) , \quad E[w(t)] = 0 , \quad E[w(t)w(t+\tau)] = q\,\delta(\tau) ,$$
$$\dot{x}_1 = x_2 , \quad E[x_1(0)] = E[x_2(0)] = 0 , \quad E[x_1(0)]^2 = P_{11}(0) ,$$
$$E[x_2(0)]^2 = P_{22}(0) , \quad E[x_1(0)x_2(0)] = 0 ,$$

which may be interpreted as Brownian motion with no drag force ($c = 0$), or the perturbation velocity and position of a rocket with uncertainty in the thrust.

Suppose that we measure both the velocity, x_2, and position, x_1; these observations are contaminated with independent, additive, white noise:

$$z_1 = x_1 + w_1 , \quad E[w_1(t)] = 0 , \quad E[w_1(t)w_1(t+\tau)] = q_1\,\delta(\tau) ,$$

$$z_2 = x_2 + w_2\,, \qquad E[w_2(t)] = 0\,, \qquad E[w_2(t)w_2(t+\tau)] = q_2\,\delta(\tau)\,.$$

Construct the optimal *steady-state* filter for this system. Note that the steady-state variance equations are too complicated to allow a closed-form solution even for this simple problem.

Problem 7. If the covariance of the error of the initial estimate is very large, that is, $P(0) \to \infty$, it is difficult to use Equation (12.5.14) as it stands. Show that Equation (12.5.14) may also be written as

$$(d/dt)(P^{-1}) = -P^{-1}F - F^TP^{-1} + H^TR^{-1}H - P^{-1}GQG^TP^{-1}\,.$$

This Riccati equation, with $[P(0)]^{-1} = 0$, is, in general, simpler to use than (12.5.14) if $P(0) \to \infty$. Note that this equation is *linear* if $Q = 0$ (no process noise).

Problem 8. Consider

$$J = E\{\|x(t_f) - \hat{x}(t_f)\|^2\} = Tr[P(t_f)]$$

for the system

$$\dot{\hat{x}} = F\hat{x} + K(z - H\hat{x})\,.$$

Determine $K(t)$ such that J is minimized for

$$\dot{x} = Fx + Gw\,.$$

This is another derivation of the Kalman filtering formula. [HINT: Define $P(t) \triangleq E(\hat{x} - x)(\hat{x} - x)^T$ and consider the differential equation governing P as the state variable equation, with $K(t)$ as a matrix of control variables.]

12.6 Optimal filtering for nonlinear dynamic processes

Most of the dynamic systems and measurement systems encountered in practice are *nonlinear*. The optimal filters developed in Sections 12.4 and 12.5 for *linear* systems may be applied to nonlinear systems with additive white noise by linearizing about a nominal path or by continually (or occasionally) updating a linearization around the current estimates, starting with an initial guess. Considerable success has been achieved with this linearization approach (e.g., see Smith, Schmidt, and McGee, 1962;† Rauch, 1965;†† Rauch, Tung and

---

†G. L. Smith, S. Schmidt, L. A. McGee, "Application of Statistical Filtering to the Optimal Estimation of Position and Velocity on Board a Circumlunar Vehicle," NASA Ames Research Center Report, NASA-TND-1205, 1962.

††H. Rauch, "Optimum Estimation of Satellite Trajectories Including Random Fluctuations in Drag," *AIAA Journal*, Vol. 3, 1965, pp. 717–722.

Striebel, 1965;† Farrell, 1966;‡ Wagner, 1966).§ However, convergence to a reasonable estimate may *not* be obtained if the initial guess is poor or if the disturbances are so large that the linearization is inadequate to describe the system. (See also Section 12.7).

Parameter estimation for the case where the measurements are related to the parameters by nonlinear relations was treated in Section 12.2 in the two examples.

Suggested filter for a nonlinear multistage process. Consider a nonlinear multistage process with additive purely random noise:

$$x_{i+1} = f_i(x_i) + \Gamma_i w_i\,, \qquad i = 0, \ldots, N-1\,, \tag{12.6.1}$$

where

$$E(w_i) = \overline{w}_i\,, \tag{12.6.2}$$

$$E(w_i - \overline{w}_i)(w_j - \overline{w}_j)^T = Q_i\,\delta_{ij}\,, \tag{12.6.3}$$

$$E(x_o) = \bar{x}_o\,, \tag{12.6.4}$$

$$E(x_o - \bar{x}_o)(x_o - \bar{x}_o)^T = P_o\,, \tag{12.6.5}$$

$$E(x_o - \bar{x}_o)(w_i - \overline{w}_i)^T = 0\,. \tag{12.6.6}$$

The measurement system is also assumed nonlinear with additive purely random noise:

$$z_i = h_i(x_i) + v_i\,, \qquad i = 0, \ldots, N\,, \tag{12.6.7}$$

where

$$E(v_i) = 0\,, \tag{12.6.8}$$

$$E(v_i v_j^T) = R_i\,\delta_{ij}\,, \tag{12.6.9}$$

$$E(x_o - \bar{x}_o)v_i^T = 0 \qquad \text{and} \qquad E(w_i - \overline{w}_i)v_j^T = 0\,. \tag{12.6.10}$$

Equations (12.6.1) through (12.6.10) are nearly identical to the corresponding equations in Section 12.4; the only differences lie in the nonlinear functions $f_i(x_i)$ and $h_i(x_i)$ in Equations (12.6.1) and (12.6.7).

One possible filter for the nonlinear system (12.6.1) through (12.6.10) is the following obvious modification of the linear filter described in Section 12.4 [Equations (12.4.11) through (12.4.15)]:

$$\hat{x}_i = \bar{x}_i + K_i(z_i - h_i(\bar{x}_i))\,, \qquad i = 0, \ldots, k\,, \qquad k \le N\,, \tag{12.6.11}$$

---

†H. Rauch, F. Tung, and C. Striebel, "Maximum Likelihood Estimates of Linear Dynamic Systems," *AIAA Journal*, Vol. 3, No. 8, August 1965, pp. 1445–1450.

‡J. L. Farell, "Simulation of a Minimum Variance Orbital Navigation," *J. Spacecraft & Rockets*, Vol. 3, 1966, pp. 91–98.

§W. E. Wagner, *Re-entry Filtering, Prediction and Smoothing*, AIAA Preprint 65–319, July 1965.

$$\bar{x}_{i+1} = f_i(\hat{x}_i) + \Gamma_i \bar{w}_i \,, \tag{12.6.12}$$

$$K_i = P_i \left(\frac{\partial h_i}{\partial x_i} \right)^T R_i^{-1} \,, \tag{12.6.13}$$

$$P_i = M_i - M_i \left(\frac{\partial h_i}{\partial x_i} \right)^T \left[\frac{\partial h_i}{\partial x_i} M_i \left(\frac{\partial h_i}{\partial x_i} \right)^T + R_i \right]^{-1} \frac{\partial h_i}{\partial x_i} M_i \,, \tag{12.6.14}$$

$$M_{i+1} = \frac{\partial f_i}{\partial x_i} P_i \left(\frac{\partial f_i}{\partial x_i} \right)^T + \Gamma_i Q_i \Gamma_i^T \,. \tag{12.6.15}$$

Here the partial derivatives $\partial f_i/\partial x_i$ and $\partial h_i/\partial x_i$ may be evaluated on a nominal path or $\partial f_i/\partial x_i$ may be evaluated with $x_i = \hat{x}_i$ and $\partial h_i/\partial x_i$ evaluated with $x_i = \bar{x}_i$. In the latter *continuous relinearization* case, P_i and M_i (and, hence, K_i) *cannot* be precalculated as they can in the *nominal path* case but must be calculated in real time, since they are coupled to the current estimates, $\hat{x}_i$, through the relinearization procedure.

Suggested filter for a nonlinear continuous process. The continuous nonlinear system corresponding to the linear system described in Section 12.5 is

$$\dot{x} = f(x,t) + G(t)w(t) \,, \tag{12.6.16}$$

where

$$E[w(t)] = \bar{w}(t) \,, \tag{12.6.17}$$

$$E[w(t) - \bar{w}(t)]\,[w(t') - \bar{w}(t')]^T = Q(t)\,\delta(t - t') \,, \tag{12.6.18}$$

$$E[x(t_0)] = \bar{x}_0 \,, \tag{12.6.19}$$

$$E[x(t_0) - \bar{x}_0]\,[x(t_0) - \bar{x}_0]^T = P_0 \,, \tag{12.6.20}$$

$$E[x(t_0) - \bar{x}_0]\,[w(t) - \bar{w}(t)]^T = 0 \,. \tag{12.6.21}$$

Similarly, the continuous nonlinear measurement system corresponding to the linear system of Section 12.5 is

$$z(t) = h(x,t) + v(t) \,, \tag{12.6.22}$$

where

$$E[v(t)] = 0 \,, \tag{12.6.23}$$

$$E[v(t)v^T(t')] = R(t)\,\delta(t - t') \,, \tag{12.6.24}$$

$$E[x(t_0) - \bar{x}_0]\,[v(t)]^T = 0 \,, \tag{12.6.25}$$

$$E[w(t) - \bar{w}(t)]\,[v(t')]^T = 0 \,. \tag{12.6.26}$$

The nonlinearities enter only in Equations (12.6.16) and (12.6.22) through the functions $f(x,t)$ and $h(x,t)$, respectively.

One possible filter for the nonlinear system (12.6.16) through (12.6.26) is the following obvious modification of the linear filter described in Section 12.5 [Equations (12.5.13) and (12.5.14)]:

$$\dot{\hat{x}} = f(\hat{x},t) + G(t)\overline{w}(t) + P\left(\frac{\partial h}{\partial x}\right)^T R^{-1}[z(t) - h(\hat{x},t)] , \qquad \hat{x}(t_0) = \bar{x}_0 , \tag{12.6.27}$$

$$\dot{P} = \frac{\partial f}{\partial x}P + P\left(\frac{\partial f}{\partial x}\right)^T + GQG^T - P\left(\frac{\partial h}{\partial x}\right)^T R^{-1}\frac{\partial h}{\partial x}P ; \quad P(t_0) = P_0 , \tag{12.6.28}$$

where the partial derivatives $\partial f/\partial x$ and $\partial h/\partial x$ may be evaluated on a nominal path *or*, for greater accuracy, they may be evaluated with $x = \hat{x}$. In the latter continuous relinearization case $P(t)$ *cannot* be precalculated as it can in the nominal path case but must be calculated in real time, since it is coupled to the current estimate, $\hat{x}(t)$, through the relinearization procedure.

Problem 1. An example of a two-dimensional estimate of position by triangulation was given in Section 12.2. Here we extend the problem to a case involving dynamics. A sketch of the problem is shown in Figure 12.6.1.

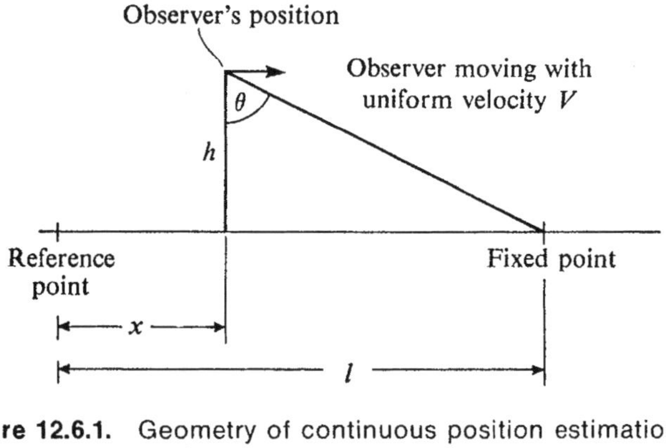

Figure 12.6.1. Geometry of continuous position estimation using continuous angle measurement.

The problem may be interpreted as an airplane, or a satellite in a circular orbit above the Earth, tracking a fixed target on Earth. The angular measurement θ is corrupted with white noise. We wish to obtain improved estimates of the altitude h and position x through measurements θ.† Formulate the nonlinear estimation problem for

†Note that this is equivalent to the two-dimensional triangulation problem in Section 12.2, where, instead of taking observations at discrete points, the observer takes continuous measurements while riding on a track.

x,h, and V and the pertinent linearized estimation equations. Show that the problem is reducible to parameter estimation if a noisy measurement of $\dot{\theta}$ is available and only V and h need to be estimated.

Problem 2. A survey team measures the elevation of a highway as a function of distance x along the highway. Let this measurement be $z_1(x) = h(x) + w_1(x)$, where $h(x)$ is the true elevation and $w_1(x)$ the error of the measurement. Let us assume that $w_1(x)$ is gaussian and

$$E(w_1(x)) = 0\,, \qquad E(w_1(x)w_1(x')) = r_1 \exp(-|x - x'|/\ell_1)\,.$$

Another survey team one year later also makes measurements of the same route and obtains $z_2(x) = h(x) + w_2(x)$, where $w_2(x)$ is a gaussian stationary purely random process with zero mean and variance, r_2 which is very much larger than r_1. Do you think it is possible to use $z_2(x)$ to improve the result of the original survey, that is, $z_1(x)$? If your answer is yes, state your scheme. If not, state your reasons.

Problem 3. An aircraft is flying a straight and level course through Colorado at a known constant speed and altitude. Suppose the elevation of the entire State of Colorado is known accurately as a function of coordinates x and y; that is, $h(x,y)$ is given. The aircraft makes a set of noisy measurements $z(0)$, $z(T)$, $z(2T), \ldots, z(kT)$ of the perpendicular distance to ground at equal intervals of time as it flies through the State. Let the initial location of the aircraft $x(0)$, $y(0)$ be $\bar{x}(0)$, $\bar{y}(0)$ with covariance

$$M(0) = \begin{bmatrix} M_{xx}(0) & M_{xy}(0) \\ M_{yx}(0) & M_{yy}(0) \end{bmatrix}$$

and assume the measurement noise is stationary, white, and zero mean.

(i) Determine the estimate $\hat{x}(kT)$, $\hat{y}(kT)$ as a result of the measurements $z(0), \ldots, z(kT)$ and the error covariance of the estimates.

(ii) What if the measurement noise has a constant but unknown bias. How do you modify your formula?

[HINT: This is a *nonlinear* but *static* estimation problem.]

Problem 4. What is the relationship between Problem 2 and 3? Can you formulate a problem involving both aspects of Problems 2 and 3? [HINT: Consider knowledge of elevation also to be in error.]

12.7 Estimation of parameters using a Bayesian approach

The estimation methods of the previous sections can be imbedded in a more general framework, where estimation is regarded as a problem of making decisions under uncertainty. This is the Bayesian decision-theoretic approach.

The parameter estimation problem. The following information is assumed given:

(a) The physical relationship governing the observation, the quantities to be estimated, and the noise:

$$z = h(x,v)\,, \tag{12.7.1}$$

where

z is the observation or measurement vector $(k \times 1)$,
x is the state vector to be estimated $(n \times 1)$,
v is the noise vector $(q \times 1)$.

(b) The joint density function of x and v, $p(x,v)$. From this, the respective marginal densities, $p(x)$ and $p(v)$, can be obtained, at least in principle.

We assume that information for (b) is available in analytical form or can be approximated by analytical distributions. Item (a) can either be in closed form or be merely computable. We wish to obtain an estimate $\hat{x}$ of x that is "best" in a sense to be defined later.

The Bayesian solution. The solution proceeds in four steps:

(a) *Evaluate* $p(z)$. This can be done analytically, at least in principle, or experimentally by Monte Carlo methods, since $z = h(x,v)$ and $p(x,v)$ are given. In the experimental case, we assume that it is possible to fit the observed distribution by a member of a family of distributions.

(b) At this point, two alternate approaches may be possible:

(i) *Evaluate* $p(x,z)$. This is possible analytically if v is of the same dimension as z and one can obtain the functional relationship $v = h^{-1}(x,z)$. In this case, by the integrand transformation Equation (10.4.5), we have

$$p(x,z) = p(x,v)|JJ^T|^{1/2}\,, \tag{12.7.2}$$

where

$$v = h^{-1}(x,z)\,, \qquad J = \frac{\partial h^{-1}(x,z)}{\partial z}\,. \tag{12.7.3}$$

(ii) *Evaluate* $p(z|x)$. This conditional density function can always be obtained, either analytically or experimentally, from $z = h(x,v)$ and $p(x,v)$.

(c) *Evaluate* $p(x|z)$ using the following relationships,

(i) Following (b)(i), we have

$$p(x|z) = [p(x,z)]/p(z)\,. \tag{12.7.4}$$

(ii) Following (b)(ii), use the Bayes formula (9.3.3):

$$p(x|z) = \frac{p(z|x)p(x)}{p(z)}. \tag{12.7.5}$$

Depending on the distributions we have assumed or obtained for $p(x,v)$, $p(z)$, this key step may be easy or difficult to carry out. Several distributions that have nice properties for this purpose can be found in Raiffa and Schlaifer, 1961, pp. 53–58. The density function $p(x|z)$ is known as the posterior density function of x. It is our state of knowledge *after* the measurements z. *By definition, it contains all the information necessary for estimation.*

(d) Depending on the criterion function for estimation we can compute the estimate $\hat{x}$ from $p(x|z)$. Some typical examples are:

(i) Criterion: Maximize the probability that $\hat{x} = x$. Solution

$$\hat{x} = \text{mode of} \quad p(x|z). \tag{12.7.6}$$

This we may call the *most probable estimate* or the unconditional *maximum-likelihood estimate.*

(ii) Criterion: Minimize $\int \|x - \hat{x}\|^2 p(x|z)\,dx$. Solution

$$\hat{x} = E(x|z). \tag{12.7.7}$$

This is the *minimum-variance estimate.*

(iii) Criterion: Minimize maximum $|x - \hat{x}|$. Solution

$$\hat{x} = \text{median of} \quad p(x|z). \tag{12.7.8}$$

This we may call the *minimum-error estimate.* The three estimates are shown pictorially in Figure 12.7.1 for a general $p(x|z)$.

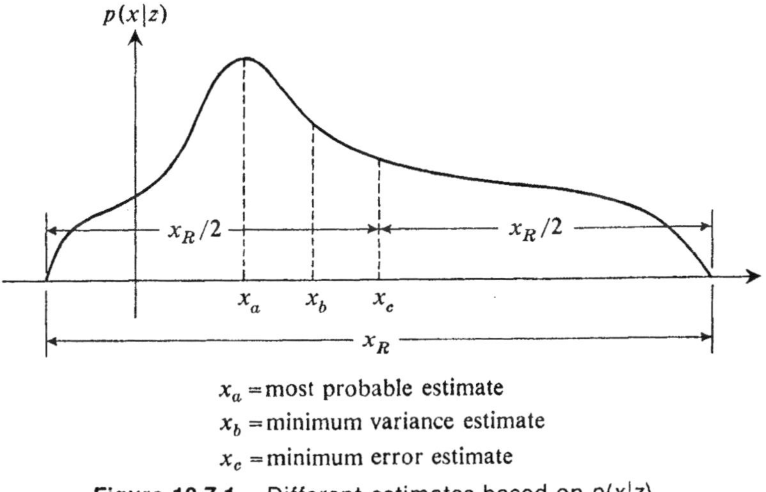

Figure 12.7.1. Different estimates based on $p(x|z)$.

Clearly, other estimates, such as confidence intervals, suggest themselves. The essential idea here is the fact that the posterior density function $p(x|z)$ is central to the estimation problem independent of the criterion we may employ.

Special case of linear measurement relations and gaussian noise. The given information in this case specializes to the following:

(a) The physical relationship

$$z = Hx + v\,. \tag{12.7.9}$$

(b) The joint density function is gaussian and independent:

$$p(x,v) = p(x)p(v)\,, \tag{12.7.10}$$

with

$$E(x) = \bar{x}\,, \qquad \operatorname{cov}(x) = M\,, \tag{12.7.11}$$

$$E(v) = 0\,, \qquad \operatorname{cov}(v) = R\,. \tag{12.7.12}$$

Following the steps for the Bayesian solution, we have

(a) *Evaluate* $p(z)$. Since $z = Hx + v$ and x,v are gaussian and independent, we immediately get (see Chapter 10)

$$p(z) \text{ is gaussian,} \qquad E(z) = H\bar{x}\,, \qquad \operatorname{cov}(z) = HMH^T + R\,. \tag{12.7.13}$$

(b) (i) *Evaluate* $p(z,x)$. We have $\partial h^{-1}/\partial z$ = identity matrix. Thus, we have

$$p(x,z) = p(x,v) = p(x)p(z - Hx)\,. \tag{12.7.14}$$

(b) (ii) *Evaluate* $p(z|x)$.† Directly, we get

$$p(z|x) = \frac{p(x,z)}{p(x)} = p(z - Hx)\,. \tag{12.7.15}$$

(c) *Evaluate* $p(x|z)$. We have

$$p(x|z) = \frac{p(x)p(z - Hx)}{p(z)}\,. \tag{12.7.16}$$

By direct substitution of (12.7.11), (12.7.12), and (12.7.13) into (12.7.16), we have

$$p(x|z) = \frac{1}{(2\pi)^{n/2}|M|^{1/2}} \exp\left\{-\frac{1}{2}(x - \bar{x})^T M^{-1}(x - \bar{x})\right\}$$

$$\times \frac{1}{(2\pi)^{p/2}|R|^{1/2}} \exp\left\{-\frac{1}{2}(z - Hx)^T R^{-1}(z - Hx)\right\}$$

†Note: Step (b) (ii) is redundant in this case.

$$\times (2\pi)^{p/2} |HMH^T + R|^{1/2} \exp\left\{ + \frac{1}{2}(z - H\bar{x})^T (HMH^T + R)^{-1}(z - H\bar{x}) \right\} \quad (12.7.17)$$

$$= \frac{|HMH^T + R|^{1/2}}{(2\pi)^{n/2}|M|^{1/2}|R|^{1/2}} \exp\left\{ -\frac{1}{2}[(x - \bar{x})^T(M^{-1} + H^TR^{-1}H)(x - \bar{x}) + z^T[R^{-1} - (HMH^T + R)^{-1}]z - z^TR^{-1}H(x - \bar{x}) - (x - \bar{x})^TH^TR^{-1}z] \right\}.$$

Now, completing the squares in the exponent of (12.7.17), we obtain

$$p(x|z) = \frac{|HMH^T + R|^{1/2}}{(2\pi)^{n/2}|M|^{1/2}|R|^{1/2}} \exp\left\{ -\frac{1}{2}(x - \hat{x})^TP^{-1}(x - \hat{x}) \right\}, \quad (12.7.18)$$

where

$$\hat{x} = \bar{x} + PH^TR^{-1}(z - H\bar{x}) \quad (12.7.19)$$

and

$$P = (M^{-1} + H^TR^{-1}H)^{-1} \quad (12.7.20)$$

or, equivalently, via the matrix inversion lemma

$$P = M - MH^T(HMH^T + R)^{-1}HM. \quad (12.7.21)$$

(d) Now, since $p(x|z)$ is gaussian, the unconditional maximum-likelihood, minimum-variance, and minimum-error estimates all coincide and are given by $\hat{x}$, the conditional mean.

This is a derivation of the relations for parameter estimation (see Section 12.2). The $(P, \hat{x})$ pair is called a *sufficient statistic* for the problem in the sense that $p(x|z) = p(x|P, \hat{x})$. It is also directly verified that x, P agree with results of Problem 4, Section 12.2.

Example. *Estimating discrete levels in gaussian noise.* A quantity x has only two possible values, x_1 and x_2, which are equally probable. A measurement of x contains additive gaussian noise, v:

$$z = x + v, \quad (12.7.22)$$

where

$$p(v) = \frac{1}{\sqrt{2\pi r}} \exp\left(-\frac{v^2}{2r} \right). \quad (12.7.23)$$

We wish to determine the conditional probabilities $p(x_1|z)$ and $p(x_2|z)$. To do this, we first determine the probability density of z:

$$p(z) = p(z|x_1)p(x_1) + p(z|x_2)p(x_2), \quad (12.7.24)$$

but

$$p(x_1) = p(x_2) = \tfrac{1}{2} \tag{12.7.25}$$

and

$$p(z|x_i) = \frac{1}{\sqrt{2\pi r}} \exp\left[-\frac{(z - x_i)^2}{2r}\right], \qquad i = 1,2. \tag{12.7.26}$$

Using Bayes's rule, we have

$$p(x_i|z) = \frac{p(z|x_i)p(x_i)}{p(z)}. \tag{12.7.27}$$

Substituting (12.7.24) through (12.7.26) into (12.7.27) gives

$$p(x_i|z) = \frac{\exp[-(z - x_i)^2/2r]}{\exp[-(z - x_1)^2/2r] + \exp[-(z - x_2)^2/2r]}, \qquad i = 1,2. \tag{12.7.28}$$

Given z, we may then calculate $p(x_1|z)$ and $p(x_2|z)$ and estimate the measured quantity as x_1 or x_2, depending on which is larger.

Further properties of the conditional mean. The conditional mean $E(x|z)$ possesses many interesting properties. If we consider the general criterion function $C(x - \hat{x})$, which is convex, symmetric, and increasing for increasing $|x - \hat{x}|$, it can be shown that the choice of $\hat{x} = E(x|z)$ will minimize the average risk defined as $\int\int C(x - \hat{x})p(x|z) p(z)\, dx\, dz$ [Deutsch (1965), Chapter 2]. Also, as was shown earlier, $E(x|z)$ is the minimum-variance estimate regardless of the distribution of $p(x|z)$. Finally, if $p(x|z)$ is gaussian, for any criterion function the optimal estimate must be a function of $\hat{x} = E(x|z)$ and $P = cov(x|z)$ since these two quantities together constitute a sufficient statistic.

Thus, just as the posterior density function $p(x|z)$ plays a central role in *estimation*, the gross characterization of $p(x|z)$, the conditional mean $\hat{x}$, plays an essential role as *estimator* for a variety of criterion functions.

12.8 Bayesian approach to optimal filtering and prediction for multistage systems

The key idea in the Bayesian approach is that of *using the measurement to update our knowledge of the state of the system*, i.e., from $p(x)$ to $p(x|z)$. In the multistage case, the updating procedure may be repeated every time a measurement is made. The posterior density function from the previous stage becomes the prior density function at the present stage. Let $Z(k)$ represent the set of measurements

$z(1), z(2), \ldots, z(k)$; then we are looking for a recursion relationship of the form

$$p[x(k+1)|Z(k+1)] = \boxed{\quad ? \quad} p[x(k)|Z(k)]. \qquad (12.8.1)$$

Assuming that $x(k)$ is a markov sequence and that $z(k)$ depends only on $x(k)$, we develop (12.8.1) as follows:

$$\begin{aligned} p[x(k+1)|Z(k+1)] &= p[x(k+1)|z(k+1), Z(k)] \\ &= \frac{p[z(k+1)|x(k+1), Z(k)]}{p[z(k+1)|Z(k)]} p[x(k+1)|Z(k)] \\ &= \frac{p[z(k+1)|x(k+1)]}{p[z(k+1)|Z(k)]} p[x(k+1)|Z(k)]. \end{aligned} \qquad (12.8.2)$$

where the markov property is invoked for the second equality. Equation (12.8.2) is *exactly the same* as Equation (12.7.5) if we identify $p[x(k+1)|Z(k)]$ with $p(x)$, the prior probability, and $p[x(k+1)| Z(k+1)]$ with $p(x|z)$, the posterior probability. If, in particular, we let $p[x(k+1)|Z(k)]$ be $N[\bar{x}(k+1), M(k+1)]$ and

$$z(k+1) = Hx(k+1) + v(k+1), \qquad (12.8.3)$$

where

$$E[v(k+1)] = 0, \qquad E[v(k)v(\ell)^T] = R\,\delta_{k\ell},$$

then it can be verified directly, in exactly the same way as (12.7.5), that $p[x(k+1)|Z(k+1)]$ is $N[\hat{x}(k+1), P(k+1)]$, where

$$\hat{x}(k+1) = \bar{x}(k+1) + P(k+1)H^T R^{-1}[z(k+1) - H\bar{x}(k+1)], \qquad (12.8.4)$$

$$P(k+1) = M(k+1) - M(k+1)H^T[HM(k+1)H^T + R]^{-1}HM(k+1).$$

The connection between $p[x(k+1)|Z(k)]$ and $p[x(k)|Z(k)]$ is a simple prediction and is given (see Section 11.6) by

$$\begin{aligned} p[x(k+1)|Z(k)] &= \int p[x(k+1)|x(k), Z(k)]p[x(k)|Z(k)]\,dx(k) \\ &= \int p[x(k+1)|x(k)]p[x(k)|Z(k)]\,dx(k), \end{aligned} \qquad (12.8.5)$$

where the markov property is invoked for the second equality. If $x(k)$ is, in addition, a gauss-markov sequence represented by

$$x(k+1) = \Phi x(k) + \Gamma w(k), \qquad (12.8.6)$$

where

$$E[w(k)] = \overline{w}(k), \qquad E[w(k) - \overline{w}(k)]\,[w(\ell) - \overline{w}(\ell)^T] = \chi\,\delta_{k\ell},$$

and $p[x(k)|Z(k)]$ is $N[\hat{x}(k), P(k)]$, then the integration in (12.8.5) can be simply carried out. In fact, we know that $p[x(k+1)|Z(k)]$ will be $N[\bar{x}(k+1), M(k+1)]$, where

$$\bar{x}(k+1) = \Phi\hat{x}(k) + \Gamma\overline{w}(k), \qquad M(k+1) = \Phi P(k)\Phi^T + \Gamma\chi\Gamma^T. \quad (12.8.7)$$

This is a Bayesian derivation of the multistage optimal linear filter for gauss-markov sequences, which we described in Section 12.3.

The answer to the more general question posed in Equation (12.8.1) can now be given by joining Equations (12.8.2) and (12.8.5):

$$p[x(k+1)|Z(k+1)]$$
$$= \frac{\int p[z(k+1)|x(k+1)]p[x(k+1)|x(k)]p[x(k)|Z(k)]\,dx(k)}{p[z(k+1)|Z(k)]} \quad (12.8.8)$$
$$= \frac{\int p[z(k+1)|x(k+1)]p[x(k+1)|x(k)]p[x(k)|Z(k)]\,dx(k)}{\int p[z(k+1)|x(k+1)]p[x(k+1)|x(k)]p[x(k)|Z(k)]\,dx(k)\,dx(k+1)}.$$

Equation (12.8.8) is the recursive relationship governing the evolution of conditional probability density of the state from one stage to another for a markov sequence *with* measurements depending only on the state; for example, $x(k)$ and $z(k)$ may obey

$$x(k+1) = f[x(k), w(k)], \quad (12.8.9)$$

$$z(k) = H[x(k), v(k)], \quad (12.8.10)$$

and $w(k)$ and $v(k)$ are purely random sequences. If one dispenses with the measurements, (12.8.8) simply reduces to (12.8.5) (see Section 11.6), the case of pure prediction. Equations (12.8.4) and (12.8.7) are extensions of (11.2.7) and (11.2.9) to the case with measurements, and Equation (12.8.8) is the extension of Equation (12.8.5).

A partial differential equation governs the evolution of the probability density function of a markov process without measurement (see Section 11.7). It is possible to derive an extension of this partial differential equation to cover the case with measurements (see Kushner,† 1964). This equation is the continuous analog of (12.8.8) and the generalization of the continuous time estimator and the continuous Riccati equation of Section 12.4. These relationships are illustrated in the table below.

†H. Kushner, "On the Differential Equations Satisfied by Conditional Probability Densities of Markov Processes With Applicators," *J. Soc. & Indust. Appl. Math.*, Vol. 2, Series A (control), 1964, pp. 106–119.

Table 12.8.1 Table of Equations Governing the Evolution of Probability Densities of Markov Processes and Sequences

Type of stochastic processes \ Information available	Without measurement	With measurement
Gauss-markov sequence	$\bar{x}_{t+1} = \Phi\bar{x}_t + \Gamma\bar{w}_t$ $P_{t+1} = \Phi P_t \Phi^T + \Gamma\chi\Gamma^T$	$\bar{x}_{t+1} = \Phi\hat{x}_t + \Gamma\bar{w}_t$ $\hat{x}_{t+1} = \bar{x}_{t+1} + P_{t+1}H^T R^{-1}(z_{t+1} - H\bar{x}_{t+1})$ $M_{t+1} = \Phi P_t \Phi^T + \Gamma\chi\Gamma^T$ $P_{t+1} = M_{t+1} - M_{t+1}H^T(HM_{t+1}H^T + R)^{-1}HM_{t+1}$
Gauss-markov process	$\dot{\bar{x}} = F\bar{x} + G\bar{w}$ $\dot{P} = FP + PF^T + GQG^T$	$\dot{\hat{x}} = F\hat{x} + PH^T R^{-1}(z - H\hat{x}) + G\bar{w}$ $\dot{P} = FP + PF^T + GQG^T - PH^T R^{-1}HP$
Markov sequence	$p(x_{t+1}) = \int p(x_{t+1}\|x_t)p(x_t)\,dx_t$	$p(x_{t+1}\|Z_{t+1}) = \dfrac{\int p(z_{t+1}\|x_{t+1})p(x_{t+1}\|x_t)p(x_t\|Z_t)\,dx_t}{\int p(z_{t+1}\|x_{t+1})p(x_{t+1}\|x_t)p(x_t\|Z_t)\,dx_t\,dx_{t+1}}$
Markov process	Kolmogorov Equation (11.7.3)	Extension of Kolmogorov Equation (Kushner, 1964)

To illustrate the use of Equation (12.8.8) for a nongaussian nonlinear case, the following example is considered.

Example 1. *Filtering of nongaussian signals.* An infrared sensor followed by a threshold detector is used in a satellite to detect warm areas on the ground. Extraneous signals, particularly reflection from clouds, enter into the measurements. The problem is to design a multistage filter to estimate the presence of warm areas on the ground using the output of the threshold detector.

Let s_k be a signal at the kth stage, that is, either one or zero. We assume that the signal sequence is a scalar Bernoulli process with

$$p(s_k) = (1 - q)\,\delta(s_k) + q\,\delta(1 - s_k) \qquad \text{(definition)}. \quad (12.8.11)\dagger$$

Let n_k be a scalar markov sequence, which, again, has only two possible values, zero or one, with

$$p(n_1) = (1 - a)\,\delta(n_1) + a\,\delta(1 - n_1)\,, \qquad (12.8.12)$$

$$p(n_{k+1}|n_k) = \left(1 - a - \frac{n_k}{2}\right)\delta(n_{k+1}) + \left(a + \frac{n_k}{2}\right)\delta(1 - n_{k+1}) \quad (12.8.13)$$

† For brevity, we use the notation

$$\delta(x) = \begin{cases} 1\,, & x = 0\,, \\ 0\,, & x \neq 0\,. \end{cases}$$

and the scalar measurement,

$$z_k = s_k \oplus n_k, \tag{12.8.14}$$

where the symbol $\oplus$ indicates the logical "or" operation.†

Essentially Equations (12.8.11) through (12.8.14) indicate the fact that, as the sensor sweeps across the field of view, cloud reflections tend to appear in groups while signals appear in isolated dots.

Initially, the probability of $s_1 = 1$ is q and that of $n_1 = 1$ is a, as given by (12.8.11) and (12.8.12). After one measurement, we calculate

$$p(n_1|z_1) = \frac{p(z_1|n_1)p(n_1)}{p(z_1)}, \tag{12.8.15}$$

$$p(s_1|z_1) = \frac{p(z_1|s_1)p(s_1)}{p(z_1)}, \tag{12.8.16}$$

where $p(z_1)$ is calculated by direct enumeration in Table 12.8.2,

Table 12.8.2 Calculation of $p(z_1)$

Values of n_1	0	0	1	1
Values of s_1	0	1	0	1
Resultant value of z_1	0	1	1	1
Probability of z_1	$(1-a)(1-q)$	$q(1-a)$	$a(1-q)$	aq

and we find via (12.8.15) and (12.8.16) that $p(s_1 = 1|z_1)$ is modified to be

$$\begin{aligned} q'(z_1) &= \frac{q\,\delta(z_1 - 1)}{(1-a)(1-q)\,\delta(z_1) + (a+q-aq)\,\delta(z_1 - 1)} \\ &= \frac{q}{a+q-aq}\,\delta(z_1 - 1) \end{aligned}$$

and $p(n_1 = 1|z_1)$ is

$$\begin{aligned} a'(z_1) &= \frac{a\,\delta(z_1 - 1)}{(1-a)(1-q)\,\delta(z_1) + (a+q-aq)\,\delta(z_1 - 1)} \\ &= \frac{a}{a+q-aq}\,\delta(z_1 - 1) \end{aligned}$$

or

$$p(s_1|z_1) = [1 - q'(z_1)]\delta(s_1) + q'(z_1)\delta(1 - s_1) \tag{12.8.17}$$

$$p(n_1|z_1) = [1 - a'(z_1)]\,\delta(s_1) + a'(z_1)\,\delta(1 - s_1). \tag{12.8.18}$$

†Another interpretation of (12.8.11) through (12.8.14) is the transmission of Morse code over a communication channel of limited bandwidth with strong burst errors. In this case, the s_k are the Morse code signals, and the n_k are the burst noises.

The important thing to notice here is that $p(s_1|z_1)$ and $p(n_1|z_1)$ have the same form as $p(s_1)$ and $p(n_1)$. Only the pertinent probabilities q and a have been modified. The next step is to compute the propagation of the densities to $p(s_2|z_1)$, $p(n_2|z_1)$. We have, by definition of the signal process, $p(s_2|z_1) = p(s_1)$; i.e., signals are independent. On the other hand, we have

$$p(n_2|z_1) = \int_{-\infty}^{\infty} p(n_2|n_1)p(n_1|z_1)\,dn_1\,. \qquad (12.8.19)$$

Using (12.8.13) and (12.8.18), we find, after some enumeration, that

$$p(n_2|z_1) = [1 - a(z_1)]\,\delta(n_2) + a(z_1)\,\delta(n_2 - 1)\,, \qquad (12.8.20)$$

where

$$a(z_1) = a + \frac{a'(z_1)}{2}. \qquad (12.8.21)$$

Now Equations (12.8.19) and (12.8.20) take the place of the original prior densities $p(n_1)$ and $p(s_1)$, and the computation for z_2 repeats the same way. In fact, the general recursion equations are

$$p(n_k|Z_k) \triangleq p(n_k|z_k,z_{k-1},\ldots) = [1 - a'(z_k)]\,\delta(n_k) + a'(z_k)\,\delta(n_k - 1)\,, \qquad (12.8.22)$$

$$a'(Z_k) \triangleq a'(z_k,z_{k-1},\ldots)$$

$$= \frac{a(Z_{k-1})\,\delta(z_{k-1})}{[1 - a(Z_{k-1})]\,(1 - q)\,\delta(z_k) + [a(Z_{k-1}) + q - a(Z_{k-1})q]\,\delta(z_k - 1)}, \qquad (12.8.23)$$

$$a(Z_{k-1}) \triangleq a(z_{k-1},z_{k-2},\ldots) = a + \frac{a'(Z_{k-1})}{2}, \qquad (12.8.24)$$

$$p(s_k|Z_k) \triangleq p(s_k|z_k,z_{k-1},\ldots) = 1 - q'(Z_k)\,\delta(s_k) + q'(Z_k)\,\delta(s_k - 1)\,, \qquad (12.8.25)$$

$$q'(Z_k) = q'(z_k,z_{k-1},\ldots)$$

$$= \frac{q\,\delta(z_k - 1)}{[1 - a(Z_{k-1})]\,(1 - q)\,\delta(z_k) + [a(Z_{k-1}) + q - a(Z_{k-1})q]\,\delta(z_k - 1)}, \qquad (12.8.26)$$

$$p(n_{k+1}|Z_k) = [1 - a(z_k)]\,\delta(n_{k+1}) + a(z_k)\,\delta(n_{k+1} - 1)\,, \qquad (12.8.27)$$

$$p(s_{k+1}|Z_k) = p(s_{k+1}). \qquad (12.8.28)$$

As a check, we considered two possible observed sequences for Z, namely, (0,1) and (1,1). With $a = 1/4$ and $q = 1/4$, we found that $p(s_2|Z_2)$ equals 0.571 and 0.337, respectively. This agrees with intuition, since the sequence (1,1) has a higher probability of being cloud reflections.

Summary. The posterior conditional density function $p(\text{state}|$ measurements) is the key to the solution of the general multistage estimation problem. Difficulties associated with the solution of the general problem appear as difficulties in the computation of $p(x|z)$. These difficulties are:

(a) *Computation of* $p(z|x)$. This problem is complicated by the nonlinear functional relationships between z and x. Except in the cases when z and x are linearly related or when z and x are scalars, very little can be done in general, either analytically or experimentally. This difficulty does not usually appear in "decision theory" since there it is assumed that $p(z|x)$ is given.
(b) *Requirement that* $p(x|z)$ *be in analytical form.* This is an obvious requirement if we intend to use the solution in real-time applications. It will not be feasible to compute $p(x|z)$ after z has occurred.
(c) *Requirement that* $p(x)$, $p(z)$, $p(x|z)$ *be conjugate or "reproducing" distributions.*† This is simply the requirement that $p(x)$ and $p(x|z)$ be density functions from the same family or that *a set of finite dimensional sufficient statistics exist.* Note that all the examples and problems discussed in this section possess this desirable property. This is precisely the reason that multistage computation can be done efficiently. In our problem, this imposed a further restriction on the functions f and h in Equations (12.8.19) and (12.8.20).

The difficulties listed above [(a) through (c)] are formidable ones. It is not likely that we can easily circumvent them except for special classes of problems, such as those discussed. This is the reason that practical progress in stochastic control beyond the linear-quadratic case is so difficult. However, the Bayesian approach offers a unified and intuitive viewpoint particularly adaptable for handling modern-day control problems in which the *state* and/or the *markov* assumptions are applicable.

12.9 Detection of gaussian signals in noise

One of the purposes of estimation is to make decisions. For example, suppose that we know that a sequence of measurements $z_1, \ldots, z_t = Z_t$ has been made on one or the other of two multistage systems, and we must decide which one it is. We know that the two sequences have the following structure:

†See Raiffa and Schlaifer (1961), Chapter III.

$$z_t^i = H^i x_t^i + v_t^i, \qquad x_{t+1}^i = \Phi^i x_t^i + \Gamma^i w_t^i; \qquad i = 0,1 . \tag{12.9.1}$$

Let H_0 be the hypothesis that $z_t = z_t^0$ and H_1 be the hypothesis that $z_t = z_t^1$. A decision procedure for deciding which hypothesis is true is to compare the likelihood-ratio function

$$L(Z_t) = \frac{p(z_1, \ldots, z_t|H_1)}{p(z_1, \ldots, z_t|H_0)} \tag{12.9.2}$$

against a threshold value. This is an optimal decision procedure for a variety of criteria, depending on the threshold value chosen [see Selin (1965), Chapter 2].†

The principal difficulty in the application of this likelihood-ratio decision procedure is the real-time generation of the joint density functions $p(Z_t|H_i)$. However, for gauss-markov sequences this can be done quite readily using procedures developed in this chapter. By Bayes's rule, we have

$$p(Z_t|H_i) = p(z_t|Z_{t-1},H_i)p(Z_{t-1}|H_i) . \tag{12.9.3}$$

Let

$$\ln[p(Z_t|H_i)] \triangleq q^i(t) .$$

Then we have

$$q^i(t) = q^i(t-1) + \ln[p(Z_t|Z_{t-1},H_i)] . \tag{12.9.4}$$

But the last term in (12.9.4) can be computed in terms of $\ln[p(z_t|\hat{x}_{t-1}^i,P_{t-1}^i)]$, since $\hat{x}_{t-1}^i$, P_{t-1}^i are sufficient statistics for Z_{t-1}. In fact, we have

$$\ln[p(z_t|Z_{t-1},H_i)] = -\tfrac{1}{2}\ln 2([\det(H^i M_{t-1}^i H^{iT} + R^i)])$$

$$- \tfrac{1}{2}(z_t - H^i\Phi^i\hat{x}_{t-1}^i)^T(H^i M_{t-1}^i H^{iT} + R^i)^{-1}(z_t - H^i\Phi^i\hat{x}_{t-1}^i) .$$

Hence, $q^i(t)$ can be computed recursively by using the filters in Section 12.3. Thus, the likelihood-ratio decision rule may be implemented recursively for a completely general class of detection problems. For more details, see Schweppe (1965).††

---

†I. Selin, *Detection Theory*. Princeton. New Jersey: Princeton University Press, 1965.
††F. Schweppe, "Evaluation of Likelihood Function for Gaussian Signals," *Trans. of IEEE on Information Theory*, Vol. IT-11, No. 1, January 1965, pp. 61–70.

Optimal smoothing and interpolation

13

13.1 Optimal smoothing for single-stage transitions

In the estimation problem considered in Section 12.3, it is obvious that improved estimates of x_0 and w_0 could be made by using the measurement z_1. In other words, measurements related to state 1 provide information about the state 0 and about the transition from state 0 to state 1, i.e., about the forcing vector w_0. We shall call these estimates *smoothing estimates* and denote them by $\hat{x}_{0/1}$ and $\hat{w}_{0/1}$ to distinguish them from the estimates $\hat{x}_0$ and $\overline{w}_0$. To be least-square estimates, the quantities $x_{0/1}$ and $w_{0/1}$ must be the values of x_0 and w_0 that minimize the quadratic form

$$J = \tfrac{1}{2}(x_0 - \hat{x}_0)^T P_0^{-1}(x_0 - \hat{x}_0) + \tfrac{1}{2}(w_0 - \overline{w}_0)^T Q_0^{-1}(w_0 - \overline{w}_0) \\ + \tfrac{1}{2}(z_1 - H_1 x_1)^T R_1^{-1}(z_1 - H_1 x_1)\,, \quad (13.1.1)$$

with the constraint

$$x_1 = \Phi_0 x_0 + \Gamma_0 w_0\,. \quad (13.1.2)$$

If x_0, w_0, v_1 are independent gaussian random vectors, the joint probability density of (x_0, w_0, v_1) is proportional to $\exp(-J)$, so that minimizing J also corresponds to maximizing the joint probability density.

Differential changes in J and x_1 due to differential changes in x_0 and w_0 are

$$dJ = dx_0^T P_0^{-1}(x_0 - \hat{x}_0) + dw_0^T Q_0^{-1}(w_0 - \overline{w}_0) - dx_1^T H_1^T R_1^{-1}(z_1 - H_1 x_1)\,, \quad (13.1.3)$$

$$dx_1 = \Phi_0\, dx_0 + \Gamma_0\, dw_0\,. \quad (13.1.4)$$

Using (13.1.4) to eliminate dx_1 from (13.1.3), we have

$$dJ = dx_0^T[P_0^{-1}(x_0 - \hat{x}_0) - \Phi_0^T H_1^T R_1^{-1}(z_1 - H_1 x_1)] + dw_0^T[Q_0^{-1}(w_0 - \overline{w}_0) \\ - \Gamma_0^T H_1^T R_1^{-1}(z_1 - H_1 x_1)]\,. \quad (13.1.5)$$

For a minimum, $dJ = 0$ for arbitrary dx_0, dw_0. This implies that the optimal estimates must be such that the coefficients of dx_0^T and dw_0^T in (13.1.5) vanish:

$$x_0 = \hat{x}_0 + P_0\Phi_0^T H_1^T R_1^{-1}(z_1 - H_1\hat{x}_{1/1}) \triangleq \hat{x}_{0/1}, \qquad (13.1.6)$$

$$w_0 = \overline{w}_0 + Q_0\Gamma_0^T H_1^T R_1^{-1}(z_1 - H_1\hat{x}_{1/1}) \triangleq \hat{w}_{0/1}, \qquad (13.1.7)$$

where, in keeping with the constraint (13.1.2), we have defined

$$\hat{x}_{1/1} \triangleq \Phi_0\hat{x}_{0/1} + \Gamma_0\hat{w}_{0/1}. \qquad (13.1.8)$$

Now, substituting (13.1.6) and (13.1.7) into (13.1.8) and solving for $\hat{x}_{1/1}$, we obtain

$$\hat{x}_{1/1} = \bar{x}_1 + P_1 H_1^T R_1^{-1}(z_1 - H_1\bar{x}_1), \qquad (13.1.9)$$

where $\bar{x}_1$ and P_1 are as given by (12.3.4), (12.3.5), and (12.3.7). It is not surprising to discover from (13.1.9) that

$$\hat{x}_{1/1} \equiv \hat{x}_1. \qquad (13.1.10)$$

By straightforward but lengthy manipulations of the previous relations, using (12.3.4) through (12.3.7), $\hat{x}_{0/1}$ and $\hat{w}_{0/1}$ may be written as

$$\hat{x}_{0/1} = \hat{x}_0 - C_0(\bar{x}_1 - \hat{x}_1), \qquad (13.1.11)$$

$$\hat{w}_{0/1} = \overline{w}_0 - B_0(\bar{x}_1 - \hat{x}_1), \qquad (13.1.12)$$

where

$$C_0 = P_0\Phi_0^T M_1^{-1}, \qquad (13.1.13)$$

$$B_0 = Q_0\Gamma_0^T M_1^{-1}. \qquad (13.1.14)$$

Let us denote the error covariance matrices for the estimates $\hat{x}_{0/1}$ and $\hat{w}_{0/1}$ as

$$P_{0/1} \triangleq E(\hat{x}_{0/1} - x_0)(\hat{x}_{0/1} - x_0)^T, \qquad Q_{0/1} \triangleq E(\hat{w}_{0/1} - w_0)(\hat{w}_{0/1} - w_0)^T. \qquad (13.1.15)$$

Using (13.1.11) and (13.1.12), it is straightforward (but tedious) to show that

$$P_{0/1} = P_0 - C_0(M_1 - P_1)C_0^T, \qquad (13.1.16)$$

$$Q_{0/1} = Q_0 - B_0(M_1 - P_1)B_0^T. \qquad (13.1.17)$$

A source of potential numerical difficulty in (13.1.11) and (13.1.12) is the term $M_1^{-1}(\hat{x}_1 - \bar{x}_1)$, since the elements of M_1^{-1} may, in some cases, be quite large and the elements of $\hat{x}_1 - \bar{x}_1$ quite small. In this case, it is better to use the identity

$$M_1^{-1}(\bar{x}_1 - \hat{x}_1) \equiv -H_1^T(H_1M_1H_1^T + R_1)^{-1}(z_1 - H_1\bar{x}_1)\,, \qquad (13.1.18)$$
$$\triangleq \lambda_0$$

so that (13.1.11) and (13.1.12) may be written as

$$\hat{x}_{0/1} = \hat{x}_0 - P_0\Phi_0^T\lambda_0\,, \qquad (13.1.19)$$

$$\hat{w}_{0/1} = \overline{w}_0 - Q_0\Gamma_0^T\lambda_0\,. \qquad (13.1.20)$$

Another source of potential numerical difficulty and complexity appears in (13.1.16) and (13.1.17) in the term $M_1^{-1}(M_1 - P_1)M_1^{-1}$, since the elements of M_1^{-1} not only have to be inverted from M_1 but also may, in some cases, be quite large while the elements of $M_1 - P_1$ are quite small. In this case, we may recast (13.1.16) and (13.1.17) in the following way:

$$\begin{aligned} P_{0/1} &= P_0 - C_0(M_1 - P_1)C_0^T = P_0 - C_0M_1C_0^T + C_0M_1C_0^T - C_0M_1C_0^T \\ &\quad + C_0P_1C_0^T = P_0 - P_0\Phi_0^TC_0^T + C_0M_1C_0^T - C_0\Phi_0P_0 + C_0P_1C_0^T \\ &= P_0(I - C_0\Phi_0)^T - C_0\Phi_0P_0(I - C_0\Phi_0)^T + C_0(P_1 + \Gamma_0Q_0\Gamma_0^T)C_0^T\,, \\ P_{0/1} &= (I - C_0\Phi_0)P_0(I - C_0\Phi_0)^T + C_0(P_1 + \Gamma_0Q_0\Gamma_0^T)C_0^T\,, \end{aligned} \qquad (13.1.21)†$$

and, similarly,

$$Q_{0/1} = (I - B_0\Gamma_0)Q_0(I - B_0\Gamma_0)^T + B_0(P_1 + \Phi_0P_0\Phi_0^T)B_0^T\,. \qquad (13.1.22)$$

Equations (13.1.21) and (13.1.22) are superior from a numerical viewpoint, since $P_{0/1}$ and $Q_{0/1}$ are now computed as the sum of two positive-semidefinite matrices (see Problem 1 of Section 12.4). On the other hand (13.1.21) and (13.1.22) still require the computation of M_1^{-1} in C_0. This can be avoided, but only at the expense of taking differences of matrices again. Consider the identity

$$M_1^{-1}(M_1 - P_1)M_1^{-1} = H_1^T(H_1M_1H_1^T + R_1)^{-1}H_1 \triangleq \Lambda_0\,. \qquad (13.1.23)$$

We may then rewrite (13.1.21) and (13.1.22) or (13.1.16) and (13.1.17) as

$$P_{0/1} = P_0 - P_0\Phi_0^T\Lambda_0\Phi_0P_0\,, \qquad (13.1.24)$$

$$Q_{0/1} = Q_0 - Q_0\Gamma_0^T\Lambda_0\Gamma_0Q_0\,. \qquad (13.1.25)$$

Equations (13.1.24) and (13.1.25) are useful in the multistage case in the next section, since the Λ matrices can be computed recursively without resorting to matrix inversion.

Problem 1. Prove Equation (13.1.9), using Equations (13.1.6), (13.1.7), and (12.3.4) through (12.3.7).

---

†This derivation is due to Donald Fraser, Ph.D. Thesis, M.I.T., 1967.

Problem 2. Prove Equations (13.1.11) through (13.1.14), using Equations (13.1.6) through (13.1.9) and Equations (12.3.4) through (12.3.7).

Problem 3. Prove Equations (13.1.16) and (13.1.17), using Equations (12.3.5) and (12.3.14).

13.2 Optimal smoothing for multistage processes

Consider, once again, the gauss-markov sequence described in Section 12.4. We would now like to find the least-square estimate of the state x_i, using *all* the measurements $z_1, \ldots, z_N$, where $i < N$. We shall call this a smoothing estimate and denote it by $\hat{x}_{i/N}$ to distinguish it from the filtering estimate $\hat{x}_i$, which uses only the measurements $z_1, \ldots, z_i$.

We obtain the smoothing estimate by appealing to our intuition and a sequential application of the results of the previous section.† Suppose that we have already obtained $\hat{x}_{i+1/N}$ and the associated $P_{i+1/N}$. They represent the best estimates we have concerning the state at $i+1$. Now we consider the smoothing problem of a single-stage transition between stage i and stage $i+1$. Here, $\hat{x}_{i+1/N}$ and $P_{i+1/N}$ play the roles of $\hat{x}_1$ and P_1 in Section 13.1, while $\hat{x}_{i/N}$ and $P_{i/N}$ play the roles of $\hat{x}_{0/1}$ and $P_{0/1}$. Hence, we have

$$\hat{x}_{i/N} = \hat{x}_i - C_i(\bar{x}_{i+1} - \hat{x}_{i+1/N}), \qquad \hat{x}_{N/N} \triangleq \hat{x}_N, \tag{13.2.1}$$

$$\hat{w}_{i/N} = \overline{w}_i - B_i(\bar{x}_{i+1} - \hat{x}_{i+1/N}), \tag{13.2.2}$$

$$P_{i/N} = P_i - C_i(M_{i+1} - P_{i+1/N})C_i^T, \qquad P_{N/N} = P_N, \tag{13.2.3}$$

$$Q_{i/N} = Q_i - B_i(M_{i+1} - P_{i+1/N})B_i^T, \tag{13.2.4}$$

where

$$B_i = Q_i\Gamma_i^T M_{i+1}^{-1}, \tag{13.2.5}$$

$$C_i = P_i\Phi_i^T M_{i+1}^{-1}. \tag{13.2.6}$$

In order to find the smoothing estimates from Equations (13.2.1) through (13.2.6), we must *first* determine the filter estimates $\hat{x}_i$, P_i, and the intermediate quantities $\bar{x}_i$, M_i. This requires a forward sweep through the measurements z_i. Since we have $\hat{x}_{N/N} = \hat{x}_N$ and $P_{N/N} = P_N$, we can then make a backward sweep with the recursion relations (13.2.1) through (13.2.6) to determine $\hat{x}_{i/N}$, $\hat{w}_{i/N}$, $P_{i/N}$, and $Q_{i/N}$. This is shown symbolically below.

†For a more rigorous formulation, see Problem 3 at the end of this section, and Section 13.3.

Forward sweep

$$\hat{x}_o \xrightarrow{\downarrow \overline{w}_o} \bar{x}_i \xrightarrow{\downarrow z_1} \hat{x}_1 \longrightarrow \cdots \bar{x}_N \xrightarrow{\downarrow z_N} \hat{x}_N ,$$

$$P_o \xrightarrow{\downarrow Q_o} M_1 \xrightarrow{\downarrow R_1} P_1 \longrightarrow \cdots M_N \xrightarrow{\downarrow R_N} P_N ,$$

Backward sweep

$$\hat{x}_{o/N} \xleftarrow{\bar{x}_1 \downarrow} \hat{x}_{1/N} \cdots \longleftarrow \hat{x}_{N-1/N} \xleftarrow{\bar{x}_N \downarrow} \hat{x}_{N/N} ,$$

$$P_{o/N} \xleftarrow{M_1 \downarrow} P_{1/N} \cdots \longleftarrow P_{N-1/N} \xleftarrow{M_N \downarrow} P_{N/N} .$$

As in the single-stage problem of Section 13.1, there are potential numerical difficulties and complexities† with the quantities $M_i^{-1}(\bar{x}_i - \hat{x}_{i/N})$ and $M_i^{-1}(M_i - P_i)M_i^{-1}$ that may be avoided by using the analogous method of Section 13.1. The smoothing estimates can then be obtained by using the following recursion relations:

$$\hat{x}_{i/N} = \hat{x}_i - P_i \Phi_i^T \lambda_i , \tag{13.2.7}$$

$$\hat{w}_{i/N} = \overline{w}_i - Q_i \Gamma_i^T \lambda_i , \tag{13.2.8}$$

$$\lambda_{i-1} = (I - P_i S_i)^T [\Phi_i^T \lambda_i - H_i^T R_i^{-1}(z_i - H_i \bar{x}_i)] , \qquad \lambda_N = 0 , \tag{13.2.9}$$

$$P_{i/N} = (I - C_i \Phi_i) P_i (I - C_i \Phi_i)^T + C_i (P_{i+1/N} + \Gamma_i Q_i \Gamma_i^T C_i^T , \tag{13.2.10}$$

$$Q_{i/N} = (I - B_i \Gamma_i) Q_i (I - B_i \Gamma_i)^T + B_i (P_{i+1/N} + \Phi_i P_i \Phi_i^T) B_i^T , \tag{13.2.11}$$

$$\Lambda_{i-1} = (I - P_i S_i)^T \Phi_i^T \Lambda_i \Phi_i (I - P_i S_i) + S_i (I - P_i S_i) , \qquad \Lambda_N = 0 , \tag{13.2.12}$$

where

$$S_i = H_i^T R_i^{-1} H_i . \tag{13.2.13}$$

The analogous equations to (13.1.24) and (13.1.25) are

$$\mathrm{P}_{i/N} = P_i - P_i \Phi_i^T \Lambda_i \Phi_i P_i , \tag{13.2.14}$$

$$Q_{i/N} = Q_i - Q_i \Gamma_i^T \Lambda_i \Gamma_i Q_i . \tag{13.2.15}$$

Smoothing estimate for fixed i and increasing N. Another form of smoothing estimate can be obtained if we consider x_i for fixed i but with N increasing. It turns out (see Problem 4 of this section) that a recursion formula for this estimate is given by

$$\hat{x}_{i/N} = \hat{x}_{i/N-1} + P_{i/N} H_N^T R_N^{-1}(z_N - H_N \hat{x}_N) , \qquad \hat{x}_{i/i} = \hat{x}_i \tag{13.2.16}$$

$$P_{i/N} = P_{i/N-1}(I - P_N H_N^T R_N^{-1} H_N)^T , \qquad P_{i/i} = P_i . \tag{13.2.17}$$

Note that simultaneous computation of $\hat{x}_i$ and P_i is required in this approach.

†The matrices M_i now have to be inverted for all i. This is extremely cumbersome.

Problem 1. Derive the recursion relations (13.2.9) and (13.2.12).

Problem 2. Show that

$$E(\lambda_i \lambda_i^T) = \Lambda_i \,.$$

Problem 3. Consider the problem of determining the values of x_0 and $w_i, i = 0, \ldots, N-1$, so as to minimize the criterion

$$J = \tfrac{1}{2}(x_0 - \hat{x}_0)^T P_0^{-1}(x_0 - \hat{x}_0) + \sum_{i=0}^{N-1} \tfrac{1}{2}(w_i - \overline{w}_i)^T Q_i (w_i - \overline{w}_i)$$
$$+ \sum_{i=1}^{N} \tfrac{1}{2}(z_i - H_i x_i)^T R_i^{-1}(z_i - H_i x_i),$$

subject to the constraint $x_{i+1} = \Phi_i x_i + \Gamma_i w_i$.

Show that the solution of this problem constitutes the smoothing estimates discussed in this section. [HINT: Use the sweep method of Chapter 5 and see Bryson and Frazier (1963).]

Problem 4. Derive (13.2.16) and (13.2.17). The Bayesian approach discussed in Section 12.8 is one way to do this:

$$P(x_i|z_N) = \frac{P(z_N|z_{N-1}, x_i)}{P(z_N|z_{N-1})} P(x_i|z_{N-1}).$$

13.3 Optimal smoothing and interpolation for continuous processes

The formulas for continuous-time smoothing can be derived formally by passing to the limit of the discrete formulas as was done in Section 12.5. However, for pedagogical reasons, let us consider a direct derivation by considering the analog of Problem 3 of Section 13.2:

Determine $x(t_0)$ and $w(t)$ so as to minimize the criterion

$$\begin{aligned} J &= \tfrac{1}{2}[\hat{x}(t_0) - x(t_0)]^T P^{-1}(t_0)[\hat{x}(t_0) - x(t_0)] \\ &+ \frac{1}{2}\int_{t_0}^{t_f} [(w(t) - \overline{w}(t))^T Q^{-1}(t)(w(t) - \overline{w}(t)) \\ &+ (z(t) - H(t)x(t))^T R^{-1}(t)(z(t) - H(t)x(t))]\, dt \end{aligned} \tag{13.3.1}$$

subject to the constraint

$$\dot{x} = F(t)x + G(t)w\,. \tag{13.3.2}$$

This is a *deterministic* optimization problem for which the standard techniques of Chapter 2 are applicable. The extremalizing condition and the Euler equations are

$$w(t) = \overline{w} - QG^T\lambda\,, \tag{13.3.3}$$

$$\begin{bmatrix} \dot{x} \\ \dot{\lambda} \end{bmatrix} = \begin{bmatrix} F & -GQG^T \\ -H^TR^{-1}H & -F^T \end{bmatrix} \begin{bmatrix} x \\ \lambda \end{bmatrix} + \begin{bmatrix} G\overline{w} \\ H^TR^{-1}z \end{bmatrix}, \tag{13.3.4}$$

$$x(t_o) = \hat{x}(t_o) - P(t_o)\lambda(t_o), \qquad \lambda(t_f) = 0. \tag{13.3.5}$$

The solution of the two-point boundary-value problem of Equations (13.3.3) through (13.3.5) will be denoted as $\hat{x}(t/t_f)$ and $\hat{w}(t/t_f)$, which are the smoothing estimates in the continuous case. Since the two-point boundary-value problem is linear, the solution can be obtained either by transition-matrix methods or by the sweep method of Chapter 5. Choosing the latter, we postulate that the solution $x(t) \triangleq \hat{x}(t/t_f)$ satisfies

$$x(t) = \hat{x}(t) - P(t)\lambda(t), \tag{13.3.6}$$

where $\hat{x}(t)$ and $P(t)$ are to be determined. Differentiating (13.3.6) and substituting (13.3.4) into the result, we obtain

$$Fx - GQG^T\lambda + G\overline{w} = \dot{\hat{x}} - \dot{P}\lambda - P(-H^TR^{-1}Hx - F^T\lambda + H^TR^{-1}z)$$

or

$$-\dot{\hat{x}} + F\hat{x} - PH^TR^{-1}H\hat{x} + G\overline{w} + PH^TR^{-1}z = [-\dot{P} + FP + PF^T + GQG^T - PH^TR^{-1}HP]\lambda.$$

Thus, if we require $\hat{x}$, P to satisfy the filtering solution of Chapter 12,

$$\begin{aligned} \dot{\hat{x}} &= F\hat{x} + G\overline{w} + PH^TR^{-1}(z - H\hat{x}); \qquad \hat{x}(t_o) = \hat{x}_o, \\ \dot{P} &= FP + PF^T + GQG^T - PH^TR^{-1}HP, \qquad P(t_o) = P_o, \end{aligned} \tag{13.3.7}$$

then (13.3.6) becomes an identity. At $t = t_f$, we have $x(t_f) = \hat{x}(t_f) \triangleq \hat{x}(t_f/t_f)$. From this we can calculate $\lambda(t)$ via

$$\dot{\lambda} = -(F - PH^TR^{-1}H)^T\lambda + H^TR^{-1}(z - H\hat{x}), \qquad \lambda(t_f) = 0, \tag{13.3.8}$$

which is a backward sweep, whereas (13.3.7) constitutes a forward sweep. The situation is exactly the reverse of Chapter 5 (see the comment on the duality principle in Section 12.4). Once $\lambda(t)$ is known, $\hat{w}(t/t_f)$ is calculated from (13.3.3) and $\hat{x}(t/t_f)$ from (13.3.6).

Problem. Let $e(t)$ = error of the smoothed estimate = $\hat{x}(t|t_f) - x(t)$

and

$$P(t|t_f) \triangleq E[e(t)e(t)^T].$$

Show that

$$P(t|t_f) = P(t) + P(t)\Lambda(t)P(t), \tag{13.3.9}$$

where†

†In Equations (13.3.10) through (13.3.12), we have that $P \equiv P(t)$, not $P(t/t_f)$.

$$\dot{\Lambda}(t) = -(F - PH^TR^{-1}H)^T\Lambda - \Lambda(F - PH^TR^{-1}H) + H^TR^{-1}H\,, \quad \Lambda(t_f) = 0\,. \tag{13.3.10}$$

Also show that

$$\frac{d}{dt}P(t|t_f) = (F + GQG^TP^{-1})P(t|t_f) + P(t|t_f)(F + GQG^TP^{-1})^T - GQG^T$$
$$-PH^THP\,, \qquad P(t_f|t_f) = P(t_f)\,, \tag{13.3.11}$$

$$\frac{d}{dt}\hat{x}(t|t_f) = F\hat{x}(t|t_f) + G\bar{w}(t) + GQG^TP^{-1}[\hat{x}(t|t_f) - \hat{x}(t)]\,, \qquad \hat{x}(t_f|t_f) = \hat{x}(t_f)\,. \tag{13.3.12}$$

Equations (13.3.9) and (13.3.10) are the continuous-time analog of Equations (13.2.10) and (13.2.12) of Section 13.2.

Continuous-time discrete-data problems: interpolation. A special case of the smoothing problem is the interpolation problem, involving a continuous-time process with discrete measurements. All that is needed is to modify the filtering equations for discrete measurements. They are

$$\dot{\hat{x}} = F\hat{x} + K_i[z(t_i^+) - H_i\hat{x}(t_i^-)]\,\delta(t - t_i)\,, \qquad \hat{x}(t_0) = \bar{x}_0\,, \tag{13.3.13}$$

$$K_i = P(t_i^+)H_i^TR_i^{-1}\,, \tag{13.3.14}$$

$$\dot{P} = FP + PF^T + GQG^T - [P(t_i^-)H_i^T(HP(t_i^-)H_i^T + R_i)^{-1}H_iP(t_i^-)]\,\delta(t - t_i)\,. \tag{13.3.15}$$

The filter estimate $\hat{x}$ and the filter covariance matrix P are discontinuous, but the interpolated estimate $\hat{x}(t|t_i)$, calculated from (13.3.12), is continuous.

Example 1.

Process

$$\dot{x} = w\,, \quad \bar{w}(t) = 0\,, \quad \bar{x}(0) = 0\,, \quad z = x + v\,, \quad P_0 = p_0\,,$$
$$Q = q\,, \quad R = r \quad \text{(all constants)}.$$

Filter

$$\dot{\hat{x}} = \frac{P(t)}{r}(z - \hat{x})\,; \qquad P(t) = \sqrt{rq}\,\frac{1 + be^{-2\alpha t}}{1 - be^{-2\alpha t}}\,;$$

$$\alpha = \sqrt{q/r}\,, \quad b = \frac{p_0 - \sqrt{rq}}{p_0 + \sqrt{rq}}\,.$$

Smoother

$$\hat{x}(t|t_f) = \hat{x}(t) - P(t)\lambda(t)\,, \qquad \hat{w}(t|t_f) = -q\lambda(t)\,,$$

$$\dot{\lambda} = \frac{P(t)}{r}\lambda - \frac{1}{r}[z(t) - \hat{x}(t)]\,, \qquad \lambda(t_f) = 0\,,$$

$$P(t|t_f) = P(t)\left[1 - \frac{(1 + be^{-2\alpha t})(1 - e^{-2\alpha(t_f - t)})}{2(1 - be^{-2\alpha t_f})}\right]$$

$$\rightarrow \tfrac{1}{2}\sqrt{rq} \quad \text{if} \quad \alpha t\,, \alpha t_f >> 1\,,$$

$$P(0|t_f) \rightarrow \frac{p_o}{1 + (p_o/\sqrt{rq})} \quad \text{if} \quad \alpha t_f >> 1\,,$$

Estimate at $t = 0$, *using later data:*

$$\frac{d}{dt}\hat{x}(0|t) = \frac{P(0|t)}{r}[z(t) - \hat{x}(t)]\,,$$

$$P(0|t) = p_o \frac{1 - b}{1 - be^{-2\alpha t}} e^{-\alpha t}\,.$$

Relationship to the Wiener filter. It is possible to express the various filters and smoothers described in this section in terms of integrals of the measurement function $z(t)$, rather than in terms of differential equations. This can be approached via the transition-matrix method of Chapter 5. Let

$$\Phi(t,\tau) = \begin{bmatrix} \Phi_{xx}(t,\tau) & \Phi_{x\lambda}(t,\tau) \\ \Phi_{\lambda x}(t,\tau) & \Phi_{\lambda\lambda}(t,\tau) \end{bmatrix} \qquad (13.3.16)$$

be the transition matrix of Equation (13.3.4). Then (see Appendix A4) we have

$$x(t_f) = \Phi_{xx}(t_f,t_o)\hat{x}_o - P_o\lambda(t_o) + \Phi_{x\lambda}(t_f,t_o)\lambda(t_o)$$
$$+ \int_{t_o}^{t_f} [\Phi_{xx}(t_f,\tau)G\overline{w} + \Phi_{x\lambda}(t_f,\tau)H^TR^{-1}z]\,d\tau\,, \qquad (13.3.17)$$

$$\lambda(t_f) = 0 = \Phi_{\lambda x}(t_f,t_o)\hat{x}_o - P_o\lambda(t_o) + \Phi_{\lambda\lambda}(t_f,t_o)\lambda(t_o)$$
$$+ \int_{t_o}^{t_f} [\Phi_{\lambda x}(t_f,\tau)G\overline{w} + \Phi_{\lambda\lambda}(t_f,\tau)H^TR^{-1}z]\,d\tau\,. \qquad (13.3.18)$$

Combining (13.3.17) and (13.3.18) and eliminating $\lambda(t_o)$, we get

$$\hat{x}(t_f) \triangleq \hat{x}(t_f|t_f) = [\Phi_{xx}(t_f,t_o) - \Phi_{x\lambda}(t_f,t_o)\Phi_{\lambda\lambda}^{-1}(t_f,t_o)\Phi_{\lambda x}(t_f,t_o)]\hat{x}_o$$
$$+ \int_{t_o}^{t_f} [(\Phi_{xx}(t_f,\tau) - \Phi_{x\lambda}(t_f,t_o)\Phi_{\lambda\lambda}^{-1}(t_f,t_o)\Phi_{\lambda x}(t_f,\tau))G\overline{w}$$
$$+ (\Phi_{x\lambda}(t_f,\tau) - \Phi_{x\lambda}(t_f,t_o)\Phi_{\lambda\lambda}^{-1}(t_f,t_o)\Phi_{\lambda\lambda}(t_f,\tau))H^TR^{-1}z]\,d\tau$$

$$\triangleq \psi_x(t_f,t_o)\hat{x}_o + \int_{t_o}^{t_f} [\psi_x(t_f,\tau)G\overline{w} + \psi_\lambda(t_f,\tau)H^TR^{-1}z]\,d\tau\,, \qquad (13.3.19)$$

where

$$\psi_x(t_f,\tau) \triangleq \Phi_{xx}(t_f,\tau) - \Phi_{x\lambda}(t_f,t_o)\Phi_{\lambda\lambda}^{-1}(t_f,t_o)\Phi_{\lambda x}(t_f,\tau),$$

$$\psi_\lambda(t_f,\tau) \triangleq \Phi_{x\lambda}(t_f,\tau) - \Phi_{x\lambda}(t_f,t_o)\Phi_{\lambda\lambda}^{-1}(t_f,t_o)\Phi_{\lambda\lambda}(t_f,\tau).$$

In the case where $\hat{x}_o = 0$, $\overline{w} = 0$, and we are interested in $\hat{z}(t_f|t_f)$, we have

$$\hat{z}(t_f|t_f) = \int_{t_0}^{t_f} H\psi_\lambda(t_f,\tau)H^T R^{-1} z(\tau)\, d\tau, \tag{13.3.20}$$

and the kernel $H\psi_\lambda(t_f,\tau)H^T R^{-1} \triangleq k(t_f,\tau)$ represents an impulse response of the time-varying Wiener filter. In steady state (that is, $t_o = -\infty$), we have $k(t_f,\tau) = k(t_f - \tau)$. Of course, the kernel for the Wiener filter is usually derived via the solution of an integral equation, and the assumption of a finite dimensional state vector is not necessary. In this sense, the Wiener approach can be said to be more general. However, computationally and conceptually, the present approach is far more practical and appealing.

Similarly, we may derive the kernel for a smoothing filter via Equation (13.3.5):

$$\begin{aligned}\hat{x}(t_o|t_f) &= \hat{x}_o - P(t_o)\lambda(t_o)\\ &= [I + P_o\Phi_{\lambda\lambda}^{-1}(t_f,t_o)\Phi_{\lambda x}(t_f,t_o)]\hat{x}_o + \int_{t_0}^{t_f} [P_o\Phi_{\lambda\lambda}^{-1}(t_f,t_o)\Phi_{\lambda x}(t_f,\tau)G\overline{w}(\tau)\\ &\quad + P_o\Phi_{\lambda\lambda}^{-1}(t_f,t_o)\Phi_{\lambda\lambda}(t_f,\tau)H^T R^{-1} z(\tau)]\, d\tau. \end{aligned} \tag{13.3.21}$$

Example 2.

Process

$$\dot{x} = w, \qquad \overline{w}(t) = 0, \qquad \hat{x}(0) = 0,$$

$$z = x + v, \qquad P_o = p_o, \qquad Q = q, \qquad R = r \qquad \text{(all constants)}.$$

Kernel function for estimate at t_f, $0 < t < t_f$

$$\hat{x}(t_f|t_f) = \frac{1}{r}\int_0^{t_f} \psi_\lambda(t_f|t)z(t)\, dt.$$

$$\psi_\lambda(t_f,t) = \sqrt{qr}\,\frac{1 + be^{-2\alpha t}}{1 - be^{-2\alpha t}}\, e^{-\alpha(t_f - t)},$$

Kernel function for estimate at $t = 0$, $0 \le t \le t_f$

$$\hat{x}(0|t_f) = \frac{1}{r}\int_0^{t_f} \psi_\lambda(t_f|t)z(t)\, dt;$$

$$\psi_\lambda(t_f|t) = \frac{1-b}{2}\, p_o\, \frac{1 + e^{-\alpha(t_f - t)}}{1 - be^{-2\alpha t}}\, e^{-\alpha t}.$$

13.4 Optimal smoothing for nonlinear dynamic processes

The concept of viewing the smoothing problem as a deterministic least-square fit problem, as described in Section 13.3, can be extended to nonlinear dynamic systems. Consider the nonlinear system

$$\dot{x} = f(x,t) + g(x,t)w \tag{13.4.1}$$

and the measurements

$$z = h(x,t) + v\,, \tag{13.4.2}$$

where w and v are zero-mean white gaussian processes with covariances Q and R, respectively. The solution of the following problem constitutes a smoothed estimate of x_0, v and w, which we shall call $\hat{x}_0$, $\hat{v}(t)$ and $\hat{w}(t)$:

Determine x_0, $v(t)$, *and* $w(t)$ $(t_0 \leq t \leq t_f)$ *to minimize*

$$J = \frac{1}{2}[(x_0 - \bar{x}_0)^T P_0^{-1}(x_0 - \bar{x}_0)]_{t_0} + \frac{1}{2}\int_{t_0}^{t_f} [v^T R^{-1} v + w^T Q^{-1} w]\, dt \tag{13.4.3}$$

subject to the constraints (13.4.1) *and* (13.4.2).

The smoothed estimate of the state $\hat{x}(t/t_f)$ obeys the differential equation (13.4.1) with $w(t) = \hat{w}(t)$ and $x(t_0) = \hat{x}_0$.

The smoothing problem as stated is a straightforward optimization problem that can be solved using the iterative methods of Chapter 7, particularly the sweep method [see Bryson and Frazier (1963)].

13.5 Sequentially-correlated measurement noise†

Up to this point we have limited ourselves to estimation problems in which the measurement uncertainty was modeled as an additive purely random sequence in the case of sequential discrete measurements, or as an additive white noise in the case of continuous measurements. Clearly, in some estimation problems the measurement uncertainty will be modeled more accurately as an additive markov sequence (sequentially-correlated noise) or, in the case of continuous measurements, as an additive markov process (time-correlated or colored noise). In principle, sequentially-correlated measurement noise can be treated by the methods of Sections 12.4 and 13.2 if one introduces multistage shaping filters driven by purely random sequences to generate the sequentially-correlated noise. (See Section 11.4 for use of this device to generate colored process noise.) However, this device will *not* work for time-correlated measurement noise in the continuous measurement case. Furthermore, even in

†This section is based on Henrikson, (1968).

the sequential measurement case, it is necessary to increase the dimension of the state vector to be estimated, which is inconvenient for a real-time filter, and, even worse, the computation of the filter gains is very likely to be ill-conditioned. Thus, it is desirable to seek better ways to handle sequentially-correlated measurement noise in estimation problems.

For simplicity, we start by considering a somewhat restricted multistage estimation problem:

$$x_{i+1} = \Phi x_i + \Gamma w_i\,, \qquad i = 0, \ldots, N-1 \qquad (x \text{ an } n\text{-vector})\,, \qquad (13.5.1)$$

where w_i is a gaussian purely random sequence with zero mean and covariance Q. The measurements are described by

$$z_i = Hx_i + v_i\,, \qquad i = 0, \ldots, N\,, \qquad (z \text{ a } p\text{-vector})\,, \qquad (13.5.2)$$

where v_i is a gauss-markov sequence that may be generated by the multistage shaping filter

$$v_{i+1} = \Psi v_i + \xi_i\,, \qquad (13.5.3)$$

where ξ_i is a gaussian purely random sequence with zero mean and covariance Q°. We wish to estimate x_i from the measurements z_i.

The augmented state approach [Kalman (1960)] treats x_i and v_i as a single larger state vector:

$$y_i \triangleq \begin{bmatrix} x_i \\ \hline v_i \end{bmatrix} \qquad (13.5.4)$$

Equations (13.5.1) and (13.5.3) can then be combined into a single larger dynamic system:

$$y_{i+1} = \Phi^a y_i + \Gamma^a \eta_i\,, \qquad (13.5.5)$$

where

$$\Phi^a = \left[\begin{array}{c|c} \Phi & 0 \\ \hline 0 & \Psi \end{array}\right], \qquad \Gamma^a = \left[\begin{array}{c|c} \Gamma & 0 \\ \hline 0 & I \end{array}\right], \qquad \eta_i = \begin{bmatrix} w_i \\ \hline \xi_i \end{bmatrix}.$$

The measurements (13.5.2) in terms of y_i are "perfect"; i.e., they contain no purely random noise:

$$z_i = H^a y_i\,, \qquad (13.5.6)$$

where

$$H^a = [H \,|\, I]\,.$$

The multistage filter for the augmented system (13.5.5) and (13.5.6) follows immediately from (12.4.11) through (12.4.15), noting that $R_i = 0$:

$$\hat{y}_i = \Phi^a \hat{y}_{i-1} + K_i^a(z_i - H^a \Phi^a \hat{y}_{i-1}), \tag{13.5.7}$$

$$K_i^a = M_i^a (H^a)^T [H^a M_i^a (H^a)^T]^{-1}, \tag{13.5.8}$$

$$M_{i+1}^a = \Phi^a P_i^a (\Phi^a)^T + \Gamma^a Q^a (\Gamma^a)^T, \tag{13.5.9}$$

$$P_i^a = M_i^a - M_i^a (H^a)^T [H^a M_i^a (H^a)^T]^{-1} H^a M_i^a, \tag{13.5.10}$$

where

$$Q^a = \left[\begin{array}{c|c} Q & 0 \\ \hline 0 & Q^\circ \end{array}\right]$$

Now, in addition to the possibly cumbersome size of the augmented state vector y_i, the computations in Equations (13.5.8) and (13.5.10) are very likely to be ill-conditioned. The covariance matrix P_i^a is positive semidefinite, since, from (13.5.10), it is easy to show that

$$H^a P_i^a (H^a)^T = 0. \tag{13.5.11}$$

This is simply a statement that, after the p perfect measurements z_i, p linear combinations of the components of the $(n + p)$-vector y are known exactly (i.e., have zero variance). Numerically, this singularity of P_i^a is quite undesirable, since P_{i+1}^a, computed from (13.5.10), can easily become indefinite due to truncation errors, and, even worse, the computation of the inverse of $H^a M_i (H^a)^T$ can be very inaccurate for cases where P_i^a approximately equals M_i^a.

These difficulties can be circumvented if we are willing to compute the following weighted difference of successive measurement vectors:

$$\zeta_i = z_{i+1} - \Psi z_i. \tag{13.5.12}$$

Using Equations (13.5.1) through (13.5.3), ζ_i may be expressed as

$$\zeta_i = H(\Phi x_i + \Gamma w_i) + \Psi v_i + \xi_i - \Psi[H x_i + v_i]$$

or

$$\zeta_i = H^\circ x_i + \epsilon_i, \tag{13.5.13}$$

where

$$H^\circ \equiv H\Phi - \Psi H \qquad \text{and} \qquad \epsilon_i \equiv H\Gamma w_i + \xi_i.$$

Notice that the difference (13.5.12) was chosen to eliminate v_i from ζ_i. Now Equations (13.5.1) and (13.5.13) constitute an estimation problem for an nth-order system (instead of an $(n + p)$th-order system) with measurements that contain a gaussian purely random sequence ϵ_i. However, there are two unusual features of this problem: (a) the derived measurement ζ_i lags one step behind the actual measurement z_i and (b) the process and measurement noises are correlated; i.e., we have

$$E[w_i \epsilon_j^T] = S\,\delta_{ij}\,, \tag{13.5.14}$$

where

$$S = Q\Gamma^T H^T\,.$$

The covariance of the measurement noise is

$$E[\epsilon_i \epsilon_j^T] = R\,\delta_{ij}\,, \tag{13.5.15}$$

where

$$R \equiv H\Gamma Q \Gamma^T H^T + Q^*\,.$$

The maximum-likelihood filter for a system with correlated process and measurement noise is easily derived (see Problem 4, Section 12.5 for the continuous case). A particularly simple derivation can be made by adjoining (13.5.13) to (13.5.1) with an undetermined matrix D:

$$x_{i+1} = \Phi x_i + \Gamma w_i + D[\zeta_i - H^* x_i - \epsilon_i] = (\Phi - DH^*)x_i + D\zeta_i + \Gamma w_i - D\epsilon_i\,. \tag{13.5.16}$$

The noise appearing in (13.5.16), $\Gamma w_i - D\epsilon_i$, is a gaussian purely random sequence, and its correlation with the measurement noise ϵ_i is given, from (13.5.14) and (13.5.15), as

$$E[(\Gamma w_i - D\epsilon_i)\epsilon_j^T] = (\Gamma S - DR)\,\delta_{ij}. \tag{13.5.17}$$

This correlation will be zero if we choose D to be

$$D = \Gamma S R^{-1}\,. \tag{13.5.18}$$

With this choice of D, the estimation problem (13.5.16) and (13.5.13) is of the type treated in Sections 12.4 and 13.2. The filtering solution is, therefore,

$$\hat{x}_i = \bar{x}_i + K_i(\zeta_i - H^* \bar{x}_i)\,, \tag{13.5.19}$$

$$\bar{x}_{i+1} = \Phi \hat{x}_i + D(\zeta_i - H^* \hat{x}_i)\,, \tag{13.5.20}$$

$$K_i = P_i (H^*)^T R^{-1}\,, \tag{13.5.21}$$

$$P_i = M_i - M_i (H^*)^T (H^* M_i (H^*)^T + R)^{-1} H^* M_i\,, \tag{13.5.22}$$

$$M_{i+1} = (\Phi - DH^*)P_i(\Phi - DH^*)^T + \Gamma Q \Gamma^T - DRD^T\,, \tag{13.5.23}$$

where S, R, and D are as defined in (13.5.14), (13.5.15), and (13.5.18).

Equations (13.5.19) through (13.5.23) constitute the maximum-likelihood filter for the system described in Equations (13.5.1) through (13.5.3). We would like to emphasize a few points about this measurement-differencing filter:

(a) This filter is of dimension n instead of $(n + p)$ as in the augmented state filter of Equations (13.5.7) through (13.5.10). This is intuitively reasonable, since the measurements z_i may be regarded as p perfect measurements involving the $(n + p)$ variables x_i and v_i.

(b) The computations for the gains K_i and D are not ill-conditioned so long as $R \triangleq H\Gamma Q\Gamma^T H^T + Q^*$ is not singular. If R is singular, this implies that the derived measurement ζ_i still contains one or more perfect measurements, and further measurement differencing is possible, which, in turn, further reduces the dimension of the filter. Alternatively, another form for the filter can be used if R is singular (see Problem 2).

(c) The estimate of the state lags one step behind the measurement. It is, in fact, a single-stage smoothing estimate, so that $\hat{x}_i$ in (13.5.19) should be interpreted as $\hat{x}_{i/i+1}$ and $\bar{x}_i$ in (13.5.20) should be interpreted as $\hat{x}_{i/i}$.

(d) Initially, when only z_0 is available, we cannot compute ζ_0. We are forced, then, to use the augmented state filter (13.5.7) through (13.5.10) for the first step. This is all right because, in general, $H^a M_0^a (H^a)^T$ is *not* ill-conditioned. After the first step, the reduced filter (13.5.19) through (13.5.23) may be used.

(e) Estimates of v_i are easily made, if desired, since we have

$$\hat{v}_i = z_i - H\hat{x}_i$$

and

$$E[(v_i - \hat{v}_i)(v_i - \hat{v}_i)^T] = HP_iH^T.$$

(f) More general cases are treated in Henrikson (1968) and in Bryson and Henrikson (1968).†

Problem 1. Show that the maximum-likelihood smoothing estimates for the system (13.5.1) through (13.5.3), with sequentially-correlated measurement noise, are given by

$$\hat{x}_{i/N} = \hat{x}_i - C_i(\bar{x}_{i+1} - \hat{x}_{i+1/N}),$$

where

$$C_i = P_i(\Phi - DH^*)M_{i+1}^{-1}, \qquad P_{i/N} = P_i - C_i(M_{i+1} - P_{i+1/N})C_i^T.$$

Problem 2. Show that alternate forms for the reduced filter of Equations (13.5.19) through (13.5.23) and the smoother of Problem 1, useful when $R \triangleq H\Gamma Q\Gamma^T H + Q^*$ is singular, are:

†A. E. Bryson and L. J. Henrikson, "Estimation Using Sampled Data Containing Sequentially Correlated Noise," *J. Spacecraft & Rockets*, Vol. 5, No. 6, June 1968, pp. 662–666.

Filter

$$\hat{x}_i = \bar{x}_i + K_i(\zeta_i - H^\circ \bar{x}_i), \quad \bar{x}_{i+1} = \Phi \hat{x}_i + \Gamma S[H^\circ M_i (H^\circ)^T + R]^{-1}(\zeta_i - H^\circ \bar{x}_i),$$

$$K_i = M_i (H^\circ)^T [H^\circ M_i (H^\circ)^T + R]^{-1},$$

$$P_i = (I - K_i H^\circ) M_i (I - K_i H^\circ) + K_i R K_i^T,$$

$$M_{i+1} = \Phi P_i \Phi^T + \Gamma Q \Gamma^T - \Gamma S[H^\circ M_i (H^\circ)^T + R]^{-1} S^T \Gamma^T - \Phi K_i S^T \Gamma^T - \Gamma S K_i^T \Phi^T.$$

Smoother

$$\hat{x}_{i/N} = \hat{x}_i - C_i(\bar{x}_{i+1} - \hat{x}_{i+1/N}), \qquad C_i = (P_i \Phi^T - K_i S^T \Gamma^T) M_{i+1}^{-1},$$

$$P_{i/N} = P_i - C_i(M_{i+1} - P_{i+1/N}) C_i^T.$$

13.6 Time-correlated measurement noise

The augmented state approach, using shaping filters, will *not* work for the continuous measurement case in which the measurement uncertainty is modeled as an additive markov process (time-correlated or colored noise). Here, we *must* go to the reduced filter and smoother, which are limits of the filter and smoother introduced in the previous section.

Again, for simplicity, we consider a somewhat restricted problem. [More general filtering problems are treated in Bryson and Johansen (1965), and the smoothing problem with time-correlated noise is in Mehra and Bryson (1968).]† We have

$$\dot{x} = F(t)x + G(t)w, \qquad t_o \leq t \leq t_f \qquad (x \text{ an } n\text{-vector}); \qquad (13.6.1)$$

where

$$E[w(t)] = 0, \qquad E[w(t)w^T(\tau)] = Q(t)\,\delta(t - \tau);$$

$$E[x(t_o)] = 0, \qquad E[x(t_o)x^T(t_o)] = P(t_o).$$

The measurements are described by

$$z(t) = H(t)x(t) + v(t), \qquad (13.6.2)$$

where $v(t)$ is a gauss-markov process that may be generated by the shaping filter

$$\dot{v} = A(t)v + B(t)\xi, \qquad (13.6.3)$$

and

$$E[\xi(t)] = 0, \qquad E[\xi(t)\xi^T(\tau)] = Q^\circ(t)\,\delta(t - \tau);$$

$$E[v(t_o)] = 0, \qquad E[v(t_o)v^T(t_o)] = V(t_o).$$

†R. K. Mehra and A. E. Bryson, "Linear Smoothing Using Measurements Containing Correlated Noise With an Application to Inertial Navigation," *IEEE Trans. on Automatic Control*, Vol. 13, No. 4, October 1968.

We may eliminate $v(t)$ by constructing a *derived measurement*, $\zeta(t)$, such that

$$\zeta(t) \triangleq \dot{z} - A(t)z . \tag{13.6.4}$$

Using Equations (13.6.1) through (13.6.4), we easily show that

$$\zeta(t) = H^*(t)x + \epsilon(t) , \tag{13.6.5}$$

where

$$H^*(t) = \dot{H} + HF - AH \qquad \text{and} \qquad \epsilon(t) = HGw + B\xi .$$

Now Equations (13.6.1) and (13.6.5) constitute an estimation problem with measurements containing white noise. However, this measurement noise is correlated with the process noise in Equation (13.6.1):

$$E[w(t)\epsilon^T(\tau)] = S(t)\,\delta(t-\tau) , \tag{13.6.6}$$

where $S(t) \equiv QG^TH^T$.

The time correlation of the measurement noise is

$$E[\epsilon(t)\epsilon^T(\tau)] = R(t)\delta(t-\tau) , \tag{13.6.7}$$

where

$$R(t) \equiv HGQG^TH^T + BQ^*B^T .$$

Using the results of Problem 4, Section 12.5, the maximum-likelihood filter for the system (13.6.1) and (13.6.5) is

$$\dot{\hat{x}} = F\hat{x} + K(\dot{z} - Az - H^*\hat{x}) , \tag{13.6.8}$$

$$K(t) = [P(H^*)^T + GS]R^{-1} , \tag{13.6.9}$$

$$\dot{P} = FP + PF^T - KRK^T + GQG^T . \tag{13.6.10}$$

In principle, it is possible to differentiate the measurements to obtain $\dot{z}$. However, it is *not* necessary in this case if we introduce an intermediate state vector $x^*(t)$ defined as

$$\boxed{\hat{x}(t) = x^*(t) + K(t)z(t) .} \tag{13.6.11}$$

Differentiating (13.6.11) and substituting for $\dot{\hat{x}}$ into (13.6.8) yields

$$\boxed{\dot{x}^* = (F - KH^*)\hat{x} - (\dot{K} + KA)z(t) .} \tag{13.6.12}$$

Equations (13.6.11) and (13.6.12) are then used in place of (13.6.8).

The *initial conditions* for the filter and for Equation (13.6.10) are rather unusual. Because of the *smoothness* of $z(t)$, the instant the measurements are started, $z(t_0)$ provides good information about the system. Using the results of Problem 2, Section 12.2, we may write

$$\boxed{\hat{x}(t_0^+) = P(t_0)H^T(t_0)\,[H(t_0)P(t_0)H^T(t_0) + V(t_0)]^{-1}z(t_0)\,,} \tag{13.6.13}$$

$$\boxed{P(t_0^+) = P(t_0) - P(t_0)H^T(t_0)\,[H(t_0)P(t_0)H^T(t_0) + V(t_0)]^{-1}H(t_0)P(t_0)\,.} \tag{13.6.14}$$

Thus, there is a *discontinuity* in the estimate and in the covariance of the error in the estimate at the initial time.

Estimates of $v(t)$ are easily made, if desired, since

$$\hat{v}(t) = z(t) - H(t)\hat{x}\,, \tag{13.6.15}$$

$$E[v(t) - \hat{v}(t)]\,[v(t) - \hat{v}(t)]^T = H(t)P(t)H^T(t)\,. \tag{13.6.16}$$

Optimal feedback control in the presence of uncertainty

14

14.1 Introduction

Having studied deterministic optimal control, random processes, and optimal filtering in the previous chapters, we are now prepared to consider stochastic optimal control, in particular, the synthesis of feedback controllers that are optimum in the ensemble average sense, in the presence of random disturbances and uncertainty in measurements and initial conditions.

We shall restrict our attention largely to linear systems with gaussian noise. We do this for two reasons: (a) Many technically important systems can be adequately represented by linear models disturbed by gaussian noise. (b) The theory for nonlinear systems with noise (and for linear systems with nongaussian noise) has not yet been developed to the point where it can be straightforwardly applied in *practical* design problems.

14.2 Continuous linear systems with white process noise and perfect knowledge of the state

A simple stochastic control problem is that of designing an optimal controller for a linear system disturbed by gaussian white noise, where the performance index is quadratic, the initial conditions are random, but perfect knowledge of the state of the system is available. Let the system to be controlled be represented by the following linear model:

$$\dot{x} = F(t)x + G(t)u + w(t)\,, \tag{14.2.1}$$

where

x = state vector with n components,
u = control vector with m components,
w = process noise vector with n components, $E[w(t)] = 0$, (14.2.2)
but $E[w(t)w^T(\tau)] = Q(t)\,\delta(t-\tau)$,

$$E[x(t_o)] = 0\,, \qquad \text{but} \qquad E[x(t_o)x^T(t_o)] = X_o\,. \tag{14.2.3}$$

Let the performance index be the ensemble average of a quadratic form like the one considered in Section 5.2:

$$J = E\left\{\frac{1}{2}(x^T S_f x)_{t=t_f} + \frac{1}{2}\int_{t_o}^{t_f}(x^T Ax + u^T Bu)\,dt\right\}, \tag{14.2.4}$$

where S_f and $A(t)$ are positive-semidefinite matrices and $B(t)$ is a positive-definite matrix. They may be arbitrarily assigned or derived from considerations of the second variation of a nonlinear problem (see Chapter 6). Clearly, we wish to *minimize* this average value.

Now, $w(t)$ represents random disturbances with zero mean and short correlation times compared to characteristic times of the system (white noise). Thus, it is impossible to predict $w(\tau)$ for $\tau > t$, even with perfect knowledge of the state for $\tau < t$. Clearly, then, the optimal controller is identical to the deterministic controller of Section 5.2 (see Problem 2 of this section):

$$u(t) = -C(t)x(t)\,, \tag{14.2.5}$$

where

$$C = B^{-1}G^T S\,, \tag{14.2.6}$$

$$\dot{S} = -SF - F^T S + SGB^{-1}G^T S - A\,, \tag{14.2.7}$$

$$S(t_f) = S_f\,. \tag{14.2.8}$$

Average behavior of the optimally-controlled system. It is often of interest to determine how the controlled system will behave, on the average. To do this, we first substitute (14.2.5) into (14.2.1):

$$\dot{x} = (F - GC)x + w\,. \tag{14.2.9}$$

This is now in the form of a gauss-markov random process like the ones considered in Section 11.4. If we let

$$\boxed{X(t) \triangleq E[x(t)x^T(t)]} \tag{14.2.10}$$

be the mean square value of $x(t)$, we have, from Section 11.4,

$$\dot{X} = (F - GC)X + X(F - GC)^T + Q\,, \tag{14.2.11}$$

$$X(t_o) = X_o\,. \tag{14.2.12}$$

The linear matrix equation (14.2.11), with the initial condition (14.2.12), allows us to predict the *mean-square histories of the state variables* and their cross-correlations. The *mean-square histories of the control variables* and their cross-correlations may be obtained from (14.2.5), using (14.2.10):

$$\boxed{E[u(t)u^T(t)] = CXC^T .} \tag{14.2.13}$$

We may also determine the average value of the performance index. Using the trace operator, Tr{ }, Equation (14.2.4) may be written as

$$J = \mathrm{Tr}\left\{\frac{1}{2} S_f X(t_f) + \frac{1}{2}\int_{t_0}^{t_f} (AX + BCXC^T)\, dt\right\} . \tag{14.2.14}$$

This expression may be transformed into a more interesting one by using Equations (14.2.7) and (14.2.11). We first add the perfect differential $(d/dt)(SX)$ into the integrand of (14.2.14), compensating by adding $S(t_0)X(t_0) - S_f X(t_f)$ outside the integral:

$$J = \mathrm{Tr}\left\{\frac{1}{2} S(t_0)X(t_0) + \frac{1}{2}\int_{t_0}^{t_f} (AX + BCXC^T + \dot{S}X + S\dot{X})\, dt\right\} . \tag{14.2.15}$$

Now, we substitute Equations (14.2.7) and (14.2.11) for $\dot{S}$ and $\dot{X}$, respectively, into Equation (14.2.15). All the terms in the integrand cancel except one, leaving

$$\boxed{J = \mathrm{Tr}\left\{\frac{1}{2} S(t_0)X(t_0) + \frac{1}{2}\int_{t_0}^{t_f} (SQ)\, dt\right\} .} \tag{14.2.16}$$

For $Q = 0$ (no process noise), $J = \frac{1}{2}\mathrm{Tr}[S(t_0)X(t_0)]$, identical to the optimal return function of Section 5.2. Thus, the presence of the process noise ($Q \neq 0$) *increases* the performance index, on the average, since $\mathrm{Tr}(SQ)$ is nonnegative if S and Q are nonnegative definite.

Statistically stationary case (a regulator). If the system to be controlled and the process noise are stationary (F, G, and Q constant) and the matrices A and B of the performance index are constant, the controller may also become stationary; that is, S and, hence, C may become constant (see Section 5.4). The mean-square value of the state, X, becomes constant and can be found from a set of linear algebraic equations obtained by setting $\dot{X} = 0$ in Equation (14.2.11):

$$(F - GC)X + X(F - GC)^T = -Q . \tag{14.2.17}$$

The mean-square value of the control also becomes constant and is found from (14.2.13), using the value of X from (14.2.17).

Example. In Example 2 of Section 5.4, consider random roll torques, $n(t)$, disturbing the missile, where $\dot{\omega} = -(1/\tau)\omega + (Q/\tau)\,\delta + n(t)$. Let $n(t)$ be white noise, with $E[n(t)] = 0$ and $E[n(t)n(\tau)] = N\,\delta(t-\tau)$. If we estimate the root mean square (RMS) value of $n(t)$ to be equivalent to 5 deg of aileron deflection and the correlation time to be .23 sec, we have, since .23 sec is much less than the 1-sec time constant of the system,

$$N \cong 2\left(\frac{Q}{\tau}\,\delta_N\right)^2 \cdot T_{\text{corr}} = 2\left[\left(\frac{10}{1}\right)(5)\right]^2 (.23) = 1150\,(\text{deg})^2\,(\text{sec})^{-3}\,.$$

From Example 2, Section 5.4, we have

$$F = \begin{bmatrix} 0, & 0, 0 \\ 10, & -1, 0 \\ 0, & 1, 0 \end{bmatrix}, \qquad G = \begin{bmatrix} 1 \\ 0 \\ 0 \end{bmatrix}, \qquad C = [27, 29, 180]\,,$$

so that we have

$$F - GC = \begin{bmatrix} -27, & -29, & -180 \\ 10, & -1\;, & 0 \\ 0, & 1\;, & 0 \end{bmatrix}.$$

Using Equation (14.2.17) with

$$Q = \begin{bmatrix} 0, & 0\;, & 0 \\ 0, & 1150\,, & 0 \\ 0, & 0\;, & 0 \end{bmatrix},$$

we then find that

$$X = \begin{bmatrix} 109, & -50, & -8.5 \\ -50, & 85, & 0 \\ -8.5, & 0, & 1.0 \end{bmatrix} \text{deg, sec units.}$$

From Equation (14.2.13), we have

$$E[u^2] = [27, 29, 180]\,X \begin{bmatrix} 27 \\ 29 \\ 180 \end{bmatrix} = 25{,}500\,(\text{deg sec}^{-1})^2\,.$$

Thus, the RMS state and control variables are

$$\sqrt{E(\delta^2)} = 10 \text{ deg}\,, \qquad \sqrt{E(\phi)^2} = 1.0 \text{ deg}\,, \qquad \sqrt{E(\omega^2)} = 9.2 \text{ deg sec}^{-1}\,,$$
$$\sqrt{E(u^2)} = 160 \text{ deg sec}^{-1}\,.$$

Problem 1. In Problem 6 of Section 5.4, consider random wind gusts disturbing the airplane. Their effect can be approximated by replacing β by $\beta + \beta_N$ in the equations of motion, where β_N is white noise with $E[\beta_N] = 0$ and $E[\beta_N(t)\beta_N(\tau)] = N\,\delta(t-\tau)$. With $N = .01$ (rad)^2 sec,† and perfect knowledge of the state, determine the steady-state mean-square values $E(\beta + \psi)^2$, $E(\phi^2)$, $E(\delta_a^2)$, $E(\delta_r^2)$.

ANSWER. Numerical computations (by R. Mehra) gave

$$\sqrt{E(\beta+\psi)^2} = .60 \text{ deg}, \qquad \sqrt{E(\phi^2)} = 2.87 \text{ deg}, \qquad \sqrt{E(\delta_a^2)} = 3.30 \text{ deg},$$
$$\sqrt{E(\delta_r^2)} = 2.48 \text{ deg},$$

where use was made of the RIAS computer program (ASP). (Reference 4. Ch. 5)

Problem 2. Consider the problem of minimizing

$$J = E\left\{\phi[x(t_f)] + \int_{t_0}^{t_f} L[x(t),u(t),t]\,dt\right\}$$

for the system

$$\dot{x} = f(x,u,t) + G(x,t)w\,.$$

Define

$$J^o(x,t) \triangleq \text{optimal value of } J \text{ starting in the state } x \text{ at time } t \text{ and using an optimal control.}$$

(a) Derive the functional equation

$$-J_t^o - \tfrac{1}{2}\mathrm{Tr}[J_{xx}^o GQG^T] = \min_u \{L + J_x^o f\},$$

satisfied by $J^o(x,t)$. Note that it is very similar to (except for the J_{xx}^o term) and reduces (in case $Q = 0$) to the Hamilton-Jacobi-Bellman equation treated in Chapter 4.

(b) Specialize the above problem to that of Equations (14.2.1) and (14.2.4) and obtain the results of Equations (14.2.5) through (14.2.8) and (14.2.16) directly. [HINT: Use the usual dynamic programming argument, but keep the second-order terms in the expansion.]

Problem 3. Consider the usual dynamic programming problem shown in Figure 14.2.1. At each point, if a particular decision is chosen

---

†Arrived at by estimating RMS gust velocity of 100 ft sec^{-1}, correlation distance of 250 ft, cruising speed of airplane = 800 ft sec^{-1} $\Rightarrow N \cong 2(100/800)^2 \times (250/800) \cong .01$ (rad)^2 sec.

(up or down), there is a probability of only $\frac{3}{4}$ that the particular decision is carried out (i.e., there is $\frac{1}{4}$ probability that the other path will be taken). Compute the minimal expected cost of traversing from path A to level 3.

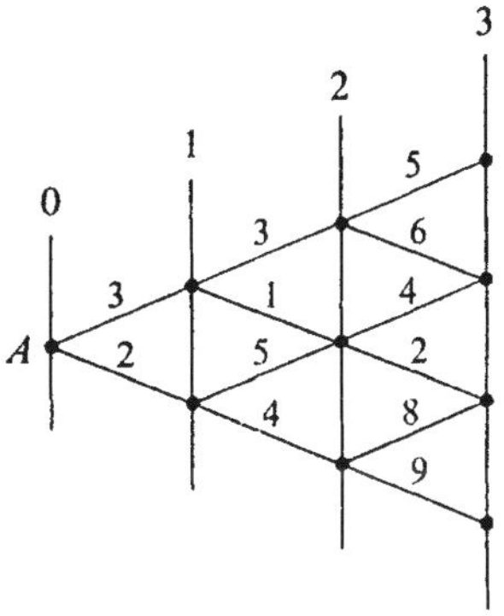

Figure 14.2.1. Path-cost grid for Problem 3.

Problem 4. Consider

$$x_{i+1} = \Phi x_i + \Gamma u_i + w_i ,$$

where

$$P(w_i/u_i) \quad \text{is} \quad N[\overline{w}_i, Q_i(u_i^T \Gamma^T \Gamma u_i)] ,$$

$$P(w_i, w_j/u_i, u_j) = P(w_i/u_i)P(w_j/u_j) ;$$

that is, the noise is *control-dependent* but independent and gaussian once given the control. If, furthermore, we assume that

$$u_i = K_i x_i ,$$

derive the mean and covariance equation of the controlled system as a function of K_i.

ANSWER. We have

$$\bar{x}_{i+1} = (\Phi + \Gamma K_i)\bar{x}_i + \overline{w}_i ,$$

$$P_{i+1} = (\Phi + \Gamma K_i)P_i(\Phi + \Gamma K_i)^T + Q_i \operatorname{Tr}[K_i^T \Gamma^T \Gamma K_i (P_i + \bar{x}_i \bar{x}_i^T)] .$$

Note that, if a criterion $J = F(\bar{x}_N, P_N)$ is selected, a completely deterministic optimization problem, using $\bar{x}_i, P_i$ as state variables and K_i as control variable, can be posed.

14.3 Continuous linear systems with process and measurements containing additive white noise; the certainty-equivalence principle

In Chapter 5, we found that the optimal control for a linear system with a quadratic performance index is a linear feedback of the state variables. In Chapter 12, we found that estimates of the state variables could be made from noisy measurements of linear combinations of the state variables, using a filter that is a "model" of the system and a feedback signal proportional to the difference between the actual and estimated measurements.

Hence, it is not surprising to find that the combination of these two logical structures, the optimal filter followed by the optimal deterministic controller, is the optimal feedback controller, in the ensemble average sense, for linear-quadratic problems with additive gaussian white noise. The conditions under which this *certainty-equivalence principle* or *separation theorem* is valid are listed below:

Given a set of measurements, $z(\tau)$, for $t_0 \le \tau \le t$, where

$$z(\tau) = H(\tau)x(\tau) + v(\tau)\,, \tag{14.3.1}$$

and a model of the dynamic system to be controlled,

$$\dot{x} = F(t)x + G(t)u + w(t)\,, \tag{14.3.2}$$

where $w(t)$, $v(t)$ are white gaussian processes,

$$E\begin{bmatrix} w(t) \\ v(t) \end{bmatrix} [w^T(t'), v^T(t')] = \begin{bmatrix} Q(t) & , T(t) \\ T^T(t) & , R(t) \end{bmatrix} \delta(t-t')\,, \tag{14.3.3}$$

$$E[w(t)] = E[v(t)] = E[x(t_0)] = 0\,, \tag{14.3.4}$$

and $x(t_0)$ is a gaussian random vector independent of $w(t)$ and $v(t)$,

$$E[x(t_0)x^T(t_0)] = P_0\,, \qquad E[x(t_0)v^T(t)] = E[x(t_0)w^T(t)] = 0\,, \tag{14.3.5}$$

find u as a functional of $z(\tau)$, $t_0 \le \tau \le t$, to minimize

$$J = E\left\{\frac{1}{2}x^T(t_f)S_f x(t_f) + \frac{1}{2}\int_{t_0}^{t_f} [x^T, u^T] \begin{bmatrix} A(t) & , N(t) \\ N^T(t) & , B(t) \end{bmatrix} \begin{bmatrix} x \\ u \end{bmatrix} dt \right\}. \tag{14.3.6}$$

The *Solution* to this problem (the optimal feedback controller) is:

$$u = -C(t)\hat{x}(t)\,; \tag{14.3.7}$$

$$\dot{\hat{x}} = F\hat{x} + Gu + K(t)\,[z(t) - H\hat{x}]\,, \qquad \hat{x}(t_0) = 0\,; \tag{14.3.8}$$

where

$$C = B^{-1}(G^T S + N^T)\,, \tag{14.3.9}$$

$$K = (PH^T + T)R^{-1}\,; \tag{14.3.10}$$

and

$$\dot{S} = -SF - F^TS + C^TBC - A\,, \qquad S(t_f) = S_f\,; \tag{14.3.11}$$

$$\dot{P} = FP + PF^T - KRK^T + Q\,, \qquad P(t_o) = P_o\,. \tag{14.3.12}$$

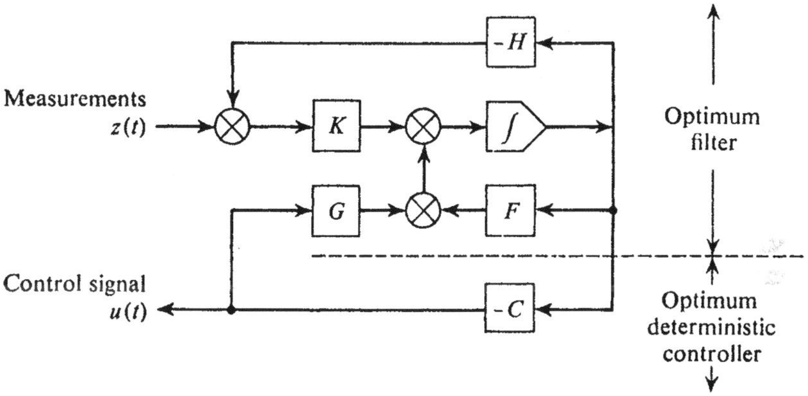

Figure 14.3.1. Flow chart of the optimal feedback controller for linear-quadratic systems with gaussian noise.

Note that $\hat{x}(t)$ is the maximum-likelihood estimate of $x(t)$, i.e., the mean of $x(t)$ conditioned on the measurements up to the time t. These estimates are used as though they were exact (certainty-equivalence) in the optimum deterministic controller.

The *certainty-equivalence principle* was stated by Simon (1956) in the field of econometrics and, in multistage form, by Joseph and Tou (1961) and Gunckel and Franklin (1963) in the automatic control field.

To better understand the certainty-equivalence principle for the control problem, we define $e(t) = \hat{x}(t) - x(t)$. From Section 12.4, we know that $E[e(t)\hat{x}(t)^T] = 0$ for all t. Hence, the criterion (14.3.6) can be rewritten as

$$\begin{aligned} J = E\left\{\frac{1}{2}\hat{x}(t_f)S_f\hat{x}(t_f) + \frac{1}{2}\int_{t_o}^{t_f}[\hat{x}\;u]^T\begin{bmatrix} A & N \\ N^T & B\end{bmatrix}\begin{bmatrix}\hat{x} \\ u\end{bmatrix}dt\right\} \\ + E\left\{\frac{1}{2}e(t_f)^TS_fe(t_f) + \frac{1}{2}\int_{t_o}^{t_f}e^TAe\,dt\right\}. \end{aligned} \tag{14.3.13}$$

Now, defining

$$\eta(t) \equiv z(t) - H\hat{x}(t) = -He(t) + v(t) = \text{correction term of the Kalman filter}\,, \tag{14.3.14}$$

we calculate

$$E[\eta(t)\eta(\tau)^T] = E[(-He(t) + v(t))(-He(\tau) + v(\tau))^T]$$
$$= HE[e(t)e(\tau)^T]H^T - HE[e(t)v(\tau)^T] - E[v(t)e(\tau)^T]H^T + E[v(t)v(\tau)^T]. \quad (14.3.15)$$

Since $e(t)$ obeys

$$\dot{e} = (F - KH)e + Kv - w, \qquad E[e(t_0)] = 0, \quad (14.3.16)$$

we have

$$E[e(t)e(\tau)^T] = \Psi(t,\tau)P(\tau), \qquad t > \tau, \quad (14.3.17)$$

$$E[e(t)v(\tau)^T] = E\int_{t_0}^{t} \Psi(t,\xi)[Kv(\xi) - w(\xi)]v(\tau)^T\, d\xi ;$$
$$E[e(t)v(\tau)^T] = \begin{cases} \Psi(t,\tau)(KR - T), & t \geq \tau, \\ 0, & t < \tau, \end{cases} \quad (14.3.18)$$

where Ψ is the transition matrix of (14.3.16).

Thus, Equation (14.3.15) becomes

$$E[\eta(t)\eta(\tau)^T] = H\Psi(t,\tau)P(\tau)H^T - H\Psi(t,\tau)(KR - T) + R\,\delta(t - \tau)$$
$$= H\Psi PH^T - H\Psi(PH^T + T - T) + R\,\delta(t - \tau) = R\,\delta(t - \tau), \quad (14.3.19)$$

that is, the correction term $K(z - H\hat{x})$ may be regarded as an *equivalent* white noise with zero mean and correlation $KRK^T\,\delta(t - \tau)$ in the equation

$$\dot{\hat{x}} = F\hat{x} + Gu + K(z - H\hat{x}), \qquad \hat{x}(t_0) = \bar{x}(t_0) = 0. \quad (14.3.20)$$

Since, in (14.3.13), the second term of the criterion is not affected by either $\hat{x}$ or the control u, the equivalent optimization problem is to minimize the first term of (14.3.13) subject to (14.3.20). But this problem was treated in Section 14.2 (see Problem 2 of that section). From this, the results of Equations (14.3.7) through (14.3.11) follow directly.†

Problem. Show that $(z_{i+1} - H\Phi\hat{x}_i)$ in the multistage case constitutes a purely random sequence.

14.4 Average behavior of an optimally-controlled system

It is possible to predict how an optimally-controlled system will behave, on the average. Obviously, this is of great interest when designing control systems.

---

†The above does not constitute a *proof* of the certainty-equivalence principle. The decomposition (14.3.13) is arbitrarily imposed. For a more complete and direct derivation, see Section 14.7.

The state of the controlled system and the state of the estimator are coupled. From Section 14.3, we have

$$\dot{x} = Fx - GC\hat{x} + w\,, \tag{14.4.1}$$

$$\dot{\hat{x}} = F\hat{x} - GC\hat{x} + K[H(x - \hat{x}) + v] \tag{14.4.2}$$

or, in terms of $e \equiv \hat{x} - x$,

$$\dot{e} = (F - KH)e + Kv - w\,, \tag{14.4.3}$$

$$\dot{\hat{x}} = (F - GC)\hat{x} - KHe + Kv\,. \tag{14.4.4}$$

Using Equations (14.3.14) and (14.3.19), the covariance matrices of e and $\hat{x}$ are given, respectively, by

$$\dot{P} = FP + PF^T - KRK^T + Q\,, \qquad P(t_o) = P_o\,; \tag{14.4.5}$$

$$\dot{\hat{X}} = (F - GC)\hat{X} + \hat{X}(F - GC)^T + KRK^T\,, \qquad \hat{X}(t_o) = 0\,. \tag{14.4.6}$$

Since $E(\hat{x}e^T) = 0$, we have

$$E[x(t)x(t)^T] \equiv X(t) = E[(\hat{x} - e)(\hat{x} - e)^T] = \hat{X}(t) + P(t)\,. \tag{14.4.7}$$

The *linear* matrix differential equation (14.4.6), with its initial conditions, allows us to predict the *mean-square histories of the state variables* and their cross-correlations. The *mean-square histories of the control variables* and their cross-correlations may be obtained from Equation (14.3.7) above:

$$\boxed{E[u(t)u^T(t)] = C\hat{X}C^T\,.} \tag{14.4.8}$$

We may also determine the average value of the performance index. Using the trace operator, Tr{ }, Equation (14.3.6) may be written as

$$J = \mathrm{Tr}\left\{\frac{1}{2}\,S_fX(t_f) + \frac{1}{2}\int_{t_o}^{t_f}\begin{bmatrix} A\ , N \\ N^T, B\end{bmatrix}\begin{bmatrix} X\ \ ,-\hat{X}C^T \\ -C\hat{X}\ , C\hat{X}C^T\end{bmatrix} dt\right\}, \tag{14.4.9}$$

where the fact $E[\hat{x}(t)e^T(t)] = 0$ was used. If we add the perfect differential $(d/dt)(SX)$ into the integrand of (14.4.9), and compensate by adding $S(t_o)X(t_o) - S_fX(t_f)$ outside the integral, we obtain

$$J = \mathrm{Tr}\left\{\frac{1}{2}\,S(t_o)X(t_o) + \frac{1}{2}\int_{t_o}^{t_f}[AX - NC\hat{X} - \hat{X}C^TN^T + BC\hat{X}C^T + \dot{S}X + S\dot{X}]\,dt\right\}. \tag{14.4.10}$$

Now we substitute into (14.4.10) for $\dot{S}$ from (14.3.11) and for $\dot{X} \equiv \dot{P} + \dot{\hat{X}}$ from (14.3.12) and (14.4.6). Most of the terms cancel, leaving

$$\boxed{J = \mathrm{Tr}\left\{\frac{1}{2}\,S(t_o)X(t_o) + \frac{1}{2}\int_{t_o}^{t_f}[SQ + C^TBCP]\,dt\right\}.} \tag{14.4.11}$$

For the deterministic case, we have $Q = 0$ and $P = 0$ and, hence, $X = \hat{X}$ and $J = \frac{1}{2}\text{Tr}[S(t_o)X(t_o)]$, identical to the optimal return function of Section 5.2. The presence of noise $(Q \neq 0, P \neq 0)$ increases the performance index, on the average, since $\text{Tr}(SQ)$ and $\text{Tr}(C^TBCP)$ are positive, in general.

14.5 Synthesis of regulators for stationary linear systems with stationary additive white noise

In Section 14.3, if F, G, Q, T, R, H, A, N, and B are all *constant* matrices, a statistically stationary state for the controlled system *may* exist for $t >> t_o, t << t_f$; that is, $\dot{P} \to 0$, $\dot{S} \to 0 \Rightarrow P, S$ constant $\Rightarrow K$ and C constant. For this technologically important case, the gains K and C, the mean-square state X, and the mean-square control may be obtained, in principle, as solutions of the following algebraic equations.†

$$0 = -SF - F^TS - A + (SG + N)B^{-1}(N^T + G^TS) \Rightarrow S$$
$$\Rightarrow C = B^{-1}(N^T + G^TS), \quad (14.5.1)$$

$$0 = FP + PF^T + Q - (PH^T + T)R^{-1}(T^T + HP) \Rightarrow P$$
$$\Rightarrow K = (PH^T + T)R^{-1}, \quad (14.5.2)$$

$$0 = (F - GC)\hat{X} + \hat{X}(F - GC)^T + KRK^T \Rightarrow \hat{X} \Rightarrow X = \hat{X} + P, \quad (14.5.3)$$

$$\Rightarrow E(uu^T) = C\hat{X}C^T. \quad (14.5.4)$$

Now we have $B \to 0 \Rightarrow C \to \infty, \hat{X} \to 0, E(uu^T) \to \infty, X \to P$; that is, the mean-square state can never be smaller than the mean-square error in estimating the state. Thus, we have

$$\boxed{X \geq P.} \quad (14.5.5)$$

Note that stability of the overall system is guaranteed if the conditions that guarantee the equilibrium solutions of $\dot{S} = 0, \dot{P} = 0$ are satisfied. They are

$$(F,G) \quad \text{controllable}, \qquad (F,H) \quad \text{observable}; \qquad (a)$$

$$B > 0, \qquad R > 0, \qquad A \geq 0, \qquad Q \geq 0. \qquad (b)$$

Example. *First-order system* (see Section 5.4, Example 1, and Section 11.4, pp. 334–5). We have

†In practice, it is usually simpler and faster to integrate the matrix differential equations until a steady-state solution is obtained.

$$\dot{x} = -\frac{1}{\tau}x + u + w\,, \qquad E[w(t)w(t')] = q\,\delta(t - t')\,, \tag{14.5.6}$$

$$z = x + v\,, \qquad E[v(t)v(t')] = r\,\delta(t - t')\,, \tag{14.5.7}$$

$$J = E\left\{\lim_{t_f \to \infty} \frac{1}{2t_f} \int_0^{t_f} (ax^2 + bu^2)\,dt\right\}, \tag{14.5.8}$$

$$0 = \frac{2}{\tau}S - a + \frac{1}{b}S^2 \Rightarrow S = -\frac{b}{\tau} + \sqrt{\frac{b^2}{\tau^2} + ab} \Rightarrow C = \sqrt{\frac{1}{\tau^2} + \frac{a}{b}} - \frac{1}{\tau}\,, \tag{14.5.9}$$

$$0 = -\frac{2}{\tau}P + q - \frac{1}{r}P^2 \Rightarrow P = -\frac{r}{\tau} + \sqrt{\frac{r^2}{\tau^2} + qr} \Rightarrow K = \sqrt{\frac{1}{\tau^2} + \frac{q}{r}} - \frac{1}{\tau}\,, \tag{14.5.10}$$

$$0 = -2\sqrt{\frac{1}{\tau^2} + \frac{a}{b}}\,\hat{X} + r\left[\sqrt{\frac{1}{\tau^2} + \frac{q}{r}} - \frac{1}{\tau}\right]^2 \Rightarrow \hat{X} = \frac{r}{2\tau}\frac{(\sqrt{1 + (q/r)\tau^2} - 1)^2}{\sqrt{1 + (a/b)\tau^2}}\,, \tag{14.5.11}$$

$$E(x^2) = \hat{X} + P = \frac{r}{\tau}\left[\sqrt{1 + \frac{q}{r}\tau^2} - 1\right]\left[\frac{\sqrt{1 + (q/r)\tau^2} - 1}{2\sqrt{1 + (a/b)\tau^2}} + 1\right], \tag{14.5.12}$$

$$E(u^2) = C^2\hat{X} = \frac{r}{2\tau^3}\left(\sqrt{1 + \frac{a}{b}\tau^2} - 1\right)^2 \frac{(\sqrt{1 + (q/r)\tau^2} - 1)^2}{\sqrt{1 + (a/b)\tau^2}}\,. \tag{14.5.13}$$

The controller is given by

$$\dot{\hat{x}} = -\sqrt{\frac{1}{\tau^2} + \frac{a}{b}}\,\hat{x} + \left(\sqrt{\frac{1}{\tau^2} + \frac{q}{r}} - \frac{1}{\tau}\right)(z - \hat{x})\,, \tag{14.5.14}$$

$$u = -\left[\sqrt{\frac{1}{\tau^2} + \frac{a}{b}} - \frac{1}{\tau}\right]\hat{x}\,. \tag{14.5.15}$$

Note that, as $(b/a) \to 0\,, \hat{X} \to 0\,, C \to \infty\,, E(u^2) \to \infty\,, E(x^2) \to P$.

If we were to specify the acceptable level of $E(u^2)$, (14.5.13) could be used to find b/a; the resulting controller gives minimum $E(x^2)$ for the specified level of $E(u^2)$.

Problem 1. In Example 2 of Section 5.4, consider white noise in the torque equation $\dot{\omega} = -(1/\tau)\omega + (Q/\tau)\delta + n(t)$, where $E[n(t)] = 0$, $E[n(t)n(\tau)] = N\,\delta(t - \tau)$, and a single measurement of the roll angle $z = \phi + v$, where $E[v] = 0\,, E[v(t)v(\tau)] = R\,\delta(t - \tau)$. Design and analyze the average performance of an optimal roll attitude regulator for this missile using quadratic synthesis.

Problems 2 and 3. Put random forcing functions into Problems 3 and 4 of Section 5.4 and assume various measurements with additive white noise. Design and analyze the statistical performance of optimal regulators (autopilots) for these airplanes.

Problem 4. *Constant altitude autopilot.* The longitudinal perturbations of an airplane in horizontal cruising flight were considered in Problem 5 of Section 5.4. There we were concerned with keeping the altitude, h, as nearly constant as possible. The relevant perturbation equations become

$$\dot{\alpha} = q - \frac{1}{\tau}(\alpha + \alpha_N), \qquad \dot{\theta} = q,$$

$$\dot{q} = \omega_0^2 Q\,\delta - \omega_0^2(\alpha + \alpha_N), \qquad \dot{h} = V(\theta - \alpha),$$

where V is the cruise velocity. The effect of vertical winds is to modify the angle of attack relative to the air by a *noise angle of attack*, α_N, which we assume is white noise.

Possible sensors and the quantities they measure are (i) an accelerometer to measure the lift perturbation, $(1/\tau)(\alpha + \alpha_N)$, (ii) a pitch-rate gyro to measure q, (iii) a free gyro to measure pitch angle θ, and (iv) an altimeter to measure h.

(a) Discuss which of these are essential to controlling h adequately and how adding "nonessential" sensors might improve the control system.

(b) Design a control system that uses only an altimeter and predict as much as you can about its behavior on the average.

Problem 5. An electric motor drives a machine at an average rotational speed p_0 with an average required torque of T_0. However, the required (or load) torque, $T(t)$, varies randomly with time in such a way that

$$E[T(t) - T_0][T(t') - T_0] = \overline{(\Delta T)^2}\exp[-\alpha|t - t'|].$$

A tachometer measures the speed of rotation p with additive white noise v; that is, we have

$$z = p + v, \qquad E[v(t)] = 0, \qquad E[v(t)v(t')] = r\,\delta(t - t').$$

The dynamics of the electric motor (direct current, separately excited field) are given approximately by

$$J\dot{p} = Ni - T(t),$$

where

$$i = \frac{e - Np}{R} = \text{current flowing through armature,}$$

N = torque per ampere ≡ back emf per rad $\sec^{-1}$,
e = applied emf (the control variable),
R = armature resistance, J = moment of inertia of rotating parts.

Using quadratic synthesis, show how to design a regulator that keeps the speed reasonably close to p_0 under the varying load torque, using minimum average power.

Note: The performance index involves the cross-product between the control and one of the components of the state. Also, a first-order shaping filter will be necessary to generate the load torque, since it has an exponential correlation.

Problem 6. Some precision components for a computer are to be kept at a constant temperature in an electrically-heated oven. The temperature of the oven is described quite accurately by

$$c\dot{T} = -h(T - T_e) + u(t),$$

where

T = oven temperature, c = heat capacity of oven and components,
h = heat transfer coefficient,
T_e = temperature of environment outside oven,
u = heat supplied per unit time.

The desired oven temperature, T_0, is chosen substantially above the mean environmental temperature, $\overline{T}_e$, because $u(t)$ can only be positive. The fluctuations of the environmental temperature are reasonably well described by

$$E[T_e(t) - \overline{T}_e]\,[T_e(t') - \overline{T}_e] = \sigma^2 \exp\left[-\frac{|t - t'|}{\tau}\right],$$

where

$$\sigma << T_0 - \overline{T}_e, \qquad \tau > \frac{c}{h}.$$

Assuming a precise thermometer for sensing the oven temperature, and a linear feedback controller with a specified constant gain k:

$$u = u_0 - k(T - T_0), \qquad \text{where} \qquad u_0 = h(T_0 - \overline{T}_e),$$

find the stationary mean-square values

$$E(T - T_0)^2 \qquad \text{and} \qquad E(u - u_0)^2.$$

14.6 Synthesis of terminal controllers for linear systems with additive white noise

In the presence of uncertainty it is *not possible,* in general, to meet terminal conditions with zero error, as in Section 5.3. Instead, we must specify tolerable *mean-square terminal errors*, using a performance index of the form considered in Section 14.3. It is clear, intuitively, that large mean-square control will be associated with smaller mean-square terminal errors; the synthesis procedure revolves essentially around this trade-off. The weighting matrices S_f and $B(t)$ in Equation (14.3.6) may be chosen to accomplish this trade-off by an iterative design procedure, as illustrated in Example 2 below.

However, there are certain important limits to this procedure. In particular, *the mean-square terminal errors can never be made smaller than the mean-square errors in estimating them* using the optimal filter, by virtue of Equation (14.4.7), and the fact that $\hat{X}$ is positive semidefinite. These latter values may be determined as soon as the measurement scheme is chosen, using the Riccati equation (14.3.12), and do not depend on the control logic used. This fact can be very useful in preliminary design of sensor-measurement schemes.

Example 1. *A simple first-order system with quadratic criteria* (see Example 1 of Section 5.2 and Problem 5 of Section 12.5).
Given $\dot{x} = u$, where x,u are scalar variables,

$$J = E\left\{\frac{1}{2} S_f [x(t_f)]^2 + \frac{1}{2} B \int_{t_0}^{t_f} u^2 \, dt\right\},$$

where

$$E[x(t_0)] = 0, \qquad E[x(t_0)]^2 = P_0,$$

and the single continuous measurement

$$z = x + v,$$

where

$$E[v(t)] = 0, \qquad E[v(t)v(t+\tau)] = R\,\delta(\tau),$$

find $u(t)$ to minimize J.

Solution. We have

$$u = -C(t)\hat{x}(t), \qquad \dot{\hat{x}} = u + K(t)(z - \hat{x}),$$

$$C(t) = \frac{S(t)}{B}, \qquad K(t) = \frac{P(t)}{R},$$

$$\dot{S} = \frac{S^2}{B}, \qquad S(t_f) = S_f \Rightarrow S = \frac{S_f}{1 + (S_f/B)(t_f - t)},$$

$$\dot{P} = -\frac{P^2}{R}, \qquad P(t_o) = P_o \Rightarrow P = \frac{P_o}{1 + (P_o/R)(t - t_o)},$$

$$\dot{\hat{X}} = -2\frac{S}{B}\hat{X} + \frac{P^2}{R}, \qquad \hat{X}(t_o) = 0.$$

This latter equation can be integrated to yield

$$\frac{\hat{X}(t)}{P_o} = \frac{\alpha(\beta - \theta)^2}{(\alpha + \beta)^3}\left\{(\alpha + \beta)\theta\left[\frac{1}{\alpha(\alpha + \theta)} + \frac{1}{\beta(\beta - \theta)}\right] + 2\log\frac{1 + (\theta/\alpha)}{1 - (\theta/\beta)}\right\},$$

where

$$\theta = \frac{t - t_o}{t_f - t_o}, \qquad \alpha = \frac{R}{P_o(t_f - t_o)}, \qquad \beta = \frac{B}{S_f(t_f - t_o)} + 1.$$

Figure 14.6.1 shows a typical case, for $\alpha = .1$, $\beta = 1.1$. Note how the mean-square state, $X = E(x^2)$, decreases monotonically, whereas the mean-square control, $E(u^2)$, increases monotonically and has a particularly sharp rise near the end of the control period.

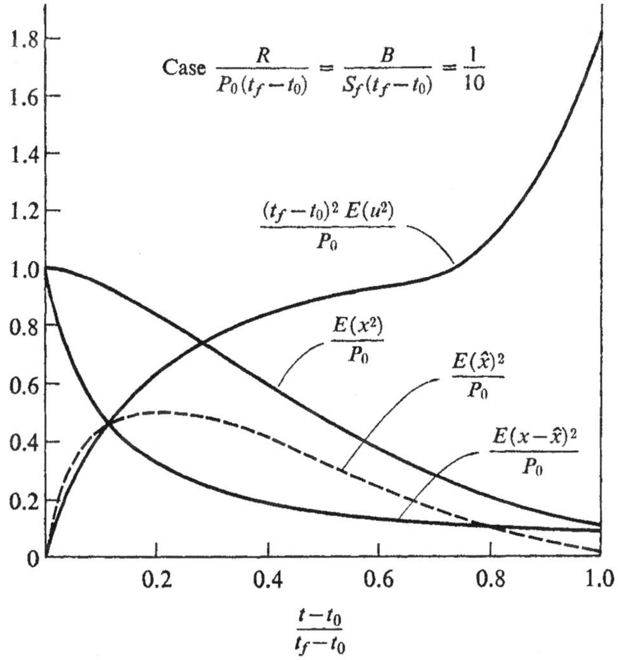

Figure 14.6.1. Histories of mean-square state and control for first-order example.

Example 2. *Intercept guidance with random target acceleration* (see Example 2, Section 5.2, with $c_1 = 0$, $c_2 = (1/b)$).

The dynamic system is

$$\dot{y} = v, \qquad \dot{v} = a_p - a_T$$

where y is the relative lateral position perpendicular to the initial line-of-sight (LOS), v is the relative lateral velocity, a_p the pursuer's lateral acceleration (the control), and a_T the target's lateral acceleration (See Figure 14.6.2).

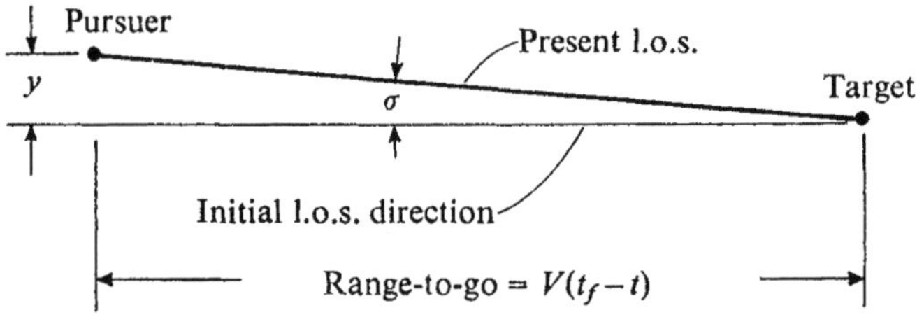

Figure 14.6.2. Nomenclature for intercept guidance example.

Here $a_T(t)$ is treated as a random forcing function with an exponential correlation:

$$E[a_T(t)] = 0, \qquad E[a_T(t)a_T(t')] = a_T^2 \exp\left[-\frac{|t-t'|}{\tau}\right]$$

The initial lateral position $y(t_0)$ is zero by definition, whereas the initial lateral velocity $v(t_0)$ is random, produced by launching error:

$$E[y(t_0)] = 0, \qquad E[v(t_0)] = 0$$
$$E[y(t_0)]^2 = 0, \qquad E[y(t_0)v(t_0)] = 0, \qquad E[v(t_0)]^2 \text{ given.}$$

The performance index is the mean square miss $E[y(t_f)]^2$ and there is a constraint on the integral mean square control. Hence

$$J = E\left\{\frac{1}{2}[y(t_f)]^2 + \frac{b}{2}\int_{t_0}^{t_f}[a_p(t)]^2\,dt\right\},$$

where b is a constant, to be selected so that the control constraint is satisfied.

The measurement is σ, the angle of the present LOS relative to the initial LOS. For $|\sigma| << 1$, $\sigma = y/[V(t_f - t)]$, where V is the relative velocity along the initial LOS direction. We assume the measurement is corrupted by fading and scintillation noise, so that

$$z = \frac{y}{V(t_f - t)} + v$$

$$E[v(t)] = 0\,, \qquad E[v(t)v(t')] = \left[R_1 + \frac{R_2}{(t_f - t)^2}\right]\delta(t - t')\,.$$

Figure 14.6.3 shows the optimal controller logic, where a first-order shaping filter with time constant τ seconds has been used to simulate the correlated target acceleration.

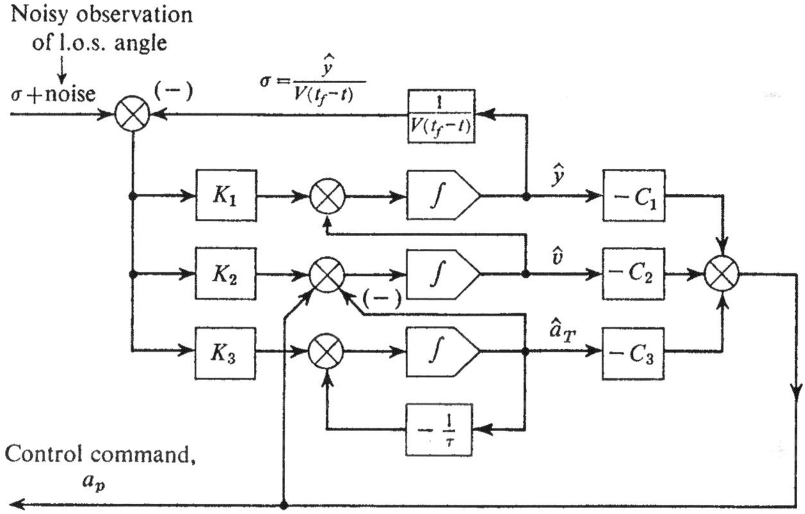

Figure 14.6.3. Optimal controller for intercept guidance example.

Figure 14.6.4 shows some computed results (courtesy W. S. Widnall, M.I.T. Instrumentation Laboratory) for parameters approximating a Falcon or Sparrow guided missile. ($V = 3000$ ft sec^{-1}, $t_f = 10$ sec, $R_1 = 15 \times 10^{-6}$ rad^{2}sec, $R_2 = 1.67 \times 10^{-3}$ rad^{2}sec^3.) A large RMS launch error was assumed (about 4° off LOS) and an RMS target acceleration of 100 ft sec.$^{-2}$ with a time correlation of 2.0 seconds was used. The RMS miss was only 31 feet (initial range was 30,000 ft), with the weighting factor b in the performance index chosen so that the maximum RMS control acceleration was 400 ft sec^{-2} ($b = 1.51 \times 10^{-2}$ sec^3). Note that the mean-square state is the sum of the MS estimate and the MS error in estimate. The terminal RMS error in the estimate of the miss is 12 ft and the RMS miss is 31 ft. No matter how much control we use, we can never reduce the RMS miss below 12 ft.

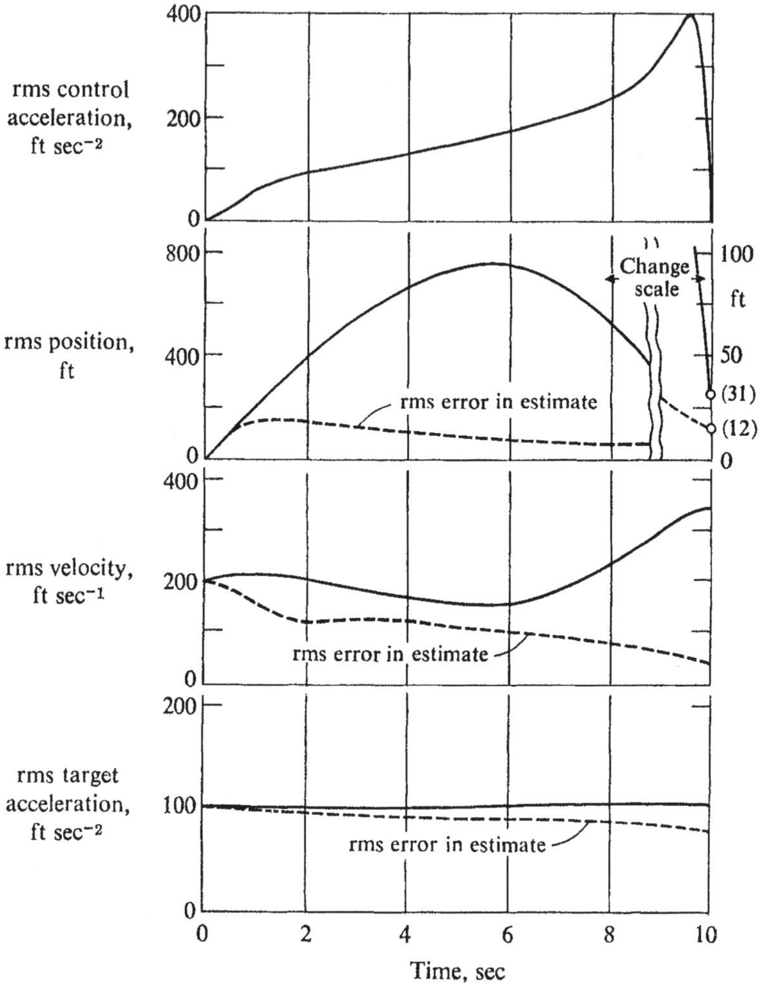

Figure 14.6.4. Numerical results for intercept guidance example.

Problem 1. Consider the system

$$\dot{x} = u + w\,, \qquad E[w(t)] = 0\,, \qquad E[w(t)w(t')] = Q\,\delta(t - t')\,;$$

$$z = x + v\,, \qquad E[v(t)] = 0\,, \qquad E[v(t)v(t')] = R\,\delta(t - t')\,,$$

where all quantities are scalars. Here w and v are independent random processes, and $x(0)$ is an independent random variable, with $E[x(0)] = 0$, $E[x(0)]^2 = P_0$. Design (i.e., solve explicitly) an optimal stochastic controller that minimizes the criterion

$$J = E\left\{\frac{1}{2} S_f x^2(t_f) + \frac{1}{2} B \int_{t_0}^{t_f} u^2(t)\, dt\right\}.$$

Draw the block diagram of the controlled system.

Problem 2. *Terminal controller for a stationary first-order system* (see Sections 5.2, Problem 2, and Section 11.4). A stationary first-order system has control $u(t)$, a random forcing function w, and random initial conditions

$$\dot{x} + \frac{1}{\tau} x = u + w\,, \qquad E[w(t)] = 0\,, \qquad E[w(t)w(t')] = Q\,\delta(t - t')\,,$$
$$E[x(0)] = 0\,, \qquad E[x(0)]^2 = P_0\,.$$

One noisy continuous measurement, $z(t)$, is made, where

$$z = x + v\,, \qquad E[v(t)] = 0\,, \qquad E[v(t)v(t')] = R\,\delta(t - t')\,.$$

Show that the terminal controller that minimizes

$$J = E\left\{\frac{1}{2} S_f [x(t_f)]^2 + \frac{1}{2} \int_0^{t_f} (Ax^2 + Bu^2)\, dt\right\}$$

is given by

$$u = -C(t)\hat{x}\, t)\,, \qquad \dot{\hat{x}} + (1/\tau)\hat{x} = u + K(t)\,(z - \hat{x})\,; \qquad \hat{x}(0) = 0\,,$$

where

$$K(t) = \frac{1}{R}\left[P_1 + \frac{P_1 + P_2}{[(P_0 + P_2)/(P_0 - P_1)]e^{2\beta t} - 1}\right],$$

$$C(t) = \frac{1}{B}\left[S_1 + \frac{S_1 + S_2}{[(S_f + S_2)/(S_f - S_1)]\, e^{2\alpha(t_f - t)} - 1}\right],$$

and

$$\left.\begin{matrix} P_1 \\ P_2 \end{matrix}\right\} = R\left(\beta \mp \frac{1}{\tau}\right), \qquad \beta^2 = \frac{1}{\tau^2} + \frac{Q}{R}\,,$$

$$\left.\begin{matrix} S_1 \\ S_2 \end{matrix}\right\} = B\left(\alpha \mp \frac{1}{\tau}\right), \qquad \alpha^2 = \frac{1}{\tau^2} + \frac{A}{B}\,.$$

Note that, for $\beta t >> 1$ and $\alpha(t_f - t) >> 1$, K and C are nearly *constant*, so that the control system is essentially a regulator in this intermediate period. (see Figure 14.6.5).

Problem 3. For the general problem in Section 14.3, let $A = 0\,, N = 0\,,$ and $S(t_f) = \mathrm{diag}(d_1, \ldots, d_q\,; 0, \ldots, 0)$. Use the concept of predicted terminal state (Equation 5.3.53) to simplify the equations of Section 14.3.

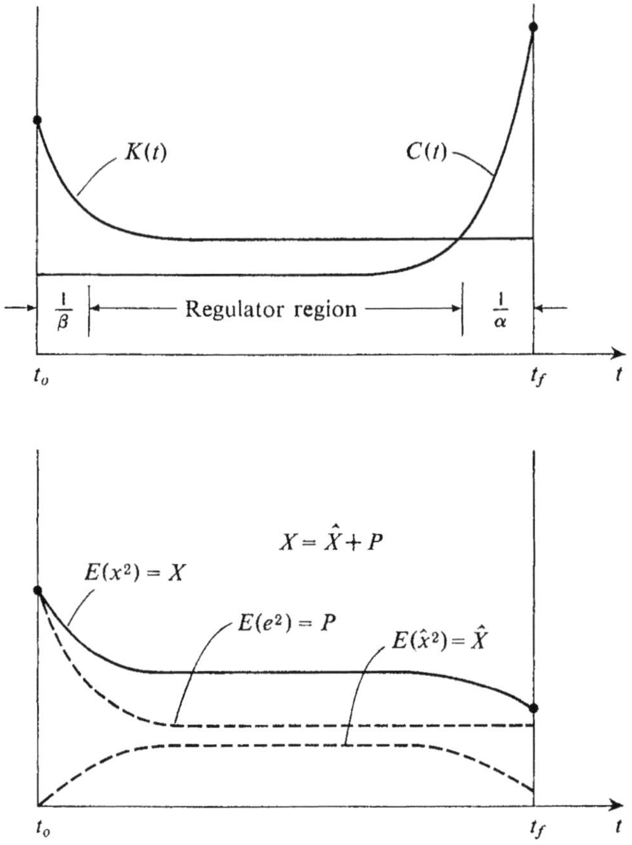

Figure 14.6.5. Behavior of terminal controller for a first-order system.

ANSWER. We have $u = -B^{-1}G^T\Phi^T(t_f,t)\bar{S}\hat{x}(t_f,t)$, where

$$\dot{\bar{S}} = \bar{S}\Phi(t_f,t)GB^{-1}G^T\Phi^T(t_f,t)\bar{S}, \qquad \bar{S}(t_f) = \text{diag}(d_1, \ldots, d_q; 0, \ldots, 0);$$
$$S_{ij}(t) = 0 \qquad \text{for all} \qquad t; \qquad i,j > q.$$

14.7 Multistage linear systems with additive purely random noise; the discrete certainty-equivalence principle

The multistage certainty-equivalence principle is very similar to the continuous certainty-equivalence principle stated in Section 14.3. It is included here for convenient reference, since it appears that digital computers will be used increasingly for both multistage and continuous-control applications.

Given a set of measurements, z_i, for $i = 0, \ldots, N-1$, where

$$z_i = H_i x_i + v_i , \tag{14.7.1}$$

and a multistage model of the dynamic system to be controlled,

$$x_{i+1} = \Phi_i x_i + \Gamma_i u_i + w_i , \tag{14.7.2}$$

where

$$E[v_i] = E[w_i] = E[x_o] = 0 , \tag{14.7.3}$$

$$E\begin{bmatrix} w_i \\ v_i \end{bmatrix} [w_j^T, v_j^T] = \begin{bmatrix} Q_i , 0 \\ 0 \ , R_i \end{bmatrix} \delta_{ij} , \tag{14.7.4}$$

$$E[x_o x_o^T] = P_o , \qquad E[x_o v_i^T] = E[x_o w_i^T] = 0 , \tag{14.7.5}$$

find u_i as a function of $z_o, \ldots, z_i$ to minimize

$$J = E\left\{\frac{1}{2} x_N^T S_N x_N + \frac{1}{2} \sum_{i=o}^{N-1} [x_i^T, u_i^T] \begin{bmatrix} A_i \ , N_i \\ N_i^T \ , B_i \end{bmatrix} \begin{bmatrix} x_i \\ u_i \end{bmatrix}\right\} . \tag{14.7.6}$$

SOLUTION. We have

$$u_i = -C_i \hat{x}_i , \tag{14.7.7}$$

$$\hat{x}_i = \bar{x}_i + K_i(z_i - H_i \bar{x}_i) , \tag{14.7.8}$$

$$\bar{x}_{i+1} = \Phi_i \hat{x}_i + \Gamma_i u_i , \tag{14.7.9}$$

where

$$C_i = (\Gamma_i^T S_{i+1} \Gamma_i + B_i)^{-1} (\Gamma_i^T S_{i+1} \Phi_i + N_i^T) , \tag{14.7.10}$$

$$K_i = M_i H_i^T (H_i M_i H_i^T + R_i)^{-1} \equiv P_i H_i^T R_i^{-1} , \tag{14.7.11}$$

and

$$S_i = \Phi_i^T S_{i+1} \Phi_i - C_i^T (B_i + \Gamma_i^T S_{i+1} \Gamma_i) C_i + A_i , \qquad S_N \text{ given,} \tag{14.7.12}$$

$$M_{i+1} = \Phi_i P_i \Phi_i^T + Q_i , \qquad M_o \text{ given,}$$

$$P_i = M_i - K_i (H_i M_i H_i^T + R_i) K_i^T \equiv (I - K_i H_i) M_i (I - K_i H_i)^T + K_i R_i K_i^T . \tag{14.7.13}$$

A block diagram of the controller is shown in Figure 14.7.1.

For interested readers, we sketch the proof of the above result for the case $N = 0$ (with no loss of generality) via dynamic programming. Define

$V_1(Z_{N-1}) \equiv$ optimal expected value of J for one-stage control process starting at $i = N-1$, knowing the measurements $\{z_o, \ldots, z_{N-1}\} \equiv Z_{N-1}$ and using an optimal u_{N-1}

$$= \min_{u_{N-1}} E[\tfrac{1}{2} x_N^T S_N x_N + x_{N-1}^T A_{N-1} x_{N-1} + u_{N-1}^T B_{N-1} u_{N-1} | Z_{N-1}] . \tag{14.7.14}$$

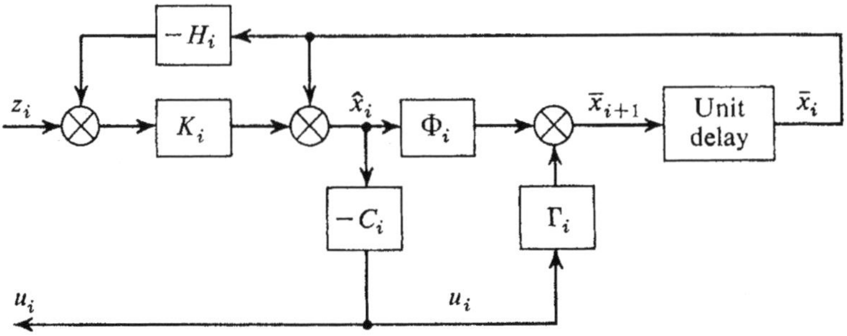

Figure 14.7.1. Block diagram of optimum stochastic controller for a multistage system.

Using the system equation (14.7.2), this becomes

$$V_1(Z_{N-1}) = \min_{u_{N-1}} E[x_{N-1}^T(\Phi_{N-1}^T S_N \Phi_{N-1} + A_{N-1})x_{N-1} + x_{N-1}^T \Phi_{N-1}^T S_N w_{N-1}$$

$$\Gamma_{N-1}u_{N-1}$$

$$+ u_{N-1}^T \Gamma_{N-1}^T S_N \Phi_{N-1} x_{N-1} + x_{N-1}^T \Phi_{N-1}^T S_N w_{N-1} \qquad (14.7.15)$$

$$+ w_{N-1}^T S_N \Phi_{N-1} x_{N-1} + w_{N-1}^T S_N w_{N-1}$$

$$+ u_{N-1}^T(\Gamma_{N-1}^T S_N \Gamma_{N-1} + B_{N-1})u_{N-1} | Z_{N-1}] .$$

Since w_{N-1} is independent of u_{N-1} and x_{N-1} and $E(w_{N-1}^T S_N w_{N-1})$ can be precomputed, we get, for the result of minimization,

$$u_{N-1} = -(\Gamma_{N-1}^T S_N \Gamma_{N-1} + B_{N-1})^{-1}(\Gamma_{N-1}^T S_N \Phi_{N-1})\hat{x}_{N-1} \equiv -C_{N-1}\hat{x}_{N-1} , \qquad (14.7.16)$$

where

$$\hat{x}_{N-1} \equiv E(x_{N-1} | Z_{N-1}) . \qquad (14.7.17)$$

If we substitute (14.7.17) back into (14.7.15), do some manipulating, and utilize the fact that $E(e\hat{x}) = 0$, we obtain

$$V_1(Z_{N-1}) = E\{x_{N-1}^T S_{N-1} x_{N-1} | Z_{N-1}\} + \text{constant term}, \qquad (14.7.18)$$

where

$$S_{N-1} = \Phi_{N-1}^T S_N \Phi_{N-1} - C_{N-1}^T (B_{N-1} + \Gamma_{N-1}^T S_N \Gamma_{N-1})^{-1} C_{N-1} + A_{N-1} . \qquad (14.7.19)$$

Now, using the principle of optimality, we have

$$V_2(Z_{N-2}) = \min_{u_{N-1}} E\{V_1(Z_{N-1}) + x_{N-2}^T A_{N-2} x_{N-2} + u_{N-2}^T B_{N-2} u_{N-2} | Z_{N-2}\} . \qquad (14.7.20)$$

But, by the definition of the conditional mean, we see that

$$E\{V_1(Z_{N-1}) | Z_{N-2}\} \equiv E\{E[x_{N-1}^T S_{N-1} x_{N-1} | z_0, \ldots, z_{N-1}] | z_0, \ldots, z_{N-2}\}$$

$$= E\{x_{N-1}^T S_{N-1} x_{N-1} | z_0, \ldots, z_{N-2}\} . \qquad (14.7.21)$$

Using (14.7.21), $V_2(Z_{N-2})$ reduces now to a form exactly identical to (14.7.14), except for indices. The cycle now repeats. Thus, the key result (14.7.16) or (14.7.7) is established. The variable $E(x_i Z_i)$ in (14.7.7) is, of course, computed via the filter of Chapter 12.

Average behavior of an optimally controlled multistage system. The state of the controlled system, x_i, and the state of the controller, $\bar{x}_i$, are coupled.† From (14.7.2) and (14.7.7) through (14.7.9), we write

$$x_{i+1} = \Phi_i x_i + \Gamma_i u_i + w_i\,, \tag{14.7.22}$$

$$\bar{x}_{i+1} = \Phi_i[\bar{x}_i + K_i(z_i - H_i\bar{x}_i)] + \Gamma_i u_i\,. \tag{14.7.23}$$

Using $e_i \equiv \bar{x}_i - x_i$ and (14.7.7) and (14.7.8), Equations (14.7.22) and (14.7.23) may be replaced by the following equivalent system:

$$e_{i+1} = \Phi_i(I - K_i H_i)e_i + \Phi_i K_i v_i - w_i\,, \tag{14.7.24}$$

$$\bar{x}_{i+1} = (\Phi_i - \Gamma_i C_i)(\bar{x}_i - K_i H_i e_i + K_i v_i)\,. \tag{14.7.25}$$

Let us define the following mean-square quantities:

$$M_i = E[e_i e_i^T]\,, \tag{14.7.26}$$

$$\overline{X}_i = E[\bar{x}_i \bar{x}_i^T]\,, \tag{14.7.27}$$

$$X_i = E[x_i x_i^T]\,, \tag{14.7.28}$$

noting that we have already defined

$$P_i = E[(\hat{x}_i - x_i)(\hat{x}_i - x_i)^T]\,. \tag{14.7.29}$$

From the fact that $E(e_i\bar{x}_i^T) = 0$ for all i [see Section 12.4, Equation (12.4.19b)], it follows that

$$\boxed{X_i = \overline{X}_i + M_i\,.} \tag{14.7.30}$$

Multiplying (14.7.24) by its transpose, taking the expected value of the result, and using $E(e_i\bar{x}_i^T) = 0$, along with

$$E[\bar{x}_i v_i^T] = 0\,, \tag{14.7.31}$$

$$E[e_i v_i^T] = 0\,, \tag{14.7.32}$$

we obtain Equation (14.7.13) as a check.

Multiplying (14.7.25) by its transpose, taking the expected value of the result, and using (14.7.31) and (14.7.32), we obtain

$$\boxed{\overline{X}_{i+1} = (\Phi_i - \Gamma_i C_i)(\overline{X}_i + M_i - P_i)(\Phi_i - \Gamma_i C_i)^T\,;} \tag{14.7.33}$$

† $\bar{x}_i$ is used in the sense of $E(x_i/z_1, \ldots z_{i-1})$; see equation (12.4.12).

and from (14.7.3), it follows that

$$\boxed{\overline{X}_o = 0\,.} \tag{14.7.34}$$

The multistage linear matrix equation (14.7.33), with the initial conditions (14.7.34), allows us to predict the ***mean-square histories of the state variables*** [through use of (14.7.30)].

Since (14.7.7) and (14.7.8) may be written as

$$u_i = -C_i[\bar{x}_i - K_i H_i e_i + K_i v_i]\,, \tag{14.7.35}$$

it follows that, using the first form of (14.7.13),

$$\boxed{E[u_i u_i^T] = C_i[\overline{X}_i + M_i - P_i]C_i^T\,.} \tag{14.7.36}$$

We may also determine the average value of the performance index. The procedure is similar to the one in Section 14.4 and is left as an exercise for the reader. The result is

$$\boxed{J = \mathrm{Tr}\frac{1}{2}\left[S_o X_o + \sum_{i=0}^{N-1} S_{i+1}(Q_i + \Gamma_i C_i P_i \Phi_i^T)\right].} \tag{14.7.37}$$

14.8 Optimum feedback control for nonlinear systems with additive white noise

Perturbation feedback control with deterministic nominal path. In Sections 6.4 and 6.8 we considered perturbation feedback control for nonlinear systems assuming perfect knowledge of the state and all forcing functions. In Section 12.6 we considered optimal filtering for nonlinear systems with additive white noise in the measurements, $v(t)$, and white noise forcing functions, $w(t)$. Obviously, these two concepts can be combined to give a quasioptimum perturbation feedback scheme for nonlinear systems with additive white noise in the measurements and white noise forcing functions. The synthesis would proceed in the following steps:

(a) Determine a nominal optimum control program, $\bar{u}(t)$, the associated state variable histories $\bar{x}(t)$, and measurement histories $h[\bar{x}(t),t]$, assuming no process noise $[w(t) = 0]$ and perfect knowledge of the nominal initial state $\bar{x}(t_o)$ and terminal conditions $\psi[\bar{x}(t_f),t_f] = 0$.
(b) Determine the covariance of the errors in estimating the state, $P(t)$, by integrating (12.6.28). Calculate the gains $P(\partial h/\partial x)^T R^{-1}$ for the linearized maximum-likelihood filter [Equation (12.6.27)].
(c) Determine the feedback control matrix, $C(t)$, of the linearized

system, using the methods of Sections 6.4 or 6.8 and a quadratic approximation to the performance index similar to (6.1.16), *with one important difference*! In the presence of noise, it is *impossible to meet the terminal conditions* $\psi[x(t_f),t_f] = 0$ *exactly*! Hence, in the place of the requirement $\delta\psi \equiv \psi_x \, \delta x|_{t=t_f} = 0$, the quadratic approximation to the performance index in the noisy case must be augmented by a quadratic form in $\delta\psi \equiv \psi_x \, \delta x$ as follows:

$$\delta^2 \bar{J} = E\left\{\frac{1}{2}\left[\delta x^T(\phi_{xx} + \nu^T\psi_{xx} + \psi_x^T N \psi_x)\,\delta x\right]_{t=t_f} + \frac{1}{2}\int_{t_0}^{t_f} [\delta x^T, \delta u^T] \begin{bmatrix} H_{xx}, H_{xu} \\ H_{ux}, H_{uu} \end{bmatrix} \begin{bmatrix} \delta x \\ \delta u \end{bmatrix} dt\right\}. \tag{14.8.1}$$

The weighting matrix, N, must be chosen to be positive definite with elements selected *by iteration* to give values of $E\{\text{diag}(\delta\psi\,\delta\psi^T)\}_{t=t_f}$ that are tolerable for the particular system. The iteration process requires the use of Section 14.4 to determine these latter quantities for a given N, since $E[\delta\psi\,\delta\psi^T] = \psi_x E(\delta x\,\delta x^T)\psi_x^T$.

Combined optimization of nominal path and perturbation feedback control. The procedure suggested in the previous subsection assumes that

$$\min\{E(\bar{J}_{\text{nom}} + \delta^2\bar{J})\} = (\min \bar{J}_{\text{nom}})_{\text{noise}=0} + \min\{E(\delta^2\bar{J})\}\,. \tag{14.8.2}$$

In general, this relation is *not* true, and the optimization of the nominal path and the perturbation feedback control should be done together. A simple parameter optimization problem illustrates this fact: Find $\bar{u} = E(u)$ to minimize

$$E[L(u)]\,, \tag{14.8.3}$$

where u is a gaussian random variable with a given variance

$$E[u - \bar{u}]^2 = \sigma^2\,, \tag{14.8.4}$$

and $L(u)$ possesses a minimum but is *not* symmetric about this minimum.

To second order we may write

$$\begin{aligned} E[L(u)] &= E\{L(\bar{u}) + (u - \bar{u})L_u(\bar{u}) + \tfrac{1}{2}(u - \bar{u})^2 L_{uu}(\bar{u}) + \cdots\} \\ &= L(\bar{u}) + \frac{\sigma^2}{2} L_{uu}(\bar{u}) + \cdots \end{aligned} \tag{14.8.5}$$

where the term in $(u - \bar{u})$ vanishes because $E(u - \bar{u}) = 0$, not because $L_u(\bar{u}) = 0$. Let u_0 be the minimizing value of $L(u)$; clearly, if $L(u)$ is not symmetric about $u = u_0$, Equation (14.8.5) indicates that

$\bar{u} = u_0$ does *not* minimize $E[L(u)]$. See Problem 1 for a specific example.

The key feature of this example is that the coefficient $L_{uu}(\bar{u})$ of the perturbation problem depends on the choice of the nominal and is *not* symmetric about the deterministic minimum. The same phenomenon can easily occur in our *nominal plus perturbation feedback* problem of the previous subsection. The coefficients ϕ_{xx}, ψ_{xx}, H_{xx}, $H_{xu}, \ldots$, etc., in Equation (14.8.1) are evaluated on the nominal path, and, if they are not symmetric about the deterministic minimum path, it is almost certain that a better choice of nominal can be made than the deterministic minimum.

The concept for the combined optimization problem, also known as the dual control problem is fairly simple, but the required analysis is rather involved [see Fitzgerald (1964),† Denham (1964),‡ and Feldbaum§ (1965)].

Optimum feedback control for nonlinear systems with perfect knowledge of the state. Consider the problem of determining $u(x,t)$ to minimize

$$J = E\left\{\phi[x(t_f)] + \int_{t_0}^{t_f} L(x,u,t)\, dt\right\}, \tag{14.8.6}$$

subject to the constraints

$$\dot{x} = f(x,u,t) + G(x,t)w(t), \tag{14.8.7}$$

where $E[w(t)] = 0$, $E[w(t)w^T(\tau)] = Q(t)\,\delta(t-\tau)$, and we assume that $x(t)$ and t are known exactly.

Let

$$J^o(x,t) \triangleq \text{optimal value of } J \text{ starting in the state } x \text{ at time } t \text{ and using optimal control.} \tag{14.8.8}$$

Using dynamic programming methods, with the added complication of including second-order terms in the Taylor series expansion, it is straightforward to show° that

$$-\frac{\partial J^o}{\partial t} - \frac{1}{2}\operatorname{Tr}\left[\frac{\partial^2 J^o}{\partial x^2} GQG^T\right] = \min_u \left[L + \frac{\partial J^o}{\partial x} f\right], \qquad J^o(x,t_f) = \phi[x(t_f)]. \tag{14.8.9}$$

---

†R. Fitzgerald, *A Gradient Method for Optimizing Stochastic Systems.* Ph.D. Thesis, M.I.T., May 1964.

‡W. Denham, *Choosing the Nominal Path for a Dynamic System with Random Forcing Function to Optimize Statistical Performance,* Tr 449, Division of Eng. & App. Physics, Harvard University, 1964.

§A. Feldbaum, *Optimal Control Systems.* New York: Academic Press, 1965.

°See Problem 2, Section 14.2. Equation (14.8.9), after u has been eliminated by the min operation, is called a *Kolmogorov backward equation.*

With $Q = 0$, Equation (14.8.9) becomes the Hamilton-Jacobi-Bellman (HJB) partial differential equation [See Equation (4.2.11)]; with $L = 0$ and $f = f(x,t)$, Equation (14.8.9) is called a Kolmogorov partial differential equation.

Because of its parabolic nature, Equation (14.8.9) is even harder to solve than the HJB equation, since the method of characteristics is no longer applicable. In other words, there is no natural analog of the Euler-Lagrange equations. Very few truly nonlinear examples have been solved thus far. For the special case in which (14.8.7) is linear and (14.8.6) is quadratic, Equation (14.8.9) reduces to the backward matrix Riccati equation (14.2.7) (again see Problem 2, Section 14.2).

Optimum feedback control for linear systems, nonlinear criteria, and noisy measurements of the state. Here we consider the case in which (14.8.7) is linear,

$$\dot{x} = F(t)x + G(t)u + w(t)\,, \tag{14.8.10}$$

and we have measurements $z(t)$ of linear combinations of the state components

$$z(t) = H(t)x + v(t)\,, \tag{14.8.11}$$

where

$$E[v(t)] = 0\,, \qquad E[v(t)v^T(\tau)] = R(t)\,\delta(t-\tau)\,.$$

However, we still have the *nonlinear* criterion (14.8.6).

For this case, $\hat{x}(t)$ and $P(t)$, as given by the linear Kalman filter of Equations (14.3.8) and (14.3.12), still constitute a set of sufficient statistics. Hence, the optimal control may be constructed as a function of $\hat{x}$ and P instead of a functional of $z(\tau)$, $t_0 \leq \tau \leq t$.

Let

$$J^o(\hat{x},t) \triangleq \text{optimal value of } J \text{ starting with estimated state } x \text{ at time } t \text{ and using optimal control.} \tag{14.8.12}$$

Again, using dynamic programming methods, one can show that

$$-\frac{\partial J^o}{\partial t} - \frac{1}{2}\,\mathrm{Tr}\left[\frac{\partial^2 J^o}{\partial \hat{x}^2} PH^TR^{-1}HP\right] - \frac{\partial J^o}{\partial x} F\hat{x} = \min_u \left[\bar{L} + \frac{\partial J^o}{\partial \hat{x}} Gu\right], \tag{14.8.13}$$

where

$$J^o(\hat{x},t_f) = \bar{\phi}(\hat{x})\,, \tag{14.8.14}$$

$$\bar{\phi}(\hat{x}) \triangleq E[\phi|\hat{x}] \equiv \int_{-\infty}^{\infty} \cdots \int_{-\infty}^{\infty} \phi(\xi)p(\xi|\hat{x})\,d\xi_1 \cdots d\xi_n\,, \tag{14.8.15}$$

$$\overline{L}(\hat{x},u,t) \triangleq E[L|\hat{x},u,t] \equiv \int_{-\infty}^{\infty} \cdots \int_{-\infty}^{\infty} L(\xi,u,t)p(\xi|\hat{x})\, d\xi_1, \ldots, d\xi_n, \quad (14.8.16)$$

$$p(\xi|\hat{x}) = \frac{1}{(2\pi)^{n/2}|P(t)|^{1/2}} \exp\left\{-\frac{1}{2}(\xi - \hat{x})^T[P(t)]^{-1}(\xi - \hat{x})\right\}, \quad (14.8.17)$$

that is, a gaussian density function.

If ϕ and L are *quadratic* functions, (14.8.13) can be reduced to the relations obtained in Sections 14.3 and 14.4. For ϕ and L not quadratic, the partial integrodifferential equation (14.8.13) is extremely difficult to solve. However, Deyst (1967)† has given a few interesting examples.

The density function for stochastic nonlinear dynamic systems. Even after solving the Kolmogorov backward equation (14.8.9) for the stochastic optimal return function $J^o(x,t)$, and in the process determining $u(x,t)$, we do not have a description of the way in which the resulting dynamic system will evolve, on the average. To obtain such a description, we must determine the forward evolution in time of the probability density function for the state $p(x,t)$. If we incorporate the feedback law $u = u(x,t)$ into Equation (14.8.7) so that $f(x,u,t) \triangleq f^o(x,t)$, and let $w(t)$ be a scalar (so that G is a vector), it can be shown that $p(x,t)$ satisfies the Kolmogorov forward equation:

$$\frac{\partial p}{\partial t} + \sum_i \frac{\partial}{\partial x_i}(pf_i^o) = \frac{Q}{2}\sum_i \sum_j \frac{\partial^2}{\partial x_i\, \partial x_j}(G_iG_jp), \quad (14.8.18)$$

where

$$p(x,t_0) \quad \text{is given.}$$

In the special case where f^o is linear, $G = G(t)$, and $p(x,t_0)$ is gaussian, the system reduces to a gauss-markov process and is completely described by the mean-value and covariance histories $\bar{x}(t)$ and $X(t)$ of Equations (11.4.6) and (11.4.17).

Additional complications occur if the state of the system at t_0 is not perfectly known and partial measurement of the state of the system is allowed as the system evolves. The density function of interest in this case is the conditional density function of the state, given all the measurement and the initial distribution of x_0. In the discrete case, the equation that governs the evolution of $p[x,t/z(\tau), t_0 \leq \tau \leq t, p(x_0)]$ is the integral-difference equation [Equation (12.8.8)] derived in Section 12.8. In the continuous case, the equation represents an extension of the Kolmogorov equation to the case involving measure-

†J. Deyst, *Optimal Control in the Presence of Measurement Uncertainties.* Ph.D. Thesis, M.I.T, February 1967.

ments [see Kushner, (1967)].† If, on the other hand, the system and measurements are linear, the problem has been completely resolved. This is, of course, the filtering solution derived in Section 12.5.

Problem 1. Find $\bar{u}$ to minimize

$$E[u(e^u - 1)]\,, \qquad \text{where} \qquad E(u - \bar{u})^2 = \sigma^2\,.$$

Problem 2. For $f^0(x,t) = F(t)x$, $G(x,t) = G(t)$, and $p(x,t_0)$ gaussian with zero mean, show that Equation (14.8.18) can be reduced to Equation (11.4.17) for the covariance $X(t)$.

†H. J. Kushner, *Stochastic Stability and Control.* New York: Academic Press, 1967.

Appendix A
Some basic mathematical facts

A1 Introduction

As mentioned in the Preface, the subject matter of this book is considered to be at the senior or first-year-graduate level of a university student. The authors presumed, therefore, that the reader would have some acquaintance with advanced calculus, ordinary differential equations, and mechanics. The purpose of this Appendix is to bring together in one place, and in one consistent notation, some basic mathematical facts currently found useful in control and optimization theory. The reader need not be intimately familiar with all the material in this Appendix, but some knowledge of the material in Sections A3 and A4 is necessary to proceed with a smooth reading of the book. To this end, we furnish proofs or detailed explanations of the material only when they form an integral part of the control and optimization theory of the book.

A2 Notation

Since we wish to analyze systems with state vectors of arbitrary dimension, the use of vector and matrix notation is very convenient. We adopt the following conventions:

(a) *Vectors*: All lower-case italic or Greek letters indicate vectors. Thus, we have

$$x = \begin{bmatrix} x_1 \\ x_2 \\ \cdot \\ \cdot \\ \cdot \\ x_n \end{bmatrix}$$

Components of vectors are denoted by subscripts, for example, $x_1, x_2, \ldots, x_n$, where x_i are real variables. These components may or may not be vectors themselves.

(b) *Matrices*: All upper-case italic or Greek letters indicate matrices. Thus, we have

$$A = \begin{bmatrix} a_{11} \; a_{12} \cdots a_{1n} \\ a_{21} & \\ \cdot & \cdot \\ \cdot & \cdot \\ \cdot & \cdot \\ a_{k1} \quad \cdots \quad a_{kn} \end{bmatrix},$$

where a_{ij} are real variables. A vector may be regarded as a $(k \times 1)$-matrix.

(c) *Scalars*: Except in examples and exercises, scalar quantities appear rarely in this book. We shall rely on the text and specific instructions to make the presence of scalar quantities clear. Either upper- or lower-case letters may be used to denote scalar quantities. Examples are: (i) ". . . consider the scalar criterion function J . . ."; (ii) "Let us consider an example of a two-dimensional system governed by the equations $\dot{x}_1 = x_2$, $\dot{x}_2 = 0$. . . ."

(d) The scalar variable t is used to denote the independent variable time, which may vary continuously or discretely. Vectors or matrices whose elements are time varying are denoted as $x(t)$ and $A(t)$, respectively. Occasionally in the book (and quite often in current literature), when there is no danger of confusion, we also write x_t and A_t for the time-varying vector and matrix, or simply omit the explicit indication of t altogether in order to simplify the appearance of some of the equations.

(e) The scalars i, j, k, ℓ, unless otherwise indicated, are used as indexing variables; i.e., they take on the values 1,2,3, ... Thus, when we say the vector x_i, we mean the vector $x_1, x_2, \ldots$.

(f) The transpose of a vector or matrix is denoted by the superscript T (the prime symbol (′) is also an accepted substitute for T in the current literature). Thus, the scalar or inner product of two vectors may be written in the following equivalent ways:

$$x^T y \equiv x'y \equiv \mathbf{x} \cdot \mathbf{y} \equiv \langle x \cdot y \rangle \equiv (x,y) \equiv [x_1, \ldots, x_n] \begin{bmatrix} y_1 \\ \cdot \\ \cdot \\ \cdot \\ y_n \end{bmatrix} = \sum_{i=1}^{n} x_i y_i .$$

The vector, dyadic, or outer product of two vectors may be written as

$$xy^T \equiv xy' \equiv \mathbf{xy} \equiv x \times y \equiv \begin{bmatrix} x_1 \\ \cdot \\ \cdot \\ \cdot \\ x_n \end{bmatrix} [y_1, \ldots, y_m]$$

$$\equiv \begin{bmatrix} x_1y_1, \ldots, x_1y_m \\ \cdot \qquad \cdot \\ \cdot \qquad \cdot \\ \cdot \qquad \cdot \\ x_ny_1, \ldots, x_ny_m \end{bmatrix}.$$

In particular, we have

$$x^Tx \equiv \|x\|^2 = \text{length of the vector } x,$$

$$xx^T \equiv \begin{bmatrix} x_1^2 & x_1x_2 \cdots & x_1x_n \\ x_1x_2 & x_2^2 \quad \cdots & \cdot \\ \cdot & & \cdot \\ \cdot & & \cdot \\ \cdot & & \\ x_nx_1 & \cdots & x_n^2 \end{bmatrix} = \text{scatter matrix of the vector } x.$$

Note that a row vector is always written as $h^T = [h_1, \ldots, h_n]$, while a $(1 \times n)$-matrix H is written without the transpose sign. The transpose rule $(AB)^T = B^TA^T$ is directly verified.

(g) When a mathematical symbol is applied to a vector or matrix, it is applied to every element of the vector or matrix. Thus, we have

$$\frac{d}{dt}x = \begin{bmatrix} \frac{dx_1}{dt} \\ \vdots \\ \frac{dx_n}{dt} \end{bmatrix}, \qquad \int A\,dt = \begin{bmatrix} \int a_{11}\,dt \cdots \int a_{1n}\,dt \\ \cdot \qquad\qquad \cdot \\ \cdot \qquad\qquad \cdot \\ \int a_{n1}\,dt \cdots \int a_{nn}\,dt \end{bmatrix}.$$

(h) The symbol $\partial/\partial x$, when applied to a scalar variable, say, J, means the row vector $[(\partial J/\partial x_1), (\partial J/\partial x_2), \ldots, (\partial J/\partial x_n)]$, the gradient of the function† J. This is also denoted by J_x or $\nabla_x J$.

†In some of the current literature, J_x or $\nabla_x J$ is considered as a column vector. The context and the rule of conformity usually make it clear as to which definition is being used.

A3 Matrix algebra and geometrical concepts

(a) *Determinant* of A is denoted by $\det(A)$ or sometimes $|A|$ and can be defined only for square matrices. It is defined by

$$\det(A) = a_{11}C_{11} + a_{12}C_{12} + \cdots + a_{1n}C_{1n},$$

where

$$A = \begin{bmatrix} a_{11} & \cdots & a_{1n} \\ \cdot & & \cdot \\ \cdot & & \cdot \\ \cdot & & \cdot \\ a_{n1} & \cdots & a_{nn} \end{bmatrix},$$

where C_{1i}, called the 1*i*th *cofactor*, is the determinant of the reduced matrix formed by crossing out the first row and the *i*th column multiplied by $(-1)^{1+i}$. Note that we have

$$\det(AB) = \det(A)\det(B), \qquad \det(A) = \det(A^T).$$

Trace of a square matrix A is defined as

$$\mathrm{Tr}(A) \triangleq \sum_{i=1}^{n} a_{ii}.$$

From above, it follows immediately that

$$\mathrm{Tr}(AB) = \mathrm{Tr}(BA).$$

Symmetric matrix is defined by $A^T = A$.
Identity matrix is defined by

$$I = \begin{bmatrix} 1 & \cdots & 0 \\ \cdot & & \cdot \\ \cdot & 1 & \cdot \\ \cdot & & \cdot \\ 0 & \cdots & 1 \end{bmatrix}.$$

(b) *Inverse matrix* A^{-1} is defined by

$$(A^{-1})_{ij} = \frac{1}{\det(A)}(C_{ji}) \qquad \text{for} \qquad \det(A) \neq 0,$$

where C_{ji} is the *ji*th cofactor of A. The inverse of A, when it exists, is unique. We have

$$A^{-1}A = AA^{-1} = I$$

and the inverse rule

$$(AB)^{-1} = B^{-1}A^{-1}.$$

Orthogonal matrix is defined by $A^{-1} = A^T$; for example, we have

$$A = \begin{bmatrix} \cos\theta & \sin\theta \\ -\sin\theta & \cos\theta \end{bmatrix}.$$

Antisymmetric matrix is defined by $A^T = -A$.

(c) *Linear Independence.* A set of n-vectors, $a_1, \ldots, a_n$, of equal dimension are said to be ***linearly independent*** if it is not possible to express any one vector as a linear combination of the other vectors. In other words, the vector equation

$$x_1 a_1 + \cdots + x_n a_n = [a_1 \,|\, a_2 \,|\, \cdots \,|\, a_n] \begin{bmatrix} x_1 \\ \cdot \\ \cdot \\ \cdot \\ x_n \end{bmatrix} = 0$$

implies that the scalars $x_1, \ldots, x_n$ are all zero. The ***rank*** of a matrix is defined as the maximum number of linearly-independent columns or rows (these two numbers are equal). A square $(n \times n)$-matrix with rank less than n is said to be ***singular***.

(d) For a square matrix A and the equation $Ax = y$, the following six statements are equivalent:

(i) $\det(A) \neq 0$;
(ii) A^{-1} exists;
(iii) $Ax = y$ has a unique solution for $y \neq 0$;
(iv) columns of A are linearly independent;
(v) rows of A are linearly independent; $Ax = 0$ has no nontrivial solutions [see (c)];
(vi) we have $\lambda_i(A) \neq 0$, $i = 1, 2, \ldots, n$, where $\lambda_i(A)$ is the ith eigenvalue of matrix A.

It follows that any matrix with two identical rows or columns has a zero determinant, and any matrix with a zero row or column has a zero determinant.

Problem 1. Verify the statements (i) through (vi).

If we consider the columns of A as individual vectors, then $Ax = y$ says that the arbitrary vector y can be expressed as a linear combination of a set of n linearly-independent vectors $a_1, \ldots, a_n$, the base vectors. Components of the solution vector x represent the particular linear combination in question. In this case, we say that the base vectors span an n-dimensional space in the sense that any vector in this space can be reproduced by appropriate vector sum of the base vectors. Usually, when we specify the components of a vector x in numerical value, we implicitly assume the unit vectors $e_i^T = (0, \ldots, 0, 1, 0 \ldots, 0)$ as the base vectors. Multiplication of x by a matrix A yields another vector y, which can be interpreted as (i) the same numerical vector x but in a different coordinate system, with the column vectors of A as base vectors; or (ii) a different vector, which has been rotated and scaled from the original vector in the same coordinate system. For example, let $x = \begin{bmatrix}1\\1\end{bmatrix}$ and $y = Ax = \begin{bmatrix}2 & -1\\1 & 1\end{bmatrix}\begin{bmatrix}1\\1\end{bmatrix} = \begin{bmatrix}1\\2\end{bmatrix}$. This is shown pictorially in Figure A.3.1

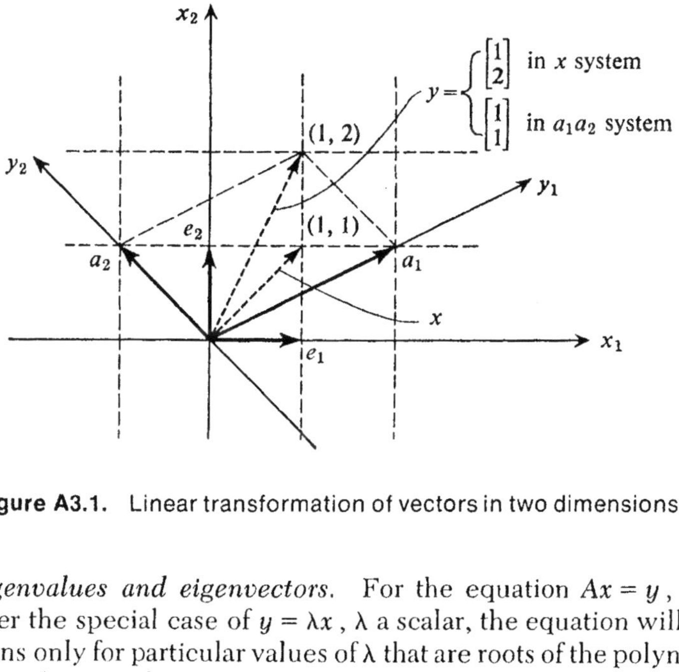

Figure A3.1. Linear transformation of vectors in two dimensions.

(e) *Eigenvalues and eigenvectors.* For the equation $Ax = y$, if we consider the special case of $y = \lambda x$, λ a scalar, the equation will have solutions only for particular values of λ that are roots of the polynomial $\det(A - \lambda I) = 0$. These values, λ_i, and the corresponding solutions, x^i, are called the eigenvalues and the eigenvectors of A. For all practical purposes [see Bellman (1960)], we can say that any square matrix A has (or can be approximated by a matrix that has) n linearly-independent eigenvectors, $x^1, x^2, \ldots, x^n$. Consequently, we can arrange the n set of eigensolutions in matrix form as

$$A\left[x^1 \,\middle|\, \cdots \,\middle|\, x^n\right] = \left[x^1 \,\middle|\, \cdots \,\middle|\, x^n\right]\begin{bmatrix}\lambda_1 & \cdots & 0\\ \vdots & \lambda_2 & \vdots\\ \vdots & & \ddots \\ 0 & & \lambda_n\end{bmatrix} \overset{\Delta}{\Longleftrightarrow} AS = S\Lambda$$

or

$$S^{-1}AS = \Lambda ,$$

which is called a *similarity transformation*. We say A is similar to a diagonal matrix† Λ. Similarity transformation has the property that, if A and B are related by $A = SBS^{-1}$, then

$$\lambda_i(A) = \lambda_i(B) , \qquad P_n(A) = 0 \Rightarrow P_n(B) = 0 ,$$

where

$$P_n(A) \triangleq a_n A^n + a_{n-1} A^{n-1} + \cdots + a_1 A + a_0 I ,$$

and a_i's are scalar constants; i.e., eigenvalues and matrix polynomial equations are preserved under a similarity transformation. It can be shown, furthermore, that

$$\text{trace of } A \triangleq \text{Tr}(A) = \sum_{i=1}^{n} a_{ii} = \sum_{i=1}^{n} \lambda_i(A) .$$

If, in addition, the matrix A is symmetric, the similarity transformation specializes to the orthogonal transformation, that is, $S^{-1} = S^T$. Finally, we have the *Cayley-Hamilton theorem*, which states that any square matrix A satisfies its own characteristic polynomial; i.e., if $P_n(\lambda) = \det(A - \lambda I)$, then $P_n(A) = 0$.

(f) *Norm* of a vector is defined as

$$\|x\|^2 = x^T x = \sum_{i=1}^{n} x_i^2 .$$

It satisfies the three axioms of a norm (length): (1) $\|x\| = 0$ if and only if $x = 0$; (2) $\|\alpha x\| = |\alpha| \cdot \|x\|$ for all scalar α; (3) $\|x + y\| \leq \|x\| + \|y\|$. Now consider any nonsingular matrix A and the vector $y = Ax$. Then, according to the discussion in (d), y is the transformed vector and has a norm that is only zero if $x = 0$. We have

$$\|y\|^2 = x^T A^T A x \triangleq \|x\|^2_{A^T A} .$$

The quantity $\|x\|^2_{A^TA}$ is called the *generalized norm* of x (it is the norm of x in a different coordinate system). The matrix A^TA is said to be *positive definite* in the sense that $\|x\|^2_{A^TA} > 0$ for all $x \neq 0$. We write, in shorthand, $A^TA > 0$. Similarly, a matrix B is said to be *positive semidefinite* if $\|x\|^2_B \geq 0$ for all $x \neq 0$ $(B \geq 0)$.

For any rectangular matrix A, $A^TA = B$ is positive semidefinite. It is positive definite if A has maximal rank and A^TA is $(n \times n)$, where n is the smaller dimension of the matrix A.

For any antisymmetric A, we have $x^TAx = 0$. Since any square

†The more correct statement would be that A is always similar to a Jordan matrix [Bellman (1960)].

matrix can be written as $A = A^a + A^s$, where $A^a = (A - A^T)/2$ and $A^s = (A + A^T)/2$, we have $x^TAx = x^TA^sx$.

Problems 2 and 3. Prove the above assertions.

The scalar function $J = x^TBx = \|x\|_B^2$ is called, also, a *quadratic form* of B, where we have assumed B to be symmetric without loss of generality (see above). If $B = S^TS$, where S is orthogonal, then $y^Ty = x^TS^TSx = x^Tx$. We thus arrive at the conclusion that orthogonal transformation does not affect the length of a vector. In other words, $y = Sx$ merely represents a rotation and/or reflection of the coordinate system. Since, in general, we have $B = S^T\Lambda S$, we can write $z = \Lambda^{1/2}Sx = \Lambda^{1/2}y$, where

$$\Lambda^{1/2} \triangleq \begin{bmatrix} \sqrt{\lambda_1} & \cdot & \cdot & \cdot & 0 \\ \cdot & \sqrt{\lambda_2} & & & \cdot \\ \cdot & & \cdot & & \cdot \\ \cdot & & & \cdot & \cdot \\ 0 & \cdot & \cdot & \cdot & \sqrt{\lambda_n} \end{bmatrix},$$

provided that λ_i are greater than or equal to zero. The linear transform $\Lambda^{1/2}$ effects a stretching or shrinking of the coordinate axes. The combined linear transform $\Lambda^{1/2}S$ simultaneously rotates, reflects, and scales, a conclusion we reached earlier in (d). If the quadratic form is also a generalized norm for x, then, clearly, $\lambda_i > 0$. Hence, an *alternate definition for positive definiteness is the positivity of the associated eigenvalues.* A practical test for positive definiteness is that the principal minors of the matrix all be positive.†

(g) Simple *geometrical figures* in n-dimensions can be expressed algebraically by vectors and matrices. The scalar equation $(a^i)^Tx - y_i = 0$, for given scalar y_i and given vector a^i, defines an $(n-1)$-dimensional flat space called a *hyperplane* in an n-dimensional space. The normal to the hyperplane is simply the vector a^i. The intersection of n such hyperplanes, $i = 1, \ldots, n$, in general determines a point that is simply the solution of $Ax = y$.

Next in order of complication are quadratic surfaces in n dimensions defined by the equation $x^TBx - c = 0$, where B is a given symmetric matrix and c a given positive scalar. If we have $B > 0$, the surface may be interpreted as the loci of the vectors whose generalized length is equal to c. According to the discussion in (f), a pure rotation and

---

†Principal minors of a matrix are the determinants of the reduced matrices formed by deleting the first row and column, the first and second rows and columns, the first, second, and third rows and columns, and so on. This test may be interpreted as applying the Routh-Hurwitz criteria to the characteristic polynomial of the matrix.

possible reflection of the coordinate system will diagonalize B. This reveals that the closed surfaces are hyperellipsoids in n-dimension, with principal axes equal to $(\lambda_i/c)^{-1/2}$. For example, with $c = 1$, and

$$B = \begin{bmatrix} \frac{3}{2} & -\frac{1}{2} \\ -\frac{1}{2} & \frac{3}{2} \end{bmatrix} = \begin{bmatrix} \cos\theta & -\sin\theta \\ \sin\theta & \cos\theta \end{bmatrix} \begin{bmatrix} 1 & 0 \\ 0 & 2 \end{bmatrix} \begin{bmatrix} \cos\theta & \sin\theta \\ -\sin\theta & \cos\theta \end{bmatrix}$$

$$\text{for} \quad \theta = 45°,$$

the curve is an ellipse with semiaxes 1 and $1/\sqrt{2}$ oriented at 45° to the coordinate axes (See Figure A3.2).

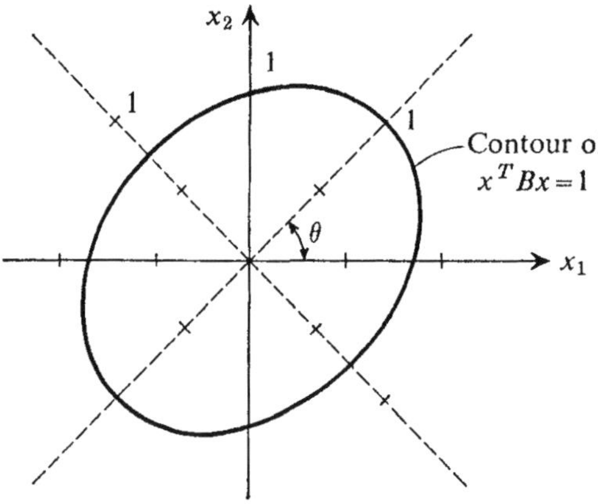

Figure A3.2. *Locus of quadratic function $x^T Bx = 1$ in two dimensions.*

The general formula for diagonalizing two-dimensional quadratic forms is

$$\left.\begin{matrix} \lambda_1 \\ \lambda_2 \end{matrix}\right\} = \frac{B_{11} + B_{22} \pm \sqrt{(B_{11} - B_{22})^2 + 4B_{12}^2}}{2},$$

$$\theta = \frac{1}{2} \tan^{-1} \frac{2B_{12}}{B_{11} - B_{22}},$$

which is easily derived from the equations

$$B_{11} = \cos^2\theta\lambda_1 + \sin^2\theta\lambda_2, \qquad B_{22} = \sin^2\theta\lambda_1 + \cos^2\theta\lambda_2,$$
$$B_{12} = \cos\theta \sin\theta(\lambda_1 - \lambda_2).$$

More generally, a scalar function of n variables, $f(x) = 0$, also defines a surface in n dimensions. At any point x_0 on the surface, we can approximate the neighborhood surface by the tangent hyperplane at x_o. This is equivalent to a first-order Taylor series expansion about x_o; that is, we have

$$f(x) \approx f(x_o) + \left.\frac{\partial f}{\partial x}\right|_{x=x_o} (x - x_o) = 0 .$$

Thus, the normal to the surface $f(x) = 0$ at x_o is simply the *gradient* of the function $f(x)$ evaluated at x_o. It represents the direction of the steepest increase of the function $f(x)$. The tangent hyperplane also separates the neighborhood of x_o into two halves, one where $f(x) < 0$, one where $f(x) > 0$. Following the same procedure, a closer approximation of $f(x) = 0$ near x_o can be attempted by including the second-order terms in the expansion; i.e., we have

$$f(x) \approx f(x_o) + \left.\frac{\partial f}{\partial x}\right|_{x_o} (x - x_o) + \frac{1}{2}(x - x_o)^T \left.\frac{\partial^2 f}{\partial x^2}\right|_{x_o} (x - x_o) = 0 ,$$

where

$$\left(\frac{\partial^2 f}{\partial x^2}\right)_{ij} = \frac{\partial^2 f}{\partial x_i \, \partial x_j} = B_{ij}$$

is a symmetric matrix of the second partial derivatives. In other words we are fitting the surface $f(x) = 0$ [or the function $f(x)$] by the quadratic surface (function) defined by the above equation. If, in particular, the matrix $B \geq 0$ (> 0), we call the corresponding function f *locally convex* (strictly locally convex) near x_o. If this is true for all x_o, the function is called convex (strictly convex).† Convex functions of this type have the obvious property

$$f(x) \geq f(x_o) + \left.\frac{\partial f}{\partial x}\right|_{x_o} (x - x_o) ,$$

which, in turn, implies either $(\partial f/\partial x) = 0$, in which case x_o is a minimum point by definition; or $(\partial f/\partial x) \neq 0$, in which case, starting from any point x, we can find a point x_o such that $(\partial f/\partial x)(x - x_o)$ is positive. Hence, $f(x_o)$ has a lower value than $f(x)$. This procedure can be repeated until either $(\partial f/\partial x) = 0$ is true or $f(x)$ becomes unbounded.

(h) Some useful formulas follow. If $f(x) = \frac{1}{2}x^T A x$, then $(\partial f/\partial x) = x^T A$, $(\partial^2 f/\partial x^2) = A$ (for $A = A^T$). If $f(x) = \lambda^T A x$, then $(\partial f/\partial x) = \lambda^T A$, $(\partial^2 f/\partial x^2) =$

†Convexity can be defined for functions that do not possess partial derivatives by the property

$$(1 - \theta) f(x_o) + \theta f(x) \geq f[(1 - \theta)x_o + \theta x] \quad \text{for all} \quad 0 \leq \theta \leq 1 \quad \text{and} \quad x, x_o ;$$

that is, linear interpolation always overestimates the value of the function.

0. If $f(\Phi) = \frac{1}{2}\|z - H\Phi x\|^2$, then $(\partial f/\partial \Phi) = -H^T(z - H\Phi x)x^T$. If $f(X) \triangleq \mathrm{Tr}(XA) = \mathrm{Tr}(AX)$, then $(\partial f/\partial X) = A$.

A4 Elements of ordinary differential equations

The class of ordinary differential equations considered in this book is of the form

$$\dot{x} = f(x,u,t), \qquad x(t_0) = x_0, \tag{A4.1}$$

where x is the state vector, u the control vector, and the form of f is given. In cases where f is not explicitly a function of time, the system of (A4.1) is said to be *stationary*. When $u = 0$ for all $t \geq t_0$, the system is said to be *free*. When both are true, the system is then *autonomous*.

Fundamental theorems. The following important theorems concerning ordinary differential equations are stated without proof.

THEOREM 1 *(Existence). If $u = 0$ and f is continuous in x and $t_0 \leq t \leq t_1$, there exists a solution $\phi(t;x_0,t_0)$ of* (A4.1), *with continuous derivatives on $t_0 \leq t < \alpha \leq t_1$, for which $\phi(t_0;x_0,t_0) = x_0$.*

THEOREM 2 *(Uniqueness). Let $u = 0$ and f be continuous in x and $t_0 \leq t \leq t_1$. Furthermore, f satisfies the Lipschitz condition*

$$\|f(x_1,t) - f(x_2,t)\| < k\|x_1 - x_2\|, \qquad k \text{ a given scalar.}$$

If ϕ_1 and ϕ_2 are any two solutions of (A4.1) *for $t_0 \leq t \leq t_1$ such that $\phi_1(t_0;x_0,t_0) = \phi_2(t_0;x_0,t_0)$, then $\phi_1 = \phi_2$ for all t.*

THEOREM 3 *(Continuity). For f satisfying conditions of Theorem 2, the solution ϕ is continuous in x_0 and t_0 as well. Hence, we are justified in writing $\phi(t;x_0,t_0)$.*

The above three theorems provide the mathematical basis for the use of (A4.1) in representing a dynamic system. When $u = u(t)$ is a continuous known function, all theorems remain valid. However, in control applications it often turns out that $u(t)$ is discontinuous. In this case, we have to consider the solution of the differential equation in pieces, each satisfying Theorems 1, 2, and 3. The pieces are joined together at the boundary by appropriate considerations of the type of discontinuities of $u(t)$.

Variational equations. Suppose that for the system of (A4.1) we have a given $x(t_0) = x_0$ and $u(t)$ for $t > t_0$; then the unique solution of $\phi_u(t;x_0,t_0)$ is determined and can be found, at least in principle. Numerically, $\phi_u(t;x_0,t_0)$ can always be determined nowadays with relative ease on an analog or a digital computer. Now, suppose that we consider small perturbations $\delta x(t_0)$ and $\delta u(t)$ in x_0 and $u(t)$. Because of Theorem 3, we shall expect small perturbations in the solution ϕ_u. Thus, we have

$$\dot{x} + \delta\dot{x} = f(x + \delta x, u + \delta u, t)\,; \qquad \delta x(t_0) = \delta x_0\,. \tag{A4.2}$$

Expanding the right-hand side of (A4.2) in a Taylor series and retaining terms only of first order, we have

$$\delta\dot{x} = F(t)\,\delta x + G(t)\,\delta u\,; \qquad \delta x(t_0) = \delta x_0\,, \tag{A4.3}$$

where

$$F_{ij} = \frac{\partial f_i}{\partial x_j}\,, \qquad G_{ij} = \frac{\partial f_i}{\partial u_j}\,, \tag{A4.4}$$

and the partial derivatives are evaluated along the known solution, $x(t) = \phi_u(t;x_0,t_0)$, and control $u(t)$. Thus, $\partial f_i/\partial x_j$ and $\partial f_i/\partial u_j$ can simply be considered as known time-varying matrices $F(t)$ and $G(t)$. The motion of a dynamic system about a known path for small perturbations is thus seen to be governed by the linear time-varying ordinary differential equations of (A4.3).

Linear ordinary differential equations. In the main, we consider the linear systems governed by

$$\dot{x} = F(t)x + G(t)u\,, \tag{A4.5}$$

where F and G are $(n \times n)$ and $(n \times m)$ time-varying matrices, respectively. We shall usually assume that F and G are continuous and bounded. If the equation is simulated on an analog computer, we need a total of n integrators. Each integrator will assume the configuration shown in Figure A4.1. Suppose that we apply a unit impulse at the input of the jth integrator at time $= \tau$ (this is equivalent

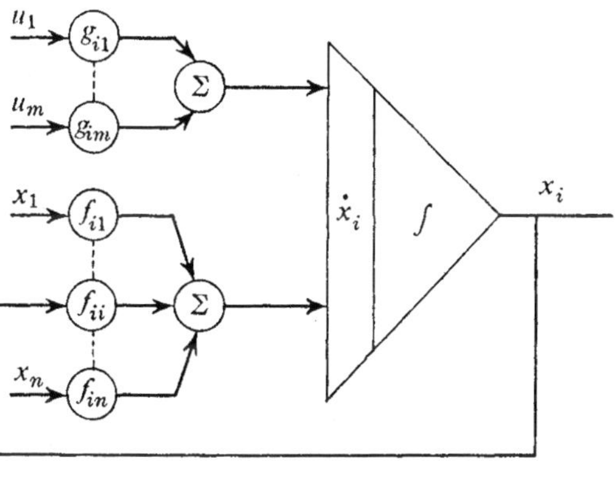

Figure A4.1. Analog computer representation of one component of Equation A4.5.

to making the initial condition at the output of the jth integrator at $t = \tau$ equal to unity), with zero initial conditions on all other integrators, and record the output of the ith integrator at time $t = t_1$. The value of this output will be designated $\phi_{ij}(t_1,\tau)$ and interpreted as the response of the ith integrator due to a unit impulse at the input of the jth integrator. If we do this for all values of i and j, there will be n^2 numbers that form a matrix called $\phi(t_1,\tau)$. By repeating the whole process for all values of τ from t_0 to t_1, we can, in principle, obtain the complete time history of the matrix $\phi(t_1,\tau)$. Now, suppose that the dynamic system has certain initial conditions $x_i(t_0)$ for $i = 1,\ldots,n$ at $t = t_0$; then the value of the output of the ith integrator at $t = t_1$ is, by linearity and superposition,

$$x_i(t_1) = \sum_{j=1}^{n} \phi_{ij}(t_1,t_0)x_j(t_0)$$

or

$$x(t_1) = \Phi(t_1,t_0)x(t_0). \tag{A4.6}$$

Similarly, we can consider the inputs $u_k(t)$, $k = 1,\ldots,m$, for $t_0 \leq t \leq t_1$ as made up of a series of impulses at all values of t, with the area of the impulse at t equal to $u_k(t)\,dt$. Then, by the principle of superposition, we find that the output of the ith integrator at $t = t_1$, due to an input $u_k(t)$, is

$$x_i(t_1) = \int_{t_0}^{t_1} \sum_{j=1}^{n} \phi_{ij}(t_1,\tau)g_{jk}(\tau)u_k(\tau)\,d\tau \tag{A4.7}$$

or, for all inputs,

$$x_i(t_1) = \int_{t_0}^{t_1} \sum_{j=1}^{n} \sum_{k=1}^{m} \phi_{ij}(t_1,\tau)g_{jk}(\tau)u_k(\tau)\,d\tau\,, \qquad i = 1,\ldots,n. \tag{A4.8}$$

Finally, superimposing the effect of initial conditions and inputs, we have, in vector matrix notation,

$$x(t_1) = \Phi(t_1,t_0)x(t_0) + \int_{t_0}^{t_1} \Phi(t_1,\tau)G(\tau)u(\tau)\,d\tau\,, \tag{A4.9}$$

where t_1,t_0 are two arbitrary time instants, with $t_1 > t_0$. Thus, the general solution of a time-varying linear differential equation is expressed in terms of the so-called *fundamental or transition matrix.* Note that the fundamental matrix $\Phi(t,t_0)$ obeys, by construction,

$$\frac{d}{dt}\Phi(t,t_0) = F(t)\Phi(t,t_0)\,, \qquad \Phi(t_0,t_0) = I\,, \tag{A4.10}$$

which is simply n sets of (A4.5) (with $u = 0$) in parallel.

Problem 1. Prove (A4.9) directly, using (A4.10) and (A4.5).

The computation of the time history of the fundamental matrix solution as described above is extremely laborious for all t,τ. However, a great simplification is possible. It turns out that $\Phi(t_1,\tau)$ for *fixed* t_1 *and variable* τ obeys the differential equation

$$\frac{d}{d\tau}\Phi^T(t_1,\tau) = -F^T(\tau)\Phi^T(t_1,\tau), \qquad \Phi(t_1,t_1) = I, \tag{A4.11}$$

which is called the adjoint equation to (A4.5). Its *real contribution* is that it enables $\Phi(t_1,\tau)$ to be computed for variable τ in a *single* integration of (A4.11). Equation (A4.11) may be derived by considering the identity

$$\Phi(t_1,\tau)\Phi(\tau,t_1) = I, \tag{A4.12}$$

where the matrix $\Phi(\tau,t_1)$ can be visualized as the initial condition matrix at τ that would produce I as the matrix response at t_1. Now differentiating (A4.12) on both sides with respect to τ, we get

$$\dot{\Phi}(t_1,\tau)\Phi(\tau,t_1) + \Phi(t_1,\tau)\dot{\Phi}(\tau,t_1) = 0.$$

Using (A4.10), this becomes

$$\frac{d}{d\tau}\Phi(t_1,\tau) = -\Phi(t_1,\tau)F(\tau),$$

which was to be proved. Several easily derived properties of $\Phi(t,\tau)$ are:

(a) The matrix $\Phi(t,t_0)$ is never singular and obeys

$$|\Phi(t,t_0)| = \exp\left\{\int_{t_0}^{t} \mathrm{Tr}[F(\tau)]\, d\tau\right\}. \tag{A4.13}$$

(b) From (A4.12), we have

$$\Phi(t,t_0) = \Phi(t,\tau)\Phi(\tau,t_0), \tag{A4.14}$$

$$\Phi(t,t_0) = \Phi^{-1}(t_0,t). \tag{A4.15}$$

Sometimes, for convenience, we write $\Phi(t,t_0)$ as $\Phi(t)$ and $\Phi(t,\tau) = \Phi(t)\Phi(-\tau)$. In particular, if the matrix F is constant, then t_0 can be taken as zero, and we have

$$\Phi(t,t_0) \triangleq \Phi(t,0) \triangleq \Phi(t) \triangleq e^{Ft} \triangleq \sum_{k=0}^{\infty} \frac{F^k}{k!} t^k, \tag{A4.16}$$

$$\Phi(t,\tau) = \Phi(t-\tau). \tag{A4.17}$$

Problem 2. Verify (A4.16) directly by substitution into (A4.5).

Equation (A4.11) is often of great use when we wish to compute the value of a particular linear combination of x at a terminal time t_1 due to input $u(t)$, $t_0 \le t \le t_1$. We have, from (A4.9),

$$h^T x(t_1) = \int_{t_0}^{t_1} h^T \Phi(t_1,\tau) G(\tau) u(\tau)\, d\tau\,.$$

If we define $\lambda^T(\tau) \triangleq h^T\Phi(t_1,\tau)$, then $\lambda(\tau)$ obeys

$$\frac{d\lambda}{d\tau} = -F^T(\tau)\lambda\,, \qquad \lambda(t_1) = h\,.$$

This equation will be encountered many times in Chapters 2 through 8 and will be referred to variously as the *adjoint equation*, the *influence function equation*, etc., or, in optimization context, as the *multiplier rule* and the *Euler equation.*

Discrete linear systems. Often, for a linear system, not only are F and G constant, but, in addition, the control u is approximated as piecewise constant:

$$u(t) = u(iT)\,, \qquad iT \le t < (i+1)T\,,\ i = 0,1,\ldots; \tag{A4.18}$$

T is usually called the *sampling period* and can be taken to be 1 by appropriate scaling. In this case, let $\Phi(T) = \Phi$ and

$$\int_0^1 \Phi(t) G\, dt = D\,.$$

It is then a simple exercise to verify that (A4.9) for $t = 1,2,3,\ldots,k\ldots,$ can be written as

$$x(k) = \Phi x(k-1) + Du(k-1) = \Phi^k x(0) + \sum_{i=0}^{k-1} \Phi^{k-1-i} Du(i)\,. \tag{A4.19}$$

The difference equation (A4.19) describes accurately the behavior of the linear system at the sampling instant and is sometimes referred to as a sampled-data system. On the other hand, if F, G and the sampling instants are time varying, (A4.19) becomes

$$x(t_k) = \Phi(t_k,t_{k-1})x(t_{k-1}) + D(t_k,t_{k-1})u(t_{k-1})\,, \tag{A4.20}$$

where

$$D(t_k,t_{k-1}) = \int_{t_{k-1}}^{t_k} \Phi(t_k,t) G(t)\, dt\,.$$

A linear discrete system may also be *defined* as one which is governed by (A4.19) or (A4.20) [see, for example, Zadeh and Desoer (1963)]. For example, a large class of digital computer circuits, such as shift registers, coders, and decoders, can be represented by (A4.19) or (A4.20), with the element x, etc., defined on a prime-modulo number field. Then all the applicable theorems from linear algebra apply directly to such systems. In the present case, where the discrete system is actually derived from a continuous linear system, we have the additional knowledge that $\det(\Phi) \neq 0$ [see Equation (A4.13)].

Matrix linear systems. Another type of linear differential equation often encountered in control and estimation theory is the linear matrix differential equation

$$\dot{P} = FP + PF^T + Q\,, \tag{A4.22}$$

where F,Q are known time-varying matrices. Of course, (A4.22) can always be rearranged into a vector linear differential equation in the form of Equation (A4.5). However, for theoretical development, it is much more convenient to be able to write down directly the general solution of (A4.22) as

$$P(t) = \Phi(t,t_o)P(t_o)\Phi^T(t,t_o) + \int_{t_o}^{t} \Phi(t,\tau)Q(\tau)\Phi^T(t,\tau)\,d\tau\,, \tag{A4.23}$$

where $\Phi(t,\tau)$ is the transition matrix associated with the system $\dot{x} = Fx$. The validity of (A4.23) can be verified directly. This formula will prove particularly useful in Chapters 11–14.

Remarks on computations. Using high-speed digital or analog computers, it is relatively straightforward to obtain a single solution for the differential equation (A4.1), even for systems of rather high order, say, $n = 10$ or higher. However, it becomes extremely inefficient to carry out the integration of (A4.1) hundreds or thousands of times. This is where the principle of superposition in linear systems can be utilized to advantage. From the theory derived for linear systems in Section A3, we know that there exists a fundamental matrix solution for a given linear system as represented by (A4.10). For any given initial conditions and inputs, the actual solution for the linear system can be expressed simply as various linear combinations of the elements of the fundamental matrix solution [incidentally, this is the reason why the determinant of $\Phi(t,\tau)$ is never zero, for otherwise, the columns will be linearly dependent, and linear combinations of them will not be able to represent any arbitrary solution for (A4.5)]. Thus, once $\Phi(t,\tau)$ is known, any other solution can be obtained simply by matrix multiplication. Furthermore, it turns out in control applications, one is often interested in the response of a system at one fixed time for initial conditions at a prior variable time, that is, $\Phi(t_1,t)$. Hence, as pointed out earlier, only one integration of the adjoint equations can furnish the complete information. These simplifications can be approximately carried over to nonlinear systems if we restrict our considerations to a small neighborhood of a given solution of (A4.1); that is, we linearize (A4.1). Then the variational equation (A4.3) applies, and we have linearity once more. Of course, we need not be permanently restricted in small neighborhood. Once a neighboring solution for (A4.1) is known, we can successively linearize and bootstrap ourselves to a distant solution by a process analogous to analytic continuation. This way we gradually sweep out a volume in the solution space of (A4.1) within which we have information for *all* solutions. On the other hand, in the case of brute-force integration of (A4.1), we would require many, many integrations to acquire the same information. This is especially wasteful if only a few of the many possible solutions are actually required.

REFERENCES

1. R. BELLMAN. *Introduction to Matrix Analysis.* New York: McGraw-Hill, 1960.
2. S. CODDINGTON and N. LEVINSON. *Theory of Ordinary Differential Equations.* New York: McGraw-Hill, 1955.
3. V. N. FADEEVA. *Computational Methods of Linear Algebra.* New York: Dover, 1959.
4. F. R. GANTMACHER. *The Theory of Matrices.* New York: Chelsea, 1959, Vols. I and II.
5. S. LEFSCHETZ. *Differential Equations—Geometric Theory.* New York: Interscience, 1957.
6. D. SOMERVILLE. *Introduction to the Geometry of n-Dimensions.* New York: Dover, 1958.
7. L. ZADEH and C. A. DESOER. *Linear System Theory.* New York: McGraw-Hill, 1963.

Appendix B
Properties of linear systems

B1 Linear algebraic equations

In this appendix we present some of the important properties of linear dynamic systems used in this book.

Many of the properties of linear systems are connected with the solution of the linear algebraic equation

$$A\alpha = \beta \,, \tag{B1.1}$$

where A is an $(m \times k)$-matrix, α a k-vector, and β an m-vector. Some well-known facts concerning (B1.1) are:

THEOREM B1. *If* $m = k$, (B1.1) *has a unique solution if* $\det(A) \neq 0$.

THEOREM B2. *If* $m < k$ *and rank* $(A) = m$, (B1.1), *in general, has an infinite number of solutions.*

THEOREM B3. *If* $m \leq k$ *and rank* $(A) = m_1 < m$, (B1.1) *has solutions if* β *belongs to an* m_1*-dimensional subspace of the* m*-space.*

THEOREM B4. *If* $m > k$ *and rank* $(A) = k$, *the least-square-fit solution of* (B1.1) *is unique and is given by* $(A^TA)^{-1}A^T\beta$.

THEOREM B5. *If* $m > k$ *and rank* $(A) = k_1 < k$, *the least-square-fit solution of* (B1.1) *will not be unique and is given by the nonunique solutions of the equation* $A^TA\alpha = A^T\beta$.

We shall see that, by proper interpretation of the matrix A and the vectors α and β, the theory of controllability and observability follows as a simple consequence of Theorems B1 through B5.

B2 Controllability

In Theorem B2, consider $m = n$ and

$$A \triangleq [\Phi^{k-1}d \,\vdots\, \cdots \,\vdots\, \Phi d \,\vdots\, d] \,, \tag{B2.1}$$

$$\alpha \triangleq u = \begin{bmatrix} u(0) \\ \cdot \\ \cdot \\ \cdot \\ u(k-1) \end{bmatrix}, \tag{B2.2}$$

$$\beta \triangleq [x(k) - \Phi^k x(0)]. \tag{B2.3}$$

Equation (B1.1) then constitutes a statement of the dynamics of an nth-order single-input linear system:

$$x(k) = \Phi^k x(0) + \sum_{i=0}^{k-1} \Phi^{k-i-1} du(i). \tag{B2.4}$$

Finding a solution to (B1.1) is equivalent to finding a scalar control sequence $u(i)$, which transfers the state $x(0)$ at $t = 0$ to the state $x(k)$ at $t = k$. For arbitrary $x(k)$ or $x(0)$ (corresponding to arbitrary β), the results of Theorems B2 and B3 require A to be of maximal rank or, what is equivalent,

$$AA^T = \sum_{i=0}^{k-1} \Phi^{k-i-1} dd^T(\Phi^{k-i-1})^T > 0. \tag{B2.5}$$

Since the pair (Φ, d) only depends on the specification of the linear discrete system, (B2.5) represents an intrinsic property of the linear system that we shall define as *controllability*, i.e., the ability to transfer an arbitrary initial state to an arbitrary final state in finite time. If (B2.5) is true, it follows that the minimum energy control sequence, in the sense that $\sum_{i=0}^{k-1} u(i)^2$ is a minimum, is given by (see Chapter 5, Section 5.3)

$$u = (AA^T)^{-1}A^T\beta. \tag{B2.6}$$

Note that, if rank (A) is not maximal, solutions may still exist to (B2.4). In other words, an uncontrollable system may still have states that are controllable. The generalization of (B2.5) to multi-input, continuous, and nonstationary systems is straightforward. For the system $x(i) = \Phi(t_i, t_{i-1})x(i-1) + d(t_{i-1})u(i-1)$, we have

$$\sum_{i=0}^{k-1} \Phi(t_{k-1}, t_i)\, d(t_i)\, d(t_i)^T \Phi(t_{k-1}, t_i)^T > 0. \tag{B2.7}$$

For the multi-input system with input matrix $D(t_i)$, we have

$$\sum_{i=0}^{k-1} \Phi(t_{k-1}, t_i) D(t_i) D(t_i)^T \Phi(t_{k-1}, t_i)^T > 0. \tag{B2.8}$$

For the continuous system $\dot{x} = Fx + Gu$, we have

$$\int_{t_0}^{t} \Phi(t,\tau)G(\tau)G^T(\tau)\Phi^T(t,\tau)\,d\tau > 0 \tag{B2.9}$$

$$\Longleftrightarrow \int_{t_0}^{t} \Phi(t_0,\tau)G(\tau)G^T(\tau)\Phi^T(t_0,\tau)\,d\tau > 0\,.$$

Various other equivalent conditions can be derived [see Kalman, Ho, and Narendra (1963)]. If we write $\Phi^T(t_f,\tau) = R(\tau)$, then $(dR/d\tau) = -F^TR$, $R(t_f) = I$, and (B2.9) can be rewritten as

$$\int_{t_0}^{t_f} R^TGG^TR\,d\tau = -Q(t_0) > 0\,,$$

where $\dot{R} = -F^TR$, $\dot{Q} = -R^TGG^TR$ (see Chapter 5, Section 5.3).

Problem. Verify (B2.9).

B3 Observability

In Theorem B4, let us consider that $k = n$ and

$$A \triangleq \begin{bmatrix} h^T\Phi^{-k+1} \\ \hline \cdot \\ \cdot \\ \cdot \\ \hline h^T\Phi^{-1} \\ \hline h^T \end{bmatrix}, \tag{B3.1}$$

$$\alpha \triangleq x(k)\,, \tag{B3.2}$$

$$\beta = \begin{bmatrix} z(1) \\ \cdot \\ \cdot \\ \cdot \\ z(k-1) \\ z(k) \end{bmatrix}. \tag{B3.3}$$

Equation (B1.1) then describes a sequence of observations on the single-output, linear, discrete system:

$$x(i+1) = \Phi x(i)\,, \qquad z(i) = h^Tx(i)\,. \tag{B3.4}$$

Finding a solution to (B1.1) is equivalent to determining $x(k)$ from the observations $z(1), \ldots, z(k)$. Since (B1.1), in this case, always has solutions by construction of (B3.4), they can be determined by the least-square fit (i.e., the least-square-fit error will be zero). In view of Theorems B4 and B5, we conclude that $x(k)$ can be uniquely determined from the observation sequence $z(i)$ if and only if (iff) A is of maximal rank, or, equivalently,

$$A^TA = \sum_{i=0}^{k-1} (\Phi^T)^{-i} hh^T \Phi^{-i} > 0 . \tag{B3.5}$$

Since the pair (h,Φ) only depends on the specification of the linear system and the output matrix, (B3.5) represents an *intrinsic* property of the observation system, which we shall define as *observability*, i.e., the ability to uniquely determine the state of a free linear dynamic system from observations of linear combinations of the output of the system in finite time. If (B3.5) is not true, we cannot uniquely determine $x(k)$. It is determined only to within a subspace of the state space spanned by the solutions of $A^TA\alpha = A^T\beta$.

The relationship (B3.5) may be generalized to multi-output, nonstationary, and continuous cases. For multiple-output $z(i) = H(t_i)x(i)$, we have

$$\sum_{i=0}^{k-1} \Phi^T(t_i,t_k)H^T(t_i)H(t_i)\Phi(t_i,t_k) > 0 . \tag{B3.6}$$

For continuous systems $\dot{x} = F(t)x$, $z = H(t)x$, we have

$$\int_{t_o}^{t} \Phi^T(\tau,t)H^T(\tau)H(\tau)\Phi(\tau,t)d\tau > 0. \tag{B3.7}$$

For further details see Kalman (1963).

B4 Stability

Consider the free dynamic system

$$\dot{x} = f(x,t) , \qquad \text{where} \qquad f(0,t) = 0 , \qquad \text{and } x \text{ is an } n\text{-vector.} \tag{B4.1}$$

We say that the system described by (B4.1) has *uniform asymptotic global stability* (UAGS) if $\|x(t)\|^2 \to 0$ as $t \to \infty$ (asymptotic), independent of t_o (uniform), and independent of direction and magnitude of $x(t_o)$ (global). The general UAGS theorem we have concerning (B4.1) is the following.

THEOREM B6. *For the system of* (B4.1), *if there exists a scalar function* $V(x,t)$ *with continuous* $(\partial V/\partial x)$, $(\partial V/\partial t)$ *such that*

(a) $\qquad V(x,t) > 0 \quad$ *for all* $\quad x \neq 0 \quad$ *and all* t,

that is,

$$0 < \alpha(\|x\|^2) < V(x,t) , \qquad \textit{where} \qquad \alpha(0) = 0 ,$$

(b) $\qquad \dot{V}(x,t) < -\gamma(\|x\|^2) < 0 \quad$ *for all* x, t,

(c) $\qquad V(x,t) < \beta(\|x\|^2)$,

(d) $\qquad \alpha(\|x\|^2) \to \infty \quad$ *as* $\quad \|x\|^2 \to \infty$,

then the system (B4.1) *has UAGS and tends toward* $x = 0$.

PROOF. *See Kalman and Bertram (1960).*

Theorem B6 gives ***sufficient*** conditions for stability. However, it does not necessarily furnish a method for determining or constructing the Lyapunov function $V(x,t)$, if one exists, for a given system. Vast amounts of literature exist dealing with partial solutions of this problem. Further details can be found in Kalman and Bertram (1960) Brockett (1966). For linear stationary systems, more specific results do exist.

THEOREM B7 *(Lyapunov). A linear stationary system $\dot{x} = Fx$ has UAGS iff, for any positive-definite matrix Q, there exists a positive-definite matrix P, which is the unique solution of the linear algebraic equation*

$$-Q = F^T P + PF . \tag{B4.2}$$

Theorem B7 states a necessary as well as sufficient condition for the stability of the stationary linear system. It also states that $V(x) = \|x\|_P^2$ is a Lyapunov function if the system has UAGS. Thus, we have an explicit construction method in this case.

Problem 1. Prove the equivalent condition

$$-Q = \Phi^T P \Phi - P \tag{B4.3}$$

for the discrete system

$$x_{t+1} = \Phi x_t .$$

Problem 2. What is the connection of (B4.2) with Equation (11.4.17)? From this, show that the variance equation (11.4.17) has a finite equilibrium solution if the system has a stable matrix, F, that is, $Re[\lambda_i(F)] < 0$. [HINT: We have that $\lambda_i(F) = \lambda_i(F^T)$.]

B5 Canonical transformations

For single-input, single-output linear stationary dynamic systems,

$$\dot{x} = Fx + gu , \qquad z = h^T x , \tag{B5.1}$$

there exists a linear transformation on the state variables that is useful in establishing relationships between state-space control theory and classical control theory based on transfer functions.

Let us assume that (B5.1) is controllable and observable. Then we have the following theorem.

THEOREM B8. *The transformation $y = Ax$ transforms* (B5.1) *to*

$$\dot{y} = F_c y + g_c u , \qquad z = h_c^T y , \tag{B5.2}$$

where

$$F_c \triangleq \left[\begin{array}{c|c} 0 & \\ \cdot & \\ \cdot & I \\ \cdot & \\ 0 & \\ \hline -a_n \quad -a_{n-1} & \cdots -a_1 \end{array}\right] \qquad g_c \triangleq Ag\,, \tag{B5.3}$$

$$h_c^T = [1,0,\ldots,0]\,,$$

and

$$A \triangleq \left[\begin{array}{c} h^T \\ \hline h^TF \\ \hline \cdot \\ \cdot \\ \cdot \\ \hline h^TF^{n-1} \end{array}\right] \tag{B5.4}$$

We have a similar transformation for the corresponding discrete system if we replace F,g by Φ and d.

PROOF. *Use direct substitution.*

The implication of Theorem B8 is that (B5.1) can always be made equivalent to the following *scalar* differential equation:

$$\frac{d^nz}{dt^n} + a_1\frac{d^{n-1}z}{dt^{n-1}} + \cdots + a_nz = b_1\frac{d^{n-1}u}{dt^{n-1}} + \cdots + b_nu\,, \tag{B5.5}$$

where

$$b_i = \sum_{k=0}^{i-1} a_{i-k}g_{c_k} + g_{c_i}\,. \tag{B5.6}$$

The numerical computation (B5.5) (or, rather, its discrete equivalent) may often be more convenient, since it involves fewer arithmetic operations.

Problem 1. Show that the generalization of (B5.6) to the time-varying case is

$$b_i(t) = \sum_{k=0}^{i-1}\sum_{s=0}^{i-k}\binom{n+s-1}{n-1}a_{i-k-s}(t)\frac{d^sg_{c_k}}{dt^s} + g_{c_i}(t)\,. \tag{B5.6$'$}$$

Problem 2. Show that the pair (F_c,g_c) is always controllable if $g_c^T = [0,\ldots,1]$. What does this imply concerning the ability of u in

(B5.5) to regulate the behavior of z and its first $n-1$ time derivatives?

Problem 3. Consider the dynamic system

$$x_{t+1} = \Phi x_t + du_t, \qquad z_t = h^T x_t,$$

where

$$\Phi = \begin{bmatrix} 0, & 1 \\ -2, & -3 \end{bmatrix}; \qquad d = \begin{bmatrix} 0 \\ 1 \end{bmatrix}, \qquad h^T = [1, 1].$$

Is the system controllable? Observable?

Problem 4. Consider the autonomous discrete dynamic system

$$x_{t+1} = \Phi x_t .$$

Assuming that x can be measured, what is the necessary and sufficient condition for determining Φ?

Now suppose that

$$z_t = h^T x_t,$$

where h is also unknown. To what extent (if at all) can we determine (Φ, h) such that, as far as output z_t is concerned, no difference can be observed? [HINT: Consider observability and canonical transformation.]

Problem 5. Given that the system

$$\dot{x} = Fx + gu$$

is asymptotically stable for $u = 0$, show that the control law

$$u = \text{sat}[-g^T Px]$$

is asymptotically stable for some $P > 0$.

REFERENCES

1. R. W. BROCKETT. "The Status of Stability Theory for Deterministic Systems," *Trans. on Automatic Control IEEE*, Vol. AC-11, No. 3, July 1966.
2. R. E. KALMAN, Y.C. HO, and K. S. NARENDRA. "Controllability of Linear Dynamic Systems," *Contributions to Differential Equations*, Vol. 1, 1963, p. 189.
3. R. E. KALMAN. "Canonical Structure of Linear Dynamical Systems," *SIAM*, Vol. A1, 1963.
4. R. E. KALMAN and J. E. BERTRAM. "Control System Analysis and Design Via the Second Method of Lyapunov," *Trans. ASME*, Vol. 82, 1960, p. 371.

References for all chapters

The aim of this book is to be as self-contained as possible. References cited in the text fall into two categories: (*i*) sources of additional information in which readers may wish to pursue further detail concerning the particular point under discussion, and (ii) general references or books treating similar material for collateral reading. References in category (i) are not necessarily the original sources, and those in category (ii) are limited to books or monographs easily available or classic references with which the authors are familiar. No attempt at a complete bibliography or an historical record is claimed. Many of the results derived in the book were obtained by the authors while teaching the material during the last five years. However, it is likely that a significant fraction of them were obtained independently by others and may even have appeared in the literature. The authors apologize in advance for omission of such references. Additional references also appear in footnotes which are not repeated here.

References for Chapter 1

1. H. KUHN and A. W. TUCKER. "Nonlinear Programming," *Second Berkeley Symposium of Mathematical Statistics and Probability.* Berkeley, Calif.: University of California Press, 1951.
2. G. ZOUTENDIJK. *Method of Feasible Directions.* London: Elsevier Pub. Co., 1961.
3. G. DANTZIG. *Linear Programming and Extensions.* Princeton, N. J.: Princeton University Press, 1963, Chapter 6.
4. G. HADLEY. *Nonlinear and Dynamic Programming.* Reading, Mass.: Addison-Wesley, 1964.
5. H. P. KUNZI *et al. Nonlinear Programming.* Waltham, Mass.: Blaisdell, 1966.

References for Chapters 2 and 3

1. C. LANCZOS. *The Variational Principle of Mechanics.* Toronto, Canada: University of Toronto Press, 1949.
2. P. CICALA. *An Engineering Approach to the Calculus of Variations.* Turin, Italy: Levrotto and Bella, 1957.
3. J. V. BREAKWELL. "The Optimization of Trajectories," *SIAM Journal*, Vol. 7, 1959.
4. D. F. LAWDEN. *Optimal Trajectories for Space Navigation.* London: Butterworths, 1963.
5. A. E. BRYSON, JR., and W. F. DENHAM. "A Steepest-Ascent Method for Solving Optimum Programming Problems," *J. Appl. Mech.*, June 1962.
6. M. ATHANS and P. L. FALB. *Optimal Control.* New York: McGraw-Hill, 1966.
7. W. F. DENHAM. "Steepest Ascent Solution of Optimal Programming Problems," Ph.D. Thesis, Harvard University, 1963.

8. L. NEUSTADT. "Synthesizing Time Optimal Controls," *J. Math. Anal. & Appl.* Vol 1, 1960, p. 484.
9. D. W. BUSHAW. "Optimal Discontinuous Forcing Terms," *Contributions to the Theory of Nonlinear Oscillations, Annals of Mathematical Studies*, Vol. 41. Princeton, N. J.: Princeton University Press, 1958.
10. A. E. BRYSON, W. F. DENHAM, and S. E. DREYFUS. "Optimal Programming Problems with Inequality Constraints. I: Necessary Conditions for Extremal Solutions," *AIAA Journal*, Vol. 1, 1963, p. 2544.
11. S. S. L. CHANG. "Optimal Control in Bounded Phase Space," *Automatica*, Vol. 1, 1963.
12. G. LEITMANN. *An Introduction to Optimal Control.* New York: McGraw-Hill, 1966.
13. E. B. LEE and L. MARKUS. *Foundations of Optimal Control Theory*, New York: Wiley, 1967.
14. G. LEITMAN (ed.). *Optimization Techniques.* New York: Academic Press, 1962.
15. G. LEITMAN (ed.). *Topics in Optimization.* New York: Academic Press, 1966.
16. L. S. PONTRYAGIN *et al. The Mathematical Theory of Optimal Processes.* New York: Interscience, 1962.
17. L. D. BERKOVITZ. "Variational Methods in Problems of Control and Programming," *J. Math. Anal. & Appl.*, Vol. 3, 1961, p. 145.
18. J. L. SPEYER, R. K. MEHRA, and A. E. BRYSON. "The Separate Computation of Arcs for Optimal Flight Paths with State Variable Inequality Constraints," *Advanced Problems and Methods for Space Flight Optimization*, Pergamon Press, Oxford, 1969, pp. 53-68.
19. E. J. MCSHANE. "On Multipliers for LaGrange Problems," *Amer. J. Math.* LXI 809-819, (1939).

References for Chapter 4

1. R. BELLMAN. *Dynamic Programming.* Princeton, N. J.: Princeton University Press, 1957.
2. S. DREYFUS and L. BERKOVITZ. "A Dynamic Programming Approach to the Nonparametric Problem in the Calculus of Variations," *J. Math. & Mech.*, Vol. 15, 1966, p. 83.
3. S. E. DREYFUS. *Dynamic Programming and the Calculus of Variations.* New York: Academic Press, 1965.
4. R. E. KALMAN. "The Calculus of Variations and Optimal Control Theory," *Mathematical Optimization Techniques*, R. Bellman (ed.). Berkeley, Calif. University of California Press, 1963.
5. R. LARSON. "Survey of Dynamic Programming Computational Procedures," *Trans. IEEE-GAC*, Vol. AC-12, No. 6, 1967, pp. 767–774.

References for Chapter 5

1. R. E. KALMAN. "Contributions to the Theory of Optimal Control," *Bol. de Soc. Math. Mexicana* (1960), p. 102.
2. R. E. KALMAN. "On the General Theory of Control," *Proc. of the First*

Int. Cong. on Automatic Control, Vol. 1. London: Butterworth Scientific Institute, 1961, p. 481.
3. R. E. KALMAN. "When Is a Linear Control System Optimal?" *Trans. ASME*, Vol. 86D, 1964, p. 51.
4. R. E. KALMAN and T. ENGLAR. "ASP—The Automatic Synthesis Program (Program C)," *NASA Contractor Report CR-475*, June 1966.

References for Chapters 6 and 7

1. H. J. KELLEY. "Guidance Theory and Extremal Fields," *Proc. IRE*, 1962, p. 75.
2. J. V. BREAKWELL, J. L. SPEYER, and A. E. BRYSON, JR. "Optimization and Control of Nonlinear Systems Using the Second Variation," *SIAM Journal*, Vol. 1, 1963, p. 193.
3. J. V. BREAKWELL and Y. C. HO. "On the Conjugate Point Condition for the Control Problem," *Intl. J. of Engineering Science*, Vol. 2, 1965, p. 565.
4. S. R. MCREYNOLDS and A. E. BRYSON. "A Successive Sweep Method for Optimal Programming Problems," *Joint Auto. Control Conf.* (1965) also *Harvard DEAP Report No. 465* (1965).
5. S. MITTER. "Successive Approximation Method for the Solution of Optimal Control Problems," *Automatica*, Vol. 3, 1966, p. 135.
6. S. R. MCREYNOLDS. "The Successive Sweep Method and Dynamic Programming," *J. Math. Anal. & Appl.*, Vol. 19, 1967, p. 565.
7. W. F. DENHAM and A. E. BRYSON, JR.. "Optimal Programming Problems with Inequality Constraints II: Solution by Steepest-Ascent," *AIAA Journal*, Vol. 2, 1964.
8. A. V. BALAKRISHNAN and L. W. NEUSTADT (eds.). *Computing Methods in Optimization Problems*. New York: Academic Press, 1964.
9. R. E. KALMAN. "Towards a Theory of Difficulty in the Computation of Optimal Controls," *Proc. 1964 IBM Symp. on Comp. and Control* (1966).
10. F. S. BECKMAN. "The Solution of Linear Equations by the Conjugate Gradient Method," *Mathematical Methods for Digital Computers*, Vol. 1, A. Ralson and H. Wilf (eds.), New York: Wiley, 1960.
11. L. S. LASDON, S. MITTER, and A. WARREN. "The Method of Conjugate Gradient for Optimal Control Problems," *IEEE Trans. Automatic Control*, April 1967.
12. D. H. JACOBSON. "Second-Order and Second Variation Methods for Determining Optimal Control: A Comparative Analysis Using Differential Dynamic Programming," *Intl. J. of Control*, Vol. 7, No. 2, 1968, pp. 175–196.
13. See also References 5 through 8, 14, and 15 of Chapters 2 and 3.

References for Chapter 8

1. K. TAIT, "Singular Problems in Optimal Control," Ph.D. Thesis, Harvard University, 1965.
2. H. J. KELLEY, R. E. KOPP, and A. G. MOYER. "Singular Extremals," *Topics in Optimization*, G. Leitmann (ed.), Vol. II, Chapter 3. New York: Academic Press, 1966.

3. H. M. ROBBINS. "Optimality of Intermediate-Thrust Arcs of Rocket Trajectories," *AIAA Journal*, Vol. 3, 1965, p. 1094.
4. C. JOHNSON and W. WONHAM. "Optimal Bang-Bang Control with Quadratic Performance Index," *Trans. ASME*, Vol. 86D, 1964, p. 107.
5. H. M. ROBBINS. "A Generalized Legendre-Clebsch Condition for the Singular Case of Optimal Control," *IBM Federal Systems Division Rept. #66–825 2043*, Sept. 1966.
6. C. JOHNSON. "Singular Problems in Optimal Control," *Advances in Control Systems*, C. T. Leondes (ed.), Vol. 2. New York: Academic Press, 1965, p. 209.

References for Chapter 9

1. R. ISAACS. *Differential Games*. New York: Wiley, 1965.
2. Y. C. HO. "Review of the Book *Differential Games*, by R. Isaacs," *IEEE Trans. on Automatic Control*, October 1965.
3. L. D. BERKOVITZ. "A Variational Approach to Differential Games, "*Advances in Game Theory, Annals of Math. Studies*, No. 52. Princeton, N.J.: Princeton University Press, 1964, p. 127.
4. S. BARON. "Differential Games and Optimal Pursuit–Evasion Strategies," Ph.D. Thesis, Harvard University, 1966; also in *IEEE Trans. Auto. Control*, Oct. 1965, p. 385.

References for Chapters 10 and 11

1. W. FELLER. *An Introduction to Probability Theory and Its Applications*, Vol. I (2nd ed.). New York: Wiley 1957.
2. E. PARZEN. *Modern Probability Theory and Its Applications*, New York: Wiley, 1960.
3. E. PARZEN. *Stochastic Processes*. San Francisco, Calif.: Holden-Day, 1962.
4. H. CRAMER. *Mathematical Method of Statistics*. Princeton, N. J.: Princeton University Press, 1946.
5. H. RAIFFA and R. SCHLAIFER. *Applied Statistical Decision Theory*. Cambridge, Mass.: Harvard University Press, 1961.
6. N. WAX (ed.). *Collected Papers on Noise and Stochastic Processes*. New York: Dover, 1954.

References for Chapter 12

1. N. WIENER. *The Interpolation and Smoothing of Stationary Time Series*. Cambridge, Mass.: M.I.T. Press, 1949.
2. R. E. KALMAN. "A New Approach to Linear Filtering and Prediction Problems," *Trans. ASME*, Vol. 82D, 1960, p. 35.
3. R. E. KALMAN and R. BUCY. "New Results in Linear Filtering and Prediction," *Trans. ASME*, Vol. 83D, 1961, p. 95.
4. R. DEUTSCH. *Estimation Theory*. Englewood Cliffs, N.J.: Prentice-Hall 1965.
5. H. SORENSEN. "Kalman Filtering," *Advance in Control Systems*, Vol. 3, C. T. Leondes (ed.). New York: Academic Press, 1966, p. 219.

References for Chapter 13

1. A. E. BRYSON and M. FRAZIER. "Smoothing for Linear and Nonlinear Dynamic Systems," *Proc. Optimum Sys. Synthesis Conf., U.S. Air Force Tech. Rept. ASD-TDR-63-119*; Feb. 1963.
2. H. E. RAUCH, F. TUNG, and C. T. STRIEBEL. "Maximum Likelihood Estimates of Linear Dynamic Systems," *AIAA Journal*, Vol. 3, 1965, p. 1445.
3. A. E. BRYSON and D. E. JOHANSEN. "Linear Filtering for Time-Varying Systems Using Measurements Containing Colored Noise," *IEEE Trans. Auto. Control*, Vol AC-10, 1965, p. 4.
4. L. J. HENRIKSON. "Sequentially Correlated Measurement Noise with Application to Inertial Navigation," Ph.D. Thesis, Harvard University, 1967; also in *Jour. Spacecraft and Rockets*, Vol. 5, 1968.

References for Chapter 14

1. H. A. SIMON. "Dynamic Programming Under Uncertainty with a Quadratic Criterion Function," *Econometrica*, Vol. 24, 1956, pp. 74–81.
2. P. D. JOSEPH and J. T. TOU, "On Linear Control Theory," *Trans. AIEE*, Part III, Vol. 80, No. 18, 1961.
3. T. F. GUNCKEL and G. F. FRANKLIN. "A General Solution for Linear Sampled-Data Control Systems," *Trans. ASME*, Vol. 85D, 1963, p. 197.
4. A. E. BRYSON. "Applications of Optimal Control Theory in Aerospace Engineering," *Jour. Spacecraft and Rockets*, Vol. 4, 1967, p. 545.

Multiple-choice examination

The following is a multiple-choice examination the reader can use to test his own knowledge on material treated in this book. This material is covered by Sections 1.1 – 1.6, 2.1 – 2.4, Chapter 4, Chapter 5; Sections 7.1, 7.2, 7.4, Chapter 10, Chapter 11, and Chapter 12. The examination should take 3 hours. The answers are given at the end.

Problem 1 (12 pt). Consider the problem of minimizing $J = \phi[x(t_f)] + \int_{t_0}^{t_f} L(x,u,t)\,dt$, subject to $\dot{x} = f(x,u,t)$ and $g[x(t_0)] = 0$ (g a vector function) by choosing $u(t)$, $t_0 \leq t \leq t_f$ and $x(t_0)$. The necessary conditions for an extremum are

(i) $\dot{\lambda}^T = -H_x$, $\quad \lambda^T(t_f) = \phi_{x(t_f)}$, $\quad \lambda^T(t_0) = -\nu^T g_{x(t_0)}$,
$\quad\dot{x} = f(x,u,t)$, $\quad H_u = 0$, $\quad g[x(t_0)] = 0$,

where $H = \lambda^T f + L$ and ν is a vector of constants;
(ii) same as (i) except that $\lambda(t_0) = \nu$;
(iii) same as (i) except that $\lambda(t_0) = 0$;
(iv) same as (i) except that $\lambda^T(t_0) = g_{x(t_0)}$.

Problem 2 (12 pt). Consider the calculus of variations problem in Chapter 2;

$$\dot{x} = f(x,u,t), \qquad J = \phi[x(t_f)] + \int_{t_0}^{t_f} L\,dt.$$

We add the scalar constraint that $g(u) = 0$ for all t(u is a vector). Necessary conditions for an extremum are

(i) $H = \lambda^T f + \mu g + L$,. $\quad \dot{\lambda}^T = -H_x$, $\quad \dot{x} = f$, $\quad \lambda^T(t_f) = \phi_{x(t_f)}$,
$\quad H_u = 0$, $\quad g(u) = 0$;
(ii) same as (i) but with addition of; $\dot{\mu} = -H_g$, $\quad \mu(t_f) = 0$;
(iii) same as (i) but with addition of $H = \text{const}$;
(iv) same as (i) except that $\dot{\lambda}^T = -H_x - \mu g_u$.

Problem 3 (12 pt). Consider the path cost problem as shown in Figure P3. Moves (up or down) at levels 1 and 3 are controlled by player A while at levels 2 and 4 player B controls. A wishes to minimize total cost while B wishes to maximize. The minimax cost starting from point (a) at level 1 to level 5 is

(i) 26, (ii) 29, (iii) 15, (iv) ∞

[NOTE: The numbers in the circles at level 5 are the terminal costs.]

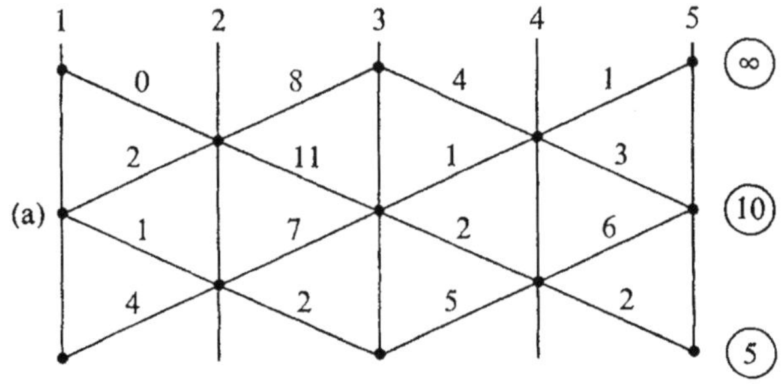

Figure P3. Cost network for Problem 3.

Problem 4 (12 pt). Consider another *minimal* path-cost problem controlled by one player only as in Figure P4. The added complication is that a given decision (up or down) has probability only 2/3 of being actually realized. The expected minimal cost of this problem is

(i) 10, (ii) 10.666666 . . . ;

(iii) a random variable J with $P(J = 10) = 2/3$, $P(J = 12) = 1/3$;

(iv) 98/9.

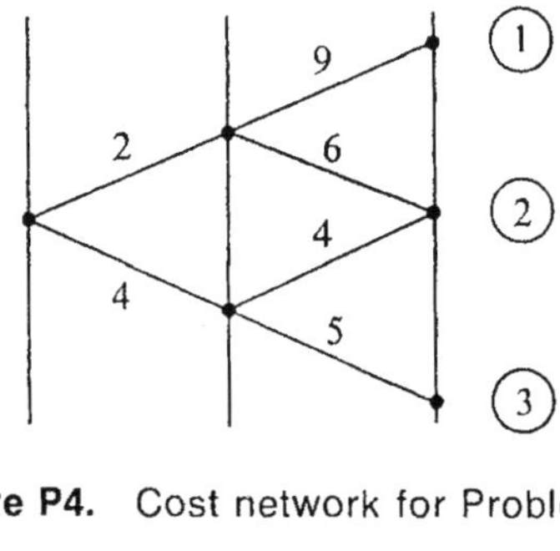

Figure P4. Cost network for Problem 4.

Problem 5 (9 pt). Choose one of the following:

(i) An unstable linear system is uncontrollable.
(ii) A stable nonlinear system is both controllable and observable.
(iii) Observability in linear systems also implies stability.
(iv) Stability, controllability, and observability are independent properties of a linear system.

Problem 6 (12 pt). For the control problem $\dot{x} = f(x,u,t)$, $J = \phi[x(t_f)]$, assume a unique optimal control exists for every starting x_0 and t_0. The value of J along an optimal trajectory is

(i) constant,
(ii) the value of the Hamiltonian function evaluated at the particular point on the trajectory,
(iii) -1,
(iv) none of the above.

Problem 7 (9 pt). Consider the linear-quadratic problem

$$\dot{x} = Fx + Gu \qquad \text{and} \qquad J = \frac{1}{2}\int_0^\infty (\|u\|_B^2 + \|x\|_A^2)\,dt\,.$$

The stability of the system under optimal control is guaranteed

(i) for any F, G, B, A;
(ii) for (F,G) controllable, $B > 0$, A anything;
(iii) for (F,G) controllable, $A > 0$, B anything;
(iv) for (F,G) controllable, $B > 0$, $A > 0$.

Problem 8 (9 pt). Consider a sequence of numbers u_t; $t = 0, \Delta, 2\Delta, 3\Delta, \ldots$, and a staircase function $u(t)$ such that $u(\tau) = u_t$, $t \leq \tau < t + \Delta$, and two constant dynamic systems

$$(^*) \quad x_{t+\Delta} = \Phi x_t + \Gamma u_t\,, x(0) = a\,,$$
$$(^{**}) \quad \dot{x} = Fx + Gu\,, x_o = a\,;$$

where $\Phi \triangleq \Phi(\Delta)$, $\Phi(t)$ is the transition matrix of F, and

$$\Gamma = \int_0^\Delta \Phi(\tau) G\, d\tau\,.$$

Then:

(i) $x(t) = x_t$ for $t = 0, \Delta, 2\Delta, \ldots$;
(ii) $x(t) \approx x_t$ for $t = \Delta, 2\Delta, 3\Delta, \ldots$;
(iii) $x(t) = x_t$ only for $t \to \infty$;
(iv) $x(t) = x_t$ only if Δ is integral multiple of the eigenvalues of F.

Problem 9 (9 pt). In Problem 8, let u_t be a purely random sequence with zero mean and covariance matrix χ and $u(t)$ a white random process with zero mean and spectral density $Q = \chi/\Delta$. Then the P_t calculated, based on the $(^*)$ model, is the same as $P(t)$ calculated based on the $(^{**})$ model at $t = 0, \Delta, 2\Delta, 3\Delta, \ldots$. This is:

(i) True.
(ii) False.

Problem 10 (9 pt). Given the dynamic system

$$\dot{x}_1 = x_2\,, \qquad \dot{x}_2 = u + 1\,,$$

and the criterion

$$J = \frac{1}{2}\int_0^\infty [a(x_1 - \tfrac{1}{2}t^2)^2 + 2b(x_1 - \tfrac{1}{2}t^2)(x_2 - t) + c(x_2 - t)^2 + du^2]\,dt\,,$$

where a, b, c, and d are constants. Then:

(i) The system can not have a stable control law.
(ii) The system can have a stable control law if $a > 0, c > 0$, $ac - b^2 > 0$.
(iii) Same as (ii) except that d > 0 also.
(iv) The control law will be stable but $u(\infty) \neq 0$.
(v) The control law will not be linear.

Problem 11 (12 pt). We have $\dot{x} = Fx + w$, where $w(t)$ is stationary zero mean white noise. If $Re[\lambda_i(F)] < 0$, that is, F is asymptotically stable, then

(i) $C(t,\tau) \rightarrow$ nonzero constant for fixed τ and $t \rightarrow \infty$,
(ii) $C(t,\tau) \rightarrow$ nonzero constant for fixed t and $\tau \rightarrow \infty$,
(iii) $C(t,\tau) \rightarrow 0$ for $(t - \tau) \rightarrow \infty$,
(iv) $C(t,\tau) \rightarrow 0$ for $(t - \tau) \rightarrow \infty$ only in steady state, where $C(t,\tau) \triangleq E[(x(t) - \bar{x}(t))(x(\tau) - \bar{x}(\tau))^T]$.

Problem 12 (12 pt). We have $\dot{x} = Fx + Gw$, F,G constant, $E[w(t)] = 0$, $E[w(t)w(\tau)^T] = Q\delta(t - \tau)$, Q constant, $E[x(0)] = x_0$, $\mathrm{Cov}[x(0)] = P_0 = 0$, $x(0)$, $w(t)$ independent and normal.

Then:

(i) $x(t), 0 \leq t \leq T < \infty$ is a stationary gauss-markov process;
(ii) $x(t), 0 \leq t < \infty$ is a nonstationary gauss-markov process;
(iii) $x(t)$ is a nonstationary, non-markov process;
(iv) $p[x(t + \Delta)/x(t)]$ is non-gaussian.

Problem 13 (12 pt). Consider the n-dimensional multistage process described by

$$x_{i+1} = \Phi_i x_i + \Gamma_i u_i\,,\ u_i \text{ is a scalar,}$$

where

$$E(u_i) = \bar{u}_i = 3i^2 - 2i + 5\,, \qquad E[(u_i - \bar{u}_i)(u_j - \bar{u}_j)] = 0 \qquad \text{if} \quad i \neq j$$

and where u_i has a "rectangular" distribution:

$$p(u_i) = \begin{cases} \dfrac{1}{2a_i} & \text{if} \quad -a_i + \bar{u}_i < u_i < a_i + \bar{u}_i\,, \\ 0 & \text{otherwise}\,, \end{cases}$$

where

$$a_i = 1 + \tfrac{1}{2}\cos(i)\,, \qquad i = 0,1,2,\ldots, \qquad E[(x_0 - \bar{x}_0)(u_i - \bar{u}_i)] = 0\,.$$

We are also given

$$\bar{x}_o \triangleq E(x_o) \qquad \text{and} \qquad P_o \triangleq E[x_o - \bar{x}_o)(x_o - \bar{x}_o)^T]\,.$$

Defining

$$\bar{x}_i \triangleq E(x_i)\,, \qquad P_i \triangleq E[(x_i - \bar{x}_i)(x_i - \bar{x}_i)^T]\,, \qquad Q_i \triangleq E[(u_i - \bar{u}_i)^2]\,,$$

we make the following assertion:

Assertion: $\bar{x}_{i+1} = \Phi_i \bar{x}_i + \Gamma_i \bar{u}_i\,, \qquad P_{i+1} = \Phi_i P_i \Phi_i^T + \Gamma_i Q_i \Gamma_i^T\,.$

Which of the following statements is true?

(i) The assertion is true.
(ii) The assertion is false because u_i is not gaussian.
(iii) The assertion is false because u_i is not stationary.
(iv) The assertion is false because u_i can take on only a limited range of values.
(v) The assertion is false, but not for reasons (ii), (iii), and (iv).

Problem 14 (12 pt.). An n-dimensional random process is described by

$$\dot{x}(t) = Fx(t) + Gu(t)$$

where F and G are constant $(n \times n)$ and $(n \times m)$ matrices, respectively, and where $u(t)$ is an $(m \times 1)$ vector with zero mean and correlation function

$$E\{u(t)u^T(\tau)\} = Q\delta(t - \tau)\,,$$

where Q is a constant positive-definite $(m \times m)$ matrix. Suppose we also have a perfect measurement

$$z(t) = H^T[Fx(t) - \dot{x}(t)]\,,$$

where H is constant, $\bar{x}(0)$ is the initial mean, and $P(0)$ the initial covariance. Define

$$I \triangleq \lim_{T \to \infty} \int_0^T [z(t)]^2\, dt\,.$$

Then which of the following is true?

(i) $I = 0$;
(ii) $I = \mathrm{Tr}(Q)$;
(iii) $I = H^T G Q G^T H$;
(iv) $I = \infty$;
(v) The value of I depends on the stability of F and/or observability of the pair (F,H).

Problem 15 (12 pt). Consider the scalar estimation problem

$$x_{i+1} = x_i + w_i\,, \qquad z_i = x_i + v_i\,,$$

where

$$v_i \text{ is } N(0,1) \text{ and white,} \qquad w_i \text{ is } N(0,q) \text{ and white,}$$

and there is no initial knowledge on x_o.

If $0 < q < \infty$, then the optimal estimate $\hat{x}_i$ is given by

(i) $\hat{x}_i = \frac{1}{i}\sum_{j=1}^{i} z_j\,,$

(ii) $x_i = z_i\,,$

(iii) $\hat{x}_{i+1} = \hat{x}_i + k_i(z_{i+1} - \hat{x}_i)\,, \frac{1}{i+1} < k_i < 1\,,$

(iv) $\hat{x}_{i+1} = \hat{x}_i + k_i(z_{i+1} - \hat{x}_i)\,, 1 < k_i < \infty\,.$

Problem 16 (9 pt). A static estimate of x is made from a measurement z:

$$z = Hx + v\,, \qquad v \text{ is } N(\bar{v},R)\,, \qquad x \text{ is } N(\bar{x},M)\,.$$

The estimate is

$$\hat{x} = \bar{x} + K(z - H\bar{x})\,,$$

and K is some constant matrix.
The estimate is

(i) unbiased,
(ii) biased with a bias of $(KH\bar{x})$,
(iii) biased with a bias of $(K\bar{v})$,
(iv) biased with a bias of $[K(\bar{v} - H\bar{x})]$.

Problem 17 (15 pt). Which of the following statements is false (in the context of estimation)?

(i) Weighted-least-square estimates are equivalent to minimum variance estimates if and only if the weighting matrix is equal to the inverse of measurement noise covariance matrix.
(ii) Mean-square error of *WLS* estimate $\geq$ mean-square error of *MV* estimate
(iii) In the gaussian case, the conditional mean is the *MV* estimate.
(iv) In the general nongaussian case, conditional mean is always the minimal Bayes-Risk estimate.

Problem 18 (9 pt). For an nth-order linear dynamic system driven by white noise, the covariance matrix of the state is

(i) dependent on the mean value of the initial state,
(ii) independent of the eigenvalues of the system transition matrix,
(iii) independent of the controllability of the system,
(iv) dependent on both F,G and the statistics of the noise.

Problem 19 (12 pt). If noisy measurements are made on some components of the state of the system in Problem 18, the mean square error of the optimal estimate based on methods in Chapter 12 is

(i) independent of the observability of the system,
(ii) independent of the value of the measurements taken,
(iii) independent of the measurement noise statistics,
(iv) dependent on the actual measurements taken.

Problem 20 (12 pt). A scalar estimate is made from the given scalar measurement z as the conditional expected value of x given z, that is, $E(x|z)$. x and z are not necessarily normal. The error of the estimation, $[x - E(x|z)]$, is

(i) independent of x,
(ii) independent of z,
(iii) uncorrelated with x,
(iv) uncorrelated with z.

[NOTE: $E(xy) \triangleq \int\int xyp(x,y)\,dx\,dy$ for any two random variables.]

Problem 21 (12 pt). Given a system

$$\dot{x} = Fx, \qquad z = Hx,$$

which is observable. Noise is added to the system as

$$\dot{x} = Fx + Gw, \qquad z = Hx + v;$$

w and v are white with zero mean and spectral densities Q and R, respectively. Then the

(i) system is observable only if $G = 0$,
(ii) system is observable only if $G = 0, Q = 0, R = 0$,
(iii) system is observable only if $Q = R$,
(iv) system is still observable for any finite G, Q, and $R > 0$.

Problem 22 (9 pts). A function L is defined as follows:

$$L = \tfrac{1}{2}[-u_1^2 + 3u_2^2 - 10u_1 - 14u_2 + 2u_1u_2 + 11].$$

(i) The function L achieves its *minimum* value at

$$u_1 = -2, \qquad u_2 = 3.$$

(ii) The function L achieves its maximum value at

$$u_1 = -2, \qquad u_2 = 3.$$

(iii) The function L has a stationary point at

$$u_1 = -2, \qquad u_2 = 3 \qquad \text{which is a saddle point;}$$

(iv) It has a singular point with eigenvalues -1 and 0.

Problem 23 (9 pt). We have

$$x(i+1) = Fx(i), \qquad \det(F) \neq 0.$$

Here F is a (3×3)-matrix. A scheme for identification of F is used in which $x(i)$ is observed for three steps, a matrix $[x(3) \vdots x(2) \vdots x(1)]$ is written as

$$[x(3) \vdots x(2) \vdots x(1)] = F[x(2) \vdots x(1) \vdots x(0)],$$

and F is determined as

$$F = [x(3) \vdots x(2) \vdots x(1)]\,[x(2) \vdots x(1) \vdots x(0)]^{-1}.$$

The only control we have in this method is in selecting $x(0)$. The method will work

(i) for all $x(0)$ that are eigenvectors of F,
(ii) for all nonzero $x(0)$ that are not eigenvectors of F and F^2,
(iii) only for all $x(0)$ of the form

$$\begin{bmatrix}1\\0\\0\end{bmatrix} \text{ or } \begin{bmatrix}0\\1\\0\end{bmatrix} \text{ or } \begin{bmatrix}0\\0\\1\end{bmatrix},$$

(iv) for all $x(0)$.

Problem 24 (15 pt). Assume all problems in an examination are weighted equally, and all problems have only four choices. Grades are assigned according to the following rule: correct choice, full credit; wrong choice, minus $\frac{1}{3}$ credit; no choice, zero credit. Note that the expected grade of a completely random choice is

$$x \cdot \frac{1}{4} - \frac{x}{3} \cdot \frac{3}{4} = 0.$$

The probability that a person who has no knowledge of the course through random choice can gain a grade of 35 (out of 100) on 25 questions is *approximately*

(i) 1/10,
(ii) 1/100,
(iii) 1/1000,
(iv) 1/10,000,
(v) $1/10^6$.

Problem 25 (12 pt). Consider the following two problems

(1) $x_{t+1} = \Phi x_t + \Gamma w_t$, w_t ; $N(0,Q)$ white, $z_t = Hx_t + v_t$,
v_t ; $N(0,R)$ white, $E(w_t v_t^T) = C$;

(2) $x_{t+1} = (\Phi - DH)x_t + Dz_t + \Gamma w_t - \Gamma \hat{w}_t$, $z_t = Hx_t + v_t$.

Problems (1) and (2) are equivalent if:

(i) $\hat{w}_t = E(w_t/v_t)$, in which case $\Gamma(w_t - \hat{w}_t)$ is uncorrelated with v_t.
(ii) Same as (i) but, in addition, $D = \Gamma CR^{-1}$ is required.
(iii) Same as (ii) but with $D = H\Gamma CR^{-1}$.
(iv) The two problems are never equivalent.

Problem 26 (9 pt). The covariance of the optimal estimate for (Problem 2) of Question 25 is governed by:

(i) $P_{t+1} = M_{t+1} - M_{t+1}H(H^T M_{t+1} H + R)^{-1} HM_{t+1}$,
$M_{t+1} = (\Phi - DH)P_t(\Phi - DH)^T + \Gamma Q \Gamma^T - \Gamma CR^{-1} C^T \Gamma^T$.
(ii) Same as (i) except that $M_{t+1} = (\Phi - DH)P_t(\Phi - DH)^T + \Gamma Q \Gamma^T + DRD^T$.
(iii) Same as (ii) except that we replace DRD^T with $DR^{-1}D^T$.
(iv) Same equation which depends on z_t since it appears in the system equation.

Problem 27 (9 pt). In the context of Chapter 12, suppose that $z = Hx + Av$, where H and A are known matrices and x, v are the usual gaussian random variables. The optimal estimation error covariance after the measurement z is given by:

(i) $P = M - MH^T A(HMH^T + R)^{-1} A^T HM$.
(ii) $P = M - MH^T(HMH^T + ARA^T)^{-1} HM$.
(iii) $P = AMA^T - AMH^T(HMH + R)^{-1} HMA^T$.
(iv) Matrix A does not affect the error covariance matrix P.

Problem 28 (12 pt). There are two identical urns. One contains red balls only, the other an equal mixture of red and white balls. Suppose one red ball is drawn from one urn. The probability that the urn contains red balls only is

(i) 1
(ii) 2/3
(iii) 1/2
(iv) 3/4

ANSWERS

Problem	*Answer*	*Problem*	*Answer*
1	(i)	3	(i)
2	(i)	4	(ii)

5	(iv)	*17*	(iv)
6	(i)	*18*	(iv)
7	(iv)	*19*	(ii)
8	(i)	*20*	(iv)
9	(ii)	*21*	(iv)
10	(iii)	*22*	(iii)
11	(iii)	*23*	(ii)
12	(ii)	*24*	(iii)
13	(i)	*25*	(ii)
14	(iv)	*26*	(i)
15	(iii)	*27*	(ii)
16	(iii)	*28*	(ii)

Index

Printed in the United States
by Baker & Taylor Publisher Services